ORGANIZATIONAL AND SOCIAL PERSPECTIVES ON INFORMATION TECHNOLOGY

IFIP - The International Federation for Information Processing

IFIP was founded in 1960 under the auspices of UNESCO, following the First World Computer Congress held in Paris the previous year. An umbrella organization for societies working in information processing, IFIP's aim is two-fold: to support information processing within its member countries and to encourage technology transfer to developing nations. As its mission statement clearly states,

IFIP's mission is to be the leading, truly international, apolitical organization which encourages and assists in the development, exploitation and application of information technology for the benefit of all people.

IFIP is a non-profitmaking organization, run almost solely by 2500 volunteers. It operates through a number of technical committees, which organize events and publications. IFIP's events range from an international congress to local seminars, but the most important are:

- The IFIP World Computer Congress, held every second year;
- open conferences;
- working conferences.

The flagship event is the IFIP World Computer Congress, at which both invited and contributed papers are presented. Contributed papers are rigorously refereed and the rejection rate is high.

As with the Congress, participation in the open conferences is open to all and papers may be invited or submitted. Again, submitted papers are stringently refereed.

The working conferences are structured differently. They are usually run by a working group and attendance is small and by invitation only. Their purpose is to create an atmosphere conducive to innovation and development. Refereeing is less rigorous and papers are subjected to extensive group discussion.

Publications arising from IFIP events vary. The papers presented at the IFIP World Computer Congress and at open conferences are published as conference proceedings, while the results of the working conferences are often published as collections of selected and edited papers.

Any national society whose primary activity is in information may apply to become a full member of IFIP, although full membership is restricted to one society per country. Full members are entitled to vote at the annual General Assembly, National societies preferring a less committed involvement may apply for associate or corresponding membership. Associate members enjoy the same benefits as full members, but without voting rights. Corresponding members are not represented in IFIP bodies. Affiliated membership is open to non-national societies, and individual and honorary membership schemes are also offered.

ORGANIZATIONAL AND SOCIAL PERSPECTIVES ON INFORMATION TECHNOLOGY

IFIP TC8 WG8.2 International Working Conference on the Social and Organizational Perspective on Research and Practice in Information Technology June 9–11, 2000, Aalborg, Denmark

Edited by

Richard Baskerville
Georgia State University
USA

Jan Stage
Aalborg University
Denmark

Janice I. DeGross
University of Minnesota
USA

SPRINGER SCIENCE+BUSINESS MEDIA, LLC

Contents

Part 3: Reforming the Classical Challenges

Part 4: Transforming Toward New Challenges

Part 5: Reformation of Conceptualizations

Part 6: Transformation of Conceptualizations

Part 7: Reforming Automation

Part 8: Transforming Automation

Part 9: Transforming into New Shapes of Technology

Part 10: Panels on Research Methods and Distributed Organizations

Foreword

The articles in this book constitute the proceedings papers from the IFIP WG 8.2 Working Conference, "IS2000: The Social and Organizational Perspective on Research and Practice in Information Technology," held June 10-12, 2000, in Aalborg, Denmark. The focus of the conference, and therefore this book, is on the basic aim of the working group, namely, the investigation of the interrelationships among four major components: information systems (IS), information technology (IT), organizations, and society. This basic social and organizational perspective on research and practice in information technology may have evolved substantially since the founding of the group, for example, increasing the emphasis on IS development.

The plan for the conference was partially rooted in the early WG 8.2 traditions, in which working conferences were substantially composed of invited papers. For IS2000, roughly half of the paper presentations were planned to be invited; the remaining half were planned to be double-blind refereed in response to a "Call For Papers." Invited papers were single-blind reviewed in order to provide the authors with pre-publication feedback and comments, along with the opportunity to revise their papers prior to its final incorporation in this book.

The IS2000 working conference was intended as a WG 8.2 milestone, and invited papers were drawn from a selected group of its pioneering members. This does not necessarily imply "old" members. We define pioneering members as those who were active in setting the intellectual direction of the group in its early years and those who have the potential to redirect the intellect of the group in the early years of the dawning century. We felt that by insuring the engagement of a dozen recognized speakers, this would motivate the remainder of the working group to submit their best work. The selection of this group of invited speakers was a difficult task, first because there were over 100 members in the pool, and second because we also wanted a fair distribution on research topics, international representation, research institutions, etc. We hope that through this process we have accomplished the engagement of our trend setters through both the invited and submitted vehicles.

Authors of invited papers were encouraged to include essays and subjective argumentation as well as reports on research, intending to provide them with the broadest leeway in getting their keenest ideas before the field. The topic and scope of their papers were largely their prerogative and it was permitted to include a historical view and/or new directions.

This book includes 12 articles that were invited and 15 articles that were submitted in response to the call for papers. Approximately 34% of the submitted papers were accepted and the blind-review scoring process provided substantial indications that the invited and submitted papers have equally high quality in scholarly interest and content. There were four panel discussions that won acceptance in the blind-review process and were offered during the conference. In the interests of providing a complete record, we have included descriptions of these panels at the end of this book.

Richard Baskerville
Jan Stage
Janice I. DeGross

Conference Chairs

General Chairs
Heinz K. Klein
Binghamton University
USA

M. Lynne Markus
Claremont Graduate University
USA

Organizing Chair
Peter Axel Nielsen
Aalborg University
Denmark

Program Chairs
Richard Baskerville
Georgia State University
USA

Jan Stage
Aalborg University
Denmark

Program Committee

Ivan Aaen	Aalborg University
David Avison	Southampton University
Michael Barrett	University of Alberta
Cynthia Beath	University of Texas
Kristin Braa	University of Oslo
Kevin Crowston	Syracuse University
Jan Damsgaard	Aalborg University
Elizabeth Davidson	University of Hawaii
Brian Fitzgerald	University College, Cork
Guy Fitzgerald	Brunel University
Bob Galliers	Warwick University
Bernie Glasson	Curtin University
Ole Hanseth	University of Oslo
Liisa von Hellens	Griffith University
Rudy Hirschheim	University of Houston
Matthew Jones	Cambridge University
Bonnie Kaplan	Yale University
Julie Kendall	Rutgers University
Ken Kendall	Rutgers University
Frank Land	London School of Economics
Tor Larsen	Norwegian School of Management
Allen Lee	Virginia Commonwealth University
Jonathan Liebenau	London School of Economics
Lars Mathiassen	Aalborg University
Erik Monteiro	Norwegian University of Science and Technology
Enid Mumford	Manchester University
Michael Myers	University of Auckland
Jaana Porra	University of Houston
Jan Pries-Heje	Copenhagen Business School
Nancy Russo	Northern Illinois University
Ulrike Schultze	Southern Methodist University
Susan Scott	London School of Economics
Erik Stolterman	University of Umeå
Burt Swanson	University of California, Los Angeles
Duane Truex	Georgia State University
Cathy Urquhart	University of the Sunshine Coast
John Venable	Curtin University
Geoff Walsham	Cambridge University

Sponsors

Department of Computer Science
Aalborg University

Computer Information Systems Department
Georgia State University

1 DISCOURSES ON THE INTERACTION OF INFORMATION SYSTEMS, ORGANIZATIONS, AND SOCIETY: REFORMATION AND TRANSFORMATION

Richard Baskerville
Georgia State University
U.S.A.

Jan Stage
Aalborg University
Denmark

This book marks the turn of the first century of the information age. The purpose of the book is to provide ideas and results from past, current and future investigations of the relationships and interactions among four major components: information systems (IS), information technology (IT), organizations, and society. These investigations share a primary focus on the interrelationships, not on the components themselves.

The contributions to the book deal with history of information systems theory and technology, with the directions faced by those sharing the concerns of the field in its future research, and with attempts to draws these two views together. The authors provide essays and research reports as well as research proposals for future work in this area. They also provide well-substantiated integrative frameworks that facilitate recognition and transfer of relevant knowledge about the roles and uses of information technology. These frameworks are based on a wide range of disciplines such as philosophy, history, sociology, political science, management, and computer science.

We have organized the contributions to this book into five discourses. Collectively, the five discourses answer the key question: "What is the status of the discipline of information systems as we stand at the juncture of the new century?" Individually, the five discourses deal with the following issues:

- *Fundamentals:* The basic concepts and models that are used for understanding and developing information systems.

- *Challenges:* The classical solutions and new problems that confront the field of information system.

- *Conceptualization:* The process of developing concepts and models that are used to understand and change information systems use and development.

- *Automation:* The interaction between humans and computers that is a key constituent part in an information system.

- *Technology:* The shape of the technology that is used to design and implement information systems.

In addition, we have aligned the contributions to each discourse with one of the following two perspectives:

- *Reformation:* The aim is to show how certain well-known issues are traditional and enduring, and that we can advance through revision and regeneration of issues from the past.

- *Transformation:* The aim is to show how other issues are entirely fresh and new, and that we can advance through generating issues that are completely unlike those of the past.

These two dimensions can be combined as illustrated in Figure 1. This table spans a framework for the contributions to this book. It should, however, be read with two limitations in mind. First, the vertical dimension does not represent a clear-cut division because the subjects are overlapping, and elements of the horizontal dimension are not always antithetical positions with an implied dialectic. Second, the positioning of a contribution in a cell of the table does not necessarily characterize the general perspective of any authors. Rather, the point is that the purpose, results, and conclusions of a particular contribution in this book align with the argument related to a specific discourse and perspective.

Based on this framework, the contributions are organized into nine parts. The first eight parts are defined by the first four discourses and each of their two perspectives. The ninth part concerns the new shapes in technology and its development (aligning generally with the transformational perspective). In the final part ten of the book, we include the descriptions of four panel discussions that regard research methods and the increasing distribution of organizations.

	Perspective	
Discourse	Reformation	Transformation
Fundamentals	*Part 1* Jones Mumford Boland	*Part 2* Davis Braa and Rolland Russo
Challenges	*Part 3* Land Mathiassen Crowston	*Part 4* Markus Kendall and Kendall Walsham
Conceptualization	*Part 5* Sawyer Hirshheim and Klein Lamb and Davidson	*Part 6* Kvasny and Truex Introna and Ilharco Venable and Travis
Automation	*Part 7* Pouloudi and Whitley Aanestad and Hanseth Rose and Truex	*Part 8* Mark Grundén Karsten
Technology		*Part 9* Wilson and Howcroft Eriksen Berg

Figure 1. Discourses and Perspectives in this Book

1. Reforming the Fundamentals

The first discourse regards the reformation or transformation in the fundamentals of the information systems discipline. In this part, we ask "Are the most fundamental concepts in the field enduring, or are these fundamental concepts evolving?" The authors primarily adopt the perspective of reformation as they build on certain fundamental concepts that were relevant in the past and show how these are just as relevant today.

In "The Moving Finger: The Use of Social Theory in WG8.2 Conference Papers 1975-1999," Jones uses citation data from the corpus of IFIP WG 8.2 working conferences to uncover indications as to how the social theory that inhabits information systems research is evolving. WG 8.2 conferences give earlier and greater attention to social theorists than is typical for the IS field, and while this theoretical base appears to have emerged, significant and irreversible shifts are not apparent.

In "Socio-technical Design: An Unfulfilled Promise or a Future Opportunity," Mumford surveys the history and purposes of the socio-technical movement in information systems to reveal its future potential. The principles of socio-technical design, and particularly its value system, are even more strongly relevant to information systems in the coming "wired world" because new organizational forms and economic processes require careful management of the increasingly fluid mutual benefits between employers and employees.

Boland, in "The Limits of Language in Doing Systems Work," uses the imagery of a roundtable discussion between respected social philosophers to frame an inventory of current assumptions about language and systems. The fundamental roadblock to the realization or successful implementation of many good ideas for information technology is language. This limit is imposed because language bounds the ability of human beings to engage in systems thinking in myriad ways. The variety of these ways is revealed in a review of some of the fundamental philosophical assumptions underlying the systems literature.

2. Transforming the Fundamentals

In contrast, the authors in this section adopt a more revolutionary perspective regarding the fundamentals of the discipline. They tend to focus on the changes that are underway in the fundamental concepts in the field of information systems. These authors concentrate on how some of the most basic ideas in the future information systems arena will be essentially different from those of today.

In "Information Systems Conceptual Foundations: Looking Backward and Forward," Davis surveys past and potential formalisms and frameworks used to define our discipline. He finds that the concept of information systems as a discipline is aggregated from a set of technologies, methods, ideas, processes, etc. The members of this set are not individually persistent, although there are currently more arrivals than departures. The effect is to make the "core" concepts of information systems more difficult to distinguish because of varying assumptions, politics, and the current free-market of ideas.

Braa and Rolland, in "Horizontal Information Systems: Emergent Trends and Perspectives," use a case study to show how contemporary trends in information systems, such as internetworking, globalization, and enterprise resource planning, represent the emergence of horizontal, rather than vertical, information systems development. Institutional responses, such as standardization and knowledge management, create tension in the communities of practice because these conflict with the need for horizontal systems to be flexible, intra-active, and emergent.

Russo argues in "Expanding the Horizons of Information Systems Development" that electronic commerce is qualitatively changing the mix of skills and knowledge needed in systems development particularly regarding the need for technical skills in telecommunications and multi-media and business skills in collaboration and flexible management.

3. Reforming the Classical Challenges

While the authors above are focused on the essential ideas underlying the perspectives we take within the discipline of information systems, others look more closely at the essential solutions and problems we are choosing. Some of the work seems to suggest that the essential problems in information systems are enduring, and that we are overlooking the most fundamental and essential solutions to these enduring problems. These authors point to a reformation of some traditional problems, and address the question "Are we neglecting the most essential information systems solutions?"

In "Evaluation in a Socio-technical Context," Land argues that the failure of socio-technical approaches to gain wider acceptance in IS practice may be largely due to the lack of acceptable ways of evaluating the organizational worth of the social elements within organizations. The discovery of practical techniques for accessing the social value of technology forms the main agenda for future socio-technical research.

Mathiassen uses the case of a large, long-term research project to construct a combined research approach to balance the diverse goals of research and practice. "Collaborative Practice Research" shows how professional work practices research is evolving into a hybrid of action research, experiments, and practice studies that enables the IS discipline to grow more closely attuned to the needs of practice.

Crowston argues in "Process as Theory in Information Systems Research" that the corpus of information systems research is ineffective because it is conceptually disconnected. Social-oriented research in information systems appears at multiple levels of analysis, e.g. at the individual level or at the organizational level. The recognized problem in linking research at these different levels of analysis can be resolved by centering the study of processes rather than the study of variance.

4. Transforming Toward New Challenges

On the other hand, other authors are suggesting that we are completely missing the right questions. These authors ask: "Are we choosing the right problems to solve?" These authors are suggesting that the essential problems have either been overlooked in the past or may have undergone revolutionary change over time. Therefore, they suggest that we are not concentrating our efforts on the real problems.

In "Toward an Integrated Theory of IT-related Risk Control," Markus argues that IS practices are growing inconsistent with recent management practice because IS has failed to achieve an integrated approach to risk management. An integrated view of IT-related risk includes system development failure, security breaches, and competitive threats, and enables intelligent, end-to-end tradeoff decisions in the management of IS.

The second chapter in this part is "Individual, Organizational, and Societal Perspectives on Information Delivery Systems: Bright and Dark Sides to Push and Pull Technologies" by Kendall and Kendall. They find that the benefits of push and pull information delivery systems may be outweighed by the social costs being unleashed on groups and individuals. The potential of each technology is described on four levels of increasing technological capabilities. The dangers include bandwidth wasted on unwanted content and the information overload, anxiety, addiction, and disorientation and the consequent stress, illness, and social dysfunction. There are remedies for these problems, ranging from unplugging to educating users and providers, but the nature and implications of the problem and the remedies are not yet understood.

Walsham in "Globalization and IT: Agenda for Research," finds us emerging into a century in which global information systems are ascendant, yet we are poorly prepared to use or understand these systems. We lack research that informs about the impact of these systems on human identity or on the effects of global groupware and intranets. We lack more critical organizational case studies and a sound multi-stakeholder perspective of the costs and benefits of global internetwork trading. We also lack a sense of the role and value of global IS in different cultures.

5. Reformation of Conceptualizations

Our authors also seem prepared to question whether our institutionalized "way of thinking" about information systems is the most useful way to approach the information systems issues facing us today. This process of conceptualization, this way of thinking, encompasses the processes that shape the way we apply fundamental concepts to finding the problems and resolving the solutions. These authors wonder whether the analytical approaches, mental models, and metaphors that inhabit our approach to the study of information systems are still right for the job. The authors in this part suggest that, in some ways, we should revise our institutionalized conceptualization processes. They are prepared to ask "Should we reform our old ways of thinking about these phenomena?"

In surveying the literature on multiple-method information systems research, Sawyer finds that the use of multiple research methodologies in each particular research study has not been well-explored in the information systems field. In "Studying Organizational Computing Infrastructures: Multi-method Approaches," Sawyer compares published examples with his own work and argues that multiple methodologies in the study of the application of any new information technology are in fact required by the need to study both the new technology and the organizational context. The technological infrastructure is too multi-faceted, i.e., unique, pervasive, context-driven, and emergent, for a monotonic study. Multi-method research suggests that multiple researchers, *a priori* theory, and an analytical dialectic should characterize information systems research.

In "Information Systems Research at the Crossroads: External Versus Internal Views," Hirschheim and Klein argue that non-IS practitioners have an unrealistic image of IS and its potential. IS academia is fragmented into non-interacting sects. The IS paradigms need to be adjusted to support better generalization and a broader view of both rigor and relevancy in order to establish the needed communication channels among IS academics, IS practitioners, and non-IS business professionals.

The final chapter in this part is "The New Computing Archipelago: Intranet Islands of Practice." By comparing the current development of organizational intranets with the past development of end user computing (EUC), Lamb and Davidson find intranet development is following the same pattern as EUC a generation earlier. Like EUC, intranets arise as islands of incompatible, independently developed technology without a shared vision or a common standard. The lure of centralized control over intranet development, standards and resources, at tension with the need to foster entrepreneurship and innovation, generates conflict over the role of IS professionals in relation to intranets, just as it did with EUC.

6. Transformation of Conceptualizations

In this part, we find authors whose work calls for entirely new conceptualization processes. These authors see the answer to this question, whether our institutionalized "way of thinking" about information systems is concordant with our needs, in a different light than the authors in part 5. They are prepared to ask instead, "Should we invent entirely new ways of thinking about information systems phenomena?"

The first chapter in this part is "Information Technology and the Cultural Reproduction of Social Order: A Research Paradigm." Using Bourdieu's theory of social practice as a framework, Kvasny and Truex show how information technology cannot seriously be used as a universal force for human empowerment because it has evolved as a cultural commodity that may only be acquired with cultural capital. Therefore, information technology can only be used to reproduce the existing social order in terms of enfranchising currently privileged actors while further disenfranchising the unprivileged.

In "The Screen and the World: A Phenomenological Investigation into Screens and Our Engagement in the World," Introna and Ilharco consider the significance of the growth to almost continuous engagement between people and their video screens. Video screens are so ubiquitous in current technology that we overlook the meaning that such screens assume for us. Screens "mediate our being in the world by presenting relevance in this world." Screens always present what matters, hide all else, and reveal a presumed *a priori* agreement about the viewer's situation in their world. Far from being simple, the concept of "screen-ness" is multi-faceted and loaded with meanings that information systems managers and designers ignore.

Venable and Travis, in "Developing a Virtual Community-based Information Systems Digital Library: A Proposal and Research Program," consider our contentment with the historically-fragmented repositories of human knowledge, calling upon us to reconsider the need for this distributed view of libraries. The information systems community should lead in developing an integrated, universal, information systems digital library. Proper development of this resource would involve a virtual community working as a group to develop a consensus regarding purpose and functionality. An amalgamation of soft systems and group methodologies may achieve this purpose.

7. Reforming Automation

The discourses move into a different plane as our authors in this part work to consider the most central facet in the interaction of information systems and organizations: automation of functions and the consequent human-machine interaction. It seems central to many of our authors that the relationships and exchanges between human beings and computing machinery are evolving. In particular, a continuous change is developing in the division of work between people and machines. In this part, we find work that reveals a deeper understanding of the interactions between people and machines, a revision and reformation of our previous thinking. The work of this group of authors collectively helps answer the question "Is the explosion of information technology actually changing the working boundaries between humans and machines?"

This part begins with "Representing Human and Non-human Stakeholders: On Speaking with Authority." Using an illustrative case study, Pouloudi and Whitley explore how an encryption algorithm became elevated to the level of other stakeholders in an information system setting. Non-human stakeholders, such as particular technologies, may become inscribed with human values. The observations of this study indicate how these values are not a singular cohesive set, but vary according to interpretation by the different human stakeholders.

In "Implementing Open Network Technologies in Complex Work Practices: A Case from Telemedicine," Aanestad and Hanseth use a case study to show how the introduction of broadband multimedia into organizations creates a hybrid *collectif* or collective of humans and non-humans. The interests of both technology and human actors must be translated for alignment through the dialog between an organization and its technology. Because this dialog is irreducibly interactional, it is impossible to plan or specify in advance, but must be cultivated in a process by which both the organization and the technology are socialized and educated as children in a new culture.

Rose and Truex, in "Machine Agency as Perceived Autonomy: An Action Perspective," consider machine agency, how technology seems to go about structuring human work. Contrasting theoretical positions, structuration and actor-network theories, can align when attributed machine autonomy is regarded. This attribution refers to the degree humans act according to a belief that the machine possesses autonomy. IT developers and managers need to consider this attribution more seriously because the closer one studies situated use, the more belief in autonomy is likely to be discovered. Consequently, agency theories can be seen themselves to be situated.

8. Transforming Automation

Another group of authors is suggesting revolutionary responses to this evolving relationship between human beings and computing machinery. These authors take a more normative stance, seeking new approaches to the design and construction of information systems that accommodate the changing boundaries. This work responds to the question, "Should we devise new ways to facilitate the interaction between humans and machines?"

In "Some Challenges Facing Virtually Colocated Teams," Mark is focussing on the challenges that arise when virtually colocated teams are to be supported by information technology. An empirical study and classical theories about interaction and group formation are used to analyze the differences between physically and virtually colocated teams, finding that technology shapes the culture of virtually colocated teams. It produces immediate and long-term behavioral effects on team development and effectiveness in terms of trust, motivation, cooperation, and patterns in participation and communication.

Grundén, in "MOA-S: A Scenario Model For Integrating Work Organization Aspects into the Design Process of CSCW Systems," concludes that work in CSCW has not developed good design methods to deal with the flexible character of either the technology or the organization, nor studied the consequences of CSCW for work processes. Using a case study approach, Grundén shows how the process of designing CSCW systems is improved by a model based on scenarios for articulating and discussing future designs of the work organization.

The final chapter in this part is "Constructing Interdependencies with Collaborative Information Technology." Karsten combines a theoretical approach with reviews of earlier empirical cases to discover that the relationships between information technology and social interdependence can be usefully interpreted by understanding four essentials. First, interdependence can originate through IT-mediated social integration. Second, time and space distanciation is enabled by stored information accessible through telecommunication. Third, the IT may situate or historically frame the institutionalization of social interdependence. Fourth, reciprocity in the storage of information can ingrain systemness, thereby transforming social integration to system integration.

9. Transforming into New Shapes of Technology

The last discourse we discover in the work of our authors regards important new ways to meet the new century of information systems development. These authors consider how the evolving worlds of technologies and social interaction inspire us to invent revolutionary new ways to shape the problems and solutions involved developing new information systems.

Wilson and Howcroft, in "The Role of Gender in User Resistance and Information Systems Failure," challenge the technical-economic view of failure and show how it is also a social construction. Using a case study of a hospital information system, they show how the role of gender in information systems failure is important because feminine values, such as hands-on caring, may be suppressed by the increase in abstract processing inherent in information technology. Discovery of gender issues may show that care can prevail and society and organizations can indeed benefit from a "technical" information system failure.

The next chapter is "Limitations and Opportunities of System Development Methods in Web Information System Design." Using a research approach that is intervention-driven and practice-oriented, Eriksen studies whether current systems development methods are appropriate for web design. He shows how newer information technologies, such as web technology, are changing social relationships by creating an "audience" for organizations. Design work in this shifting social context requires a distinction between producer and consumer communities and between interior and exterior design. Further, design methods must be concerned with dividing operational responsibility and maintaining the site information. Hypermedia methodology is shown to be useful to determine input specifications instead of formally described work tasks.

The final chapter is, perhaps appropriately, "Lessons from a Dinosaur: Mediating IS Research Through an Analysis of the Medical Record." Berg takes up the classical technical approach that still underlies information system development and thereby influences the way we shape the interaction between human beings and computers. The case of a paper-based versus electronic patient record is used to illustrate the inadequacy of the technical approach and the need for an interpretive approach in developing a satisfactory electronic solution.

10. Panels on Research Methods and Distributed Organizations

Two additional issues of a very basic nature inhabit the chapters in the first nine parts of the book. These issues are research methodology and new organizational forms. The first issue regards the problems involved in studying the complex social context of technology in human work. The second issue regards both the new ways in which humans can interact using technology and the ways this technology limits human action. In addition to the research papers presented in the other parts, this section includes brief descriptions of four panel discussions that deal exclusively with these two themes.

11. Conclusion: Directions of the Collective Work

Collectively, the work in the ten parts of this book comprises a discourse on a new horizon for our understanding of the interaction between human beings in organizations and society and their information systems. There are essays, theoretical papers, and empirical research reports about the changing roles and new impacts of information technologies in specific organizational contexts. These contributions suggest the new ways and means by which organizations can improve the design, implementation, and maintenance of information technologies. These contributions reveal the changing role that information technology plays in the lives of people as individuals and as members of complex social institutions such as government, community, business, professional societies, and other forms of social associations.

These changes are complex and dramatic. Much worth is discovered in our traditional thinking about information systems. The contributions to this book show exactly how to regenerate some well-known thinking in terms of fundamental concepts, solutions to known problems, conceptualization processes, and the interaction between people and machines. Worth is also discovered in revolutionary thinking for some issues. We call for completely new thinking in terms in other fundamental concepts, totally new problems, new conceptualization processes, new boundaries between people and machines, and new ways of shaping technologies.

What is our story as we turn the first century of the information age? We believe that certain important fundamental concepts remain relevant. Significant and irreversible shifts in social theory have not appeared, information technology is still central in the social interaction in organizations, and indeed has become more critical because the traditional mutual benefits between employers and employees has become more fluid. As always, human language barriers remain the fundamental roadblocks to IT implementation. We believe other transformational concepts are arising, and indeed our set of concepts is growing because there are more useful ideas arriving than are departing. For example, we need to distinguish the newer horizontal, rather than vertical, forms of information systems development and recognize the changing mix of skills and knowledge required in this area.

Many traditional problems remain to be solved, like finding acceptable ways of evaluating the organizational worth of social elements, developing better research methods, and linking our discoveries into a system. Possible solutions include developing hybrid methods that link research and practice, or centering processes for integrating theories. We are also becoming aware of new or unrecognized problems, such as the lack of any integrated approaches to risk management, the social costs being unleashed on people by the wiring of society, and our headlong rush into globalized systems without a proper understanding.

We believe our conceptualization processes are inadequate. Some need revision, such as accepting that multiple methodologies are not just desirable, but necessary for information systems research. We need to adjust our paradigms to establish better communication channels throughout the field. We have to realize that the old end user tension between central control and innovation has reappeared in the intranet world. Innovative conceptualization processes are also needed. The video screen is becoming the central means for discovering our relevance to our universe, and now we must be prepared to approach information technology as a cultural commodity that may only be acquired with cultural capital. We are at the threshold of integrated, universal digital libraries where access to all of humanity's culture, our art, history, and technology, is neatly rolled and accessed in a virtual Library of Alexandria.

There has been an incredible evolution in the relationships and exchanges between human beings and computing machinery. Machines are acquiring human status, not through robotics, but rather through reification. Computers have become inscribed socially with autonomy and human values, and organizations have become hybrid collectives of humans and non-humans. Consequently we need new norms for defining the relationships and exchanges between human beings and computers. We need methods that permit designers to appreciate how technology shapes the culture of virtually colocated people, deals with flexibility that is absolutely necessary in both information technology and organizations, and allows designers to interpret information technology in relation to the social interdependence of organizational members.

Information technology has not only broken through technical barriers, but through social ones as well. We must raise our awareness of the social issues to the same new plane as that reached by technology. For example, we can see clearly the role of gender in defining information systems success or failure. We can see how web design implies changing an incredible web of social relationships. We can also see how rational-technical information systems design has endured even more unreasonably than rational-technical information systems management, and that the need for interpretive approaches for systems development is paramount.

We believe the new century raises on our horizon other complicated issues that we were unable to deal with in this volume. One of these issues is the ethics of these new IS and IT concepts and practices. The changes suggested above reveal to us many raw ethical dilemmas which arise in the development, use, and consequences of information technology, or in research about such technology. Many questions about the new century of information systems are answered in the work presented in this volume. However, we recognize that the ethical problems deserve more future space in our thinking and our research than was afforded here, and levy this as a challenge to future authors in this series.

About the Authors

Richard Baskerville is an associate professor in the Department of Computer Information Systems at Georgia State University. His research focuses on security and methods in information systems, their interaction with organizations and research methods. He is an associate editor of *The Information Systems Journal* and *MIS Quarterly*. Richard's practical and consulting experience includes advanced information system designs for the U.S. Defense and Energy Departments. He is a former chair of the IFIP Working Group 8.2, and a Chartered Engineer under the British Engineering Council. Richard holds MSc and Ph.D. degrees from the London School of Economics. He can be reached by e-mail at baskerville@acm.org.

Jan Stage is an associate professor in Computer Science at Aalborg University. His research interests include methods for analysis and design in software engineering, object-oriented analysis and design, and analysis and design patterns. His articles have appeared in *MIS Quarterly, Information and Software Technology*, *Information Technology and People*, and *The Scandinavian Journal of Information Systems*. He is a co-author of *Object-Oriented Analysis and Design* (in Danish and Swedish) and holds a Ph.D. in Computer Science from Oslo University. Jan can be reached by e-mail at jans@ intermedia.auc.dk.

Part 1:

Reforming the Fundamentals

2 THE MOVING FINGER: THE USE OF SOCIAL THEORY IN WG 8.2 CONFERENCE PAPERS, 1975-1999

Matthew Jones
Judge Institute of Management Studies
University of Cambridge
England

The moving finger writes; and, having writ,
Moves on: nor all thy Piety nor Wit
Shall lure it back to cancel half a line
Nor all thy Tears wash out a Word of it.
Fitzgerald, *The Ruba'iyat of Omar Khayyam*

1. Introduction

The remit of Working Group 8.2 (WG 8.2) is officially identified by the International Federation for Information Processing (IFIP) as "the interaction of information systems and the organization" and its "Scope and Aims" statement (http://www.ifipwg82.org/) talks of "building theories about the role and impact of IT in specific organizational contexts." Thus, while WG 8.2 may not be the only group of IS researchers concerned with understanding the relationship between social context and the development and use of information systems, for example, WG 8.1 ("Design and Evaluation of Information Systems") and WG 9.1 ("Computers and Work") may be expected to share similar interests, the Group would seem a potentially important forum for research that seeks to address the social dimension of IS. To the extent, moreover, that interest in social context may be seen to be characteristic of the concerns of interpretative research methods, the growing acceptance of such methods in mainstream journals such as *MIS Quarterly* could be seen to place WG 8.2 in the vanguard of the IS field.

Before WG 8.2 members become too conceited about themselves as brave pioneers at the frontier of IS research, however, it may be worth considering the record of the Group in manifesting an appreciation of the social dimension of IS in its own work. One measure of such appreciation, it may be argued, might be taken to be the use of social theory: seeing how far the frameworks and theories developed by Group members actually draw on the disciplines that most directly address the social issues with which it purports to be concerned. In this paper, therefore, an analysis will be presented of the proceedings of WG 8.2 conferences to examine the extent to which social theorists have been explicitly cited. The analysis will also seek to explore the pattern of citations over time to see how this may have changed. Has interest in social theory grown over the course of the Group's history? Which particular theorists have been most cited? Is there any evidence of a cumulative tradition (Keen 1991) or of transient fashions (Abrahamson 1996)?

The structure of this paper is as follows. In the following section, the case for considering use of references to social theorists as a measure of the appreciation of the social dimension of IS will be discussed. This is followed by a description of the methodology employed in analyzing the WG 8.2 conference proceedings. The results of this analysis are then presented and comments on these findings are made. The paper concludes with a discussion of the implications of the findings for IS research in general and WG 8.2 in particular.

2. Social Theory and IS Research

A number of objections could be raised to the use of references to social theorists as a measure of the appreciation of the social dimension of IS and it would seem important to address these before proceeding with the analysis. The first objection might be that the IS field is, or should be, theoretically self-sufficient and that theories from other domains are irrelevant to an understanding of IS phenomena. WG 8.2's "Aims" statement, however, describes the Group as seeking to develop "integrative frameworks...based on a wide range of disciplines." The contribution of other disciplines to IS research would seem to be recognized, therefore, within WG 8.2 at least.

A refinement of this initial objection might be that even if other disciplines can usefully contribute to IS research, the theories and concepts from these disciplines are of limited direct use as they require significant adaptation to IS-specific conditions. As Keen (1991) and Jones (1997) have argued, however, many, if not all, of the issues relating to the social aspects of information systems are already staple elements of the concerns of other disciplines. The case for the IS field developing an idiosyncratic theoretical basis would, therefore, seem weak.

The desirability of IS researchers being aware of, and drawing on, theory from other disciplines in their work, moreover, need not be justified solely in terms of economy of effort. An orientation that is receptive to theoretical ideas from sources outside the IS field would also seem likely to promote a richer appreciation of IS issues. By putting these issues in a larger perspective, they may be understood as particular cases of broader phenomena, and as located within wider contexts, whether social, historical, or economic. Explicitly seeking to connect IS work with that in other domains also offers the opportunity for exchange and, as Keen argues, potential influence. If IS researchers only cite each other, then the field risks becoming narrow and hidebound. Measuring IS research against the standards of other fields may also help to encourage sophistication and innovation in IS research practice.

A third objection might be that focusing on the contribution of social theory is casting the net too widely and overlooking a field more evidently applicable to the interests of WG 8.2, namely organizational theory. Despite the argument of writers such as Donaldson (1985) that organizational theory is a distinctive area of research separate from social theory, however, this view is not widely shared. More typically (e.g., Burrell and Morgan 1979) organization theory is seen as a specialized subset of social theory. This is not to say that organizational research does not have a distinctive domain of study or particular issues that it addresses more than most, but its main underlying concerns may be seen to be addressed in other fields. In this respect, the relationship of organizational research to social theory may be seen to be analogous to that of IS, rather than being an equivalent "reference discipline" in its own right. This would seem particularly the case given the rather blurred boundary between the organizational and IS literatures, with significant IS articles being published in organizational journals such as *Administrative Science Quarterly, Organization Science*, and *Organization Studies*.

Finally, it should be noted that in analyzing the contributions from social theory to the work of WG 8.2 the aim is not to suggest that other areas, such as systems or computer science, have not made significant contributions to IS research. Rather, since, as was argued above, a concern with social issues may be seen to be an important differentiation for the work of WG 8.2, the extent to which social theory has been drawn on in papers presented at its working conferences would seem deserving of particular attention.

3. Research Methods

An analysis was conducted of the references cited in papers included in the published proceedings of 15 WG 8.2 conferences from 1979 to 1999, as shown in Table 1. It is not certain, however, whether Table 1 includes all of the WG 8.2 conference during this period. The Group itself does not have records, let alone copies, of the proceedings of all its conferences and the IFIP Secretariat records only go back to 1986. Despite inquiries to long-standing WG 8.2 members, some of whom had recollections of earlier conferences but could find no references for any proceedings, therefore, it is possible that Table 1 is incomplete. It does, however, match the British University Libraries' combined catalogue records of IFIP-related conferences, so any earlier conferences may not have had published proceedings or did not identify themselves as having been organized by IFIP WG 8.2. Table 1 also concurs with a note by Hank Lucas, the first chair of WG 8.2, in *Oasis* (Lucas 1994), which reported that the Group's first meeting was probably in Amsterdam in autumn 1975, and that the first WG 8.2 working conference was held in Bonn from 11-13 June 1979, with the proceedings being published as Lucas *et al* (1980). There then appears to have been a gap of four years until the second conference in Minneapolis in 1983, since which time they have continued every nine to 18 months to date.

Table 1. IFIP WG 8.2 Conferences, 1979-1999

Date	Location	Proceedings
11-13 June, 1979	Bonn, Germany	Lucas et al. (1980)
22-24 August, 1983	Minneapolis, USA	Bemelmans (1984)
1-3 September, 1984	Manchester, UK	Mumford et al. (1985)
27-29 August, 1986	Noordwijkerhout, The Netherlands	Bjørn-Andersen and Davis (1988)
29-31 May, 1987	Atlanta, USA	Klein and Kumar (1989)
2-4 July, 1989	Ithaca, USA	Kaiser and Oppeland (1990)
14-16 December, 1990	Copenhagen, Denmark	Nissen, Klein, and Hirschheim (1991)
14-17 June, 1992	Minneapolis, USA	Kendall, Lyytinen, and DeGross (1992)
17-19 May, 1993	Noordwijkerhout, The Netherlands	Avison, Kendall, and DeGross (1993)
11-13 August, 1994	Ann Arbor, USA	Baskerville et al. (1994)
7-9 December, 1995	Cambridge, UK	Orlikowski et al. (1996)
26-28 August, 1996	Atlanta, USA	Brinkkemper, Lyytinen, and Welke (1996)
31 May - 3 June, 1997	Philadelphia, USA	Lee, Liebenau, and DeGross (1997)
10-13 December 1998	Helsinki, Finland	Larsen, Levine, and DeGross (1998)
21-22 August, 1999	St Louis, USA	Ngwenyama et al. (1999)

The analysis carried out involved the identification of *all references that might be broadly defined as being to works of social theory in all the submitted papers in all the conference proceedings listed in Table 1*. Keynote papers were excluded, where it was possible to identify these, as they are often given by people from outside the WG 8.2 community and may, therefore, be considered un-representative of its views or, if by WG 8.2 members, typically offer an overview of a particular topic that might be expected to adopt a different approach to theory than a general research paper. The paper by Hirschheim (1985) was also excluded for this latter reason. As descriptions of panels were not always included in proceedings, and their use of references was also not consistent, these were excluded too. This gave a total of 293 papers over the period.

A very inclusive definition of "social theory" was adopted. Indeed, it was more a matter of recording all references apart from those that could be clearly identified as *not* being the work of social theorists. This resulted in a list of 154 names including not just sociologists, such as Bourdieu and Giddens, but also economists, such as Coase and Williamson, and philosophers, from Plato to Rorty.

The range of social theory analyzed thus included most of the "reference disciplines" specifically identified by Klein et al. (1996) in the first draft of the WG 8.2 "Aims" statement, i.e., "philosophy, history, sociology, political science, management and computer science," with the exception of the last two. Computer science was excluded as not being a social science, even broadly defined, while management was seen to be in the same position with respect to IS as was described above in relation to organization theory.

In most cases, the inclusion criterion proved relatively simple to apply, but, with a few management theorists, the boundary was not always so clear. For example, F. W. Taylor, and relatedly Braverman, were excluded as being primarily organizational theorists despite the sociological import of their work. On the other hand Weber, and perhaps more questionably Crozier, were included. In practice, these decisions made relatively little difference to the results (excluding them would only reduce the total number of authors cited by about six), but the existence of an element of subjective judgement in setting the boundary may be noted. A related decision was whether to include management or organizational works by authors who might otherwise be described as social theorists. An example is Kolb, whose learning circle is arguably a work of social psychology, but who has also written on its management implications in *Sloan Management Review*. The principle adopted was to include all works of such authors. Conversely, sociological works that were primarily concerned with methodological, rather than theoretical, issues, such as Miles and Huberman (1984) and Yin (1984), were excluded. The possible exception to this rule was the inclusion of Burrell and Morgan (1979) on the grounds that its survey of social and organizational theory may be a significant, if not always beneficial (Jones 1999a), source of influence on the understanding of social theory in the IS field.

Each work of social theory, identified according to these criteria, that was cited in each paper was recorded. Thus the number of works by any theorist cited in a particular paper, as well as the numbers of papers citing a particular social theorist, were identified. The country of the institution identified as the location of the first-named author for each paper was also recorded, unless this author explicitly identified themselves as a visitor, in which case the location of their "home" institution was recorded.

As a comparison, a search was also made of the ProQuest bibliographic database, which includes the full text of about 500 management journals. This sought to identify all articles referring to four of the social theorists most widely-cited in the WG 8.2 proceedings in combination with the term "information system" or "information systems."

3. Results

Of the 154 social theorists identified, 89 were cited in more than one paper. Only 14, however, were cited in more than 10 papers. Table 2 lists these authors and the number of citations to their works (a paper could cite more than one work by a particular author) at WG 8.2 conferences between 1979 and 1999 and also since 1992. The other authors, cited in less than 10 papers, are listed in Appendix 1. The maximum number of social theorists cited in any one paper was 22, and the maximum number of works of any social theorist cited in one paper was nine for Habermas. More than three quarters of the references, however, were to only one work by a particular theorist and only just over a third of the papers cited more than one social theorist.

Table 2. Social Theorists Cited in More than 10 Papers at WG 8.2 Conferences, 1979-1999 and 1992-1999

	1979-1999		1992-1999	
	Number of papers citing	**Number of citations**	**Number of papers citing**	**Number of citations**
Giddens	34	44	24	33
Habermas	27	56	17	29
Burrell and Morgan	24	24	10	10
Berger	23	25	13	13
Latour	18	35	18	35
Foucault	15	37	14	36
Geertz	15	21	10	16
Glaser	15	17	10	12
Rogers	15	19	12	15
Popper	11	14	3	3
Williamson	11	16	8	12
Callon	10	19	10	19
Gadamer	10	15	4	6
Law	10	16	10	16
Total	**238**	**358**		

As Table 2 indicates, the "popularity" of particular theorists has varied over time. For example, all of the citations of Actor Network Theorists (Latour, Callon, and Law), and all but one of Foucault's, have occurred after 1992, while the majority of references to Popper and Gadamer occurred before that date. Figure 1 illustrates this in more detail for four of the most widely-cited authors. While Giddens has been the most frequently-cited theorist overall, citations of the work of Latour have grown rapidly since 1994, such that he was the most-cited theorist at the 1995, 1997, and 1998 conferences. Conversely, the works of Habermas appear to be relatively less cited in recent conferences.

Another view of the change in citations over time is provided by an analysis of a "social citation density" (the total number of social theorists cited in all papers at a particular conference divided by the total number of papers at that conference) of the different conferences as shown in Table 3. This indicates that this value has generally increased over time, but that certain conferences were notable for the large number of references to social theory, especially Manchester 1984, and that others have had a much lower proportion of references to social theorists. Table 3 also indicates the proportion of papers at each conference not including any references to social theorists.

From Table 3, we can see that the citation density for conferences that were held in the USA was lower than for those held in Europe (1.8 compared with 2.3) and the proportion of papers at U.S.-based conferences not citing any social theorists is also slightly higher (46% compared to 40%). Considering the institutional affiliation of first-named authors of papers, we find that papers by U.S.-based authors also showed a slightly lower citation density (1.8 compared with 2.0) than authors from other countries, but that there was no difference in the proportion of papers not citing any social theorists between these two groups of papers (both had 44% of papers with no citations).

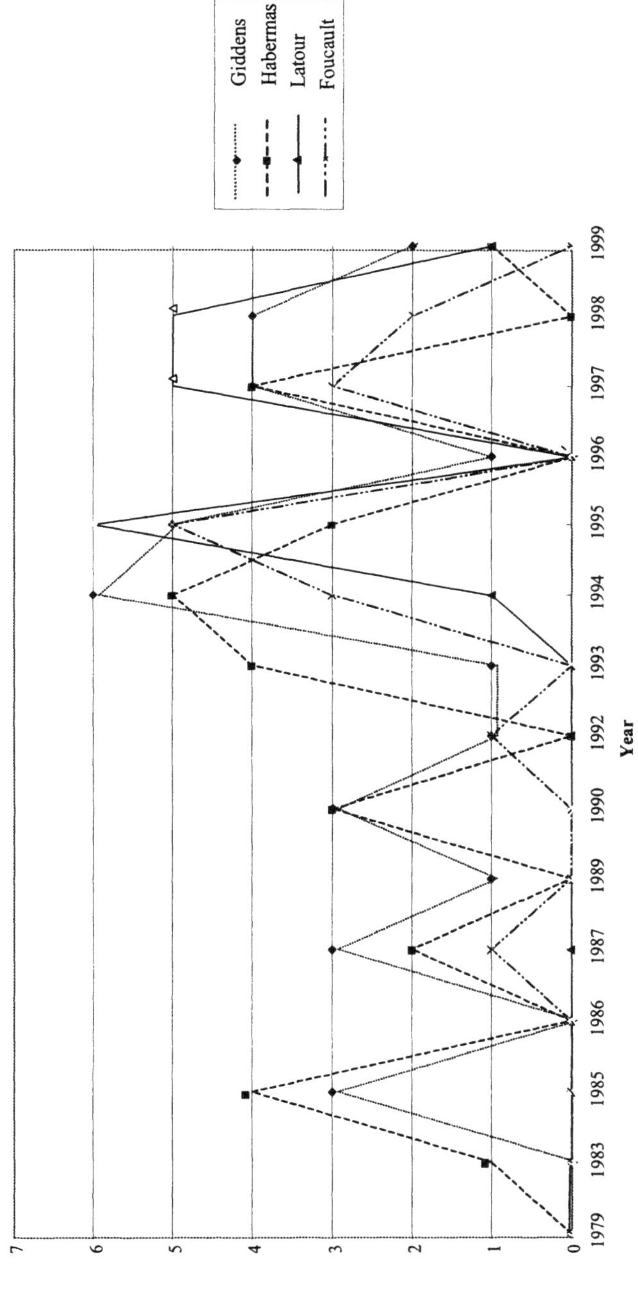

Figure 1. Frequency of Citation of Social Theorists in Papers at WG8.2 Conferences, 1979-1999

Table 3. Rates of Citations of Social Theorists at WG 8.2 Conferences, 1979-1999

Conference		Number of papers	Total citations	Papers with no citations	Citation density	No citations %
1979	Bonn	24	8	18	0.3	73
1983	Minnesota	23	12	19	0.5	83
1984	Manchester	12	58	1	4.8	8
1986	Noordwijkerhout	21	11	15	0.5	71
1987	Atlanta	13	28	6	2.2	46
1989	Ithaca	20	16	11	0.8	55
1990	Copenhagen	27	102	5	3.8	19
1992	Minneapolis	14	16	7	1.1	50
1993	Noordwijkerhout	22	21	10	1.0	45
1994	Ann Arbor	18	46	4	2.6	22
1995	Cambridge	18	75	5	4.2	28
1996	Atlanta	17	4	14	0.2	82
1997	Philadelphia	22	92	2	4.2	9
1998	Helsinki	29	80	7	2.8	24
1999	St Louis	14	39	2	2.8	14
Total/*Average*		**294**	**608**	**126**	*2.1*	*43*

Table 4 presents the results of the journal analysis. This indicates that the four widely-cited social theorists in the WG 8.2 conference proceedings were also regularly cited in conjunction with the term information system(s) in management journals. Not all of these citations were in Information Systems journals, however. Thus, looking just at the numbers of times these authors were cited in *MIS Quarterly* and *Journal of Management Information Systems* (the only two IS journals picked out by the ProQuest search), we find that these journals account for less than half of the total citations and, in the case of Foucault, for less than 15%.

4. Discussion

With references to 154 different authors, nearly 90 of whom were cited more than once, and with economists and philosophers as well as sociologists receiving more than 10 citations, the WG 8.2 conferences may be seen to illustrate a considerable breadth of theoretical interest. At the same time, however, it could be argued that the analysis demonstrates a fairly selective approach to social theory. Thus, the works of Giddens, Habermas, and, more recently, Latour have perhaps received disproportionate attention relative to other theorists such as Bourdieu, Garfinkel, Strauss, or Schutz (to pick a few

Table 4. Frequency of Citation of Four Social Theorists in Association with the Term "Information Systems" in Certain Management and IS Journals, 1992-1999

All journals	Giddens	Habermas	Foucault	Latour
1992	0	1	1	0
1993	0	0	0	0
1994	0	2	2	2
1995	7	0	2	2
1996	0	0	0	0
1997	2	1	0	2
1998	7	2	6	4
1999	7	3	4	3
Total	23	9	15	13
IS Journals only				
1992	0	0	0	0
1993	0	0	0	0
1994	0	2	1	0
1995	0	0	0	0
1996	5	0	0	0
1997	1	0	0	2
1998	0	0	0	0
1999	5	1	1	1
Total	11	3	2	3

examples of authors receiving only three or four citations). Unless it is felt that the most widely-cited authors have a unique insight on IS phenomena, then there would seem to be considerable opportunity for extending the Group's theoretical resources through an exploration of such comparatively-neglected authors' works.

To suggest that WG 8.2 research might benefit from drawing on a wider range of social theory, however, is not to belittle the current level of usage. As a comparison of Tables 2 and 4 indicates, papers at WG 8.2 conferences generally cite social theorists more frequently than is typical in the IS field as a whole (assuming that *MIS Quarterly* and *Journal of Management Information Systems* are not unrepresentative). The total number of citations of the four social theorists in WG 8.2 conference proceedings also exceeds that for all articles in management journals discussing information systems covered by the ProQuest search.

Some indication that this is a distinctive characteristic of WG 8.2 is given by a comparison of the citation density of proceedings for conferences held jointly with other IFIP working groups (Atlanta 1996 with WG 8.1 and Helsinki 1998 with WG 8.6). This shows that the number of social theorists cited in papers at these conferences was lower than the preceding or following WG 8.2 conferences. Without access to membership lists for the respective Groups, it is not possible to confirm whether this is due to a lower number of social theorists being cited in papers by members of other Groups, but these results would suggest that this was the case. The observations of Keen (1991) in comparing the WG 8.2 and ICIS (International Conference on Information Systems) conferences held in Copenhagen in December 1990 would also seem to support the view of WG 8.2 as giving considerably greater attention to social theory than is typical in the IS field.

The citing of social theorists is not a universal characteristic of all WG 8.2 conference papers, however. With an average of just over two references to social theorists per paper, but more than 80% of papers at some conferences having no such citations, references to social theorists are spread fairly thinly, with the majority of references provided by only a few papers. Less than 10% of papers, for example, cited more than five social theorists.

Nor are high rates of citation of social theorists typical of all WG 8.2 conferences, even if the effect of joint conferences is removed. While the "social citation density" has generally increased over time, some conferences stand out as having had particularly high rates. In this respect, the legendary status of the 1984 Manchester conference would seem deserved, at least in terms of the interest in social theory shown in the papers presented there. This may be related to the conference theme, as the other "methodological" conferences, Philadelphia and, to a slightly lesser extent, Copenhagen also showed a relatively high level of citation of social theorists. The high citation rates at the Cambridge conference may reflect the presence of Bruno Latour as a plenary speaker, which may have been expected to attract submissions from authors interested in, and hence citing, his work and that of other Actor Network theorists.

Following comments by Fitzgerald et al. (1985) and by Keen, it would also seem that there may be a geographical difference in the rate of citation of social theorists. Thus there is a slightly lower rate of citation of social theorists in papers presented at WG 8.2 conferences held in the USA, which typically attract a higher proportion of U.S.-based authors, and in papers by U.S.-based authors. Whether this reflects a more powerfully-institutionalized orthodoxy in the IS research community in the USA, however, or a greater receptiveness to social theory amongst non U.S.-based researchers cannot be decided from this analysis. It is also the case that a number of US-based authors have been among those most frequently and consistently citing social theorists in their WG 8.2 conference papers, so any restrictive effect of theoretical orthodoxy in the USA is clearly not universal.

As quite a few researchers have papers in more than one volume of proceedings, it is also evident that certain individuals, although not necessarily all those who have presented more than one paper, have made a distinctive contribution to the citation pattern of the WG 8.2 conferences. While it would not be appropriate to identify specific individuals, it is possible to discern at least two types of such contribution. One demonstrates an admirable, if sometimes lonely, dedication to work drawing on the same few theorists over a number of years, while others show a greater eclecticism. It is not clear, however, that either type could be said, on the basis of the citation analysis, to be significantly more influential than the other. Thus, while the iconoclasts have not always been successful in promoting wider recognition of the theorists on whom they have focused, the eclectics might be seen as simply following research fashions (Abrahamson 1996).

Relatedly, it may be questioned to what extent citation is an appropriate measure of influence. An Actor Network analysis, for example, might identify certain references as being constructed as *obligatory points of passage* for the WG 8.2 community, or particular sub-groups within it. Citations thus become not a source of insight, but a badge of membership. Although the citation count on its own does not provide enough evidence to substantiate more than a general impression, it would appear that this may be particularly the case with certain methodological references, especially Yin (1984). Although, as explained, such references were not included in the citation count, 25 references to Yin were found (which would make him the third most cited author), all of which were to the 1984 book on case study research (or later editions). Given the particular positivist approach to case research this advocates, which would not seem always to sit easily with the interpretive orientation of some of the papers in which it is cited, there may be thought to be an element of tokenism in citing the work in some cases. A similar process of obligatory citation may also be conjectured for a number of other authors included in the analysis. For example, almost all citations of Berger and Glaser were to *The Social Construction of Reality* (Berger and Luckmann 1967) and *The Discovery of Grounded Theory* (Glaser and Strauss 1967) respectively.

The use of citations as an indication of the influence of particular authors would also seem to assume that the referenced works have contributed significantly to the research. Even if citing a work is not just tokenism, however, it is not clear that the inclusion of specific references necessarily means that they have always been understood, or perhaps even read. Given the breadth of potential sources for IS research, it would seem likely that some authors are probably encountered primarily through, sometimes excellent, secondary works. In the absence of the time and resources to pursue all of the material back to the originals, there may be a temptation to assume that the secondary works provide an accurate understanding, especially if this helps to simplify the argument. This is not to say that the reading of an original source ensures that it is understood either. As Jones (1999b) has argued in relation to Giddens's Structuration Theory, for example, some of its uses in the IS field appear to conflict with key aspects of Giddens's writing.

On the other hand, depending on citations misses out on two potentially important forms of influence: via other sources or in terms of a general appreciation whose origins may not be specifically acknowledged, or necessarily even recognized. In the first respect, the exclusion of management and IS references from the analysis may have disguised significant indirect influences. For example, the contribution of some secondary sources—Boland (1985) on phenomenology or Lyytinen and Klein (1985) on Habermas, to pick two examples from WG 8.2 proceedings—should not be under-rated. Perhaps more significantly, the unacknowledged or unrecognized contribution of theorists may indicate a more profound degree of influence than the citation of original works. Thus when the ideas of a social theorist have been sufficiently institutionalized that their origins are no longer considered worthy of note, they become part of the tacit knowledge of the field (Latour 1987).

Such a development could be seen as a welcome sign of maturity in the IS field, indicating the establishment of at least some element of the cumulative knowledge tradition that authors such as Checkland and Holwell (1998) suggest is essential to its future as a discipline. Thus the decline in references to Habermas could indicate that his ideas are now "taken as read" by WG 8.2 members. The explicitly critical tone of parts of the Group's "Scope and Aims" statement, for example, the reference to critical and ethical discourses, might be seen to support this view.

As has been noted, however, the evidence of the citation analysis could equally lend support to a rather different view of the field, in which the absence of reference to primary social theoretic literature indicates lack of awareness and faddishness rather than institutionalization. Moreover, whether it is desirable, let alone possible, to establish a cumulative knowledge tradition in the IS field is itself open to debate, as the parallel discussion in the organization theory field indicates (Pfeffer 1995; Van Maanen 1995a, 1995b). The continuing diversity of the pattern of citations at WG 8.2 conferences and the absence of any single dominant approach may be seen as a sign of health, as much as a cause for regret.

That there is no single dominant approach, however, does not imply that there may be no groupings of citations. Thus, accepting that the citation data is not able to assess the indirect influence of social theorists, a number of different "schools" of WG 8.2 research may be identified. The primary grouping from this point of view is, evidently, between those WG 8.2 researchers who cite social theorists in their work and those who do not, with the former being slightly in the majority over the course of the Group's history. Among the former, the first significant grouping from an historical viewpoint was of those drawing on the work of Habermas. Interest in the work of Giddens emerged soon afterwards. Although co-cited in some papers, it is possible to distinguish between authors whose work has predominantly cited one or the other, with the latter being the slightly larger group. Foucault was first cited in a WG 8.2 conference paper in 1987, but it was not until 1993 that a significant number of papers drew on his work. Perhaps the clearest indication of a new grouping is that of the Actor Network theorists. First cited only in 1994, Latour has now become the third most cited social theorist. This may, as has already been noted, be attributable, in part, to his plenary paper at the Cambridge conference, but the interest has been maintained subsequently. References to the work of Latour are also associated with significant co-citation of other Actor Network theorists, especially Callon and Law, such that both have now received 10 citations in WG 8.2 conference papers. In no case, however, are either Callon or Law cited in papers not also citing Latour. Such a clear pattern of co-citation is not found with any other social theorists.

Given the evidence of the journal analysis that WG 8.2 conferences appear to have given greater, and earlier, attention to social theorists than is typical in the IS field, therefore, the Group's position as pioneers in the use of social theory would seem to be supported. As the citation analysis does not consider subsequent citation of WG 8.2 conference papers at later conferences, however, it is not possible to substantiate the claims of Klein (1999) regarding the influence of particular WG 8.2 papers or whole conferences on the pattern of subsequent research. Moreover, even if the analysis may be seen to indicate that WG 8.2 conferences have been successful in creating a focus for discussion of social theory in IS research, it is not clear whether any influence that this may have had on the field has been through the conferences themselves or through the writing of WG 8.2 authors in other venues. The success of WG 8.2 members in winning best paper awards at ICIS and *MIS Quarterly* may be more effective in creating wider awareness of social theory than discussion within the WG 8.2 community.

5. Conclusions

The reference to *The Ruba'iyat of Omar Khayyam* in the title of this paper was chosen, long before the citation analysis was completed, because of its ambiguity. Depending on where you end the quotation, it can be seen to comment on ephemerality of interest in social theory or on the enduring contribution of earlier work. As is so often the case, however, the evidence of the analysis provided no conclusive support for either position. In part, this reflects inherent limitations of such bibliographic measures as indicators of influence (see, for example, Johnson and Podsakoff 1994), but also the absence of an obvious pattern in the data.

Clearly, moreover, the present analysis offers only a partial, and arguably over-simplistic, view of the theoretical influences on WG 8.2 research, and there is undoubtedly considerable scope for its refinement and extension. For example, this might involve the further analysis of proceedings to consider the disciplinary basis of all references; a co-citation analysis to identify theoretical clusters; a similar detailed analysis of a number of leading IS journals to enhance the comparative exercise; or an analysis of the disciplinary backgrounds of paper authors and plenary speakers. While such additional investigations might provide a more complete picture of WG 8.2 research, however, it would be surprising if they were to alter significantly the broad outline revealed by the present study.

This suggests that WG 8.2 conferences have been notable for the extent to which the papers have drawn on social theory and that concern with social theory has generally grown over time. Whether the citations indicate a cumulative tradition, or perhaps a number of such traditions, or whether they illustrate a faddish fluctuation of interests over time is much more difficult to assess. Without detailed analysis of individual papers, for example, it is not possible to identify whether a decline in citations reflects amnesia rather than institutionalization, or an increase reflects fashionable tokenism or a significant growth in influence.

The variation in citations over time also shows no consistent trend. Thus, while the relative decline in the number of citations of Habermas and the recent surge of interest in Actor Network theorists might be seen as evidence of a shift in fashion, it is not as if Habermas is no longer cited or individual Actor Network theorists dominate the citations in an unprecedented way. Interest in different social theorists, at least as indicated by citations of their work, have clearly changed over time, but not in a systematic way, nor such that it is possible to conclude that there has been a significant and irreversible shift in the theoretical interests of WG 8.2.

If there are some very general patterns to be identified from the analysis, they might be that WG 8.2 research has drawn most strongly on social constructionism, broadly defined, with a significant input from Critical Social Theory, or Habermas at least, and, lately, Actor Network Theory. Alternatively it may be noted that there are some areas of social theory, such as the psychoanalytical literature (only one reference each to Freud and Jung, none to Kristeva, Lacan, or Levinas), that appear to have had a perhaps surprisingly limited influence in WG 8.2. As with the less-cited authors discussed above, this would seem to suggest opportunities for the Group's future theoretical extension, the pursuit of which could contribute to the maintenance and enhancement of WG 8.2's reputation as a source of theoretical innovation in the IS field.

As a final note, the whole of this paper may be seen to be predicated on the assumption that the citation of social theorists by IS researchers is "a good thing" and that higher social citation densities at WG 8.2 conferences, to the extent that they are indicative of a greater awareness and understanding of social theory, should be welcomed. While the analysis presented in this paper suggests that this view may be accepted by at least some significant proportion of the WG 8.2 community, it might be argued that there are more important issues for IS researchers to attend to. Recent concerns with "relevance" to business practitioners (Senn 1998) and "practice driven research" (Zmud 1998), for example, might seem to question the value of researchers devoting their attentions to (obscure) social theorists, especially those critical of mainstream management thinking. While, as Keen (1991) argues, there need not be a conflict between relevance and social theory and that practitioners are not necessarily antipathetic to ideas that challenge their assumptions, the priorities and concerns of practitioners may not typically embrace a broad interest in social theory.

WG 8.2 need not be defensive about its use of social theory, though, but should regard it as a valuable resource in the promotion of informed and constructive debate with both practitioners and other IS researchers. Engagement with theoretical debates would also seem necessary if IS research is to be seen as having a contribution to make to other disciplines. The question for IS researchers, therefore, is not whether they should engage with social theory, but how to do so. The evidence of this paper would suggest that WG 8.2 has been a conducive environment for such engagement and has every opportunity to continue this distinctive role in the future.

Acknowledgments

I would like to thank Eija Karsten for the "Inter-Researcher Loan" of two volumes of proceedings, David Avison, Richard Baskerville, and Rudi Hirschheim, as well as Dorothy Hayden of the IFIP Secretariat for their help in identifying the publication details for the proceedings.

References

Abrahamson, E. "Management Fashion," *Academy of Management Review* (21:1), 1996, pp. 254-285.

Avison, D., Kendall, J. E., and DeGross, J. I. (eds.). *Human, Organizational and Social Dimensions of Information Systems Development.* Amsterdam: North-Holland, 1994.

Baskerville, R., Smithson, S., Ngewnyama, O., and DeGross, J. I. (eds.). *Transforming Organizations with Information Technology.* Amsterdam: North-Holland, 1994.

Bemelmans, Th. M. A. (ed.). *Beyond Productivity: Information Systems Development for Organizational Effectiveness.* Amsterdam: North-Holland, 1984.

Berger, P. L., and Luckman, T. *The Social Construction of Reality.* Harmondsworth: Penguin, 1967.

Bjørn-Andersen, N., and Davis, G. B. (eds.). *Information Systems Assessment: Issues and Challenges.* Amsterdam: North-Holland, 1988.

Boland, R. J. "Phenomenology: A Preferred Approach to Research on Information Systems," in *Research Methods In Information Systems*, E. Mumford, R. Hirschheim, G. Fitzgerald, and T. Wood-Harper (eds.). Amsterdam: North-Holland, 1985, pp193-201.

Brinkkemper, S., Lyytinen, K., and Welke, R. J. (eds.). *Method Engineering: Principles of Method Construction and Tool Support*. London: Chapman & Hall, 1996.

Burrell, G., and Morgan, G. *Sociological Paradigms and Organizational Analysis*. Portsmouth, NH: Heinemann, 1979.

Checkland, P., and Holwell, S. *Information, Systems and Information Systems: Making Sense of the Field*. Chichester: Wiley, 1998.

Donaldson, L. *In Defence of Organisation Theory: A Reply to the Critics*. Cambridge: Cambridge University Press, 1985.

Fitzgerald, G., Hirschheim, R. A.., Mumford, E., and Wood-Harper, A. T. "Information Systems Research Methodology: An Introduction to the Debate," in *Research Methods In Information Systems*, E. Mumford, R. Hirschheim, G. Fitzgerald, and T. Wood-Harper (eds.). Amsterdam: North-Holland, 1985, pp. 3-9.

Glaser, B., and Strauss, A. *The Discovery of Grounded Theory*. Chicago: Aldine, 1967.

Hirschheim, R. A. "Information Systems Epistemology: An Historical Perspective," *Research Methods In Information Systems*, E. Mumford, R. Hirschheim, G. Fitzgerald, and T. Wood-Harper (eds.). Amsterdam: North-Holland, 1985, pp. 13-35.

Johnson, J. L., and Podsakoff, P. M. "Journal Influence in the Field of Management: An Analysis Using Salancik's Iindex in a Dependency Network," *Academy of Management Journal* (37:5), 1994, pp. 1392-1408.

Jones, M. R. "It All Depends What You Mean by Discipline...," in *Information Systems: An Emerging Discipline?*, J. Mingers and F. Stowell (eds.). Maidenhead: McGraw Hill, 1997, pp. 97-112.

Jones, M. R. "Mission Impossible: Pluralism and Multi-paradigm IS Research," in *Information Systems: The Next Generation*, L. Brooks and C. Kimble (eds.). Maidenhead: McGraw Hill, 1999a, pp. 71-82.

Jones, M. R. "Structuration Theory," in *Re-thinking Management Information Systems*, W. J. Currie and R. Galliers (eds.). Oxford: Oxford University Press, 1999b, pp. 03-135.

Kaiser, K. M., and Oppeland, H. J. (eds.). *Desktop Information Technology*. Amsterdam: North-Holland, 1990.

Keen, P. G. W. "Relevance and Rigor in Information Systems Research," in *Information Systems Research: Contemporary Approaches and Emergent Traditions*, H-E. Nissen, H. K. Klein, and R. Hirschheim (eds.). Amsterdam: North-Holland, 1991, pp. 27-49.

Kendall, K. E., Lyytinen, K., and DeGross, J. I. (eds.). *The Impact of Computer Supported Technologies on Information Systems Development*. Amsterdam: North-Holland, 1992.

Klein, H. K., Lyytinen, K., Orlikowski, W., and Pentland, B. "Draft Scope and Aims of WG 8.2," *Oasis* (30:2), 1996, p. 3.

Klein, H. K. "Knowledge and Methods in IS Research: From Beginnings to the Future," in *New Information Technologies in Organizational Processes: Field Studies and Theoretical Reflections on the Future of Work*, O. Ngwenyama, L. D. Introna, M. D. Myers, and J. I. DeGross (eds.). Boston: Kluwer Academic Publishers, 1999, pp. 13-25.

Klein, H. K., and Kumar, K. (eds.). *Systems Development for Human Progress*. Amsterdam: North-Holland, 1989.

Larsen, T. J., Levine, L., and DeGross, J. I. (eds.). *Information Systems: Current Issues and Future Challenges*. Laxenburg, Austria: International Federation for Information Processing, 1998.

Latour, B. *Science in Action*. Milton Keynes: Open University Press, 1987.

Lee, A. S., Liebenau, J., and DeGross, J. I. (eds.). *Information Systems and Qualitative Research*. London: Chapman & Hall, 1997.

Lyytinen, K. J., and Klein, H. K. "The Critical Theory of Jurgen Habermas as a Basis for a Theory of Information Systems," in *Research Methods In Information Systems*, E. Mumford, R. Hirschheim, G. Fitzgerald, and T. Wood-Harper (eds.). Amsterdam: North-Holland, 1985, pp. 219-236.

Lucas, H. "Historical Record," *Oasis* (28:1), 1994, p. 7.

Lucas, H., Land, F. F., Lincoln, T. J., and Supper, K. (eds.). *The Information Systems Environment*. Amsterdam: North-Holland, 1980.

Miles, M. B., and Huberman, A. M. *Qualitative Data Analysis: A Sourcebook of New Methods*. London: Sage, 1984.

Mumford, E., Hirschheim, R., Fitzgerald, G., and Wood-Harper, T. *Research Methods in Information Systems*. Amsterdam: North-Holland, 1985.

Ngwenyama, O., Introna, L. D., Myers, M. D., and DeGross, J. I. (eds.). *New Information technologies in Organizational Processes: Field Studies and Theoretical Reflections on the Future of Work*. Boston: Kluwer Academic Publishers, 1999.

Nissen, H-E., Klein, H. K., and Hirschheim, R. (eds.). *Information Systems Research: Contemporary Approaches and Emergent Traditions*. Amsterdam: North-Holland, 1991.

Orlikowski, W. .J., Walsham, G., Jones, M. R., and DeGross, J. I. (eds.). *Information Technology and Changes in Organizational Work*. London: Chapman & Hall, 1996.

Pfeffer, J. "Mortality, Reproducibility, and the Persistence of Styles of Theory," *Organization Science* (6:6), 1995, pp. 681-691.

Senn, J. "The Challenge of Relating Information Systems Research to Practice," *Information Resource Management Journal* (11:1), 1998, pp. 23-28.

Van Maanen, J. "Fear and Loathing in Organizational Studies," *Organization Science* (6:6), 1995a, pp. 687-692.

Van Maanen, J. "Style as Theory," *Organization Science* (6:1), 1995b, pp. 133-143.

Yin, R. K. *Case Study Research: Design and Methods*. London: Sage, 1984.

Zmud, R. "Conducting and Publishing Practice-driven Research," in *Information Systems: Current Issues and Future Challenges*, T. J. Larsen, L. Levine, and J. I. DeGross (eds.). Laxenburg, Austria: International Federation for Information Processing, 1998, pp. 1-33.

About the Author

Matthew Jones is a University Lecturer in Information Management at the Department of Engineering and the Judge Institute of Management Studies at the University of Cambridge. He previously held postdoctoral positions at the universities of Reading and Cambridge, where he was involved in the development of computer-based models for public policy decision making. His current research interests are concerned with the social and organizational aspects of the design and use of information systems and the relationship between technology and social and organizational change. Matthew can be reached by e-mail at m.jones@jims.cam.ac.uk.

Appendix 1

Frequency of Citation of Other Social Theorists in WG 8.2 Proceedings,1979-1999

Number of papers citing	
1	Aristotle; Arrow; Axelrod; Barnes; Bartlett; Bauman; Bergson; Bloor; Capra; Carnap; Chardin; Dandeker; Danzinger; Deleuze and Guttari; Dilthey; Dubinskas; Eagleton; Elias; Freeman; Freire; Freud; Friedmann; Fromm; Gleick; Greimas; Grice; Goleman; Hesse; Hirsch; Ihde; Illich; Jung; Koestler; Laclau; Laing; Leibniz; Locke; Lukacs; Lefebvre; Malinowski; McLuhan; Mumford, L; Myers; Nagel; Parsons; Pascal; Perez; Piaget; Pinch; Plato; Prigogine; Rawls; Rousseau; Ryle; Soja; Teece; Thomas, WI; Varela; Vygotsky; Waddington; Whitehead; Whyte; Zerubavel; Znaniecki; Zukav
2	Arendt; Baudrillard; Becker; Bhaskar; Boguslaw; Coase; Csikszentmihalyi; Evans-Pritchard; Galtung; Gramsci; Hacking; Hempel; Horkheimer; James; Kant; Knight; Lakatos; Luhmann; Machlup; Marcus; Marx; Merleau-Ponty; Radnitzky; Toulmin; Weber; Winch
3	Agar; Austin; Bernstein; Blumer; Boyer; Bruner; Chomsky; Eco; Garfinkel; Gergen; Haraway; Hughes; Leontjev; Marcuse; Mead; Strauss; von Hippel
4	Akrich; Bourdieu; Bunge; Derrida; Douglas; Kelly; Maturana
5	Bell; Bijker; Clifford; Collins; Ellul; Feyerabend; Heidegger; Husserl; Lyotard; Ricoeur; Rorty; Schutz
6	Apel; Goffman; Lakoff; Mackenzie; Searle
7	Bateson; Crozier
8	Kolb; Wittgenstein
9	Kuhn; Polanyi; Winner

3 SOCIO-TECHNICAL DESIGN: AN UNFULFILLED PROMISE OR A FUTURE OPPORTUNITY?

Enid Mumford
Emeritus Professor
Manchester University
England

1. Early History

Socio-technical design is now more than 50 years old. It began with the desire of a group of therapists, researchers, and consultants to use more widely the techniques they had developed to assist war damaged soldiers regain their psychological health and return to civilian life. This group, most of whom had been associated with the London Tavistock Clinic before the war and some of whom were medically qualified, believed that the therapeutic tools and techniques they had developed could usefully be applied to the organization of work in industry. They saw this as restricting and degrading many lower rank employees who were forced to spend their days carrying out simple, routine tasks with no possibility of personal development or job satisfaction.

The Tavistock Institute of Human Relations was founded by this group in London in 1946 with the aid of a grant from the Rockefeller Foundation. It was set up to bring together the psychological and social sciences in a way that benefitted society. In 1948, when the Tavistock Clinic became part of the Health Service, the Institute became a separate organization (Trist and Murray 1993).

Because many of the original members were psychiatrists, all early members of staff were required to undergo psychoanalysis. There was a belief that they had to understand themselves before they could assist with the problems of others. Both the Clinic and the Institute focused on group rather than individual treatment. This was partly because of a shortage of staff but also because group therapy was a recognized and successful method of helping with problems. This therapeutic background meant that staff were interested in results as well as theories. This led them in the direction of "action research," in which analysis and theory is associated with remedial change. The Institute believed that there should be "no therapy without research and no research without therapy." Today this could be restated as "no theory without practice, no practice without research." In 1947, a publishing company, Tavistock Publications, was founded and a new journal, *Human Relations*, was created in association with a research group led by Kurt Lewin and located at the Centre for Group Dynamics at the University of Michigan.

In 1972, the socio-technical movement was formally internationalized by the creation of a Council for the Quality of Working Life, which had members, usually academics, from many countries throughout the world. A number of academic groups became actively interested in socio-technical research. These included the Work Research Institute, Oslo, and groups at the University of Pennsylvania in the United States, York University in Toronto, Canada, and the Centre for Continuing Education in Canberra , Australia. Kurt Lewin, at the University of Michigan in Ann Arbor, also had a considerable influence on thinking and action.

2. Promises and Possibilities

When socio-technical design was first developed it was seen by its creators as a means for optimizing the intelligence and skills of human beings and associating these with new technologies in a way that would revolutionize how we live and work.

The socio-technical school believed in flexibility and intellectual growth: that individuals and groups could reorganize and redevelop to meet new challenges in changing environments and that this change process need not be too demanding and difficult. In the 1970s, many companies accepted this message and tried to restructure their procedures and change their cultures to meet new kinds of objectives, both human and technical. Unfortunately, few of these endeavors had any long term success. The attraction and validity of bureaucracy was seen as stronger and safer and the new humanistic approaches as over-risky. This paper will trace the history of socio-technical design as it moved from success to failure, attempt to find some explanations for why an approach that seemed to offer so much never realized its potential in the past, and make some predictions about its relevance for the future.

Socio-technical theory has been continually developed and tested since the Tavistock Institute was founded. Throughout its history, its practitioners have always tried to achieve its two most important objectives: the need to humanize work through the redesign of jobs and democracy at work. In order to realize these goals, the objective of socio-technical design has always been "the joint optimization of the social and technical systems." Human needs must not be forgotten when technical systems are introduced. The social and the technical should, whenever possible, be given equal weight. Over the years, this objective has been interpreted in many different ways but it is still an important design principle.

The technical system was seen as covering technology and its associated work structure. The social system covered the grouping of individuals into teams, coordination, control and boundary management. It also covered the delegation of responsibility to the work group and a reliance on its judgment for many operational decisions. A distinction was made between semi-autonomous groups and self-managing groups. The former are given authority for decision making but may lack the means to achieve this; for example, an effective information system. The latter have both authority and the necessary knowledge to control their own activities.

3. The Evolution of Socio-technical Concepts

Albert Cherns, an Associate of the Tavistock Institute, described the socio-technical design principles in an article in *Human Relations* (Cherns 1976). These were:

Principle 1. **Compatibility**. The process of design must be compatible with its objectives. This means that if the aim is to create democratic work structures, then democratic processes must be used to create these.

Principle 2. **Minimal Critical Specification**. No more should be specified than is absolutely essential. But the essential must be specified. This is often interpreted as giving employee groups clear objectives but leaving them to decide how to achieve these.

Principle 3. **The Socio-technical Criterion**. Variances, defined as deviations from expected norms and standards, if they cannot be eliminated must be controlled as close to their point of origin as possible. Problems of this kind should be solved by the group that experiences them and not by another group such as supervision.

Principle 4. **The Multifunctionality Principle**. Work needs a redundancy of functions for adaptability and learning. For groups to be flexible and able to respond to change, they need a variety of skills. These will be more than their day-to-day activities require.

Principle 5. **Boundary Location**. Boundaries should facilitate the sharing of knowledge and experience. They should occur where there is a natural discontinuity—time, technology change, etc. in the work process. Boundaries occur where work activities pass from one group to another and a new set of activities or skills is required. All groups should learn from each other despite the existence of the boundary.

Principle 6. **Information** must go, in the first instance, to the place where it is needed for action. In bureaucratically run companies, information about efficiency at lower levels is collected and given to management. It is preferable for it to go first to the work group whose efficiency is being monitored.

Principle 7. **Support Congruence**. Systems of social support must be designed to reinforce the desired social behavior. If employees are expected to cooperate with each other, management must also show cooperative behavior.

Principle 8. **Design and Human Values**. High quality work requires:

- jobs to be reasonably demanding;
- opportunity to learn;
- an area of decision making;
- social support;
- the opportunity to relate work to social life;
- a job that leads to a desirable future.

Principle 9. **Incompletion**. The recognition that design is an iterative process. Design never stops. New demands and conditions in the work environment mean that continual rethinking of structures and objectives is required.

William Pasmore, writing in *Human Relations* (1985), provides a positive assessment of what the socio-technical approach has achieved over the years. He describes the key insights provided by the early researchers as a recognition that the work system should be seen as a set of activities contributing to an integrated whole and not as a set of individual jobs. As a result, the work group becomes more important than individual job holders. Control should be devolved downwards with the work system regulated by its members, not by external supervisors. This would increase both efficiency and democracy. At the same time, flexibility and the ability to handle new challenges would be enabled through a work design philosophy based on skill redundancy. Work group members should have more skills than normal production required. (Today this is called multi-skilling.) Work activities should not be restricted to routine tasks. Work group members should have as many discretionary as prescribed tasks to perform. And, most importantly, the individual member of any team must be seen as complimentary to any machine, not subordinate to it. This would remove the dictatorship of the moving assembly line. Finally, because an important objective of the socio-technical approach is to increase knowledge, the design of work should lead to an increasing amount of variety for the individual and group so that learning can take place.

4. International Developments in the 1960s and 1970s

In Europe in the 1950s and 1960s , industry was weak and was being rebuilt. The strength and productivity of the United States was greatly envied and believed to be a product of better management. European industry was seen as centralized and authoritarian while American industry was becoming more democratic through the influence of the human relations movement. The principal initiators of socio-technical design were the Scandinavian countries. Their approaches had marked similarities. Norway, Sweden, and Denmark, although using different methods and emphasizing different aspects of work, all had a common set of values on what they hoped to achieve (Cooper and Mumford 1979). These values were made explicit in legislation, and management and trade unions were required to cooperate in achieving improvements in the work situation. Work design, although an early manifestation of the desire to improve the quality of working life, was only one aspect of the process of joint decision taking.

In academic circles, a great deal of optimism was associated with these new ventures. Geert Hofstedte, a Dutch expert, believed the humanization of work could become the third industrial revolution. He saw the first as the move from muscle power to machinery in the 19th century, the second as the arrival of information technology, and the third as these new approaches to work (Hofstedte 1979).

Let us now examine the experiences of the principal participating countries in more detail.

4.1 Norway

Norway was a major pioneer in the humanization of work. In 1962, a group of Norwegian researchers, headed by Einar Thorsrud, who was assisted by Fred Emery then at the Tavistock Institute, initiated what was called "The Norwegian Industrial Democracy Programme." This was a three phase program focusing on, first, creating improved representative systems of joint consultation. These involved the creation of worker directors. Next the program progressed to workplace democracy with employees gaining the authority, power, and resources to change their own work organization, when and where this was appropriate. This led to four major experiments in work restructuring in Norwegian industry.

A national strategy for the humanization of work was a product of these initiatives. This incorporated a Norwegian law on working conditions which gave workers the right to demand jobs conforming to socio-technical principles of good work practice—variety, learning opportunity, own decision power, organizational support, social recognition, and a desirable future. Following on and responding to this came a program for increasing trade union knowledge about technology and, as a result, union bargaining power. This program was led by a group at the Norwegian Computing Centre headed by Christen Nygaard (Eldon 1979). The industrial democracy project was stimulated by the fact that, in the 1970s, much of Norwegian industry was being taken over by multinationals and the environment had become very turbulent.

Although the work design experiments were generally successful, Norway experienced two kinds of resistance to the democratization of work. There was a general belief on the part of workers that any management inspired change must be for the worse, while engineers and technologists saw some of the changes as threatening to their positions and status. These problems have dogged many other change programs.

4.2 Sweden

Sweden was in the same situation as Norway and copied its example. By 1973, between 500 and 1,000 work improvement projects were taking place in Swedish industry. Sweden had made its first efforts toward the democratization of working life through the establishment of joint industrial councils in 1946. In the 1970s, the Swedish Government took this further by introducing a "Joint Regulation of Working Life Act." This was implemented in 1977. Both management and unions now needed some guidance on how to proceed in the new areas of codetermination. These were wide ranging covering, the interests of employees, with an emphasis on self-managing groups. They also included better personnel management, better strategic planning and increased productivity

(Apslund and Otter 1979). A program was agreed that encouraged unions and management to broaden the activities of joint councils so that these could develop new strategies for organizational redesign and business improvement. It was also agreed that unions did not have to rely on the goodwill of management. If management did not make sufficient progress with implementation, then the unions could apply pressure.

A major breakthrough was a move from job design to organizational design. It was in the later 1970s that Per Gyllenhammer created his new "dock assembly" work system at Volvo's Kalmar Plant. This removed the traditional flow line system of car production and substituted group working, with a single group assembling an entire car (Lindholm and Norstedt 1975). The project also developed the idea of worker directors, which the Swedish government required in state enterprises.

An important piece of knowledge acquired during this project was that self-managing groups separated by space and time have more difficulty in coordinating and controlling their activities than those organized bureaucratically. They require excellent information systems to assist their self management. These groups must also be able to set clear production objectives that are acceptable to management. Another problem is how to manage the interface between the workers and the technical systems when there are no foremen, production planners, or quality controllers. The group has to manage all these activities itself. Negotiation now has to replace orders as the primary tool of management and this in itself is very difficult to manage. Success with these new work systems requires the enthusiasm of both management and unions.

4.3 Denmark

Formal management/worker cooperation on job content and job design began in Danish companies after the second world war. An agreement in 1947 led to Consultative Committees with equal numbers of employer and employee representatives being set up in a number of large companies (Larsen 1979).

In 1970, a new agreement was made between the Danish Employers Confederation and the Danish Federation of Trade Unions. This required a focus on both production and job satisfaction. It also gave employees the opportunity to become decision partners in the design of their own work situations. A number of factors influenced this move toward work humanization. They included increased interest from management and unions who both saw advantages in a more contented work force. Stable conditions of employment also played their part.

The results, although encouraging, indicated that work humanization could not be achieved without overcoming a number of difficulties. Not all groups of employees had the same interests and wanted the same solutions. A lack of support from senior management or from the trade unions could also slow down progress, as could changes in a company's marketing or economic situation. Danish experience suggested that certain conditions were required for success. These included company stability and financial health. Change was extremely difficult if workers were being laid off. As in Sweden, good relationships and a history of cooperation together with an enthusiastic top management and positive union officials were also necessary. Technology must not act as a design constraint and there must be a wage payment system that reinforces group working. Employees should also have a good level of education.

4.4 France

In the 1970s, France too was interested in the humanization of work. A survey of 18 companies in 1975 and 1976 showed that a great many jobs had now been enlarged, enriched, or rotated (Trepo 1979). The principal reasons for this effort were a search for production gains together with a recognition of the need to reduce labor problems, which included absenteeism, industrial conflict, and poor quality work. In an attempt to overcome these, the French government introduced legislation requiring employers to demonstrate how they had improved working conditions and how they proposed to improve these further. But the French trade unions were suspicious of these job design efforts, seeing them as yet another possible means to exploit workers.

4.5 Italy

Italy was a rather different situation from France. In Italy, the existing rigidly structured and tightly controlled form of work organization, often called Taylorism, was seen as a product of Fascism. The Italian unions, in contrast to unions in other countries, were prepared to fight against this and were determined to secure control over the organization of work (Rollier 1979). The initiative for change, therefore, came from the unions with management as reluctant partners. The union became a major force pressing for change and also the focal point for the promotion and spread of organizational research. Agreements in the early 1970s with companies such as Olivetti and Fiat paved the way for experiments similar to those at Volvo with "production islands" and flexible work cycles. As might be expected, there was resistance from employers, although Olivetti was an exception. The company was converting from engineering to electronics and needed new forms of work organization.

All large Italian companies were afraid of the trade unions and most produced suggestions for work changes, but there was little conviction that the new work system would lead to increases in production. In 1974, Italy had a major economic crisis. Management became frightened of the economic situation and started reshaping their production systems with the aim of breaking the unions. This meant restoring the old Taylorist model and abandoning the proposed changes.

4.6 Germany

Strategies to improve the humanization of work in West Germany began in the early 1970s. These were strengthened, in 1973, by a major strike in I. G. Metall over the humanization of work and worker participation. The result of this was that Works Councils now had a say in corporate development and that these subjects became a part of collective bargaining. They also led to discussions between parliament, government, and the trade unions (Leminsky 1975). It was increasingly recognized that work was of central importance to a satisfactory life and that rewarding work must contain opportunities for autonomy, freedom, and choice.

This meant that the content of work had to be changed. There must be better training, job enrichment, and the organization of work around groups. Production, repairs, and control would now all be carried out by these groups. These reforms were implemented through new laws and by making Works Councils responsible for their introduction and for monitoring their effectiveness. A program for the humanization of work was introduced by the Federal Ministries of Labor and of Science and Technology in May 1974. This program had three components. First, the development of standards and minimum requirements for machines and workplaces. Second, the development of technologies to meet human requirements. This included computers. Third, case studies and models for the organization of work, based on the socio-technical analysis used in Britain and in Norway. Firms that were willing to introduce new forms of group work, which included more job variety, would receive subsidies to meet part of the cost of these experiments. These changes were facilitated by new legislation, which formalized and ratified workers' rights.

Works Councils were the principal change agents and any plans for reorganization made by the employer had to be agreed by the Works Council. This meant that the trade unions had to train their Works Council members in the management of change and in how to influence policy. The unions also succeeded in gaining Mitbestimmung—the equal representation of labor on supervisory boards and labor directors on executive boards. These became the new worker directors.

This humanization of work program continued successfully for some years but was criticized by socio-technical consultants in other countries for excluding the worker on the shop floor from discussions. Everything was left to the trade unions.

4.7 Netherlands

The Netherlands has always taken a lead in work humanization and a major European pioneer in socio-technical design in the 1960s and 1970s was Philips in Eindhoven. The company had many programs that incorporated what the firm called work restructuring and work consultation (Mumford and Beekman 1994). Today we might call these work design and participation. These programs were the responsibility of a special department called Technical Efficiency and Organization.

The commitment of this department to technical change began in the 1960s when the company first noted signs of unrest among blue collar workers who were doing boring and monotonous jobs. Management, and in particular the Director of TEO, were determined to overcome this. Philips believed strongly in the socio-technical principle that the social must have the same importance as the technical and they also understood the relevance of the social sciences to good management.

Philips recognized that work restructuring and participation required major changes in attitude from both management and workers. This new perspective was achieved through meetings, discussions, and lectures, all of which included the Works Council and the trade unions. Although in the 1980s many of these high hopes for the spread of job enrichment and employee participation diminished for harsh economic reasons, in the 1970s Philips was providing an inspiring example of socio-technical design (Mumford and Beekman 1994).

4.8 United Kingdom

In 1949, the Tavistock Institute pioneered two action research projects. One was a study of joint consultation at the Glacier Metal Company, the other was an investigation of the organization of work in the newly nationalized Coal Board (Jacques 1951). The chief researcher in the first project was Elliott Jacques and, in the second, Ken Bamforth, who had worked as a miner and found many ideas for the redesign of work in his mining experiences (Trist and Murray 1963; Scott, McGivering, and Mumford 1963).

These projects were both successes and failures. New patterns of consultation worked successfully at Glacier but were restricted by the authoritarian attitudes of senior management . Jacques eventually left the Tavistock as he came to believe that the authority structure of British industry, supported by a legal framework, made any fundamental employee democracy difficult if not impossible. The coal mine research had a mixed reception. Group work involving multi-skilling and a degree of self management worked well on experimental faces but was not viewed favorably by the trade union as it conflicted with wage negotiations, which were based on traditional work structures. The Coal Board was not enthusiastic, either, as it did not want trouble with the unions (Mumford 1997).

In 1965, a large scale socio-technical project took place in Shell UK with the assistance of the Tavistock. Shell UK was interested in a new management philosophy that incorporated the idea "that the resources of a company are also the resources of society" (Hill and Emery 1971). The company set out to redefine its objectives in terms of this philosophy. It was decided that these social and business objectives could best be achieved through the use of socio-technical concepts. The Tavistock principle of seeking to achieve the joint optimization of technical and human factors was to guide implementation of the program. This project lasted for four years in the UK and the experiments then continued in Shell plants in Austria, Holland, and Canada. They are still taking place.

4.9 United States

In the 1960s and 1970s, the notions of organizational development and the human relations model were extremely popular in the United States but, as the business environment changed, these became less relevant. In 1972, interest in the socio-technical approach was awakened. A decline in productivity was associated with unhappy employees who were alienated from their work. At the same time, competition from Japan and West Germany was increasing. Socio-technical projects in the United States were usually initiated by management without union or worker participation and were directed at increasing organizational effectiveness as well as the quality of working life. Most unions viewed these new policies with suspicion, seeing them as an attempt to undermine their interests or to increase productivity to the disadvantage of the worker (Davis and Cherns 1975, p. 5). But there were exceptions. The United Automobile Workers' Union negotiated contracts with General Motors, Ford, and Chrysler in which clauses were included establishing joint management-union committees to improve the quality of working life and to encourage and monitor experiments in job redesign. These projects continued for a number of years.

In the 1980s, an influential group of American researchers, managers, and consultants formed themselves into the Socio-technical Round Table. This group was originally sponsored by the Society of Manufacturing Engineers and managers from both the Digital Equipment Corporation and General Motors played a major part in its early activities. Socio-technical researchers and practitioners from other countries were invited to join. It played a major role in communicating the socio-technical message to American industry. This group is still active today.

Socio-technical projects were not restricted to Europe and the United States. India was one of the pioneers in work redesign. An early project was carried out in a cotton mill in Ahmedabad, where a group of workers became responsible for a group of looms, work was reorganized and an increase in productivity occurred. These new methods did not last and a visit to the firm by Tavistock researcher A. K. Rice in 1963 found that the old methods had returned. A new management was reluctant to give up power (Rice 1953). However socio-technical initiatives continued, led by an Indian supporter of the Tavistock approach, Processor Nitish De.

5. Why was Socio-technical Design So Popular in the 1970s?

By the end of the 1970s, there was evidence that socio-technical ideas were becoming accepted. The reasons for this interest were similar in all participating countries. Industry was expanding and many firms had labor difficulties. There were problems in obtaining staff and firms were scared of losing those they had.

Projects were spreading from manufacturing to service industries and it appeared that workers were becoming increasingly dissatisfied with the old methods of production. The socio-technical supporters believed that "quality of working life" was an emergent value and that human development could be fostered through work. In their view, the technical imperative would eventually fade away and labor and management would not continue to operate in an adversarial mode. They must and could collaborate. But the socio-technical group was over optimistic. Progress was not as great as its members believed. Initiatives usually came from individuals at the top of a company anxious to achieve stability and harmony and, even more important, to reduce labor shortages. These initiatives would become fewer once the labor market changed and many were seeking work. A major difficulty during this period was that few trade unions embraced the socio-technical concept. Many saw this as a threat to their power and influence.

A group that acted as an effective communicator and facilitator for socio-technical design at this time was the Quality of Working Life Council. This international group was drawn from many different European countries as well as India, North America, and Australia. It was chaired by Einar Thorsrud, a leading Norwegian academic, and spread the quality of work message throughout the world through meetings, training sessions, books and articles. Its members worked with many different companies, initially helping them to introduce socio-technical projects onto their shop floors and later into offices. This group was very influential. It had a common purpose and a strong network of relationships. The members acted as information conduits in their respective countries and through attendance at international conferences.

6. The 1980s

Strategies which work well at one time may not be successful at another. Both culture and the business climate can change. Many researchers have seen the 1980s as a disappointing time for organizational innovation. Industry came under pressure to cut costs and socio-technical approaches were increasingly seen as expensive and risky. Computer-assisted clerical and production systems were becoming very popular and an era of what has been described as "computer aided neo-Taylorism" arrived (Moldaschl and Weber 1998). The work of many clerks was routinized as computers moved into offices and a new shop floor technology called lean production took over the car plants. Lean production involved team work of a limited kind, also multi-skilling, direct feedback, and continuous improvement, but work was not made more flexible and interesting. It became faster, more streamlined, and more stressful (Stace 1995). The principal differences between socio-technical design and lean production were the methods for controlling and coordinating work. Socio-technical design created decentralization of control and coordination by the user group. In contrast, lean production focused on the standardization of work processes (Niepce and Molleman 1988).

Although there were few socio-technical initiatives in Britain during this period, a number of researchers, including the author, successfully carried out projects to assist the introduction of new computer systems. All of these followed the socio-technical model. They were participative in that future users at all levels played a major role in the design task, in particular rethinking the design of jobs and work processes for their own departments before new systems were installed. These user design groups, aided by systems analysts who acted as advisers on technical issues, tried to give equal weight to technical and human concerns and introduced team work, multi-skilling and a degree of self management (Mumford 1995, 1996a). The projects included large companies, such as Rolls Royce and ICI, and a number of major banks and hospitals in the UK, together with the Digital Equipment Corporation in the United States. In both countries, these socio-technical design projects were brought to a successful conclusion and implemented. One of the largest and most significant of these was the participative design of XSEL, one of Digital's first expert systems. This was developed to assist the sales force to configure VAX computers and was designed for worldwide implementation (Mumford and MacDonald 1989).

The socio-technical initiative now became dispersed and centered on smaller groups in different countries. The Tavistock retained its influential role in the UK, projects in Scandinavia continued, Eric Trist was influential in the United States and Fred Emery in Australia, the American Socio-technical Round Table was created, and Federico Buttero set up a consultancy in Italy. But the international impact was now greatly reduced. Noone was seriously pushing an integrated message internationally.

In the 1980s, industry's principal objective became cutting costs to compete in increasingly challenging international markets and maintaining or raising the price of their shares. Reducing costs through reducing staff numbers was one way of doing this and socio-technical approaches were seen as having little to offer (Mumford 1996).

7. The 1990s

The 1990s proved very frustrating to the exponents of socio-technical design. Companies recognized the need for change and were motivated to make changes but chose methods such as lean production and business process reengineering that took little account of employee needs and did not produce good human results. There were, however, exceptions. Despite difficult economic circumstances, a number of companies in the United States, Europe, and Australia continued with socio-technical projects, remodeling these to fit changing economic and social conditions. Today, the emphasis in Australia is on participative design, Scandinavia favors a democratic dialogue between management and workers, and the expert group of socio-technical consultants belonging to the Socio-technical Round Table assists American companies. Many U.S. projects are based on the development of high commitment and high performance work groups based on the cooperative sharing of power between workers and management.

8. What Can Socio-technical Design Contribute in the Future?

Socio-technical theory continues to be of interest to researchers. The Dutch are now developing an approach called Modern Socio-technical Theory, which focuses on production structures as the main determinant of any socio-technical program. The theory behind this approach is that most production systems are over complex and cannot be easily controlled; they need to be simplified (Eijnatten and Zwaan 1998).

Sweden has also been developing the socio-technical concept by bringing the company's business environment into the redesign task. Volvo now uses the phrase "Delivery, Quality and Economic Results" (DQE) to describe its objectives, which are primarily related to cost control. Results are achieved through achieving direct contact between work groups and groups in the external market such as customers and suppliers. The proposed next step is to develop socio-technical systems for business. Adler and Docherty (1998) suggest that the dominant socio-technical research tradition has shifted over time from a social dimension in the 1970s to a technical dimension in the 1980s, greatly influenced by the Dutch, and a business dimension in the 1990's developed by research groups in Scandinavia.

Despite these initiatives in Scandinavia and The Netherlands, few companies in other countries have been interested in extending the use of socio-technical design as a general design principal. The prevalence of down sizing in the 1990s has led to flatter hierarchies in many firms and it has been recognized that innovative companies require highly skilled groups who can work as members of high performance teams. These teams give their members responsibility and autonomy but they are usually privileged groups in senior positions, often working in high stress conditions.

Industry is now moving into turbulent waters as globalization increases, technology produces new organizational forms, and an underprivileged section of the world population finds that employment is not available. All of these are a recipe for conflict and possible disaster. The most important contribution socio-technical design can make to this situation is its value system. This tells us that, although technology and organizational structures may change in industry, the rights and needs of all employees must always be given a high priority. These rights and needs include varied and challenging work, good working conditions, learning opportunities, scope for making decisions, good training and supervision, and the potential for making progress in the future. The socio-technical principles of quality of life and personal control must also be applied to those that are not privileged to have paid employment and who rely on the state for security.

Opportunities for a socio-technical revival may soon be arriving. Millennium society is unlikely to be contented and placid and there are already signs of major conflicts ahead. Commercial success in tomorrow's world requires motivated work forces who are committed to the interests of their employers. This, in turn, requires companies and managers who are dedicated to creating this motivation and recognize what is required for this to be achieved. A return to socio-technical values, objectives, and principals may provide an answer to many of our future problems.

References

Adler, N., and Docherty, P. "Bringing Business into Socio-technical Theory," *Human Relations* (51:3), 1998, pp. 319-345.

Apslund, C., and Otter, C. V. "Codetermination through Collective Effort," Chapter 12 in *The Quality of Working Life in Western and Eastern Europe*, C, L Cooper and E. Mumford (eds.). London: Associated Business Press, 1979.

Cherns, A. "Principles of Socio-technical Design," *Human Relations* (2:9), 1976, pp. 783-792.

Cooper, C. L., and Mumford, E. "Introduction," in *The Quality of Working Life in Western and Eastern Europe.* London: Associated Business Press, 1979.

Davis, L., and Cherns, A. *Quality of Working Life* (Volume 2). Free Press of Glencoe. 1975.

Eijnatten, F. M. v., and Zwaan, A. v. d. "The Dutch Approach to Organizational Design: An Alternative Approach to Business Process Reengineering," *Human Relations* (51:3) , March 1998, pp. 289-318.

Eldon, M. "Three Generations of Work Democracy Experiments in Norway: Beyond Classical Socio-technical System Design," Chapter 11 in *The Quality of Working Life in Western and Eastern Europe*, C. L. Cooper and E. Mumford (eds.). London: Associated Business Press, 1979.

Hill, P., and Emery, F. *Towards a New Philosophy of Management.* London: Gower Press, 1971.

Hofstedte, G. "Humanization of Work: The Role of Values in a Third Industrial Revolution," Chapter 12 in *The Quality of Working Life in Western and Eastern Europe*, C. L. Cooper and E. Mumford (eds.). London: Associated Business Press, 1979.

Jacques, E. *The Changing Culture of a Factory.* London: Tavistock Publications, 1951.

Larson, H. H. "Humanization of the Work Environment in Denmark," Chapter 7 in *The Quality of Working Life in Western and Eastern Europe*, C. L. Cooper and E. Mumford (eds.). London: Associated Business Press, 1979.

Leminsky, G. "Trade Union Strategies for the Humanization of Work in the FRG," in *Human Choice and Computers*, E. Mumford and H. Sackman (eds.). Amsterdam: North Holland, 1975.

Lindholm, R., and Norstedt, J. P. *The Volvo Report*. Swedish Employers Confederation, 1975.

Moldaschl, M., and Weber, W. G. "The Three Waves of Industrial Group Work: Historical Reflections on Current Research on Group Work," *Human Relations* (51:3) , March 1998 , pp. 347-388.

Mumford, E. "Assisting Work Restructuring in Complex and Volatile Situations," in *Developing Organizational Consultancy,* J. E Neumann, K. Kellner, and A. Dawson-Shepherd (eds.). London: Routledge, 1997.

Mumford, E. *Effective Systems Design and Requirements Analysis*: *The ETHICS Method.* London: Macmillan, 1995.

Mumford, E. *Ethical Tools for Ethical Change.* London: Macmillan, 1996.

Mumford, E. "Risky Ideas in the Risk Society," *Journal of Information Technology* (11), 1996, pp. 321-331.

Mumford, E., and Beekman, G. J. *Tools for Change and Progress.* Amsterdam: CSG Publications, 1994.

Mumford, E., and MacDonald, B. *XSEL's Progress.* New York: Wiley, 1989.

Niepce, W., and Molleman, E. "Work Design Issues in Lean Production from a Socio-technical Perspective: Non-Taylorism or the Next Step in Socio-technical Design?" *Human Relations* (51:3), March 1988, pp. 250-287.

Pasmore, W. A. " Social Science Transformer: The Socio-technical Perspective," *Human Relations* (48:1), January 1985, pp. 1-22.

Rice, A. K. "Productivity and Social Organization in an Indian Weaving Shed," *Human Relations* (6), 1953, pp. 297-329.

Rollier, M. "Taylorism and the Italian Unions," Chapter 10 in *The Quality of Working Life in Western and Eastern Europe,* C. L. Cooper and E. Mumford (eds.). London: Associated Business Press, 1979.

Scott, W., McGivering, I., and Mumford, E. *Coal and Conflict.* Liverpool: Liverpool University Press, 1963.

Stace, D. A. "Dominant Ideologies, Strategic Change and Sustained Performance," *Human Relations* (48:1), January 1995, pp. 553-570.

Trepo, G. "Improvement of Working Conditions and Job Design in France," in *The Quality of Working Life in Western and Eastern Europe,* C. L. Cooper and E. Mumford (eds.). London: Associated Business Press, 1979.

Trist, E, and Murray, H. *The Social Engagement of Social Science, Volume 2: The Socio-technical Perspective.* Philadelphia: University of Pennsylvania Press, 1993.

About the Author

Enid Mumford is an Emeritus Professor of Manchester University, England. She was formerly Professor of Organizational Behavior at the Manchester Business School. Her latest books are *Effective Systems Design and Requirements Analysis*: *The ETHICS Method* (London: Macmillan 1995), *Ethical Tools for Ethical Change* (London: Macmillan, 1996), and *Dangerous Decisions: Problem Solving in Tomorrow's World* (New York: Kluwer Academic/Plenum, 1999). Enid can be reached by e-mail at enid@em.u-net.com.

4 THE LIMITS OF LANGUAGE IN DOING SYSTEMS WORK

Richard J. Boland, Jr.
Weatherhead School of Management
Case Western Reserve University
U.S.A.

Abstract

Doing systems work brings us to the limits of language as few human activities do. It uniquely joins the empathetic reading of human motivations, desires, and needs with a creative envisioning of new socio-technical arrangements in hopes of transforming the world. It is at once humble and audacious, finely detailed and grandly epic. Fundamental notions of goodness, truth, and beauty are relied upon in ways that forever challenge our ability to justify.

This paper sets five voices in dialogue to explore the limits of language in doing system work. The five voices, C. West Churchman, Sir Geoffrey Vickers, Richard Rorty, Bruno Latour, and Pierre Bourdieu, represent a wide range of 20th century traditions in system thinking, philosophy, sociology of technology, and social theory. Their dialogue is animated, conflictual, melodic, and unnerving, much like system work itself. Instead of a consensus on language, limits, or systems, they provide us a landscape and some paths for future exploration in our own dialogues.

This is a fictional account of an imaginary, virtual meeting in which the voices of Pierre Bourdieu, C. West Churchman, Sir Geoffrey Vickers, Bruno Latour, and Richard Rorty create a roundtable discussion on the topic "Are there limits to language which affect the design of information systems?" The pretext for this virtual meeting is a funding initiative by the Millennium Technology Committee, an arm of the European Union Millennium Celebration Council, to conduct a series of studies on the current state of information technology deployment in industrial society. One of the studies, of which this roundtable is a part, is to revisit the age old questions of system implementation. Why do so many good ideas for information technology go unrealized? Why are so many systems not successfully implemented? Why are systems often not used to their full potential? In this roundtable on the limits of language and the doing of systems work, four of the participants are imagined to be participating in a human form, while Geoffrey Vickers is imagined to be in a cyber form. He is being represented in the roundtable discussion by a conversational system that has been created from Sir Geoffrey's writings and those of his favorite authors.

The format is thus somewhat unusual, but this performative approach to writing was chosen as a way to make the ideas of these authors come alive just a bit more than in a normal paper. Also, it is an attempt to allow ideas to emerge in the process of writing the voices of the paper, allowing each voice to have its way, and see where it leads. The things these characters say are not intended to represent anything their real counterparts, living or dead, actually have said, although they are intended to represent what the characters might have said. So there are no quotations from their writings where these statements are to be found. Rather, this virtual meeting is put forward as a modest attempt by one person, who has spent some time thinking about their individual writings, to set them in conversation and see what happens.

The format is a video conference in which Geoffrey Vickers' synthesized voice is accompanied by animation of him based on a film produced during his presentation at the Second International Symposium on Communication Theory and Research in March, 1966.

Geoffrey Vickers: This workshop was called because you are some of my favorite thinkers on the questions relevant to the design of information technology in organizations. West Churchman was chosen for his work in system thinking and the systems approach to inquiry. Bruno Latour was chosen for his studies of the way we put our interests to work in creating and deploying technologies, the study of how systems come to be as a continuing accomplishment. Pierre Bourdieu was chosen for his unique approach to the study of the social world and the process of generative structuralism that produces and reproduces organizational practices. Richard Rorty was asked to join us because of his beautiful writings as an American pragmatist, in the tradition of Dewey. So you might see that I have chosen this group as a reflection of how I define myself. I am a pragmatist who is focused on practice and recognizes the socially imbued quality of the multivalued judgements we must make in designing and managing organizational systems.

The topic of "The Limits of Language" was proposed because of my own position that the language and cognitive schemas, which we as humans have evolved over the last million years or so, are now encountering a new kind of environment, one of our own making, to be sure, but one in which we are increasingly blinded to the limits of thought and action that we have created for ourselves. The limits of language are thus, in a sense, of our own making in the context of our evolving state of societal development. I would like to start our discussion by exploring those ideas a bit more fully and then open it up to your criticism and your own position on language, limits, and design. But first, I want to thank you all for joining me in this virtual roundtable tonight.

Pierre Bourdieu: I am willing to participate in this experiment, but I think the questions of organization, computers, and technology that you pose are much more complex than the way you have represented them.

West Churchman: You haven't, for instance, mentioned ethics as an essential feature of any such discussion.

Bruno Latour: I don't know what I can contribute to this complicated topic you have chosen. I know so little about organization, but then none of us does. Still, I will do what I can.

Richard Rorty: Well, I'm looking forward to an interesting evening and will try to help the conversation along.

Geoffrey Vickers: Let me open the discussion by summarizing why I have posed the topic as the "Limits of Language." I don't have a specific limit in mind, but rather a belief that all systems have in them an inherent set of self-generated limits and that those limits are associated with the characteristic dynamics of that system as those dynamics affect its ability to evolve and adapt to changing circumstances. That, of course, is a general system belief, and in a moment I will identify some ways in which I see that systemic principle applying to language. But another reason I choose that title is Wittgenstein's haunting phrase from the Tractatus, "The limits of my language are the limits of my world." That line is so evocative for me, that I find it popping into my thoughts all the time. I was reminded of these lines just recently while reading an interview with Frank Gehry on the expression of ideas in architecture and he said, "You cannot escape your language." So in both those senses, I felt that "Limits of Language" would be an intriguing topic for us.

The systemic limits of language that concern me most are, first, the limits related to the collapse of the multivalued experience of human judgement into a single valued language of policy discourse; and second, the limits resulting from a language that has evolved over many millennia of rather slow change, confronting a world of exceedingly rapid change. Let me discuss each of them briefly. First, we are, I believe, limited in our ability to reason at the policy level because our language confuses questions of value with questions of efficiency. Our language collapses judgements of what we value and of what constitutes a betterment for us given those values, with judgements of what constitutes an expansion of our resources or an increase in efficiency in our use of these resources. Judgements of betterment are political judgements, and judgements of expansion are economic judgements, each having their own language and logic. Yet, we replace a judgement of betterment with a judgement of expansion by allowing an economic language to be the sole language for thinking through organizational and technological design questions. In organization design, we don't use a language of politics to discuss our values and our judgements of betterment, and our organizations are the worse off for it—especially in the uses of technology.

Pierre Bourdieu: If by your preference for a language of politics you are saying that the field of organization is fundamentally a field of power struggles, then we have a common reference point.

Geoffrey Vickers: I'm not just saying that—I am saying that our language is inadequate to our task as responsible actors in today's social world. Perhaps the best way to map it into your work is to say that our habitus and language practice present us with objectified structures that collapse judgements of betterment into an economic vocabulary of expansion. This fools us into using a trajectory-like image for guiding our thinking about design, management, and organization rather than a more appropriate image of balance. We are dominated by images of directionality, especially upward thrust, rather than images of the give and take in evolutionary adaptation. Our guiding imagery is based on increase and forward motion rather than balance and adjustment.

In this way, our language and its related mental schemata hides from us the systemic properties of self generated limits in social systems and of the need for more cybernetically sensitive vocabularies for use in policy discourse. We go around changing the world to suit ourselves and mistakenly believe that we are expanding our opportunities rather than limiting them. We believe our increasing use of technology is giving us increased power and control. This expectation is part of our habitus from which we cannot easily escape. We don't see that it is really creating new forms of instability in our social systems, and that those instabilities become the source of new problems, which we address with the same misguided logics of expansion and control.

Pierre Bourdieu: I would like to pose right at the beginning that much as I appreciate your use of my ideas and the generative cycle of practices you have just sketched for us, I take the limits of language to be quite different than what you have proposed, especially if we mean to reflect on the practice of system work. If system work means the construction of information technologies in organizations, then I think it is much more appropriate for us to step back from suppositions about language structures and look carefully at the field in which that language operates, especially the struggle for power in that field and the ways in which system work is a structuring structure. By that I mean that we must attend to the ways in which system work is the reproduction of power relations and the redistribution of capital in organizational fields. System work and its use of language has a structure that shapes the ongoing creation of structures. Systems of information technologies then themselves become structuring structures and in so doing they objectify the subjectivity, which you seem to be so fascinated by.

If there is a limit to language, it is found in the ways in which individuals come to believe that their habitus—their orientations, expectancies, and readiness to act—are a universal subjectivity rather than a localized field of practice in which their particular subjectivity is generated. A generative structuralism of binary oppositions that characterizes a domain of action marks positions in their field and determines the forms and distribution of its capital. This is all relational and must be approached as such. In other words, it isn't the logic of the actor, but the logic of the practice in a field that is going to generate the objective probabilities of that field that we must consider if we are to explore limits of language in system work.

C. West Churchman: You are at once too close to the ground with your talk of local practice in fields, and too far from the real problem with your talk of structured structuring structures. Yes, we have habitus and fields, or as I prefer to say, "Weltanschauung," but that is a condition for the operation of reason, and it is the operation of reason that we must pay attention to here. If there are limits to language, they will make differences in our choice decisions, or they will make no difference at all. And the central requirement for reason is to have a guarantor. The guarantor of reason in language use will ultimately have to do with finding a way to, as it were, swallow the whole. By this I mean that reasoned choices, to be rational in any meaningful sense, must be choices that consider the full, relevant system and chose from the full set of available alternatives. So it is not just the thing that someone does or chooses to do that determines its rationality; but we must also include a consideration of all the things that they do not do or choose to do. It is the construction of the alternatives that we choose among that we should be paying attention to, if we are serious about the real limits of language. Looking at practices may be of some relevance in the modeling of systems, but it is on the modeling of our world, and of worlds other than ours, that we must focus, not on practice as it is today. Modeling of possible worlds and inquiry into their functioning as wholes is what will enable us to create a more enduring and beautiful world, which is what doing systems work is all about.

Pierre Bourdieu: All the other possible worlds and their alternatives! Worlds other than our own! Such bizarre and uniquely American ideas of "boundless frontiers" and fascination with an expansive destiny. The logic of practice is always very heavily constrained when our habitus, as memory, meets the objectivity of a field. Our history is brought forward into the present in our habitus, and the field we encounter has an objectified structure that we cannot reinvent. The result is an objective set of probabilities for practice, not an open world of "anything goes." It is because we carry the full burden of our past that we experience the limits of language.

Bruno Latour: Excuse me, but if I may sneak a word in edgewise here, I am fascinated and also humbled by the incredibly detailed knowledge and precise expression you two possess about something I find utterly mysterious. The words you have been using with such abandon—words like society, structures, policy, cybernetic systems, and so on—are completely beyond my humble ability to do tricks with and to juggle in the ways that you do. My vocabulary in comparison is quite poor. I don't know how to begin dealing with these invisible things you seem to take for granted. Where can I go and observe them? What door of what room in regular or cyber space should I open so that I can learn to know them in this amazing way?

I know it's tempting to speak of such imaginary things as if they had a certain causal power in your life. These kinds of fetishes can perhaps bring you some piece of mind or at least provide a handy stopping point for your arguments, but they should be resisted if we are to say something interesting about technology and organizing.

When I told my friend Barbara Czarniawska that I was going to participate in this cyber salad on the limits of language, she immediately exclaimed, "There are no limits to language! We are always saying new things, always inventing new words, creating new forms of expression and new genres!" And, of course, she is right in a certain sense, but I do see even in our little discussion here this evening how particular ways in which the participants use language sets limits for our thinking. Sir Geoffrey, for instance, in his strangely disembodied form, argues that our language tricks us into thinking about trajectories when we really should be thinking about evolutionary adaptability. But either way, whether it is a mythical journey to Nirvana or whether it is sexual reproduction making us ever more fit as a species, you end up fixated on an imaginary essence without realizing how that essence is necessarily tangled up in all manner of mysterious, invisible causal explanations.

Professor Bourdieu, on his part, is limited by the very precision of his language, so that fields, habitus, forms of capital, positions, and power struggles become a landscape from which he cannot escape to simply look around and consider what other things might be going on in creating an organization and its technologies. What a dreary, predetermined world this language of practice becomes. We can use it to explain everything, yet we understand nothing. I realize that Professor Bourdieu is somewhat aware of how our apparatus for viewing the world limits us, and that he asks his devotees to purify their predetermined analysis with reflexivity on their own reflexivity. But these reverberating reflexive shadows will not help their eyesight, and it is a humble, open set of eyes and ears we need in studying system work.

West Churchman wants us to be able to see wholes—but the social world we live in is flat. There is no place one can stand to see wholes or anything like them in this flat world of ours. There are no lumpy, abstract high points in the landscape: a closely limited horizon of local situations is all we have to see and navigate within. So immediately his language limits us to the unseeable, the unspeakable, and the unreachable. This, I hope you will all agree, is a pretty severe limit.

Pierre Bourdieu: Mr. Latour is a bit too flamboyant for my taste and his false modesty of how little we know of organizations is not well founded. I do know something about information technology design and organizations. I believe it would be clear to anyone who takes the trouble to collect and quantify and categorize the data in the exhaustive and careful manner that I have, that we can know certain things about the fields, habitus, and practice involved. The field of the technology analyst is a globalized professional field in which consultant/designers struggle for the cultural capital of intellectual achievement and for economic capital. The field of the worker is, in contrast, a local organizational field in which workers struggle primarily for the social capital of affiliation. From this view, the problem of implementation is readily apparent. The logic of practice of the consultant/designer expects that workers should be readily willing to make some simple change in daily routines because it is a rational response to the functional requirements of accumulating intellectual capital. For the worker, embedded in a generative cycle of social capital based practices, a change that is considered minor by the designer is in fact a threatening disruption of their life and very position in the field. A disruption of the very relations and conditions on which their social capital is based.

I further know that the consultant/designer and the worker are in fields with different temporal rhythms. For the worker, it is a rhythm of short cycles and many repetitions per day. For the designer, it is a long cycle rhythm with weeks or months between milestones or repetitions. For the worker, it is a rhythm of familiarity; for the designer, a rhythm of novelty. The designer moves freely through a global professional space while the worker is generally confined to a local market of limited movement. And I could go on and on with these things we know quite clearly about information technology and organizations. It is not so mysterious as you claim.

C. West Churchman: I agree with you on that last point, at least. And in my own defense, I know that the designer has to act. As befuddling as the situation might seem, designers have to muster the courage and the moral judgement to model the whole as best they can, recognizing the inferential leaps involved, and deducing a choice. Imperfect as this sounds, and difficult as the limits of language we have all identified make it, the designer has to act and act responsibly. And because I know that, I know that reason as a guarantor for the designer's action is the foundational language problem we must address.

Richard Rorty: And I know that we have a striking diversity of vocabularies here: flatland, generative structuralism, schematas, deductive logics, and all the subtexts they proliferate. There is a peculiar sense in which all of you are using vocabularies that are implicitly claiming to let us hear or see organizations as they really are. Of course, you know I reject that claim, but what are we left with? What kind of conversation do we make with all these vocabularies overturning and undermining each other? I think of Wittgenstein's comment on language from the *Philosophical Investigations*: "And how many kinds of sentences are there? Countless kinds. Think of the tools in a tool box." I would like us to think of these different vocabularies we have displayed here tonight not as contestants in a competition to see who is right or who is closest to getting the correct description of technology or organization or system work, but as tools available to a discerning crew of workers doing system work. Or perhaps we should think of them as voices in a chorus where the thing we try to get right is the harmonious blending of voices—the aesthetics of representation that we can interweave with these diverse voices. The "cash value" of what we can do with them. And also, how these voices in this chorus can open us to invite other vocabularies of representation including sound, visual imagery, art, and even dance into our conversations.

I like Barbara's strong intuition about the question we are discussing. There are limits to language only if we let there be limits, through shutting off other voices and closing off our conversations, because we fear they might lead to dead ends. But, of course, this leaves us in our self made dead end, conversing only with those who prefer our own preferred vocabulary. Keeping the conversation going with an open and changing chorus of vocabularies is the best way I know of to keep language subject to limits.

Geoffrey Vickers: This notion of music, imagery, and art is one point I strongly agree with. I've always felt there were multiple forms of consciousness at work in the way judgements were actually made in organizations. I've always thought that multiple kinds of sensations were involved in what I call appreciate judgements, or the process in which our judgements of reality and our judgements of value are brought together in a multi-valued judgement of fit. We know something is the correct thing to do not because of logic alone, but because of an appreciative judgement of appropriate balance between the multiple values we hold and the reality we face. Appreciative judgement rests on aesthetics and on all the senses, which are subject to aesthetic judgements of fitness, appropriateness, and desirability. Appreciative judgement is what drives action when it is good, true, and beautiful, not rational choice.

Richard Rorty: There you go again, down a path of searching for an ideal manager's own preferred vocabulary.

Bruno Latour: But you know, Richard, as I sit here listening to you, it suddenly hits me how you yourself are trapped in a limited vocabulary of associating thinking with something that happens in our heads and something that takes place in words. I much prefer to think of thinking as something we do with our hands. As Hutchins shows so beautifully, in his *Cognition in the Wild,* our cognition is a distributed process in which humans and artifacts together create calculation and intelligent performance. So rather than confusing vocabulary with thinking, as you seem to do when you ask us to keep our thinking open by keeping our vocabularies open, I would much rather have us think about thinking as something we make with artifacts, handwork, physical motion, tactile manipulation, inscriptions, and, of course, words as well. We make ideas, systems, and minds as we work with objects and words—we don't have understandings first and then merely put them in words. We should focus then on the making of cognition in as open a way as possible, not the blending of voices as if they were ready made. Think, Richard, of how you treat language as a ready made "thing" for workers to pick up and use. When you ask us to search for new languages or vocabularies, you ask us to "cast about" for them as if we could just fish a new vocabulary out of a stream.

To pick up on the tool metaphor you started to develop but then dropped, I would like us to think of system work as tool based work. The tools that they use in doing system work are their language, and the kinds of tools they are able to put to hand are their vocabularies. It is not just the words they use, but all the diagram techniques, interviewing strategies, and the ways they use their bodies and their hands that are the thinking and the language of system work. Maybe if we could see a kind of continuous motion picture of the system worker close up, it would help us see how unique and singular each site of system work is. If there are limits to this hand-work language, they are to be found in each filming location where the actors are using both words and artifacts in making their own contexts as they make the organization and its information systems. So there are only local limits, and local limits will prevail in language use.

Pierre Bourdieu: Well, perhaps there are local limits, but this is only true in a trivial sense, a sense in which the langue and parole of language are confused. The local, situated use of langue, its parole, will always be a limit in a superficial sense. Of course, in understanding the logic of social practice, they are profoundly important, but in a generative sense, which seems to be the focus here, the structuring structure of language, its langue, is the determining language operation. Langue and the structure of its binary oppositions set the limits of language that will prevail in doing system work.

West Churchman: I feel that in my saying what I conclude about language, I have ended up being mischaracterized here, and perhaps at some later time that is a part of the limits of language that I will think about further. But for now, what I must point out is how my conclusions about reason, guarantors, and the need for a sense of the whole system have turned out to mask something even more important in my views on language and their limits—something that serves as my ontological grounding and is in stark contrast to Bourdieu and Latour. Reason, guarantors, or a sense of whole are not the wellspring of my thinking—they are merely the best conclusion I can reach, using both logic and emotion. The wellspring for my thinking is the individual human being. The lonely, isolated, mortal, struggling, flesh and blood human being who acts: that is what requires reason, sense of whole, and a guarantor. And that foundation of the singular, passionate, morally responsible, and often anguished human from which I draw my conclusions is missing from all these arguments.

I am a humanist, pure and simple, and I am proud of it. I reject what both Latour and Bourdieu have said—realizing that they disagree between themselves quite strongly, but seeing each of them as losing sight of the primacy of the individual. Latour accuses me of looking to the imaginary and the unknowable for the operation of reason. But he, in turn, has made the individual disappear in favor of a circulating network in which artifacts are as important as humans, any node of which is subject to the same types of mediation and translations of interests. Bourdieu makes the individual disappear into recursively reproduced practices where habits replace the passion and will of the singularly potent person. Give me the flesh and blood, the agonizing existential reality of the human being facing the dread of everyday responsibilities. That's where I want to start. That is what is real: the individual human actor answering to God and the future of humankind for her actions.

Richard Rorty: That's good, very good—a nice dramatic move, West. It really got me on my feet and dancing and that's important after so much solemn celebration of the cerebral. Seriously, though, what you say is important because it brings us back to the way that our existence in a human community activates our emotions, our sense of affiliation, our sense of moral responsibility, and all the many uniquely human qualities that should, I think, be central to this discussion of language and limits. It seems to me that West's passion for the morally responsible individual is important if we are to bring ourselves as human actors and our own limits into this discussion, which I think we must. Literature, of course, is another way to do that.

Geoffrey Vickers: Your own writing has prompted me to reread Wittgenstein and I must say he seems to be very central to what West Churchman is saying about the flesh and blood person. But I'm not sure he comes to the same conclusion, if it could be said that he comes to any conclusion at all. Wittgenstein's own search for the ideal in language and for logic as a guarantor of the truth of our statements seems to be the place Churchman is ending up, with both of them having become aware of the limits of logical propositions and reasoning. But Wittgenstein later abandoned that search for logic and embraced the importance of language use in forms of life, or as he called them, language games. He warned us against letting our language "go on holiday" as it does in most theorizing, losing touch with the "rough ground" of engagement in the world. Rather than searching for a guarantor, making a careful description of the actual use of language in its multiplicity of possible meanings is all we can hope to achieve. There is no higher question about language. Appeals to theory are, therefore, irrelevant.

Bruno Latour: Finally, something I can agree with. You cannot get too close or too detailed in your description of social life or organizational work. And maybe that is the curse of system development. It is, of necessity, a search for the abstract and the general in the face of the details of life as it unfolds. If system designers could abandon their models and keep their noses closer to the ground, we would all be better off.

Geoffrey Vickers: I see we are running out of time here. We seem to have covered a lot of ground in our discussions, but I'm wondering just how far we have really come. I have told the Millennium Technology Committee that we would prepare some kind of statement summarizing our position on the limits of language in doing system work, and I do hope we can achieve that; perhaps in a subsequent session.

Richard Rorty: Yes, I see we have about one minute left, enough for me to make a final point. I have enjoyed Bruno trying to find a weak spot—or should I say a strong spot— in my ontology of language. But I think it's safe to say that I can match him or anyone step for step in following a Wittgensteinian path of appreciating that "knowing how to go on" is about all we can hope for when it comes to theorizing organizational work. Let me close, though, by pointing out the one limit no one has made clear, and that is the limit of being in language and at the same time trying to talk about language and its limits. Bruno thinks my metaphor of casting about for new vocabularies is ontologically misguided, but at least I don't presume we can somehow step outside of language. Whether you hope for a rational guarantor, a careful description, a reflexively cleansed approach to studying fields of practice, or a greater control over language construction and language choice, you are relying on a rather amazing capacity to use language to observe language as if from afar. I don't hold any such pretension and I am skeptical of any claims that rely on such an ability.

Geoffrey Vickers: I'm going to have to let that be the last word in this conversation. I think it's safe to say that many interesting and challenging ideas have come out of this roundtable. I'm not surprised we couldn't reach some agreement in such a short time, but I am hopeful that the conversation we have started can be continued by us and by others who are as fascinated by language and language use as I am. Thank you for participating.

Acknowledgments

The author appreciates the constructive criticism made by Geof Bowker and Ulrike Schultze on an earlier version of this paper, and the helpful discussion she had with Barbara Czarniawska while it was being written.

References

Bourdieu, P. *Outline of a Theory of Practice.* Cambridge, UK: Cambridge University Press, 1997.

Bourdieu, P. *The Logic of Practice.* Stanford: Stanford University Press, 1990.

Bourdieu, P., and Wacquant, L. J. D. *An Invitation to Reflexive Sociology.* Chicago: University of Chicago Press, 1992.

Churchman, C. W. *The Design of Inquiring Systems: Basic Concepts of Systems and Organization.* New York: Basic Books, 1971.

Churchman, C. W. *The Systems Approach.* New York: Delacorte Press, 1968.

Churchman, C. West. 1968. *The Challenge to Reason.* New York: McGraw-Hill

Latour, B. *Science in Action.* Cambridge, MA: Harvard University Press, 1987.

Latour, B. *Aramis or the Love of Technology.* Cambridge, MA: Harvard University Press, 1986.

Latour, B. *We Have Never Been Modern.* Cambridge, MA: Harvard University Press, 1993.

Rorty, R. *Philosophy and the Mirror of Nature.* Princeton , NJ: Princeton University Press, 1979.

Rorty, R. *Consequences of Pragmatism.* Minneapolis, MN: University of Minnesota Press, 1982

Rorty, R. *Contingency, Irony, and Solidarity.* Cambridge, England: Cambridge University Press, 1989.

Vickers, G. *Value Systems and Social Process.* Harmondsworth, UK: Penguin Books, 1970.

Vickers, G. *The Art of Judgement.* New York: Basic Books, 1965.

Vickers, G. *Freedom in a Rocking Boat.* Harmondsworth, UK: Penguin Books, 1970.

About the Author

Richard J. Boland, Jr. is Professor and Chair of the Department of Information Systems at the Weatherhead School of Management at Case Western Reserve University. Previously he was Professor of Accountancy at the University of Illinois at Urbana-Champaign. He has held a number of visiting positions, including the Eric Malmsten Professorship at the University of Gothenburg in Sweden in 1988-89, and the Arthur Andersen Distinguished Visiting Fellow at the Judge Institute of Management Studies at the University of Cambridge in 1995. His major area of research is the qualitative study of the design and use of information systems. Recent papers have concerned sense making in distributed cognition, hermeneutics applied to organizational texts, and narrative as a mode of cognition. Richard is Editor-in-Chief of the research journal *Accounting, Management and Information Technologies*, and co-editor of the Wiley Series in Information Systems. He serves on the editorial board of six journals, including *Information Systems Research* and *Accounting, Organizations and Society.* He can be reached by e-mail at rjb7@po.cwru.edu.

Part 2:

Transforming the Fundamentals

5 INFORMATION SYSTEMS CONCEPTUAL FOUNDATIONS: LOOKING BACKWARD AND FORWARD

Gordon B. Davis
University of Minnesota
U.S.A.

Abstract

The academic field of information systems has developed because organizations use a specialized body of knowledge about information and communications systems. Teaching and research support these organization needs. The field may be defined in terms of observed information systems in organizations and also in terms of the function or field of activity for system planning, development, management, and evaluation. Since the systems deal with capture, repositories, processing, and communication of data, information, and knowledge, these are also defined.

Conceptual foundations for the field are the set of concepts and propositions that explain why structures are designed the way they are, tasks are scheduled and accomplished in the way they are, and activities are performed the way they are. There are three approaches to conceptual foundations: an intersection approach that accepts any concept from any field if it appears to add insight and explanation to information systems practice and research, a core approach that seeks to define those ideas that characterize the discipline and make it distinct, and an evolutionary approach that seeks a cohesive set of concepts by combining the concepts from the core approach with concepts from other fields that over time are found to be especially useful to information systems.

At this time, there is significant variety and a number of concepts that are said to be useful in research and practice. In the long run, the evolutionary approach relative to conceptual foundations will probably prevail and reduce the scope and variety somewhat. It is a mixed strategy that fits the diversity inherent in a worldwide community of scholars. As the core concepts are developed and clarified, the core will be strengthened. However, there will still continue to be strong use of other bodies of knowledge containing concepts that support explanation and research relative to information systems.

1. Introduction

For both academic and research purposes, the field of information systems deals with systems for delivering information and communications services in an organization and the activities and management of the information systems function in planning, designing, developing, implementing, and operating the systems and providing services. These systems capture, store, process, and communicate data, information, and knowledge. The systems combine both technical components and human operators and users. The environment is an organization or a combination of organizations. Participants tend to describe the organization in terms of purposive, goal directed behavior, but in practice the organization also reflects personal agendas, power issues, prejudices, misunderstandings, etc. To explain this combination of technology, human participants, rationality, and other behaviors requires a rich set of concepts.

Starting in the 1960s and 1970s, research in information systems looked to bodies of knowledge that contained concepts and research results or research methods relevant to the study of information systems and the activities of information systems personnel. The bodies of knowledge most often used were system concepts (both soft and hard), information concepts, humans as information processors, organization behavior, management, and decision making.

The body of concepts (the conceptual foundations) has grown, and subspecialties are emerging. The paper explores three views relative to the growth of conceptual foundations. The first is an open view, that the intersection of disciplines provides rich opportunities and, therefore, the growth should be tolerated and perhaps encouraged. The second view is that this growth leads to a chaotic field that has difficulty coexisting with fields that have drawn tight boundaries around their disciplines. The proponents of the second view propose an emphasis on the essential core, that which differentiates information systems from other disciplines. The third, the evolutionary view, believes that the field will be somewhat more bounded because many conceptual foundations are not sufficiently robust relative to information systems to maintain themselves as part of the field. In other words, the field will naturally begin to be more selective. This view encourages more emphasis on the core but resists excluding bodies of knowledge that enrich the explanations of the field.

The paper presents some assumptions about the field of information systems, develops a definition of the field, describes three approaches to conceptual foundations, describes the current situation relative to conceptual foundations, and presents the author's view of future development of conceptual foundations for the field and its effect on research.

2. Some Assumptions about the Field of Information Systems

The paper rests on some assumptions about the nature of the field of information systems (by whatever name), the field being applied rather than basic science, the necessary practitioner connection, the nature of organizations employing information systems, and the complementary rather than alternative nature of different views of information systems. Although the field theoretically may develop a research agenda and conceptual foundations separate from its education mission, there is likely to be interaction.

2.1 A Field by Whatever Name

In North America, the terms **information system (IS)** and **management information system (MIS)** are identical in meaning and interchangeable in use. They refer to the system providing information technology-based information and communication services in an organization. These terms, and similar terms such as **information management**, also refer to the organization function that manages the system. The system terms and function names are broad in scope and encompass information technology systems and applications for transactions and operations, support of administrative and management functions, organizational communications and coordination, and for adding value to products and services. The academic field may be termed **information systems** (or IS), **management information systems** (MIS), **information management,** or **management of information systems** (MoIS). In other countries, there may be variations, such as **informatics** (often modified by organization, administration, or a similar term to differentiate from informatics as computer science).

The changes in terminology in the field reflect changes in the scope and consequently the research agenda. When computers were first utilized in organizations in the mid-1950s, the applications were primarily simple processing of transaction records and preparation of business documents and standard reports. This use was termed **data processing (DP)** or **electronic data processing (EDP)**. The business function for developing and managing the processing systems was also termed data processing. By the mid-1960s, many users and builders of information processing systems developed a more comprehensive vision of what computers could do for organizations. This vision was termed a **management information system (MIS).** It enlarged the scope of data processing to add systems for supporting management and administrative activities including planning, scheduling, analysis, and decision making. The business function to build and manage the management information system was often termed MIS.

In the 1980s and 1990s, there was a merging of computer and communications technologies in organizations. The organizational use of information technology was extended to internal and external networks, systems that connect an organization to its suppliers and customers, and communications systems that enablepeople in organizations to perform work alone or in groups with greater effectiveness and efficiency. Many organizations were able to achieve competitive advantage by the use of information and information technology in products, services, and business processes. Innovative applications based on information technology created value by providing services any time, at any location, and with extensive customization. Web-based communication and transaction applications became common. Information technology-based systems were employed to change organization structures and processes. There emerged a tendency to employ simple, general terms such as information systems or information management to identify both the multifaceted information technology systems and the corresponding organization function.

2.2 Information Systems as an Applied Academic Field Must Connect to Practice

Information systems is a relatively new organization function and academic field. Although there have been some changes in other business functions and related academic fields, the set of organization functions has remained reasonably stable since the advent of modern management and organization theory and practice. What then is the basis for a new function and a new academic field?

Organizations have separate functions because of the benefits of specialization and the limits of humans in dealing with specialized bodies of knowledge and practice. There is a separate marketing function because there is a specialized body of marketing knowledge and specialized marketing activities that are performed best by specialists. The entire organization needs some understanding of marketing, accounting, finance, etc., but not everyone can be expected to have sufficient depth of knowledge and skill to perform all activities. Using accounting as an example, everyone uses accounting reports and provides input into accounting, but end-user accounting in which each person decided on the chart of accounts to use and the rules and procedures for accounts and reports would result in confusion and lack of performance of vital functions. The accounting function has specialists who deal with the chart of accounts, financial reporting, reports to governments and regulators, analysis of financial results, etc.

Information systems emerged as a separate organization function because of the need for specialized development and operational activities and specialized management procedures. It is possible to outsource many technical activities, but the core activities of strategic planning for information systems, determining requirements, obtaining and implementing systems, providing support, evaluation, and so forth require technical and managerial specialists.

Academic fields emerge when there is a body of specialized knowledge and practice that can be provided by an academic discipline. There is a strong mapping of organizational and societal needs to the fields of study in colleges and universities. The observed systems and activities of organization functions provide the basis for research. This logic is demonstrated in the development of the academic field of information systems.

There is a direct relationship between the activities of the information systems function in an organization and the academic field of information systems. The academic field describes the structure and activities of the function and explains "why" they are needed, "why" they are organized and conducted the way they are, and alternatives that may be applied and conditions suggesting their use. The academic body of knowledge not only describes and explains but also guides the development and application of practice by suggesting concept-based improvements. The rationale for the explanations, suggestions, and alternatives are derived from concepts and theories of human organization, communications, decision making, human capabilities, and so forth. The concepts employed are selected from large bodies of underlying discipline knowledge; selection is based on relevance to explaining or guiding practice. The debate over research rigor versus relevance to practice is ongoing. A recent issue of the *MIS Quarterly* (Vol. 23, No 1, March 1999) presented many facets of this debate with responses from well-known scholars. The issue also contained articles on qualitative, interpretive, and case research in information systems.

Information systems in organizations
IS practice
↕
Academic body of knowledge describing, explaining, and guiding IS practice
Includes concepts developed especially for IS or appropriated and specialized for information systems (body of information systems theory)
↓
Relevant selections from bodies of underlying discipline knowledge

Figure 1. Relationship of Practice to Field Theory to Underlying Disciplines

2.3 Information Systems Support Organization Objectives and Organizational Rationality

It is clear that the normative view of organizations as having clear objectives, pursuing these objectives with rationality, and employing information system to support analytical processes is not a complete picture of how organizations operate in practice. However, this normative view (typified by Simon) is a useful presumption. It supports the design of an information system meeting ideal requirements. Since organizations never function according to the ideal, why design for it? Because it provides a coherent model of organizations and information systems. A view that organizations are chaotic in nature and irrational in operation provides an unstable basis for development and implementation of systems.

2.4 Complementary Nature of Different Views of Information Systems

The orderly, rational view of organizations provides the basic model and basic assumptions for the design and development of information systems. Alternative views provide a basis for adjustments to processes and procedures in order to deal with organizations as they are. A view of information systems research (see Boland and Hirschheim 1987; Cotterman and Senn 1992; Nissen, Klein, and Hirschheim 1991) suggests the scope of these alternatives. They enlarge the ability of analysts and users to improve design and use of systems. Two examples are a socio-technical perspective and a human-centered perspective. The socio-technical perspective is described in a number of publications, especially by Mumford. I have used Nurminen's humanistic perspective as the basis for a human-centered view (Nurminen 1988).

The socio-technical perspective does not reject the idea of rationality in organizations and the existence of organizational objectives to be met. The perspective emphasizes the fact that technology affects the nature of work, and there are alternative ways to incorporate technology into work design. Those who are affected by the introduction of technology should be included in the design process. The final design reflects the social nature of work as well as the efficient use of technology. The socio-technical perspective complements the technical systems perspective.

In the rational, technical perspective, tasks are performed by dividing functions and activities between humans and technology. Humans are assigned functions and activities requiring the unique abilities of humans; computers and other information technology perform functions and activities to which they are suited. To the designer, humans and machines are alternative objects to be designed into the system. An alternative perspective is to view humans as being able to construct work activities using tools provided to them. The emphasis of the designer is to provide a set of information technology functions that a human user can employ in performing a task. Again, the underlying assumption of organizations with purpose and rationality is not eliminated. The rational process of combining capabilities of humans and machines is the starting point for thinking about the design of a system. The human-centered view complements the technical view by introducing the notion that improvement in organizational systems can be achieved if humans are given tools to support self-design of activities to accomplish tasks.

2.5 A Teaching Perspective on Different Views

There are teaching implications to how academics formulate the field and deal with its complexity. Those who teach information systems across the spectrum from beginning survey courses for all students to advanced courses for majors in information systems may have noticed that entry-level textbooks and other course materials portray information systems as part of organizational rationality. Some simple concepts emphasizing the human element relative to systems may be introduced. Advanced courses may explain difficulties with this portrayal and suggest that simple, rationality-based methods for doing systems analysis and design may be deficient. Other fields have somewhat the same problem. Elementary accounting presents accounting processes, reports, and concepts in a very rational context of measurement of the financial consequences of organization activities. Intermediate accounting explains that the elementary accounting presentations did not deal with difficult measurement and reporting problems.

One can argue that a rational, orderly presentation of the structure and purpose of the information system of an organization helps students to develop a useful mental model of the system. Likewise, there are pedagogical reasons for describing systems analysis, development, and implementation as a simple rational, step-by-step process with goals, objectives, and deliverables. However, those students who will become practitioners in information systems should gain a richer view of the complexity of systems and system development processes. The problem may be that we have not agreed on the difference between the descriptive, rationality-based material to be presented in the overview course and the rich explanation required by those doing the work or supervising it. The idea of a first survey course followed by a richer, in-depth advanced course is not well defined.

3. Definitions of the Field

Although there is no agreed-upon definition for the field, most definitions converge quite well based on the need for two definitions. One definition is based on the observed system. If an organization describes its information system to an observer, using goal-directed language, the system can be explained in terms of its various elements (technical and human) and in terms of the organizational activities served. An additional definition describes the organization and activities of the information system function. Both definitions are required to define information systems as they exist, since both system and function are required. An additional set of definitions deals with the nature of data, information, and knowledge, because these elements are captured, stored, processed, moved, combined, communicated, and so forth by information systems.

3.1 Definition of Information System Based on the Observed System

A system-oriented definition describes the observed system and identifies its boundaries within the structure and operations of organizations. This matches the historical development of information technology within organizations. A simple definition might be that an information system is a system in the organization that delivers information and communication services needed by the organization. This can be expanded to describe the system more fully.

> The **information system or management information system of an organization** consists of the information technology infrastructure, application systems, and personnel that employ information technology to deliver information and communications services for transaction processing/operations and administration/ management of an organization. The system utilizes computer and communications hardware and software, manual procedures, and internal and external repositories of data. The systems apply a combination of automation, human actions, and user-machine interaction.

This definition is based on observations of the technical and procedural components of information systems in organizations and the structures and activities that make it work. The structure of the information system for an organization consists of the hardware/software infrastructure, repositories, and two broad classes of application software: transaction processing/operations and administration/management.

- Infrastructure. The information technology infrastructure consists of the computer and communications hardware and software and the repository management software. It provides processing, communications, and storage capabilities required by application software systems and user activities.

- Repositories. The repositories store data required for transactions, operations, analysis, decision making, explanations and justifications, and government/legal requirements. Repositories have varying scopes such as enterprise, parts of the organization (divisions, offices, departments, etc.), groups, and individuals. The stores include data about entities relevant to the organization; text and multimedia stores of analyses, reports, documents, data search results, e-mails, faxes, conversations, etc.; stores of procedures and directions for performing organizational activities including models for analysis and decision making. The repositories are also termed databases, files, data warehouses, knowledge bases, and model bases.

- Transaction processing/operations applications. Transaction processing applications record and process business transactions such as accepting a customer order, placing an order with a vendor, making a payment, and so forth. These applications range from periodic transaction processing to online immediate processing. They include web-based applications that link an organization with its customers and suppliers. Operations applications schedule and direct the operations of the organization as products are produced and distributed and services are scheduled and performed. Transaction processing and operations are increasingly integrated in enterprise systems as a continuous flow from transactions to operations that they initiate.

- Administration/management applications. These applications support clerical and knowledge workers in performing tasks individually and collaboratively. They support management requirements for data, analysis, reports, and feedback for operational control, management control, and strategic planning. Areas of application include decision support systems, executive support systems, knowledge management systems, and online analytical processing.

It is important to the field of information systems to understand and explain the characteristics of the observed systems. The infrastructure, repositories, and two broad classes of application systems—transaction processing/operations, and management/ administration—can be studied in terms of form, function, behavior of personnel using and operating the systems, behavior relative to organization activities, and value added by their use.

The definition of the observed system was developed very early. The pioneers in computing in organizations incorporated all of these ideas in their plans and visions. Early plans included all of the above applications. The ability to deliver these concepts grew over the years, but the ideas were there from the beginning. For example, my own definition of the observed system from the 1974 edition of *Management Information*

Systems: Conceptual Foundations, Structure, and Development contained the basic concepts:

> [An] integrated, man/machine system for providing information to support the operations, management, and decision-making functions in an organization. The system utilizes computer hardware and software, manual procedures, management and decision models, and a data base. [Davis 1974, p. 5]

Early methodologies for doing information systems work were based on a technical view with the designer dividing work between technology and humans. However, very early in the development of the field, there was recognition of the importance of understanding the way the users understand the outputs of the system. One illustration of this early development was Professor Börje Langefors, holder of the first chair in Sweden for information processing. His seminal book, *Theoretical Analysis of Information Systems*, was published in 1966. He distinguished between infological and datalogical work areas. Infological concepts and methods relate to the information to be provided to an organization to meet user needs. Datalogical concepts and methods define the organization of data and technology in order to implement an information system. His infological equation was insightful: $I = i(D,S,t)$.

I = the information produced by the system
D= the data made available by system processes
S = the recipient's prior knowledge and experience (world view)
t = the time period during which interpretation process occurs
i = the interpretation process that produces information for a recipient based on both the data and the recipient's prior knowledge and experience

In the infological equation, information is not just the result of algorithmic processing but is also the result of the prior knowledge and experience of the person receiving the results of processing data. Therefore, no two individuals receive the same information from this processing. However, users in common problem domains and similar data uses have prior knowledge and experience that is sufficiently similar to allow shared use of data and meaningful communication of interpretations. One of the important tasks of system developers for structured reporting and analysis applications is to elicit and document shared concepts within a domain of practice. In some cases, change processes are incorporated in system development to ensure that the recipients have a shared knowledge of the concepts and rules underlying the application and a shared understanding of the reports and analyses provided to them.

3.3 Definition of Information System Function and its Activities

Organizations are human artifacts designed and built to achieve human organization objectives. Information systems are human artifacts needed by organizations. The needs and requirements must be identified and systems must be planned and built. They are the product of human imagination and human development processes. The requirements reflect not only technical capabilities but also social and behavioral considerations. Systems are built through a combination of information technology and development procedures. The system procedures include software, human procedures, and procedures incorporated in forms and other non-technical mechanisms.

The domain of information systems as a function or field of activity and study includes activities for system development and system management and evaluation:

- Strategic planning for information and communication systems. There is a co-alignment of the organization strategy with information and communication system strategy. Technology capabilities provide opportunities for the organization strategy, and the organization strategy defines requirements for information technology infrastructure and systems. For example, the capabilities of the Internet provide opportunities for the IS function to suggest new ways of doing business, and the organization's strategic decisions to deploy web-based applications define elements of the information systems strategy.

- Management of the information system function. This includes unique problems of management of IS activities and resulting unique measurement and evaluation issues. Management issues include evaluation of outsourcing for various activities and supervision of outsourcing contracts.

- Information systems personnel. There are unique positions such as systems analyst, programmer, and network designer. Selecting, motivating, training, managing, and evaluating these personnel employ both general human resources methods and unique factors related to information systems employees.

- System development processes. Requirements determination and development processes ranging from structured development cycles to rapid prototyping and end user systems are part of these processes. Unique methods and tools are employed, such as development methodologies, CASE tools, and diagraming notations and processes. Information systems change organizations. They reflect management decisions about how the organization will interact with customers, suppliers, personnel, etc. Implementation of new systems is a change process with significant organizational effects.

- Evaluation. Evaluation of results includes measurement of satisfaction with systems and economic/organizational effects. Understanding both development successes and failures is useful.

3.4 Data, Information, and Knowledge

Information systems provide capture, repositories, processing, and communication of data, information, and knowledge. The definitions of these three terms is made difficult because of the lack of precision in everyday conversation and because one person's data may be another person's information (Buckland 1991). However, there is a convergence relative to the meaning of the terms:

> **Data** consists of representations of events, people, resources, or conditions. The representations can be in a variety of forms, such as numbers, codes, text, graphs, or pictures.
>
> **Information** is a result of processing data. It provides the recipient with some understanding, insight, conclusion, decision, confirmation, or recommendation. The information may be a report, an analysis, data organized in a meaningful output, a verbal response, a graph, picture, or video.
>
> **Knowledge** is information organized and processed to convey understanding, experience, accumulated learning, and expertise. It provides the basis for action. Knowledge may be procedural (how to do something), formal (general principles, concepts, and procedures), tacit (expertise from experience that is somewhat hidden), and meta knowledge (knowledge about where knowledge is to be found).

An information system captures data based on information system design decisions. Not everything can be captured, so someone makes a decision. If all needs for data and uses of information were known in advance, the decisions about the data to capture and store would be simple. However, we do not have foreknowledge. Also, there is a cost of capture and storage, so decisions must be made. The tendency is to capture easily measured characteristics of events. For example, in a retail purchase transaction, item number, price, date, etc. are captured, but potentially vital data items are not captured, for example, the mood of the customer, whether the item was the one wanted or purchased as a second choice, whether for own use or a gift, and so forth.

Capturing knowledge has both conceptual and practical problems. The employees of an organization may develop habits and informal procedures that provide high levels of service and performance. The procedural knowledge is not codified and, therefore, not stored by the organization. Tacit knowledge of how to do things is stored in the minds of workers but not in the manuals or training courses of the organization. There is typically no organizational memory for tacit knowledge. Capturing and codifying procedural knowledge and the tacit knowledge of valuable long-term employees is now a major information systems issue.

4. Approaches to Conceptual Foundations

The conceptual foundations for a field are the set of concepts and propositions that explain why structures are designed the way they are, tasks are scheduled and accomplished in the way they are, and activities are performed the way they are. For example, maintenance of application systems can be explained by a few underlying concepts, such as:

- Open systems decay over time as the environment changes; therefore, the system no longer fits the altered environment.

- When users employ an application, they appropriate the technology and alter the way it is used from that envisioned by the developers.

The concepts or propositions employed to explain or guide information systems design and its development processes come from fields that typically have bodies of knowledge related to the concepts or propositions used. These are, therefore, termed underlying disciplines.

For an observer who takes a normative standpoint, information systems is fairly straight forward (as are most applied fields). Organizations have transaction processing and operations requirements and requirements related to administration, management, analysis, and decision making. These require an information technology infrastructure that must implemented and managed. Based on requirements, application systems are either acquired or built. Systems must be designed, maintained, and updated. Training and support must be provided. The observer may conclude the field is simple and its concepts are simple.

In practice, infrastructures and applications are not just technology and software. There is a complex interaction with technology, application software, and users. Requirements are not obtained by simply asking. There is a process of discovery for both users and developers as the requirements emerge. Strategic applications, productivity improvement, reduced cycle time, user friendly systems, quality improvement, and so forth are the result of innovative thinking that comes from dialogue among participants who have trust both in each other and also in the processes of requirements determination and system development. The field is, therefore, complex and its conceptual foundations have emerged from the intersection of information systems problems with principles, concepts, and prescriptions from a number of fields (Davis 1992).

There are three approaches to conceptual foundations and underlying disciplines for an applied academic field. One is to be open to ideas from many other disciplines; any time there is an interesting intersection with concepts in another discipline, the concept and related disciplinary knowledge is added to the set of conceptual foundations. Conceptually, the entire set of useful concepts defines the boundaries of the field as an academic discipline. The second approach is to focus on a core set of conceptual foundations. Other ideas may be appropriated for information systems use, but the core set defines the field as an academic discipline and not the entire set of useful concepts. The intersection approach and the core approaches are at two radically different ends. A third alternative is an evolutionary view that the information systems field will become more bounded as some concepts are dropped as not being useful enough to stay in the set of important concepts.

4.1 The Underlying/Intersection Approach to Conceptual Foundations

The conceptual foundations for the emerging field of information systems in organizations started to develop in the 1960s. Scholars in North America and Europe were the most active in the early developments. Early conceptual definitions of information systems (or management information systems) focused on the elements making up the system of information storage and processing and the applications supported by the system. The conceptual foundations that emerged were based on the interaction of information technology, information systems, organizational systems, and individuals and groups employing or affected by the systems. The key concepts or underlying conceptual foundations were defined as concepts of information, humans as information processors, system concepts, concepts of organization and management (relevant to information systems), decision making, and value of information. Soft systems and socio-technical concepts were introduced to counter-balance a strong tendency to view information systems from an engineering rationality and not consider the views and perceptions of all stakeholders.

The boundaries of the field of information systems from the mid-1960s to the mid-1980s were characterized by expansion of infrastructure, applications, and conceptual foundations. Infrastructure changes were the combining of communications systems with computing systems and the emergence of end-user computing and personal computers. Applications expanded in support of collaborative work and individual and group decision making. The role of information systems in organizational communications introduced organizational communications as an underlying set of concepts. Databases were conceptualized as repositories of data (attributes) about things (entities) important to the organization and its processes. Organization power and politics considerations emerged as important concepts. Strategic value of information technology began to be studied. Adoption of new technology became an important topic. Some concepts of interorganizational systems were introduced.

From the mid-1980s to the year 2000, reengineering emerged. Although presented as a revolutionary idea, it is based on the fundamental system concept that organizational systems decay (entropy) and should therefore periodically be reengineered, sometimes radically. The radical idea of artificial intelligence achieved some practical results with expert systems, thereby bringing expertise and expert systems into the set of concepts underlying system design. Information systems had been justified on the basis of economic value to the firms adopting them; value to the economy was assumed. Under questioning relative to the economic value of information technology in improving productivity, analytical modeling and economic analysis emerged as a part of the information systems field. As the percentage of knowledge workers increased, concepts began to emerge about how knowledge work quality and productivity are improved by information technology. Recognition that information technology had the power to remove time and location constraints to organizations focused attention on the value of knowledge resources in an organization, leading to knowledge management as a subarea in the field. The Internet and the technology for the world wide web changed the nature of information storage, search, and access. Web technology changed both business to business and business to consumer applications. Search strategies and knowledge acquisition (long reserved for librarians and similar experts) became part of the field of information systems.

The intersection approach looks for concepts and principles from other fields that may apply to problems in the information systems field. One of the most important reasons to keep the intersection approach is that more powerful ideas and innovations are likely to arise at the intersection of two fields. The thinking of information systems personnel can be enriched by encouraging exploration in other fields rather than looking inward to the body of knowledge accepted by the IS field.

4.2 The "Core" Approach to Conceptual Foundations

An objection to the observed system and information system function as the basis for defining the field and its research boundaries is that the definition is not stable. As a new technology or new area of application emerges, it is pasted onto the definition of the observed system and the functional activities. To illustrate, mobile communication/computing devices and electronic commerce are examples of new technology and new applications. Both appear to be important to observed systems, and both have an effect on the activities of the IS function. The new systems lead to new sets of concepts and related disciplinary knowledge, partly because there is no set of core concepts that can be applied to all new technologies and new application areas. There is no way to constrain the growth of concepts borrowed from other discipline.

With unconstrained growth in intersections, a field may become unfocused. If there were agreement on core concepts, they might better define the information systems discipline within the context of other organization disciplines. The core concepts explain why information systems as a field differs from other fields. Also, core concepts can be the basis for cumulative research that is not constrained by the changing landscape of technology innovations and new applications.

Weber (1997; see also Wand and Weber 1995) argues that deep structure information systems phenomena are the core of information systems as an academic discipline. The deep structure of the information system consists of those characteristics of the information system that capture the meaning of the real-world system as perceived by users. An information system is a system that represents objects and activities in the real world. It codes, stores, receives and transmits, and processes representations of the real world. It also should be able to track events in the system it represents. The representation should communicate the structure of the system in terms of its behavior, including subsystems that make it easier for users to understand the system and deal with it. The representation should be simpler and more efficient for communication and reasoning than the system being described. Weber argues that developing a better understanding of the core phenomena will provide a conceptual foundation for how well an information system represents user perceptions of the real-world system. To clearly explain the deep structure will provide a unique information system contribution to theory.

> A core serves to *characterize* the discipline. It represents the essence of the discipline—the body of knowledge that leads others to recognize it and to acknowledge it as being *distinct* from other disciplines and not just a pale imitation of them....I can see only three ways in which the core of a discipline can be teased out. The first is to identify a body of phenomenon that is not accounted for by theories from other disciplines and to build novel theories to account for these phenomena....The second way is to take phenomena that *are* purportedly accounted for by theories from other disciplines and to again build novel theories to account for these phenomena....The third way is to look for *breakdowns* in theories borrowed from reference disciplines when they are applied to IS-related phenomena. [Weber 1997, pp. 27-28, emphasis in original.]

The core approach positions the information systems function as principal providers of information technology infrastructure, application systems and information technology services. This suggests a stronger emphasis on system principles (both hard and soft), matching technology infrastructure to organization structure, technology implementation and system change management, stability and quality in system operations, information technology strategy planning, and evaluation of value added. Under this core view, IS is not the principal mover for web-based applications, e-commerce, knowledge management, etc. but is the development partner with others in the organization.

Falkenberg and Lindgreen (1989) take a different approach to information system concepts. They tend to focus on conceptual models, axioms, taxonomies, levels of abstraction, etc. This effort fits into the core approach. A subsequent effort by Falkenberg et al. (1996) produced a framework for information systems concepts as part of the FRISCO Task Group of IFIP. This group continues to meet and hold conferences on information system concepts.

4.3 The Evolutionary Approach to Reducing the Set of Conceptual Foundations

Without deciding on the question of whether there should be tighter boundaries around the concepts and theories dealt with by information systems, there are natural evolutionary tendencies toward tighter boundaries. Given the large number of interesting concepts related to information systems (along with an underlying body of knowledge), there is a natural tendency to constrain the field in order to be more coherent, focus on the key elements, etc. The second evolutionary tendency for reducing the set of conceptual foundations comes from the fact that some interesting ideas, concepts, theories, and practices are discarded from the set of conceptual foundations because they are not useful enough to continue in use or to continue research based on them.

The stream of research described as cognitive style is an example of dropping a topic that had consumed significant resources in the IS field. The basic proposition is that people differ in their cognitive abilities and information systems should be designed to match them. A good system design/cognitive ability match presumably results in improved performance. There is a significant body of knowledge about cognitive styles. Three problems with the cognitive matching proposition finally lead to the demise of this research.

1. People are not either/or relative to cognitive style; there is a distribution. For example, heuristic and analytic styles range from highly one or the other to slightly more one than the other. There is no method for calibrating information system design to match the variety of cognitive styles.

2. People are adaptable. They can adapt to systems that are not designed explicitly to their intuitive style. Training will help people adapt the system to their natural cognitive style. A person with a heuristic style can adapt to a system designed for an analytical style.

3. Very few applications are designed for a single person. Not only is it costly, but an individual may move to another position and the next person, with perhaps another style, will need to use the system. Training the new user is generally less costly than creating a new system. Also, applications may be used by group, so the system needs to be useable by all members of the group.

5. The Current and Possible Future Status of Conceptual Foundations for Information Systems

Looking at the academic field of information systems in the year 2000, its scope in terms of technology, development processes, and applications has expanded dramatically in the past 30 plus years. This expansion covers the time when information systems emerged as an academic field (in the mid-1960s) to the present. The scope is so large in the year 2000 that subfields have begun to emerge.

As a check against my observations of conceptual foundations and topics in the field, I reviewed articles in the completed research and research in progress for ICIS 1998 and ICIS 1999 and articles in eight issues of the *MIS Quarterly* from December 1997 through September 1999. The underlying bodies of concepts and methods for the articles having a declared or implied concept/theory were as follows:

Underlying Bodies of Concepts and Theories	Number of Uses
Psychology	5
Cognitive Psychology	17
Sociology/Organization Behavior	49
Management Strategy	19
Economics	20
System Concepts and Principles	4
Communications	2
Decision Making	6
Information Concepts	2
Total	124

I also tallied the articles in terms of concepts, theories, processes, and applications systems that are unique or somewhat unique to information and communications systems in organizations:

Bodies of Concepts, Theories, Processes, and Application Systems Unique or Somewhat Unique to IS	Number of Uses
Information systems management processes	15
Information system development processes	30
Information system development concepts	20
Representations in information/communication systems (databases, knowledge bases, etc.)	8
Application systems (somewhat unique because of information technology)	59
Total	132

These illustrate the use of underlying disciplines but do not disclose the variety of concepts and theories from these disciplines or the variety of unique IS processes and applications. An expanded view for underlying disciplines and unique IS processes and applications is found in Tables 1 and 2. These are extensive but may not include all concepts, theories, processes, and applications that are part of the field.

5.1 Information Systems Conceptual Foundations in the Future

There are three possibilities in the next decade or so relative to conceptual foundations for the academic field of information systems:

- *A continued expansion of conceptual foundations as more intersections develop with other disciplines.* The interesting problems and issues for a field such as information systems are at the intersection with other disciplines and bodies of knowledge. As examples, group decision systems can be better implemented based on research that considers underlying research on group decision processes. Information technology systems for knowledge management are improved and IS research on knowledge management is more insightful when research in cognitive science is incorporated. Consumer psychology research becomes important when researching e-commerce systems. Under this scenario, each new area of application of information systems in organizations may bring with it underlying concepts and a body of research, so there will be continued expansion of conceptual foundations.

- *A dramatic redrawing of the map of conceptual foundations to emphasize the core.* Given the pressure from academic colleagues to define a core for information systems and proposals for this core from respected IS colleagues, the field might decide to define the field in terms of conceptual foundations at the core. Many fields in the university have done this. They define their field narrowly in terms of core activities and unique contributions. They exclude many interesting intersections with other fields.

Table 1. Underlying Disciplines for Information Systems and Concepts/Theories Used

Psychology
- Theories of human behavior
- Motivation theories
- Theory of reasoned action

Cognitive Psychology
- Human information processing
- Human cognition
- Expertise
- Artificial intelligence
- Cognitive style
- Creativity
- Knowledge
- Cognitive representations/visualization
- Human-machine interfaces

Sociology/Organization Behavior
- Nature of work (knowledge work, clerical work, etc.)
- Governance theories
- Organization design concepts
- Process models
- Culture

Technology Adoption/Diffusion
- Adaptive structuration
- Social network theory
- Actor network theory
- Social influence
- Organization change
- Organization learning
- Trust
- Ethics

Management/Strategy
- Strategy
- Innovation
- Competitive advantage
- Resource view of firm
- Knowledge management
- Risk management
- Evaluation
- Outsourcing

Economics
- Principal-agent theory
- Transaction cost economics
- Productivity
- Information economics
- Social welfare
- Adverse selection
- Value of information
- Incomplete contracting
- Intermediation

System Concepts and Principles
- Artificial systems
- Requisite variety
- Soft systems
- Complexity
- Control theory-cybernetics
- Socio-cybernetic theory of acts
- Task/technology fit (equifinality)
- System economics (reuse)
- Maintenance of systems (negative entropy)
- Process theory
- System models

Communications
- Media choice
- Collaborative work
- Speech acts theory

Decision making
- Behavioral decision making
- Normative decision models
- Group decision making
- Neural networks/genetic algorithms

Information concepts
- Mathematical theory of communications
- Quality, errors, and bias concepts
- Value of information
- Semantics
- Semiotics (theory of signs)

Table 2. Bodies of Concepts, Theories, Processes, and Applications Unique or Somewhat Unique to Information Systems

Information systems management processes
- Strategic planning for infrastructure and applications
- Evaluation of IS/IT in the organization
- Management of IS personnel
- Management of IS function and operations

Information system development processes
- IS project management
- IS project risk management
- Organization/participation in projects
- Requirements—technical and social
- Acquisition of applications
- Implementation of systems
- Training/acceptance/use

Information system development concepts
- Concepts for methods
- Socio-technical concepts
- Speech acts theory
- Rational decomposition concepts for requirements
- Social construction for requirements
- Concepts of errors and error detection
- Testing concepts for complex socio-technical systems
- Quality concepts for information/ communications systems

Representations in information/communication systems (databases, knowledge bases, etc.)
- Representations of the "real" world
- Coding of representations
- Storage, retrieval, and transmission of representations
- Tracking events
- Representing changes in events
- Representing structure of system

Applications systems (examples)
- Knowledge management
- Expert systems
- Neural networks
- Decision support systems
- Collaborative work systems/virtual teams
- Group decision support systems
- Telecommuting systems/distributed work
- Supply chain systems
- ERP systems
- Inter-organizational systems
- Organization communications systems: internet, intranet, e-mail, etc.
- Training systems
- E-commerce applications
- Customer support systems

- *Narrowing of focus with more emphasis on the core, but still including important intersections with other fields.* A redrawing of the map of conceptual foundations can define a core but still include bodies of knowledge that clearly underlie information systems. There can be a recognition that the core can incorporate clearly understood concepts without reference to other disciplines that may also use the same concepts.

My view is that the current set of concepts is too large, because some of them are not robust in providing explanations. There needs to be some pruning. The field has tended to ignore some of the core concepts and issues in favor of proven concepts from other disciplines. It will be profitable to remedy this neglect and strengthen the core concepts both by research and by explicating the concepts and their applicability. However, the field of information systems has natural overlap with other disciplines, and these intersections should remain part of the domain of the information systems discipline.

This evolutionary view of the field can be implemented if leaders in the field identify unprofitable concepts that can be dropped, strengthen the core concepts, and remain open to new intersections if there is good evidence to support their inclusion.

5.2 Some Comments about Research Methods in Information Systems

Much of the discussion about the field has contrasted views that Checkland and Holwell (1998) have termed functionalist (hard) versus interpretive (soft). The functionalist, hard system view tends to focus on the goals of organizations and how information systems should be designed to support these rational goals. The interpretive, soft systems view of organizations is multi-faceted with conflict and social relationships dominating. In the functionalist view, information systems are designed to aid rationalized activities and rational decision making. In the interpretive view, information systems provide data and communication facilities used by organization participants in making sense of the world and negotiating actions to be taken. Checkland and Holwell present these two views as opposing and leading to confusion and lack of coherence and stability in an emerging field.

My own experience is that information systems as a field is ahead rather than behind other fields in management and administration. Because information systems began fairly early to become an international discipline, a variety of views about the field and its research were encouraged. The field has a richer set of views than other fields because the positivist philosophy that dominated the American research and the phenomenology philosophy that tended to dominate in Europe were both supported by the worldwide community. The IFIP 8.2 Manchester working conference demonstrated the willingness of researchers in the IS field to appreciate the different approaches to research (Mumford et al. 1985). The *MIS Quarterly*, which began with an espoused policy of positivist research, demonstrated in practice a willingness to accept interpretive research. The current Editor-in-Chief, Allen Lee, of the *MIS Quarterly* is known as an advocate of qualitative methods.

In other words, the confusion often cited by those examining the state of the field can be interpreted as a coming together of world views and research views. The field seems to value diversity of methods. To some, the lack of a sparse set of methods and a restricted, accepted vocabulary demonstrates an immature field. I make the counter argument that they demonstrate a field that is incorporating a rich set of methods and vocabulary to make sense of a complicated world. There is a clear trend to an acceptance of positivist and interpretive methods as being complementary. Other fields such as accounting, finance, and marketing are less international and less open to a variety of research methods and world views.

I have often been characterized as a positivist. Actually, my world view of research was altered by the IFIP 8.2 Manchester conference. I believe a world-class scholar must be competent in both hypothesis testing using quantitative data and qualitative, interpretive methods using observations, interviews, and participation. My *preference* for a hypothesis testing dissertation for entry-level students is pragmatic rather than dogmatic. Such dissertations tend to be more tractable and provide good grounding in data analysis. A student should also have doctoral studies experience in qualitative research. The point is that the best scholars in the field will have an ability to employ both methods. I observe the European doctoral students becoming better trained in hypothesis testing methods and American students receiving some training in interpretive methods. There is a coming together rather than a splintering apart.

As a check against my observations of diversity in research methods, I tallied research methods for articles in the completed research and research in progress for ICIS 1998 and ICIS 1999 and articles in eight issues of the *MIS Quarterly* from December 1997 through September 1999.

Research Method	#	%
Survey	37	26
Case/cases	34	24
Model without data	15	11
Model with data	14	10
Experiments	14	10
Design/prototype	12	09
Framework	06	04
Other	08	06

6. Summary and Conclusions

The essence of the paper is that information systems intersects with many other disciplines. Some view this dependence with alarm. I view it as an opportunity. I agree with Banville and Landry (1992), who state:

> The field is attractive to many, including the authors, because of its great variety of approaches and their potential and actual cross-fertilization....Members of the MIS field should not refuse any help from other disciplines, given the richness and complexity of their main research object—management information systems—and their numerous facets.

Definitions of the information systems field and IS function tend to converge because practice can be observed and described. The main issue for conceptual foundations in the next decade is whether to focus on a narrow core set of concepts or to continue in the current free market for concepts that are useful and meaningful. Some critical events in the past years suggest that some concepts and related bodies of knowledge will be discarded as not being sufficiently useful and others will be added. We have probably neglected the core and, therefore, it may be useful to define it more clearly and precisely. This effort, however, will not preclude the inclusion of a rich set of intersections with other disciplines.

References

Banville, C., and Landry, M. "Can the Field of MIS be Disciplined?," in *Information Systems Research: Issues, Methods and Practical Guidelines*, R. Galliers (ed.). London: Blackwell Scientific Publishers, 1992, pp. 61-88.

Buckland, M. *Information and Information Systems*, New York: Praeger Publishers, 1991.

Boland, R. J., Jr., and Hirschheim, R. A. (eds.). *Critical Issues in Information Systems Research*. Chichester, UK: John Wiley & Sons, 1987.

Checkland, P., and Holwell, S. *Information, Systems and Information Systems—Making Sense of the Field*. Chichester, UK: John Wiley & Sons, 1998.

Cotterman, W. W., and Senn, J. A. (eds.). *Challenges and Strategies for Research in Systems Development*. Chichester, UK: John Wiley & Sons, 1992.

Davis, G. B. "Systems Analysis and Design: A Research Strategy Macro-analysis," in *Challenges and Strategies for Research in Systems Development*, W. W. Cotterman and J. A. Senn (eds.). Chichester, UK: John Wiley & Sons, 1992, pp. 9-21.

Davis, G. B. *Management Information Systems: Conceptual Foundations, Structure and Development*. New York: McGraw-Hill Book Company, 1974.

Falkenberg, E. D., Hesse, W., Lindgreen, P., Nilsson, B. E., Oei, J. L. H., Rolland, C. Stamper, R. K., Van Assche, F. J. M., Verrijn-Stuart, A. A., and Kos, K. *FRISCO: A Framework of Information Systems Concepts*. The IFIP WG 8.1 Task Group FRISCO, December 1996. Available from http://ftp.leidenuniv.nl/pub/rul/fri-full.zip.

Falkenberg, E. D., and Lindgreen, P. (eds.). *Information System Concepts: An In-depth Analysis*. Amsterdam: North Holland, 1989.

Langefors, B. *Theoretical Analysis of Information Systems*. Lund, Sweden: Studentlitteratur, 1966.

Mumford, E., Hirschheim, R. A., Fitzgerald, G., and Wood-Harper, A. T. (eds.). *Research Methods in Information Systems*. Amsterdam: North-Holland, 1985.

Nissen, H.-E., Klein, H. K., and Hirschheim, R. (eds.). *Information Systems Research: Contemporary Approaches and Emergent Traditions*. Amsterdam: North-Holland, 1991.

Nurminen, M. I. *People or Computers: Three Ways of Looking at Information Systems*. Lund, Sweden: Studentlitteratur, 1988.

Wand, Y., and Weber, R. "On the Deep Structure of Information Systems," *Information Systems Journal* (5), July 1995, pp. 203-223.

Weber, R. *Ontological Foundations of Information Systems*. Melbourne, Australia: Coopers & Lybrand, 1997.

About the Author

Gordon B. Davis is the Honeywell Professor of Management Information Systems at the Carlson School of Management, University of Minnesota. He received his MBA and Ph.D. from Stanford University. He also holds honorary doctorates from the University of Lyon, the University of Zurich, and the Stockholm School of Economics. He is a Fellow of the Association for Computing Machinery. He is the U.S.A. representative of IFIP Technical Committee 8 (Information Systems) and has served as chairman of TC8. He serves on the editorial boards of major journals in the field. He has published extensively and written 20 books in the MIS area. His areas of research include MIS planning, information requirements determination, conceptual foundations for IS, control and audit of information systems, quality control for user-developed systems, in-context assessment of information systems, and management of knowledge work. Gordon can be reached by e-mail at gdavis@csom.umn.edu.

6 HORIZONTAL INFORMATION SYSTEMS: EMERGENT TRENDS AND PERSPECTIVES

Kristin Braa
University of Oslo
Norway

Knut H. Rolland
University of Oslo
Norway

Abstract

At the brink of the new millennium, emerging trends like globalization and the Internet—as well as the buzzword "knowledge management"—have profound impacts on how business organizations design and deploy their IT solutions. Standardization and integration seem to be the common strategy—whether ERP systems, middleware-based IS, intranets, or IT infrastructures. However, in practice these systems are often heterogeneous and constrained by various socio-technical aspects. In focusing on this phenomenon, the concept of a "horizontal information system" is introduced. Drawing from examples from a maritime classification company, we take a closer look at the phenomenon and some challenges for design and deployment of such systems are discussed.

1. Introduction

In this paper, we investigate trends in knowledge-intensive and globally dispersed organizations in using IT for standardizing and integrating knowledge, work, infrastructure, and information systems. The term "horizontal information system" is introduced to underscore the distinct challenges facing the design, implementation and use of large-scale information systems that cut across different communities-of-practice.[1] Despite enabling technologies, including the success of Internet-based technologies, the deployment of a horizontal information system is likely to be constrained by installed-base issues, social and political aspects of knowledge sharing, and the increasing socio-technical interdependencies created. We see a tendency of moving from vertical information systems to horizontal and integrated systems that cut across the organization. Trends emerging from Internet-based technologies, such as intranets for internal organizational communication, enhance this. Other trends, such as globalization and knowledge management, support this tendency in deploying large-scale infrastructure-like information system for the entire enterprise. The general term globalization has been used to describe the increasing economic and political interdependence in the world society. These contemporary trends seem to some extent to mobilize organizations into focusing on integration of work and knowledge, and on standardization of both technology and work. These attempts at integration and standardization will involve pitfalls and challenges and the result of striving for increased control might be losing control.

A case that put into focus the challenges in designing and implementing horizontal information systems is described. The company is a maritime classification company (MCC), operating world wide as an independent foundation working with the objective of "safeguarding life, property and the environment." The MCC is a global company that comprises 300 offices in 100 countries, with a total of 5,500 employees. The horizontal information system under implementation is a classification support system designed for supporting surveyors in their inspection of ships throughout the world. We will use this case to emphasize characteristics and possible pitfalls in implementing such systems. From this rich case, we mainly emphasize aspects we find relevant for discussing challenges related to implementation of horizontal information systems. For a more detailed reading of the case, see Rolland (1999).

This paper is organized in the following way. First, we discuss trends that "drive" the deployment of horizontal information systems, the Internet, globalization, and the focus on utilizing IT for the management of knowledge. Then, the characteristics and challenges concerned with horizontal information systems are discussed in light of the above mentioned trends. Next, some of the challenges that are related to horizontal information systems are identified from a study of the deployment of a large-scale IS in MCC. Using examples from this particular case, some challenges implementing such systems are briefly outlined.

2. Emergent Trends and Perspectives

2.1 The Internet Factor

The explosive adoption of Internet technologies during the 1990s has woven local networks into a global network, making up the infrastructure of the information society. The telephone took 37 years to acquire 50 million users, the television needed 15 years to get the same amount of viewers, while the World Wide Web managed to reach 50 million

---

[1]In IT architecture terminology, horizontal/vertical systems denote ho deeply the system penetrates the architecture; e.g., a word processor is a horizontal system.

surfers in about three years (*Observer* 1999). Nobody knows exactly how many people are connected to the Internet, but it has been estimated at between 120 million and 150 million people, and, more importantly, the number continues to grow exponentially. Our discipline is faced now with interesting challenges that must be met by both existing and new research paradigms. The Internet was initially an experimental network between contractors and computer science researchers working for the U.S. Department of Defense. From the mid-1980s and until 1990, it proved very successful as a world wide information infrastructure for faculty, staff, and students at universities and research centers. In 1991, the restrictions against commercial use of the Internet were removed. The same year, World Wide Web software was released. The Web is one of the main driving forces of the Internet, where it is being used widely by large and small businesses, by private citizens, in schools, and by consumers (Guice 1998). The Internet as a unifying concept for the development of open and simple standards has proven to be a strong force in setting the agenda for the development of commercial software. Public and private organizations recognize that they need to have an opinion about how the technology affects their business. The question is not if but how the Internet can be utilized as interaction and integration media internally in organizations and externally in interaction with customers (Braa and Sørensen 1999). The Internet as a global infrastructure plays an increasingly important role in both information systems practice and research. The ability of the Internet as a common platform to build services upon also creates expectations. Standalone information systems are expected to integrate with the global network. Internet technology supports horizontal solutions involving a variety of actors, both those behind the service and those using the service. Thus, these systems become large, heterogeneous networks that need to be aligned with an installed base of existing systems as well as practices. In this way, the Internet serves as an important integrating technology.

2.2 Knowledge Management: Social and Political Aspects

One of the motivations for developing large-scale IS that cut across organizational departments and functions comes from the assumption that these systems will enable knowledge sharing and thereby serve as an important tool for the establishment of an organizational memory. In management science, economics, and recently within information systems, "knowledge" has been put forward as the most valuable asset for organizations in the "knowledge-based economy" (Neef, Siesfeld, and Cefola 1998). IT has been expected to play an important role in the management of knowledge in organizations (e.g. Earl 1996; Liebowitz 1999). Several frameworks for this have been proposed in the literature. For instance, Earl (1996) defines knowledge management as consisting of knowledge systems, networks, knowledge workers, and the learning organization. Earl draws on two case studies to illustrate how IT has enabled knowledge-based strategies. However, in order to establish a knowledge-based strategy, Earl refers to challenges concerned with (1) organizational collaboration; (2) training and personal development, and (3) organizational incentives to support knowledge sharing and collaboration. These three preconditions comes close to the challenges well known from decades of research within the field of CSCW (e.g., Grudin 1994; Markus and Connolly

1990; Orlikowski 1992b). Thus, one could argue that knowledge management comprises nothing new—it is just a rewrapping of the social and political issues involved in using IT for supporting collaboration and sharing information between different user groups. However, knowledge management as a trend can mobilize deployment of horizontal information systems in organizations. In this way, knowledge management becomes important for understanding different organizational actors' motivations and intentions and the rationale behind the design of a particular system. In our case of a major maritime classification company, one of the main objectives for developing a horizontal IS was to increase sharing and creation of knowledge in the organization. The view that knowledge can be treated as a commodity makes the state-of-the-art technologies unlimitedly enabling—downplaying the constraining factors illustrated by recent research in CSCW and IS. Even though the Internet factor and other information technologies making it technically "easier" to develop large-scale information systems, it is less evident that these systems will be successful in terms of knowledge sharing and creation.

2.3 The Consequences of Globalization for the Design of IS

The general term *globalization* has been used to describe the increasing economic and political interdependence in the world society. More specifically, Giddens (1991) describes the globalization phenomenon as time-space distanciation. In the conceptual framework of time-space distanciation, the attention is directed to the complex relation between local involvement and interaction across distance. The level of time-space distanciation is much higher now than in any other previous period, thus the relation between local and distant social forms and events becomes correspondingly stretched. This stretching process is what Giddens refers to as globalization, in the sense that the modes of connection between different social contexts or regions become networked across the earth surface as a whole. Globalization is thus defined as the intensification of worldwide social relations that link distant localities in such a way that local happenings are shaped by events occurring many miles away and vice versa. Globalization is to be understood as a dialectical phenomenon, in which events at one pole of a distanciated relation often produce divergent or even contrary occurrences at another. For Giddens, modernization and globalization are closely connected. Globalization is the most visible form modernization is taking today and risk society is emerging (Beck 1992). Everything is connected to everything, the interdependency increases and control decreases. Increasing risk means decreasing control. Traditionally, modernization implied increased control in line with Beninger's (1986) outline of the "control revolution." More knowledge and more and better technology implied increased control. In the age of high modernity and globalization, more knowledge may just as well lead to more unpredictability, more uncertainty, and less controllability (Hanseth and Braa 2000). This shift, which may appear contradictory, can be explained by the increasing role of side-effects (Beck, Giddens, and Lash 1994). Globalization means integration. At the same time, all change—new technologies introduced, organizational structures and work procedures implemented, etc.—has unintended side-effects. Any change may affect those interacting with processes being involved in the change. Side-effects of local events often have global consequences. And the more integrated the world becomes, the longer and faster side-effects travel and the more significant their consequences will be. Globalization also means globalization of side effects.

In the so-called information economy it has been argued that IT and globalization are reinforced by each other (Bradley, Hausman, and Nolan 1993; Castells 1996), and that these processes will shape markets and the way businesses compete. Interestingly, this will also change the way organizations use IT—how information systems are designed —and the motivations for developing these systems. In this context, the IS-related literature seems to recommend that global organizations utilize IT for increasing control and coordination (e.g., Ives and Jarvenpaa 1991). Earl and Fenny (1996) suggest that global and large-scale information systems have the potential to contribute to the global efficiency, local responsiveness, transfer of learning, and making global alliances. The role of IT as a key factor to bring these changes about is often thought of as an opportunity to increase control and enhance coordination, while opening access to new global markets and businesses (Ives and Jarvenpaa 1991). Bartlett and Ghoshal (1998) claim that firms operating in this global markets will increasingly be at a serious strategic disadvantage if they are unable to firmly control their worldwide operations and manage them in a globally coordinated manner. Within this model, corporations are focusing on more close coordination of increasingly more complex and global processes. At the same time, globalization is experienced as creating an increasingly more rapidly changing, dynamic, and unpredictable world.

In a variety of businesses and organizations, there seems to be a growing trend to build large-scale horizontal information systems. More specifically, these can be categories of systems such as enterprise resource planning (ERP) systems, in-house developed client/server systems based on middleware architectures, or large intranets based on Internet technologies and standards. Typically, these are systems that cut horizontally across the organization aiming at integrating and standardizing the organization's business processes. IS research on these topics, focusing on the social-technical processes that take place when organizations are deploying large-scale information systems, are few. Davenport (1998), who has surveyed the recent trends in enterprise systems, notes that information integration and standardization may reduce flexibility by imposing their own logic on the company's strategy, culture, and organization. On the other hand, some organizations may well succeed in implementing such systems. Similarly, in the contemporary discussions around information infrastructures, it has been shown how design and redesign of such large-scale systems are constrained by an installed base of systems, standards and practices (Star and Ruhleder 1996).

Thus, globalization and growing unpredictability, uncertainty, and less control cause profound consequences for how organizations use IT and deploy large-scale information systems. IT and information systems are not unlimitedly enabling technologies that corporations can deploy to increase strategic advantages in terms of information integration and a standardized IT infrastructure. These technologies are inevitably connected to larger social systems, which in turn impose a variety of socio-technical constraints on the use of IT. For instance, IT plays a key role in the implementation of "flexible specialization" models by enabling more flexible production systems. On the other hand, as seen in the case of implementing SAP in a global organization (Hanseth and Braa 1998), large and complex IT infrastructures may block the changes in organizational structures and processes necessary for a global company to excel in the global market. Thus, in this perspective, technology becomes an actor, which may decrease the number of possible redesigns and hence, in this way, technology in general becomes both enabling and constraining (Orlikowski 1992a). This insight suggests IT will be both constraining and enabling for global organizations in increasing their control and coordination. In addition, since information technologies and systems become an integral part of almost any work process, this ultimately increases the interdependencies between different work processes and between those practices and the technologies involved.

3. Horizontal Information Systems

In the 1970s, Galbraith (1973) claimed that the uncertainty faced by organizations was due to insufficient information. Uncertainty was defined as the difference between the information needed for the successful execution of an organizational task and the information available in the organization. Consequently, the information processing abilities of the organization had to be increased as the organization faced increasing uncertainty. In Galbraith's information processing model of the organization, one design strategy could be to deploy vertical information systems in order to increase the information processing abilities and avoid an overheated hierarchy. As illustrated in our case of a maritime classification company, information technologies and standards for interoperability and computer networking, combined with visions of (global) knowledge sharing and information integration—represent a shift toward deploying horizontal information systems. This shift from vertical and local information systems—to horizontal and global information systems comes with a range of new business opportunities as well as distinct challenges and pitfalls.

Horizontal information systems are different from traditional information systems in how they handle typical support for different communities in the organization or between organizations. Typically, traditional information systems focus on feeding the upper levels (i.e., strategic management) of the organization with relevant information for making decisions. Moreover, in this world, it was relatively easy to point to the typical users, making it possible to design the system for a special group of users (i.e., managers and secretaries). In short, the focus was on automating vertical information processing through a transaction processing system. A typical example is a payroll system used by the administrative staff for information on employees, salaries, and the production of payslips. In addition, management might use the system for planning staff levels and promotions, or for reporting to the tax office.

3.1 Characteristics and Challenges

In describing and understanding large-scale information systems as a phenomenon, we draw from insights given by theories of globalization (Beck 1992; Giddens 1991). The term horizontal information system is used to denote the distinct characteristics and challenges concerned with deploying large-scale information systems. However, horizontal information systems are neither a clear-cut concept nor a solution for how to deploy large-scale information system, but rather a perspective in order to offer a way of understanding this contemporary phenomenon. Horizontal IS imply that work practices as well as different technologies become increasingly interconnected and integrated, and accordingly, these systems become more vulnerable to unintended side-effects. Therefore, systems that are deployed for increased control can ultimately turn out to imply less control, because of the side-effects introduced. Since horizontal IS typically cut across functions, stakeholders, and communities of practice within an organization, the deployment of such systems faces some distinct challenges:

- *Increasing the interdependencies in the organization.* The deployment of a horizontal IS in an organization implies that different communities of practice will be connected more closely. This implies increasing the number of interdependencies between technologies and work practices in the organization. The growing socio-technical interdependencies make it almost impossible to distinguish between the technical and the non-technical issues, which in turn constrains the use of IT and the deployment of a horizontal IS in the organization. Linking different communities of practice also has social and political aspects, as for instance how employees share their knowledge through a horizontal IS depends on non-technical structures in the organization (e.g., reward systems, professions).

- *Undermining the interpretative flexibility of artifacts.* In real-time and real-life work practices, artifacts can have multiple meanings according to context and the situation. Artifacts are a profound part of work practices, and following Law (1992), the social is made up of heterogeneous networks of both materials and humans. Thus, artifacts and work practices are intrinsically linked in heterogeneous networks that constitute the focal social system, which coordinates and ensures a smooth flow of work. As a horizontal IS is used in different local contexts where artifacts are embedded in different practices, discovering the different roles and meanings of the artifacts becomes increasingly important for not undermining the "workflow from within" (Bowers, Button, and Sharrock 1995) or establishing a new workflow from within through a horizontal IS.

- *Lack of control because of unintended side effects created.* Unintended side-effects can be caused by both human conduct and non-humans, as for instance software and hardware. In the case of large software systems, for example, it is conventional wisdom that maintenance and correction of errors, in fact, often introduces new errors. When interdependencies are established through a horizontal IS, side-effects will not be isolated but distributed. Thus, the processes involved when deploying horizontal IS would be less controllable and involve negotiations for aligning the actors' interests.

- *Necessity of continuously negotiating and maintaining interfaces.* Horizontal IS will typically provide interfaces to other systems. For instance, a large-scale intranet may have interfaces to old legacy systems and databases. Horizontal IS are often deployed to replace the organization's fragmented way of storing information and the current information systems. At the same time and to a certain degree, the horizontal IS must extend the old information infrastructure.

- *Aligning the variety of different communities of practices.* A horizontal IS will, in most cases, not be perfectly aligned with different practices in different communities. However, this does not imply that the system is a failure or that improvisations and work-arounds are done by the users in order to compensate for "bad" design. Improvisations and work-arounds are part of human conduct for securing a smooth flow of work. On the other hand, the design must not undermine the users' possibility for improvising and adjusting the system to their work situation. Thus, the challenge is to balance between standardization and flexibility—not to describe a "correct" set of requirements for supporting special work practices.

- *Horizontal IS constitute an installed base.* When horizontal IS become stabilized and institutionalized in the organization, they tend to have infrastructural characteristics—e.g., they become increasingly hard to change and, at the same time, the pressure for doing necessary changes increases (Monteiro 1998). In this way, a horizontal IS in an organization will constitute a powerful installed base.

These are all challenges that have to be considered to some extent when implementing a horizontal information system. There will, of course, be variations according to, for instance, how deeply the horizontal IS penetrates the work practice. An intranet service providing biographical data of the employees will not have the same implication as implementing a patient record system at a hospital.

3.2 Related Research

3.2.1 Information Infrastructure

The term *information infrastructure* has been used to describe large-scale networked structures that often cut across work-practices, departments, functions, and organizational borders (e.g., Bud-Frierman 1994; Monteiro and Hanseth 1995; Rolland 1999; Star and Ruhleder 1996). Hence, any horizontal information system could also be defined as an information infrastructure, but not necessarily the other way around, as a horizontal information system will focus more on supporting more or less specific activities for different communities-of-practice. Information infrastructures, however, as the term is used in the literature, span from tailor-made large-scale collaborative systems (Star and Ruhleder 1996), large EDI networks (Monteiro and Hanseth 1995), national information infrastructures (Branscomb and Kahin 1995), to the Internet (Monteiro 1998). Bud-Frierman states that the concept of an information infrastructure is a potentially useful unit of discourse, being both a historical and cultural entity in addition to being used to describe both micro- and macro-level structures.

In general, an information infrastructure can be understood as a term for describing the heterogeneous, dispersed, complex, and interdependent components, which our "work" relies on to collaborate and coordinate activities through sharing and interchange of information in a given context. Along these lines, an information infrastructure becomes a socio-technical phenomenon. Information infrastructures are always more than cables, communication protocols, routers, and computers. More specifically,

Hanseth (2000) emphasizes that an information infrastructure is evidently an enabling, shared, open, socio-technical, and heterogeneous installed base. An information infrastructure is never built from scratch, and there will always be an installed base in terms of a heterogeneous social system consisting of technical as well as non-technical components. An infrastructure is a set of connected and interconnected components, which can be conceived as "ecologies of infrastructures" (Hanseth 2000). One infrastructure consist of ecologies of sub-infrastructures by:

- building one infrastructure as a layer on top of another;
- linking logical related networks; and
- integrating independent components, making them interdependent.

In this way, a horizontal information system could be understood as a component in a larger information infrastructure. Similarly, according to Star and Ruhleder (1996), an information infrastructure cannot be understood as pure technology, but an Information Infrastructure is always embedded in a larger social structure. Moreover, Star and Ruhleder emphasize that an infrastructure is something that develops in relation to practice, it is not to be conceived as a "thing" or a static technical structure, and the question becomes "When is an infrastructure?" not "What is an infrastructure?" Consequently, "an infrastructure occurs when the tension between local and global is resolved" (Star and Ruhleder 1996, p. 114). Thus, this information infrastructure discussion focuses on some interesting aspects that increasingly are met when designing and deploying large-scale information systems. For instance, the focus on the installed base, that is, the understanding that you can never develop a system from scratch, there is always something there in the form of social practices, artifacts, and very often a heterogeneous collection of different information systems.

In using the term horizontal information systems, we are interested in discovering the socio-technical processes surrounding the alignment between different practices, artifacts, the old information systems, and new systems and technologies. For instance, why is an information system successfully aligned within one context, whereas it can be totally misaligned in a different context? How do we design and implement information systems that cut across different contexts? Furthermore, how is this integration process shaped by the existing artifacts (i.e., paper documents) and work practices, and in what ways are an installed base enabling and constraining for a certain information system to be implemented?

3.2.2 Communities of Practice and Artifacts

One aspect of designing large-scale information systems is that they cut across several communities-of-practice. The term communities-of-practice has been used to denote a social group where a certain practice is common, coordinated, and reproduced (Brown and Duguid 1991). Thus, in any large organization, there will typically exist numerous communities: an organization can be described as a community-of-communities. Artifacts, whether information systems or paper documents, play important roles in a community-of-practice were they mediate relations and coordinate activities, both within the community and between different communities. Usually, artifacts are not universally interpreted among different communities-of-practice. When deploying horizontal information systems, information provided by the system can be interpreted very differently within different communities-of-practice. Furthermore, when information, which earlier existed on paper documents, as for instance standardized reports or checklists, becomes part of a horizontal information system, the information provided can be interpreted differently. Artifacts like paper documents have been recognized for having material and social aspects that are important for the meaning of the information inscribed on them (Braa and Sandahl 1998; Brown and Duguid 1994). For instance, Braa and Sandahl describe an attempt at a news agency, to implement paper-based TV schedules into a document information system. The design of the new document information system failed, because one of the resources that the users relied on in their work practice was not considered relevant for the design. At the news agency, faxes and shelves indicated progress and states of the work process and, since this work process was visible to all workers at the office, the artifacts played an important role in coordination of the work. In the document information system, this coordination mechanism did not exist and subsequently the system broke down.

4. From Local and Vertical to Global and Horizontal

4.1 The Case of MCC

During the 1990s, Maritime Classification Company (MCC) has been challenged through increased global competition and swift changes that have effected their business environment. An important part of MCC's strategy to meet these challenges has been to deploy IT with the intention of reengineering their way of working and become a "learning organization." This alignment of the business strategy and the IT strategy indicated a shift from a local and vertical IS toward a more global and horizontal information system. In 1997, as a part of this strategy, MCC invested approximately US$ 52 million in common infrastructure and a large-scale information system. In this paper, the large-scale information system implemented will be referred to as the horizontal information system. The common IT infrastructure was launched in 1997-98 under the mantra "one world—one MCC" and comprised a WAN that links 300 offices, common NT servers, office applications, common e-mail system, and shared document databases.

In addition to the global infrastructure campaign, MCC had, since 1993, been working on the horizontal information system for supporting the work of the surveyors as well as the information requirements of managers and customers. The vision was that the horizontal information system would enable knowledge sharing and transparent access to all relevant information on vessels, certificates, surveys, etc., regardless of roles, departments, and positions in the organization. The prominent idea was to integrate all relevant information for classification of vessels in a common product model. A product model is a standardized representation of all parts of a ship and the relationships between those parts. This common product model was developed using the UML modeling language and additional CASE tools and serves as the common standard for the horizontal information system. In short, the horizontal information system is a state-of-the-art client/server system built on Microsoft's COM architecture as middleware and a common SQL-based relational database as a server. The system was planned to be implemented in December 1997; however, due to the complexity of the technological solution and changing requirements, it was not put in use until early in 1999.

4.2 Local Variations and Standardization

4.2.1 The Horizontal Information System

MCC has systematically worked for streamlining and standardizing their work processes and several projects have been undertaken to define new work processes. As an overall strategy, MCC has emphasized standardization on three different levels: (1) common work processes; (2) common product model serving as a standard for the horizontal information system; and (3) common IT infrastructure (Figure 1).

Sharing knowledge through the horizontal information system implied, to a certain extent, that the terminology and the representation of knowledge used were agreed upon. To solve this problem, MCC developed a large product model as a standardized information model for all applications comprising the horizontal information system. The idea was to represent product data, as well as work tasks, on a standardized form to make it possible to share knowledge through the system between the different offices. However, system developers and others soon realized that the challenges with developing

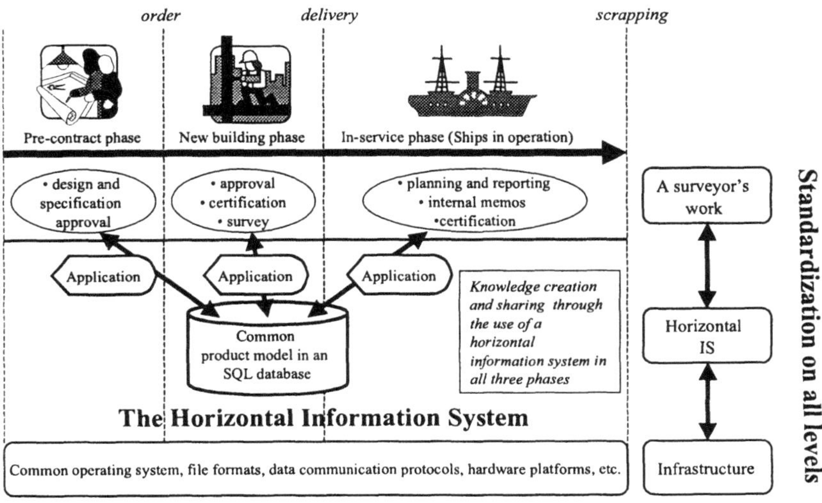

Figure 1. Life-Cycle Information Management in MCC

such models were not only of a technical character, but organizational and political as well. A manager from the software development project pointed at the fact that, historically, different departments and groups in the organization had used different terminology and that a "stiffener" was not a "stiffener" throughout the organization.

The knowledge-intensive nature of the surveyors' work made it considerably challenging to predefine and standardize this work and to design IT systems that would not pose too many constrains for their situated and context dependent work. The surveyors play an important role in the organization by conducting various types of surveys on vessels. Their work involves both practical work (e.g., investigating machinery and technical equipment on a vessel) and office work (e.g., writing technical reports, communicating with customers, using computer applications, etc.). In this way, the surveyors' work can be characterized as knowledge-intensive. Surveyors have to keep up with changing rules and regulations concerning certification as well as technical knowledge within a variety of disciplines (e.g., materials engineering, propulsion systems, hydrodynamics, electronics). MCC is authorized for doing surveys and certification on behalf of more than 130 national administrations. In addition, surveyors do surveys based on MCC's own classification rules and IMO (International Maritime Organization) regulations. In doing their daily work, the surveyors draw upon experience-based and tacit knowledge. For instance, the surveyors draw from their tacit knowledge to intuitively find those spots on a vessel's hull that could have cracks or rust. Similarly, they do considerable work before the survey to gain as much explicit knowledge of the vessel as possible. For instance, they have to know the status of the ship, in terms of length, tonnage, flag, and information on the owner of the ship. Thus, the surveyors' tacit and explicit knowledge as well as their communication skills are all factors that determine MCC's accumulated capability for safeguarding life and property.

According to the surveyors interviewed, the implementation of the new horizontal information system led to considerable changes in the role of the surveyor. Some meant that this new system would make the surveyors more or less "data collectors" for MCC. They would spend considerably less of their time "out in the field" doing practical engineering work as more of their time would be occupied doing office work.

4.2.2 Communities-of-practice within MCC

As a global organization, MCC consists of several different communities-of-practice. The differences between these communities were recognized during the design, implementation, and use of the new horizontal information system. In particular, the different interpretations and interests were visible when designing and implementing the horizontal information system with the intention of sharing knowledge throughout the global organization and standardizing the work done in different communities. This posed special challenges and problems for the design as well as the implementation of the system.

The surveyors are not one homogenous group, but more or less autonomous engineers that work in different offices around the world. Surveyors in the MCC organization are working with different kinds of surveys according to where the ship is in its life cycle (Figure 1). Even before a ship is designed and constructed, the ship owner decides to classify the new ship according to the MCC classification rules. An MCC surveyor must certify all components that are to be installed on the ship and their manufacture. In this process, the MCC surveyor and the yard that is building the ship would benefit from information on previous ship designs and components. It would be extremely helpful for a surveyor to know if other surveyors have recognized any typical failures or safety hazards concerning a specific design or component. Other surveyors specialize in doing surveys on ships that are in operation. Different communities of surveyors have different views on, for, instance how different components of a ship are related. These issues made it extremely difficult for the system developers to describe the "correct" requirements for the horizontal information system.

MCC is a global organization where the different local stations are embedded in cultural and institutional environments that have different degrees of similarity. For instance, at one MCC office in Germany, the engineers insisted on writing additional comments in German instead of using "standard texts" in English. At this office, all of their customers have similar requirements and their primary focus is on delivering results in the form of technical reports as efficiently as possible. At this office, the requirement was to have an integrated IS where they avoided entering the customer's address and name more than once. In a small Norwegian office, however, the surveyors' work is more varied and this functionality is not required. On the contrary, they focus on flexibility in the IT support and that the different IT applications they use should have a consistent user interface.

Cultural differences in the division of labor make it difficult to design a system that standardizes work processes. In Eastern countries, the surveyor role was different. For instance, in Asia it was culturally determined that doing office-like work was the job of a secretary. Consequently, the users of the horizontal information system in Eastern countries would most likely be secretaries and not surveyors.

Regarding the design of a horizontal IS, these examples illustrate the challenges involved when developing a system to be used by different communities of practice. As noted in section 3, this underscores that a horizontal IS will, in most cases, not be perfectly aligned with different practices in different communities. But, on the other hand, such systems could be successful if they link different communities of practice without undermining the communities' internal practices.

4.3 Standardization and Flexibility

4.3.1 Negotiating and Maintaining Interfaces

The surveyors have established a system of different paper-based checklists for supporting the different types of surveys conducted by a surveyor in MCC. There are a total of 74 different checklists to be used with different kinds of surveys and types of vessels. The checklists were constructed by different people for different contexts and environments. Thus, there is no standard representation or common use of terminology, and these checklists have not been a part of the official documentation given to the customers. On the other hand, these checklists have been most helpful for inexperienced surveyors who use them down to every detail, and in this way they are learning what to focus on when conducting a survey of a vessel. More experienced surveyors usually do not follow the items in the checklists when conducting a survey—they only use it in a very limited way.

The initial plan for the horizontal information system was to include a standardized version of these paper-based checklists, in order to structure the input of information. This standardized way of reporting the survey information was required, because this information was used in the generation of various survey reports. This created several dilemmas for the implementation and use of the new information system. The strategy was to include a very standardized set of checklists in order to be able to generate statistical information from the surveyors reporting through the standardized checklists. In the later stages of development and during implementation, this strategy was abandoned due to the organizational and technical complexity. Different groups of surveyors and system developers had to agree upon a common terminology and a general breakdown structure of a ship, which turned out to be a longitudinal and complex process. In addition, the systems developers had already programmed a version according to the product model philosophy that for various reasons did not meet some of the surveyors' requirements. This meant that the complex product model had to be changed, which in turn required an effort of modeling and programming.

This exemplifies the necessity of negotiating with different interest groups in order to obtain the needed flexibility in the design. The interdependencies created through a horizontal IS, and the interconnectivity between technical and non-technical issues, imply that many seemingly technical design decisions become an issue for negotiation. This makes such systems difficult to plan and increasingly difficult to change as the development proceeds.

4.3.2 Aligning the Variety of Communities of Practice

One of the main reasons for standardizing the checklists was to create a standardized set of data to support the automatic production of reports. Some of these reports are used in communication with officers and crew on a ship. For instance, when the surveyor has conducted a survey on a vessel, a preliminary survey report and a memo to the owner of the ship are given to a member of the crew on the ship. These reports summarize the job that has been done and what the surveyor found during this particular survey. The surveyor fills in the reports onboard the ship and then their meaning is carefully explained to the ship officers or other members of the crew. It is of profound importance that any "Conditions of Class" are fully understood, so that the crew is able to do the required repairs and adjustments in order to maintain the safety for crew and cargo on the ship. In this context, the reports are artifacts that act as mediators in the communication between the surveyor and different members of the crew. Some of the surveyors stated that the information on the reports generated by the horizontal information system did not have a meaningful structure for the surveyors using the system. For instance, one of the reports was structured according to the alphabetical order of the codes related to the different surveys. The surveyors, on the other hand, were used to categorize the information according to different components of the ship (e.g., hull, machinery, propellers, thrusters). Changing the structure on the reports also has implications for how the surveyors communicate with the crew on a ship. According to the surveyors, some crew members found it easier to understand concrete things like hull and propellers compared to MCC's abstract four letter codes. The surveyors had different ways of compensating for avoiding this situation. For instance, the surveyors took their time in explaining every detail in the reports for the crew. In addition, attempts were made to change the standardized templates for the reports included in the horizontal information system. Since these templates were plain Microsoft Word files, it was possible for the surveyors to locally modify the templates and, in this way, restructure the contents of the automatically generated reports. However, since it was not possible to save a new version of a template into the system, this work-around created some problems, and the users were strictly prohibited from modifying the templates. By modifying the templates for the reports, there would exist two (or more) versions of the same report: one for the user that had modified the local template and one for all others who accessing that particular report through the horizontal information system. Thus, this created a serious dilemma: the reports should be tailored to different customers' needs, but at the same time, only one version of the same report. should exist

This underscores why it becomes increasingly important to allow flexibility in use in supporting knowledge-intensive work in different communities-of-practice with a horizontal IS. Prior to the implementation of the system, the survey reports were often tailored according to whom the surveyor was going to meet on the ship. In addition, numerous types of surveys were conducted during the visit on the ship, leading to complex survey reports supposed to support the communication between the surveyors and the crew. Thus, how these reports are structured is extremely context dependent, and the need for flexibility for the surveyors to modify the generated reports becomes a prerequisite for useful reports. But, with the implementation of the horizontal information system, this flexibility vanished, leading to several work-arounds and potentially different versions of the same report.

4.4 Installed Base Issues: Unintended Side Effects

The design and implementation of the horizontal information system was considerably constrained by an installed base. The mainframe system that had to be used in parallel with the new system especially affected not only the design and implementation processes, but also how the surveyors used the system. In this way, the mainframe system became an actor, which had to be considered in all phases of development and use. However, interestingly, this mainframe system was at first regarded as a resource and not as a constraint.

Clearly, for both technical and organizational reasons, it was impossible to implement the system in all 300 offices simultaneously. Hence, for a period, the old mainframe system had to be used in parallel with the horizontal information system. Offices using the old Mainframe System and those using the Horizontal Information System are dependent upon having correct and updated information when planning and reporting surveys. In order to update the common database used by the horizontal information system with data from the mainframe system and vice versa, various scripts were made. In other words, the installed base made it necessary to develop a gateway, because a discrete transition from the old system to the new system was impossible. Due to the complexity of the product model, the technically different databases used, as well as several adjustments in the design, it was difficult to ensure perfect updates between the mainframe system and the horizontal information system. Thus, the mainframe system represented an important part of the installed base that had to be considered in the design and implementation processes. However, the installed base issues were not considered until the new horizontal information system was tested with what was considered as relevant data. The data in the mainframe system, which had at first been considered as an enabling resource to be included in the new horizontal information system, became a constraint for the design and implementation. Furthermore, this had unintended side-effects that increased the surveyors' distrust of the new horizontal information system. In using the new system, some of the surveyors experienced losing some of their information because of the imperfect gateway between the two systems' databases. The surveyors had to enter the information into the system several times, and hence, this made their office work more time consuming and stressful. This increased the distrust of the horizontal information system, and thereby created work-arounds. Some of the surveyors stated that they were more careful not to enter too much data into the system at a time. At the same time, one of the intentions with the horizontal information system was to support more detailed, consistent, and a larger amount of information than before. In fact, the unintended side effects of the horizontal information system may have led to the opposite; namely, that the surveyors report less information than before. This distrust toward the new system caused the surveyors to double check the information provided. For instance, they constantly used a large book containing information on all vessels classified by MCC and compared the information in this book with the information on the screen.

5. Challenges in Implementing Horizontal Information Systems

The possibility of gaining benefits of integration suffers from the complexity created by increasing the interfaces that need to be negotiated and maintained. Thus, side effects may be difficult to control. Developing and deploying the horizontal information system at MCC illustrate how seemingly technical issues are inherently interconnected with non-technical issues such as work practice of the different communities; the various cultural and institutional environments; distrust toward the system, and so forth. The involvement of different communities-of-practice, an installed base, and the somehow fluctuating requirements for the system made the implementation process a dynamical and complex process of negotiating and adjusting current designs. According, implementation of an information system that cuts horizontally across practices, departments, and cultures is considerably more time consuming than in the case of the more traditional systems. Drawing from actor-network theory (e.g., Callon 1991; Latour 1991; Law 1992), it can be argued that, in the deployment of such horizontal information systems, one is actually trying to change considerably larger networks, compared to traditional information systems. Thus, it is difficult to implement a horizontal IS all in at once. The old system, which had to be used in parallel, not only affected the design and implementation process, but also how the surveyors used the system. Further, the horizontal IS itself becomes an installed base that could constrain redesigns as well as further development of new systems. Huge resources are invested and interdependencies created and thus it becomes impossible to reverse the process.

The challenges concerning the design of large-scale information systems are neither local nor global, they are, rather, horizontal, and thus the question is not how to achieve a seamless integration between existing local practices—or a global and all embracing standard. Using IT in a flexible way that enables knowledge sharing in communities-of-practice, as well as linking the various communities in ways that do not undermine local work practices, is challenging. Emerging trends such as Internet technologies, globalization, and knowledge management are influencing the way information systems are designed. In describing and understanding horizontal information systems as a phenomenon, we draw from insights given by theories of globalization (Beck 1992; Giddens 1991). The concept of horizontal information systems is introduced in order to emphasize challenges to be met in implementing large scale information systems that cut across communities-of-practice. Such systems do not exist in isolation, but interact with various other systems, artifacts, and practices; relations are continuously negotiated and almost never reach a stable state as a typical "infrastructure." As shown in this case, there is a need for flexibility and the variety of communities-of-practice will "fight" the standardization attempt by means of their local practice. However, a certain degree of standardization is needed in order to communicate across practices and borders.

References

Bartlett, C. A., and Ghoshal, S. *Managing Across Borders: The Transnational Solution*. Boston: Harvard Business School Press, 1989.

Beck, U. *Risk Society: Towards a New Modernity*. London: Sage, 1992 (first published in German as *Risikogesellscahft,* 1988).

Beck, U., Giddens, A., and Lash, S. *Reflexive Modernizations—Politics, Tradition and Aesthetics in the Modern Social Order*. Cambridge, UK: Polity Press, 1994.

Beniger, J. R. *The Control Revolution. Technological and Economic Origins of the Information Society*. Cambridge, MA: Harvard University Press, 1986.

Bowers, J., Button, G., and Sharrock, W. "Workflow from Within and Without: Technology and Cooperative Work on the Print Industry Shopfloor," in *Proceedings of the Fourth European Conference on Computer-Supported Cooperative Work, ECSCW'95,* H. Marmolin, Y. Sundblad, and K. Schmidt (eds.), Stockholm, 1995, pp. 51-66.

Braa, K., and Sandahl, T. I. "From Paperwork to Network: A Field Study," in *Proceedings from the Third International Conference on the Design of Cooperative Systems,* Cannes, France, May 26-29 1998.

Braa, K., and Sørensen, C. "The Internet Factor," *Scandinavian Journal of Information Systems* (10:12), 1986.

Bradley, S. P, Hausman, J. A., and Nolan, R. L. (eds.). *Globalization, Technology, and Competition*. Boston: Harvard Business School Press, 1993.

Branscomb, L. M., and Kahin, B. "Standards Processes and Objectives for the National Information Infrastructure," in *Standards Policy for Information Infrastructure*, B. Kahin and J. Abbate (eds.). Cambridge, MA: MIT Press, 1995.

Brown, J. S., and Duguid, P. "Organizational Learning and Communities-of-practice: Toward a Unified View of Working, Learning, and Innovation," *Organization Science* (2:1), 1991, pp. 40-57.

Brown, J. S., and Duguid, P. "Borderline Issues: Social and Material Aspects of Design," *Human-Computer Interaction* (9), 1994, pp. 3-36.
Bud-Frierman, L. (ed.). *Information Acumen – The Understanding of Knowledge in Modern Business*. London: Routledge, 1994.
Callon, M. "Techno-economic Networks and Irreversibility," in *A Sociology of Monsters – Essays on Power, Technology and Domination*, J. Law (ed.). London: Routledge, 1991.
Castells, M. *The Rise of the Network Society*. Oxford: Blackwell Publishers, 1996.
Davenport, T. H. "Putting the Enterprise into the Enterprise System," *Harvard Business Review*, July-August 1998.
Earl, M. J. "Knowledge Strategies: Propositions From Two Contrasting Industries," in *Information Management—The Organizational Dimension*, M. J. Earl (ed.). Oxford: Oxford University Press, 1996.
Earl, M. J., and Fenny, D. F. "Information Systems in Global Business: Evidence from European Multinationals," in *Information Management—The Organizational Dimension*, M. J. Earl (ed.). Oxford: Oxford University Press, 1996.
Galbraith, J. *Organization Design*, Reading, MA: Addison-Wesley, 1973.
Giddens, A. *Consequences of Modernity*. WHERE: Polity Press, 1991.
Grudin, J. "Groupware and Social Dynamics: Eight Challenges for Developers," *Communications of the ACM* (37:1), 1994, pp. 92-105.
Guice, J. "Looking Backward and Forward at the Internet," *The Information Society* (14:3), 1998.
Hanseth, O. "Infrastructures and the Economy of Standards," in *From Control to Drift: The Dynamics of Corporate Information Infrastructures*, C. U. Ciborra, K. Braa, A. Cordella, B. Dahlbom, A. Failla, O. Hanseth, V. Hepsø, J. Ljungberg, E. Monteiro, and K. A. Simon (eds.). Oxford: Oxford University Press, 2000.
Hanseth, O., and Braa, K. "Technology as a Traitor: Emergent SAP Infrastructure in a Global Organization," in *Proceedings of the Nineteenth International Conference on Information Systems*, R. Hirschheim M. Newman, and J. I. DeGross (eds.), Helsinki, December, 1998, pp. 188-197.
Hanseth, O., and Braa, K. "Globalization and 'Risk Society'," in *From Control to Drift: The Dynamics of Corporate Information Infrastructures*, C. U. Ciborra, K. Braa, A. Cordella, B. Dahlbom, A. Failla, O. Hanseth, V. Hepsø, J. Ljungberg, E. Monteiro, and K. A. Simon (eds.). Oxford: Oxford University Press, 2000.
Ives, B., and Jarvenpaa, S. L. "Applications of Global Information Technology: Key Issues for Management," *MIS Quarterly*, March 1991, pp. 32-49.
Latour, B. "Technology is Society Made Durable," in *A Sociology of Monsters—Essays on Power, Technology and Domination*, J. Law (ed.). London: Routledge, 1991.
Law, J. "Notes on the Theory of the Actor-Network: Ordering, Strategy, and Heterogeneity," *Systems Practice* (5:4), 1992, pp. 379-393.
Liebowitz, J. (ed.). *Knowledge Management Handbook*. Washington, DC: CRC Press, 1999.
Markus, M. L., and Connolly T. "Why CSCW Applications Fail: Problems in the Adoption of Interdependent Work Tools," *Proceedings of the Conference on Computer-Supported Cooperative Work*. New York: ACM, 1990, pp. 371-380.
Monteiro, E. "Scaling Information Infrastructure: The Case of Next-Generation IP in the Internet," *The Information Society* (14:3), 1998, pp. 229-245.
Monteiro, E., and Hanseth, O. "Social Shaping of Information Infrastructure: On Being Specific About the Technology," in *Information Technology and Changes in Organizational Work*, W. Orlikowski, G. Walsham, M. R. Jones, and J. I. DeGross (eds.). London: Chapman & Hall, 1995.
Neef, D., Siesfeld, G. A., and Cefola, J. (eds.). *The Economic Impact of Knowledge*. London: Butterworth-Heinemann, 1998.
Observer. "Guide to the Internet," *The Observer*, January 17, 1999, pp. 32.

Orlikowski, W. J. "The Duality of Technology: Rethinking the Concept of Technology in Organizations," *Organization Science* (3:3), 1992a, pp. *398-427*.

Orlikowski, W. J. "Learning from Notes: Organizational Issues in Groupware Implementation," in *Proceedings of the Conference on Computer-Supported Cooperative Work*. New York: ACM, 1992b, pp. 362-369.

Rolland, K. "Information Infrastructure Transition: Challenges with Implementing Standardised Checklists," in *Proceedings of the Twenty-second Information Systems Research Seminar in Scandinavia*, T. K. Käkölä (ed.), Keuruu, Finland, 1999, pp. 95-110.

Star, S. L., and Ruhleder, K. "Steps Toward an Ecology of Infrastructure: Design and Access for Large Information Spaces," *Information Systems Research* (7:1), 1996, pp. 111-134.

About the Authors

Kristin Braa is an associate professor at the Department of Informatics at the University of Oslo, Norway. She has for several years done research within participatory design, digital documents, interdiciplinarity in research, integration, and geographically dispersed information systems. She has edited the books *Net Society* (in Norwegian) and *Planet Internet* (with Bo Dahlbom and Carsten Sørensen). Kristin can be reached by e-mail at kbraa@ifi.uio.no.

Knut H. Rolland graduated with a degree in informatics from the Norwegian University of Technology and Science in 1997. He has since 1998 been a Ph.D. student in informatics at the University of Oslo. His main interests include the interplay between IT and organization, particularly the design and implementation of large-scale information systems and infrastructures. Knut can be reached by e-mail at knutr@ifi.uio.no.

7 EXPANDING THE HORIZONS OF INFORMATION SYSTEMS DEVELOPMENT

Nancy L. Russo
Northern Illinois University
U.S.A.

Abstract

Advances in information systems technologies and applications and new realities in the business and economic climate are changing the nature of information systems development. In particular, the areas of electronic commerce and Web-based applications challenge our notions of the role of the information systems function in the development process. This essay explores the impact of several of these changes on information systems development, and on developers themselves, and suggests how developers must adapt to meet these new requirements.

1. Introduction

Much has been written about the continual changes experienced by the information systems field and the fact that these changes are occurring at an ever more rapid pace. According to Donald J. Listin, of Cisco Systems, "technology is moving so quickly that products are dead in 18 months" (Byrne 1999).

Organizations are changing to adjust to these dynamic business conditions. Changes are driven by commercial technology, particularly the demands of e-business, the globalization of markets, and the pressure for reengineered, quality-oriented organizations (Truex, Baskerville, and Klein 1999). These change drivers are pushing organizations to find new ways of doing business. Organizations are moving from functional, hierarchical management structures to team-based, networked structures. They are moving from a product focus to emphasis on targeted services that meet market demand. Strategic alliances and integration across the supply chain are becoming commonplace.

Information technology is the strategic enabler of these changes. The vast majority of the new ways of doing business could not be accomplished without information technology. In fact, in some cases, the information systems are not merely *supporting* the organization, they *are* the organization (*e.g.*, Amazon.com). It is difficult to identify any area of the lives of those living in industrialized nations that has not been impacted by information technology. Information technology is central to healthcare, education, government services, manufacturing, transportation, financial services, entertainment, and retailing. More and more services and products are becoming information-based and there is no reason to expect this trend to end. The need to design the information technology components of these products and services, as well as designing the means to integrate them, not only puts great pressure on developers, but raises ethical issues as well. As more and more of our interactions, both business and social, take place in the virtual world of the Web, who are we: the flesh and blood being sitting at the keyboard, or the virtual presence on the network? And what happens to those who aren't "on the net"? Will we eventually have two parallel societies? Obviously there are many issues surrounding the trends in information technology that beg consideration. This essay addresses just a small segment of this; namely, that of the impact of these trends on business information systems development.

The purpose of this essay is to describe a number of the changes taking place in the information systems development environment and to discuss how these changes can be addressed in the education and training of systems development professionals. The remainder of the paper discusses the drivers of change in business information systems development, the impact of these changes on development and developers, and the manner in which educational systems can better prepare future developers.

2. Pressures on Systems Development

In addition to pressures from the business environment, there are technological and application-oriented advances that are changing both the process and product of systems development. Y2K preparations have drained substantial IS resources for the past several years. Coming out of this, IS departments are faced with backlogs of development projects, as well as with a shortage of IS personnel. Thus there is pressure to produce more with less resources, and to produce it quickly.

This push means that development groups are facing shortened time horizons. The focus is on small-scale, rapid development (Fitzgerald 1997, 1998). This new reality challenges some of the traditional assumptions about information systems development (Truex, Baskerville, and Klein 1999). Detailed, lengthy planning and analysis phases are no longer feasible in rapidly changing environments. Projects of long duration are not tolerable because the fast pace of change means that the underlying business could change dramatically during the time between analysis and implementation.

In addition to pressure to do more with less, developers in many organizations are responsible for applications on a growing number of platforms (Fitzgerald 1998; Russo, Hightower, and Pearson 1996). For example, IBM currently develops applications for

32 different hardware and software platforms. This adds not only to the difficulty of initial development, but also adds considerably to testing resource requirements. Additional complexity is added when applications and data are distributed across homogeneous or heterogeneous networks. Even with open systems interconnectivity, organizations still face tremendous problems with connecting and integrating disparate systems. This becomes an even greater challenge on a global scale. The Internet and the Web have served to stimulate greater global business activity. For IT departments in multinational companies, this means greater worldwide distribution of computing platforms, vastly increased global information flows, and many other technical, legal, and cultural issues (Stephens 1999).

Changes are occurring in the nature of software applications as well. Software vendors are moving away from selling shrink-wrapped software toward selling (leasing) applications as services on the Web. Through the influence of object-oriented methods and tools, development is becoming much more component-based. It is predicted that some developers will be component makers, while others are component users (Lyytinen, Rose, and Welke 1998). Web application development has had a tremendous impact on the field because it is different from traditional IS development, not only in terms of the characteristics of the applications themselves, but also in terms of the people involved in Web application development.

In a general study of Web development (Russo and Graham 1998), fewer than one third of the Web developers were employed in an information systems department. Most did not come from traditional computer science/IS backgrounds, although a few had some prior programming knowledge (not necessarily work experience). Less than 15% of the respondents received formal training in using hypertext markup language or other Web development tools or languages. Few of the developers worked as part of a formal development team in developing their Web applications. Most worked alone, and others worked informally with users. Web development in many organizations is decentralized and, because it is often not the responsibility of any one group, it is very difficult to monitor and control. In this study, over half of the respondents indicated that their organizations had no standards or guidelines for Web development. Although these results must be interpreted with some caution, due to the changes that have occurred since the data was collected (1996-1997), they do tell us that our concept of a "developer" must be broadened beyond the boundaries of the IS function and beyond the traditional end-user concept.

The hottest trend in the field, e-commerce, is a specialized type of Web application development. In e-commerce development, we see the convergence of cross-functional resources: telecommunications, databases, interfaces, marketing, graphic design, and others. These systems may appear simple to set up, but the cost to establish a functional e-commerce system can be quite high. Estimates range from $10,000 to $100,000 for an electronic catalog to $1 million to $10 million for a dynamic, interactive application (Forester Research).

As organizations move away from a brick and mortar existence, and into the virtual space of the Web, traditional approaches to systems development become less viable. When the goal is to create something brand new, rather than to build something to support an existing function or process, priorities change. Analyzing a current system in the "real" space, when designing a system for "virtual" space, is not very helpful because the virtual space is not subject to the same limitations. When designing in the virtual, Web-based environment, the focus moves from merely meeting user requirements to devising new, innovative things that users haven't seen before. Not only do Web-based systems have to be functional and easy to use, they must also be creative, interactive, fun, and exciting.

Another trend is an increased level of package customization vis-à-vis new development (Fitzgerald 1997). Enterprise resource planning packages such as SAP and PeopleSoft are replacing in-house development. This move toward ERP systems raises questions about the impact that such a move will have on organizations' internal IS functions. Many organizations found that when they outsourced functions in the past, they lost access to a valuable resource. Organizations face a similar problem when moving to ERP. When one software product replaces all or most of the legacy systems in a firm, what happens to the IS department? Are fewer people needed? Are people with different skills required? Can maintenance be performed in-house, or will consultants from the software vendor be required? If the organization loses the flexibility of changing or replacing system functions, what will this mean to the organization's ability to respond to changing market needs? If the majority of organizations standardize on a few ERP packages, what will this mean for the ability to develop and use information systems for competitive advantage?

A small study (Russo, Kremer, and Brandt 1999) looked at some of these issues. The study found that in the firms examined, it was nearly as likely for the IS staff to grow as to decrease. The biggest change was in IS staff functions (and this was typically underestimated). Little streamlining in the operations of the IS department was realized and operating costs of the IS function were more likely to increase than to decrease. Productivity and efficiency benefits were slow to be realized. No overall conclusions can be drawn yet, but the impact of these packages should be monitored.

Most of the major ERP vendors reported slower growth in the second quarter of 1999. It may be that this represents a reversal of the trend, or it may simply mean that organizations are waiting for the Web-enabled versions of the packages to be released, or that they are switching to smaller vendors who have already made the migration to Web-enabled ERP systems. In any case, these enterprise-wide packages are a reality and they have had and will continue to have an impact on the IS function.

Table 1 summarizes the impact these factors are having on the information systems development environment. The following section will expand on these and other changes happening in IS development.

3. Changes in ISD Skill Requirements

Changes in the development environment call for changes in the skill sets of systems developers. Most of the skills described, which include both "hard" and "soft" skills, are not new. However, the importance of these skills is increasing and developers who can integrate them will be in high demand. This discussion addresses primarily developers of new business information systems, rather than those involved in maintenance of legacy systems or those developing large scale control systems, although some of the concepts may apply to those areas as well.

Table 1. Factors Impacting the ISD Environment

Economic Factors	Organizations continue to be pressured to do more with less. This is acerbated by the shortage of technical personnel with the necessary skill sets. As organizations complete Y2K projects, there will be a backlog of other projects waiting to be addressed.
Globalization	Both the integration of mature markets and newly emerging markets are facilitated by the Internet, resulting in a wide variety of computing platforms and increased global information exchange. For IT departments in multinational organizations, this introduces a number of technical, legal, and cultural issues that must be addressed.
Technology	Development on diverse platforms is becoming the norm. Technology is changing rapidly and continues to provide capacity for faster, more powerful systems.
Applications	Applications development is becoming increasingly component-based and media-based rather than function-oriented. Rather than developing applications software, vendors are providing software services via the Web.
Business Requirements	The expectation is that applications will be available anytime, anywhere. Greater integration within and among organizations is necessary.
Pace of Change	Growth is more rapid. We have seen more change in the last few years, primarily related to the Internet, than we have seen in the last 20 years.

The technologies used to develop the application types most in demand today, Web-based and e-commerce systems, require developers do be knowledgeable in two relatively diverse areas: telecommunications and multi-media (Lyytinen, Rose, and Welke1998). On the one hand, developers need technical telecommunications skills to design and implement communications networks to support distributed, Web-based applications. But developers must also be able to design visually pleasing, fun to use, multi-media interfaces that effectively communicate the organization's message and make it easy for users to navigate and use the site. All of this is in addition to providing the necessary functionality and security behind the scenes.

On-line collaboration can speed software development and reduce costs. Collaboration may be between project team members, with external business partners, and with outside service providers. The Web can provide a place to post coding standards, the project repository, and up-to-date system requirements, and thereby enable faster and easier communication between clients and developers. The ability to update this project information simultaneously for all development partners should reduce delays and errors. The ability to integrate the work of developers from many different locations helps organizations to cope with personnel shortages and to utilize highly-skilled developers located in other parts of the world (often at a lower cost).

All of this collaboration, however, will require that developers have good teamwork skills. Not only do developers have to be able to communicate with other developers, but they also have to be able to work with other business professionals, graphic designers and artists, and others both inside and outside the organization.

In today's fast-paced environment, developers must be able to deal with ambiguity. Organizational culture, relationships, and decision processes are continually changing, following no predefined pattern (Truex, Baskerville, and Klein 1999). Therefore, it becomes impossible to define a static set of system requirements; instead, the point of departure must be incomplete and partially defined specifications. Developers (and their organizations) must be proactive, flexible, and adaptive to respond to the changing needs of the marketplace. The fact that many of the systems developed today are not merely extensions of the existing functions of organizations, but are in fact the totality of the organization, means that the creativity, aesthetic sense, and judgement of the developer become more important than it has been viewed in the traditional development arena (Stolterman 1999; Stolterman and Russo 1999).

There is a high demand for professionals with both business and IT skills (Mateyaschuk and Jaleshgari 1999). Bridging the gap between functional business areas and information technology requires a renewed focus on developing knowledge, skills, and abilities that will enable information systems developers to interface both with the technical specialists and with the users and to understand each group's problem-solving processes and structures (Hale, Sharpe, and Hale 1999.) The ability to leverage business knowledge and integrate it with IT knowledge can result in distinctive organizational competencies (Butler and Murphy 1999).

These skills are summarized in Table 2. The final section of the paper will discuss how educational institutions can help developers build these skills.

4. Building New Development Skill Sets

Admittedly, there are many developers today who have well-honed skills, including those discussed above. However, it does not appear to this author (nor to the recruiters with whom she has discussed this issue) that all new systems developers being "produced" by academic institutions are well prepared in all of these areas. It may be that no one developer can master all of the skills required. It seems imperative, though, that to meet the IS development needs of organizations today and in the future, educational institutions should seek to provide experiences that will develop these important skills.

Providing up-to-date courses on telecommunications technology is obviously required. The ability to integrate multiple platforms is essential. Many programs include some type of multi-media experience; however, this is typically taught from a technical "how to" perspective, rather than utilizing knowledge from the graphic design field. Broadening the focus of these courses could make our graduates much more effective, whether they are doing the graphic design themselves or interacting with graphic designers on the development team.

Cross-functional business knowledge should be a part of the education of any developer who intends to work in the business environment. The use of real-world cases and projects can aid in integrating business knowledge with technology knowledge.

Table 2. Developer Skills Required

Teamwork	Developers can expect to work with multi-skilled teams, with members drawn from not only functional business areas, but also from the art and graphic design fields. Collaboration will take place not only within organizations, but across organizational boundaries on a global scale.
Adaptability	Development organizations will need to become more proactive to keep up with the changing needs of the marketplace. This will require maximum flexibility and minimal bureaucracy. In a rapidly changing environment, it is impossible to know everything with certainty. Developers will have to accept ambiguity as they move away from the traditional notions of well-planned and thoroughly analyzed projects and into a world where change is constant and traditional development approaches may not apply.
Business Knowledge	It is the ability to leverage business organizational knowledge that will provide organizations with competitive advantage. Broad organizational knowledge, integrated with technical skills, provides the basis necessary for identifying current and future opportunities.
Multi-media Skills	In Web-based applications, such as e-commerce, the design of the user interface is critical. It is more, however, than just a nice looking screen. These applications are complex, multi-media structures.
Telecommunications Skills	As more and more systems go on the Web, developers will need to be well-versed in various telecommunications technologies.

It has become quite common for students to work on project teams. It is even more useful when these teams are geographically dispersed and from different educational and cultural backgrounds. This makes communication and cooperation more difficult, and thus forces teams to develop effective communication skills.

Flexibility, adaptability, and the ability to deal with ambiguity are harder to teach. However, by exposing students to a wide variety of development approaches and situations, and stressing the dynamic nature of the field, we can help prepare them. It is also important that we help our students develop their innovativeness, creativity, and judgement skills. We can move in this direction by having students evaluate different information systems and explore ways to identify what is good and what is bad in an information system design (Stolterman 1999).

How well do existing curricula meet these needs? As examples of IS curriculum models in the U.S., we can look at the IS'97 Model Curriculum (Davis et al. 1997) and the Information Systems-Centric Curriculum'99 Program Guidelines (Lidtke et al. 1999).

In both of these curriculum models, we see the traditional fundamentals of IS, programming, database, telecommunications, and analysis and design courses. Both models stress the importance of interpersonal (teamwork) skills, communication skills, and analytical skills (problem solving) in addition to the technical skills The ISCC'99 model has in addition two ethics components, a course on the dynamics of change, and a comprehensive collaborative project (ideally in cooperation with industry). If all IS programs would move in the direction of these models, we would be closer to meeting the needs of today's organizations. However, some gaps remain. In particular, the weakest area of these curricula, and in most current IS programs, is in the area of design. We teach our students the proper way to use various design techniques to produce particular design documents, based on the results of the analysis of the existing system. Unfortunately, for many of them, the systems they will be designing aren't based on any existing system and traditional design techniques, even object-oriented ones, don't help very much when the most important part of the system is the look and feel of the interface.

To prepare our developers for the environment they will face in the future, we should strive to go beyond the technical skills, which are still required, and beyond the collaborative and communication skills, which have been added only in the past decade or so, to the creative realm, where judgement, innovativeness, and the ability to adapt and apply techniques and knowledge in new ways become possible.

5. Conclusions

Changes in technology, business functions, and market conditions are fueling tremendous advances in the systems development arena. A more diverse group of individuals are being drawn into information system development, particularly in the area of Web applications and electronic commerce, and a more diverse skill set is required of those traditional information systems professionals who are moving into these integrative areas. It is becoming more important for those involved in system development to be able to integrate business knowledge and technical knowledge, to possess strong technical skills in the areas of telecommunications and multi-media design, and to have "soft" skills in terms of collaboration, flexibility, creativity, and judgement.

It is becoming more and more difficult to strictly segregate information systems development itself from other activities. For example, as more and more people use information systems on a regular basis, and adapt them to meet their own specific purposes, the distinction between user and developer blurs. The development tools that allow us to move directly from our design to the production system have dissolved the border between design and delivery of a finished product. The whole area of Web application development challenges our notions of what is an information system life cycle, how users influence the development process, and what is the role of the IS function in the development process. Even now, our notion of who or what is an information system "developer" is changing as the process of system development becomes part of nearly all other activities.

References

Butler, T., and Murphy, C. "Shaping Information and Communication Technologies Infrastructures in the Newspaper Industry: Cases on the Role of IT Competencies," in *Proceedings of the Twentieth International Conference on Information Systems*, P. De and J. I. DeGross (eds.), Charlotte, NC, 1999, pp. 364-377.

Byrne, J. "The Search for the Young and Gifted: Why Talent Counts," *Business Week*, October 4, 1999, pp. 108-116.

Davis, G. B., Gorgone, J. T., Couger, J. D., Feinstein, D. L., and Longnecker, H. E., Jr. *IS'97 Model Curriculum and Guidelines for Undergraduate Degree Programs in Information Systems*, Association of Information Technology Professionals, 1997.

Fitzgerald, B. "The Use of System Development Methodologies in Practice: A Field Study," *Information Systems Journal* (7:3), 1997, pp. 201-212.

Fitzgerald, B. "An Empirical Investigation into the Adoption of ISD Methodologies," *Information & Management* (34:6), 1998, pp. 317-328.

Hale, D. P., Sharpe, S., and Hale, J. E. "Business-Information Systems Professional Differences: Bridging the Business Rule Gap," *Information Resources Management Journal*, April-Jun 1999, pp 16-25.

Lidtke, D. K., Stokes, G. S., Haines, J., and Mulder, M. C. *ISCC'99: An Information Systems-Centric Curriculum '99*, National Science Foundation, 1999.

Lyytinen, K., Rose, G., and Welke, R. "The Brave New World of Development in the Internetwork Computing Architecture (InterNCA): Or How Distributed Computing Platforms will Change Systems Development," *Information Systems Journal* (8:4), 1998, pp. 241-253.

Mateyaschuk, J., and Jaleshgari, R. P. "The New CIOs," *Informationweek*, August 16, 1999, pp. 18-20.

Russo, N. L., and Graham, B. R. "A First Step in Developing a Web Application Design Methodology: Understanding the Environment," in *Methodologies for Developing and Managing Emerging Technology Based Information Systems* A. T. Wood-Harper, N. Jayaratna. and J. R. G. Wood (eds.). London: Springer, 1998, pp. 24-33.

Russo, N. L., Hightower, R., and Pearson, J. M. "The Failure of Methodologies to Meet the Needs of Current Development Environments," in *Lessons Learned from the Use of Methodologies*, N. Jayaratna and B. Fitzgerald (eds.). London: British Computer Society, 1996, pp. 387-393.

Russo, N. L., Kremer, A. D., and Brandt, I. "Enterprise-Wide Software: Factors Affecting Implementation and Impacts on the IS Function," *Proceedings of the Thirtieth Annual Meeting of the Decision Sciences Institute*, November 1999, pp. 808-810.

Stephens. D. O. "The Globalization of Information Technology in Multinational Corporations," *Information Management Journal* (33:3), 1999, pp. 66-71.

Stolterman, E. "The Design of Information Systems: *Parti*, Formats, and Sketching," *Information Systems Journal* (9:1), 1999, pp. 3-20.

Stolterman, E., and Russo, N. L. "IS Design Methodologies and the Networked Organization," presented at the IFIP Working Group 8.2 Workshop, Charlotte, NC, December 12, 1999,

Truex, D. P., Baskerville, R., and Klein, H. "Growing Systems in Emergent Organizations," *Communications of the ACM* (42:8), 1999, pp. 117-123.

About the Author

Nancy L. Russo is an associate professor of Information Systems at Northern Illinois University. She received her Ph.D. in Management Information Systems from Georgia State University in 1993. In addition to on-going studies of the use and customization of system development methods in evolving contexts, her research has addressed IT innovation, research methods, and IS education issues. Her work has appeared in *Information Systems Journal*, *Journal of Information Technology*, *Journal of Computer Information Systems*, *System Development Management*, *Journal of Systems and Software*, and various conference proceedings. Nancy can be reached by e-mail at nrusso@niu.edu.

Part 3:

Reforming the Classical Challenges

8 EVALUATION IN A SOCIO-TECHNICAL CONTEXT

Frank Land
London School of Economics
England

1. Introduction

The socio-technical approach to managing business and organizational change has been around for about half of the 20th century. Ever since the pioneers of the approach at the Tavistock Institute for Human Relations published the outcome of their study of the attempts by the National Coal Board in the UK to improve productivity by the introduction of mechanization (Trist 1981; Trist and Bamforth 1951; Trist et al. 1963), socio-technical methods have been discussed and used in the implementation of change and in particular for the introduction of new technologies. Advocates of the socio-technical approach can be found over the entire industrialized world (Coakes, Lloyd-Jones, and Wills 2000). Indeed, the philosophy that underlies much of the thinking of IFIP's Working Group 8.2 rests firmly on socio-technical foundations.

Nevertheless, as Enid Mumford, one of the pioneers, laments in her paper at this conference, socio-technical design is an enigma. It has offered so much and produced so little and we need to know why. There are many case studies demonstrating both the successful use of socio-technical methods and studies which show clearly that paying attention to the social issues as a complement to the techno-economic issues produces results that satisfy managerial aspirations (Land, Detjearuwat, and Smith 1983; Mumford and Henshall 1979; Mumford and MacDonald 1989). Her paper discusses the problem and searches for fresh approaches to revitalize the unfulfilled promise of socio-technical design to come to fulfilment in the 21st century. Those who believe in the market system and its ability to select from a portfolio of innovations the ones that will lead to competitive success will argue that if, over a 50 year or so life span, the market system has failed to select socio-technical design methods, than perhaps those methods have less to offer than their protagonists claim.

This paper attempts to make a contribution to the discussion by offering its own diagnosis, suggesting some research agendas and concentrating on the role of evaluation.

2. The Socio-technical Dilemma

The socio-technical philosophy rests on two perhaps contradictory premises.

The first can be called the *humanistic welfare paradigm*. Socio-technical methods focus on design of work systems to improve the welfare of employees. The prime aim of redesigning work systems is the improvement of the quality of working life (Cherns 1976; Davis and Taylor 1972). Designers seek to develop ways of organizing work that result in improvements of job satisfaction in a number of ways often based on contradictions to the design precepts of the Tayloristic School of Scientific Management. High on the list of desirable attributes can be found concepts like autonomy, self-actualization, self-regulating teams, empowerment, and reducing stress at the work place.

Diagnostic analysis of existing work situations by socio-technical researchers found evidence linking productivity and performance failures to the neglect of many of the attributes listed above. There is much anecdotal and case study evidence demonstrating that there is a link between concern for employee welfare and the effective operation of the organization including the smooth implementation of, often, far reaching change programmes (Land, Detjearuwat, and Smith 1983; Mumford and Henshall 1979; Mumford and MacDonald 1998). Hence those who uphold the Humanistic Welfare Paradigm could claim that although they valued improvements in the quality of working life, the realization of these improvements had a direct beneficial impact on the performance of the organization reflected in its bottom line.

The value system based on ethical principles is illustrated by Mumford's often repeated statement that she would not work with an organization that included forced redundancy among the targets of a change in system (Mumford 1981).

The second can be called the *managerial paradigm*. All change (designed change) is instrumental and serves to improve the performance of the organization. The performance measures are illustrated by the concepts used: adding to shareholders' values, making the business more competitive, improving the bottom line, making the organization more responsive to changing circumstances.

To the extent that socio-technical by its concentration on concepts such as the quality of working life leads to a more contented workforce, and a more contented workforce leads to improved performance as measured by the above attributes, socio-technical methods will be used. And many advocates of the socio-technical approach came to it via the managerial paradigm rather than the social welfare paradigm. Nevertheless, the burden of proof rests with the advocates of socio-technical design methods.

3. The Problem of Evaluation

The general problem of evaluation can be illustrated by the difficulties encountered by researchers and practitioners in accounting *ex post* for the benefits accruing to the organization from the deployment of information systems based on computer and communication technology. For many years researchers failed to find any positive correlation between investment in such systems and improvements in profitability, competitiveness, and productivity (Brynjolfsson and Hitt 1993; Landauer 1995; Strassmann 1985, 1990). Only recently have researchers been able to provide evidence that investments in information and communications technology (ICT) yield positive increases in productivity far beyond that produced by alternative investment strategies (Brynjolfsson and Hitt 1996). Even these findings are still subject to argument and controversy.

Researchers have tried to overcome the problem by using surrogate measures. Of these, the most widely accepted are measures that attempt to assess user satisfaction with the implemented system (DeLone and McLean 1992; Garrity and Sanders 1998). However, user satisfaction is not an output measure and it is doubtful whether is would convince hard-nosed accountants that it proves the value of an investment.

The special case of evaluating the contribution of social elements presents an even greater challenge.

Neither classical economics nor traditional accounting practice recognize that the social elements of a business have a value and contribute to the worth of the business, or more widely to the worth of society. Labor, for example is treated merely as a factor of production, homogenous, replaceable or substitutable for or by other factors. Neo-classical economics or more modern approaches such as those based on transaction costs (Williamson 1986) or those using the theory of games (Binmore and Dasgupta 1987; Kay 1993) take us no further. Accounting practice has concentrated on the recording of directly accountable elements of cost and revenue, although more difficult elements such as "goodwill" have become components of balance sheets. More recent attempts to incorporate notions of human resource accounting are also limited in their recognition of social values. In the past, welfare economics recognized the importance of providing a balance between the costs and benefits attributable to the individual organization from an activity and the costs and benefits of that activity to the larger society, but today managerial economics pays little or no regard to the importance of "externalities."

Nor do the writers on industrial and corporate success, or those who write about business strategy pay much attention to social issues. The most widely read and espoused views on what makes a corporation successful, for example John Kay's (1993) Foundation of Corporate Success, do not include factors such as having a contented workforce as a critical enabling factor. Of the nearly 500 cited references in his book *The Foundations of Corporate Success*, there are none that relate to the importance of socio-technical change management. Porter's (1979, 1980) analysis of the forces which make a business competitive do not include the workforce and its attitudes and behavior. The furthest most writers go is to point to "resistance to change" as an inhibiting factor to "progress," which has to be dealt with in one way or another (Keen 1981). Business school courses, which indoctrinate large numbers of the decision makers of the future , tend to pay scant attention to social factors nor do they attempt to debate values other than those stemming from the managerial paradigm.

Hence it is not surprising that the advocates of the socio-technical approach are faced with problems of evaluation not necessarily faced by the advocates of alternative approaches. Other approaches to organizational change, for example, the more recent phenomenon of business process reengineering (BPR) (Davenport 1993; Hammer and Champy 1993), by appealing directly to the values inherent in the managerial paradigm, have had little difficulty in gaining acceptance. Despite a reported failure rate of between 50% and 70%, a very large proportion of corporations in the industrialized world have made some attempt to use BPR. The promise of BPR can be demonstrated directly using conventional and well understood tools of accounting and evaluation.

To gain a wider acceptance of the socio-technical approach, the evaluator has to demonstrate its value or worth in terms relevant to the managerial paradigm. The evaluation takes two forms:

> *Ex ante* to demonstrate that the expected outcome of a socio-technical inspired change program meets the instrumental expectations of the business and does so with less risk and more certainty than alternative approaches.

> *Ex post* to provide evidence acceptable to the financial management that the desired outcomes have been achieved.

But the evaluation method, too, has to prove acceptable to the decision makers within the organization. And in the typical organization, evaluation is regarded as a purely technical process, carried out according to rules that make it possible to compare evaluations of very different projects on a single scale. Hence the introduction of methods of evaluation deemed suitable for socio-technical designs, which attempt to define the value of other sets of variables previously neglected, face severe difficulties. The problem is perhaps exacerbated by the socio-technical premise that the evaluation process itself should be a socio-technical process and not merely a technical exercise.

To satisfy the humanistic welfare paradigm, the evaluator has to demonstrate that the proposed changes will result in improvements of individual and societal welfare and, once the changes have been implemented, that the improvements have been attained. As has been suggested above, it may not be possible for any proposed change to meet both managerial and humanistic criteria.

An early set of case studies sounds a more optimistic note. In 1957, the London based think tank, PEP (Political and Economic Planning), published an account of three case studies in automation: the manufacture of bearing tubes, an oil refinery process control application, and the application of the LEO computer to business data processing (PEP, 1957). The overall assessment concludes,

> Perhaps the most important theme running through these three case studies is, in fact, the new type of team work which is needed when automatic methods are used. In this there is hope that automation may result not in social loss for the sake of economic gain but rather in social gain hand-in-hand with economic advance.

It is interesting to note that, in the case study of LEO computers, the study cites management's motivation for deploying the computer as including "the hope that something might be done to minimize the drudgery of clerical work." The detailed costings published as part of the case study, however, provide no entry for reduction of drudgery.

4. Evaluation Methods

Conventional evaluation methods provide limited possibilities for including social elements in cost/benefit assessments. These require the evaluation of the second order impacts of social changes. For example, if a socio-technical design reorganizes the work situation to provide, as a direct outcome (first order impact), an increase in job satisfaction, the expected second order consequence might be a reduction in absenteeism, an improvement in health, and hopefully an increase in productivity. Each of these has a measurable impact on the cost/benefit equation. But the *ex ante* assessment of the scale of these effects is difficult and tends to rely on an act of faith by the evaluator rather than a rational calculation. Relying on the usual statistical standby of prior experience does not work well because the impact of the changes is highly situational. In principle it should be possible to check *ex post* the extent to which predictions of second order effects have been realized. But few organizations carry out rigorous *ex post* studies (Kumar 1990).

Evaluation methods are themselves assessed in a number of studies (Farbey. Land and Targett 1993; Hirschheim and Smithson 1988; Wilcocks, 1994).

A promising approach is that based first on the recognition that organizational change, and in particular large scale change, addresses a range of problems and targets a number of objectives (Kenney and Raiffa 1976; Land 1976; Zangemeister 1970). Some outcomes follow directly from the change, for example, the saving of staff; others are the second order consequence of the direct impact. For example, a second order consequence of making staff redundant may be resistance to further change. In that case, the second order effect is a negative one. The economic consequences may be readily measurable by conventional costing techniques, i.e., they are tangible, as is the saving of staff, or more difficult to measure, i.e., they are intangible, as is the increased resistance to change in the example above.

Second, it is recognized that different stakeholders can attach quite different values to the objectives even though the objectives themselves may be shared. But there are objectives which are not shared and some stakeholders attach a negative value to them while others regard them as beneficial. Thus to the senior management in the example above, the saving of staff is ranked as an important objective, while to the personnel department it may be regarded as dysfunctional.

Third, values are measured in the natural units of the goals. Thus a measure of improved responsiveness will be the expected change of response time (*ex ante*) or the achieved change of response time (*ex post*). This is also one of the major drawbacks of this type of evaluation. Instead of reducing all values to the commonly accepted money value, multi-objective, multi-criteria methods reduce all values to a common utility function.

Fourth, the evaluation process is ideally a socio-technical one. That is, it is an iterative process of discovery involving all classes of stakeholders. Technical and social considerations are equally acceptable. Evaluation is regarded as a mutual exploration of the issues, not as a mere recording of technical data. It is recognized that evaluation is a political process (Hawgood and Land 1988) and is seen as an arena for fighting for cherished objectives or alternatively for denying other's objectives that are seen as harmful to one's own interests. Potential conflicts' are exposed and steps can be taken to resolve difficulties arising from the conflict.

A number of evaluation methods are based on the articulation of the multiplicity of objectives that lie behind the designed change and the multiplicity of values that are attached to each objective by different stakeholders. The generic term for such methods are multi-objective, multi-criteria (MOMC) methods. Information economics (Parker and Benson 1987) and the balanced score card (Kaplan and Norton 1992) are variations of the MOMC concept. Some of these will be discussed below.

In the 1970s, an action research project involving a group a savings banks in the UK used MOMC techniques to help the banks evaluate alternative strategies for reorganization. Acting as facilitators, the researchers introduced MOMC to the banks. The outcome was that the banks selected and implemented a program of reorganization that, up to that point, had been rejected because it did not meet conventional cost/benefit criteria. The program incorporated a number of objectives derived from a socio-technical design exercise. Despite the apparent success of the evaluation and the subsequent reorganization, the new methods did not become accepted as part of the normal apparatus of design and evaluation in the banks. Without the presence of the research team, the new methods were seen as outside the accepted norms of accepted (or acceptable) practice in the business.

There are many similar examples of apparently successful socio-technical interventions (Land, Detjearuwat, and Smith 1983; Mumford and Henshall 1979; Mumford and MacDonald 1989). Nevertheless, socio-technical methods failed to get imbedded in those organization as part of standard practice.

6. Other Evaluation Methods

Information economics recognizes that the benefits (and costs) from IS based organizational change include elements with which conventional cost/benefit analysis cannot deal with. It permits the evaluator to "account" for a range of intangibles such as improved customer service and the predicted consequential changes following from the initial change—"value linking" in the terminology of information economics. Benefits and risks are separated into two domains, a technical domain and a business domain, which are evaluated independently. No explicit guidance is given relating to the benefits, costs, and risks associated with the social component of the organization. But the conceptual basis of information economics and the suggested evaluation process could be adapted to incorporate an explicit valuation of the social elements. However, although acknowledged for its potential in assessing the value of IS, information economics has not been widely implemented. Perhaps its focus on IS evaluation rather than as a general tool of evaluating change projects has inhibited its acceptance by management.

A general evaluation methodology that has gained a measure of acceptance from corporations is the balanced scorecard (Kaplan and Norton 1992, 1993; 1996a, 1996b), first reported in *Harvard Business Review* in 1992. Because it appears to be congruent with the requirements of the corporate financial establishment, it might prove to be an acceptable method for evaluating the worth of the socio-technical approach. More recently, a version tailored specifically for the evaluation of IS projects has been suggested in decision support systems (Martinsons. Davison, and Tse 1999). However, the methodology in the form in which it has been presented both by its originators and by its adapters does not explicitly include the value of social elements in the scorecard.

The balanced score card sets out to overcome the weakness of traditional return-on-investment measures of performance by adding measures that reflect customer satisfaction, internal business processes, and the ability to learn and grow. Its orientation is toward future potential as a complement to measures of historic performance. The scorecard is designed to maintain a balance "between short- and long-term objectives, between financial and non-financial measures, between lagging and leading indicators, and between internal and external perspectives" (Kaplan and Norton 1996b).

Kaplan and Norton (1992, 1996a, 1996b) propose the following four perspectives in a balanced scorecard:

1. The customer perspective: Are we satisfying customers needs? How do we look to customers? **Mission:** To achieve our vision, by delivering value to our customers.

2. The financial perspective: How do we look to shareholders? **Mission:** To succeed financially by delivering value to our shareholders.

3. The internal perspective: Are we working effectively and efficiently? What must we excel at? **Mission:** To satisfy our shareholders and customers by promoting efficiency and effectiveness in our business process.

4. The learning and growth perspective: How can we continue to improve and to create value? How can we serve customers better in the future? **Mission:** To achieve our vision by sustaining our innovation and change capabilities through continuous improvement and preparation for future challenges.

The four perspectives are linked. Thus internal efficiency (perspective 2) plus customer satisfaction (perspective 3) leads to financial success (perspective 1).

For each perspective, the evaluator is expected to draw up a table of goals and their appropriate measures. Thus under the customer perspective, a goal might be "reduce delay between an order arriving and it being delivered." The appropriate measure might be the average time taken to make a delivery, and perhaps a measure of the standard deviation. Selecting the measures is itself a process in which the evaluator consults relevant stakeholders. The goals within each perspective have to be ranked in accordance with both their perceived importance to the organization and the degree of risk involved.

Critics can point to the absence of a number of other important perspectives. Thus it would be possible to introduce a perspective relating to the relationship between the business and its partners: how do we look to our suppliers, subcontractors, and so on? As set out above, these perspectives are designed for the business corporation operating in a competitive market. However, it is relatively straightforward to define perspectives that are more suited to a public service enterprise, or indeed to the peculiarities of specific business units such as an IS department (Martinsons, Davison, and Tse 1999) or a research department.

The balanced score card approach lends itself to the addition of a further (socio-technical) perspective:[1]

5. The employee perspective: Are we improving the quality of working life? How do we look to our employees? **Mission:** To achieve a contented, highly motivated workforce at all levels in the organization.

Achieving the mission links with all the other perspectives. Thus a highly motivated, enthusiastic workforce links with the customer perspective in that customers are more likely to want to do business with the firm in question if they are met with an enthusiastic service. Note that the contrary also applies. Unmotivated staff are less likely to encourage customers to return. Again, a highly motivated workforce is more likely to contribute to a learning organization.

The balanced scorecard can also be used to arbitrate between the working of the managerial and the humanistic welfare paradigm by making explicit the values attached to both sets of precepts.

The socio-technical experience will help to define the goals and measures relevant to the employee perspective. These will include goals such as improving job satisfaction, reducing stress, increasing autonomy, increasing participation by the work force, becoming more open with the work force about future plans, and many others. The measures can include direct first order measures such as changes in job satisfaction, or indirect second order measures such as the reduction in absenteeism and ill health.

The process of defining goals and measures can be treated as a socio-technical process, that is, as an exploratory learning activity involving stakeholders and attempting to resolve conflicts through a process of negotiation.

7. Research Agenda

The challenge for those of us who believe that socio-technical design methods are valid, and that they uniquely have the capability of satisfying the value criteria of both the humanistic welfare paradigm and the managerial paradigm, is to provide the evidence in a manner that will convince all affected stakeholders. There have been ample demonstrations that methods which offend specific groups of stakeholders are more likely to fail to achieve their goals. The following research strategies are designed to help in establishing a new case for the socio-technical approach:

1. Identify organizations that in the past have used socio-technical design methods. Trace the history of the application and its operation. Gauge the perceptions of stakeholders on how they viewed socio-technical methods then and now. What caused perceptions to change? Evaluate the degree of success of the application, perhaps by using the modified form of the balanced scorecard outlined above, both in term of meeting techno-economic objectives and social objectives. Find out if socio-technical design methods have been used elsewhere in the organization, whether they are still used, and if not, why not. The research methodology selected

---

[1] A fuller exploration of the use of the balanced score card method is under preparation by B. Farbey, F. F. Land ,and D. Targett, "The Balanced Score Card as a Basis for Socio-technical Evaluation."

for such studies can include interpretive and hermeneutic case studies and techniques from the classical school of empirical research.

2. Search for organizations willing to try out socio-technical design for a real change program. The research team can act as tutor to the organization and observe and monitor the progress of the experiment. Or it can intervene actively, as part of an action research program, and like Mumford in her numerous cases (for example Mumford and Henshall 1979; Mumford and MacDonald 1989) act as the facilitator. Ideally the research team should chronicle the history of the project through all its phases including operation. Stakeholder perceptions and achievements in relation to techno-economic and social goal should be monitored.

3. Although there is little evidence that socio-technical methods as such have been widely used, there is also some evidence that socio-technical precepts have become a part of the language related to the management of change. Socio-technical ideas may have become incorporated in other approaches. Thus, much is made in the literature on business process reengineering that one of the explicit goals of business transformation is "employee empowerment." A research project could look at the most widely lauded methods of strategic management, management of change, and IS development methodologies to check the extent to which these incorporate explicit socio-technical concepts, how these are developed in actual projects, and how they are perceived by stakeholders.

4. A related research project could examine the possibility of relevant aspects of other approaches being merged into socio-technical design methods. In other words, ask the questions "What can the socio-technical school learn from other schools?" and "How can the socio-technical school influence other schools to adopt socio-technical principles?"

5. A major focus of this paper is on finding methods of evaluation acceptable to decision makers—methods that are capable of valuing both social and techno-economic goals. Such methods must fit in with the preconceptions of the established financial community. This paper suggested the possibility of using a modified version of the balanced scorecard. More work needs to be done on the definition of the new form. This will need to be tested in a number of situations. This suggests action research projects that aim at introducing organizations to the new form of the scorecard, monitoring its use, testing its effects, and checking on stakeholder perceptions. Ideally there should be a number of such projects working in different environments, and a meta project that carries out comparisons.

6. Research into other types of evaluation methods and procedures may also be investigated, including in particular methods used to evaluate "problem" areas such as the value of human life in a hospital or on the battlefield.

To those of us who have faith in the value and values of the socio-technical approach, a rich seam of research findings based on the above agenda would be most welcome.

8. Conclusions

Many reasons have been suggested for the failure of the socio-technical approach to gain more general acceptance. It is well established in academic circles, including the academic IS community, but it is not well known among practitioners or even consultants. In business schools, it may make an appearance, but it tends to be a token appearance. This paper attempts to make some contribution to the discussion of these failures and to suggest in particular the importance of finding acceptable ways of evaluating the worth of the social elements to the running of an organization. It proposes the modification of an established evaluation methodology and further research into evaluation methods and evaluation practice. Finally, the paper outlines the kind of research that is needed to bridge the gap between the academic and the practitioner.

References

Binmore, K., and Dasgupta, P. (eds.). *The Economics of Bargaining*. Oxford: Basil Blackwell, 1987.

Brynjolfsson, E., and Hitt, L. *New Evidence on the returns to Information Systems*. Cambridge, MA: MIT Press, 1993.

Brynjolfsson, E., and Hitt, L. "Paradox Lost? Firm-level Evidence of the Returns on Information Systems Spending," *Management Science* (42), 1996, pp. 541-558.

Cherns, A. "Principles of Socio-technical Design," *Human Relations* (2:9), 1976, pp. 783-792.

Coakes, E., Lloyd-Jones, R., and Willis, D. (eds.). *The New Sociotech: Graffiti on the Long Wall*. London: Springer Verlag, 2000.

Davenport, T. *Process Innovation: Reengineering Work Through Information Technology*. Boston: Harvard Business School Press, 1993.

Davis, L. E., and Taylor, J. G. *The Design of Jobs*. Harmandsworth, UK: Penguin Books, 1972.

Delone W., and McLean, E. "Information Systems Success: The Quest for the Dependent Variable," *Information Systems Research* (3:1), 1992, pp. 60-95.

Farbey, B., Land, F. F., and Targett, D. *How to Assess Your IT Investments: A Study of Methods and Practice*. Oxford: Butterworth-Heinemann, 1993.

Garrity, E. J., and Sanders, G. L. *Information Systems Success Measurement*. Hershey, PA: Idea Group Publishing, 1998.

Hammer, M., and Champy, J. *Reengineering the Organization: A Manifesto for Business Revolution*. New York: Harper Business , 1993.

Hawgood, J., and Land, F. F. "A Multivalent Approach to Information Systems Assessment," in *IS Assessment, Issues and Challenges*, N. Bjørn-Andersen and G. B. Davis (eds.). Amsterdam: North Holland, 1998.

Hirschheim, R., and Smithson, S. "A Critical Analysis of Information Systems Evaluation," in *IS Assessment, Issues and Challenges*, N. Bjørn-Andersen and G. B. Davis (eds.). Amsterdam: North Holland, 1988, pp. 17-37.

Kaplan, R., and Norton, D. "The Balanced Score Card: Measures that Drive Performance," *Harvard Business Review* (70:1), 1992, pp. 71-79.

Kaplan, R., and Norton, D. "Putting the Balanced Scorecard to Work," *Harvard Business Review* (71:5), 1993, pp. 134-142.

Kaplan, R., and Norton, D. "Using the Balanced Score Card as a Strategic Management System," *Harvard Business Review* (74:1), 1996a, pp. 75-85.

Kaplan, R., and Norton, D. *The Balanced Scorecard: Translating Strategy into Action*. Boston: Harvard Business School Press, 1996b.

Kay, J. *Foundations of Corporate Success: How Business Strategies Add Value*. Oxford: Oxford University Press, 1993.

Keen, P. G. W. "Information Systems and Organizational Change," *Communications of the ACM* (24:1), 1991.

Kenny, R. L., and Raiffa, H. *Decisions with Multiple Objectives: Preferences and Value Tradeoffs*. New York: John Wiley, 1976.

Kumar, K. "Post-implementation Evaluation of Computer Based IS: Current Practice," *Communications of the ACM* (33:2), 1990, pp. 203-212.

Land, F. F. "Evaluation of Systems Goals in Determining a Design Strategy for a Computer-based Information System," *Computer Journal* (19:4), 1976.

Land, F. F., Detjearuwat, N., and Smith, C. "Factors Affecting Social Control," *Systems, Objectives, Solutions* (3:5 and 3:6), 1983.

Landauer, T. K. The Trouble with Computers: Usefulness, Usability and Productivity. Cambridge, MA: MIT Press, 1995.

Martinsons, M., Davison, R., and Tse, D. "The Balanced Scorecard: A Foundation for the Strategic Management of Information Systems," *Decision Support Systems* (25), 1999, pp. 71-88.

Mumford, E. "Socio-technical Design: An Unfulfilled Promise or a Future Opportunity?" in *Organizational and Social Perspectives on Information Technology,* R. Baskerville, J. Stage, and J. I. DeGross (eds.). Boston: Kluwer Academic Publishers, 2000.

Mumford, E. *Values, Work and Technology*. The Hague: Martinus Nijhoff, 1981.

Mumford, E., and Henshall, D. *A Participative Approach to Computer Systems Design*. London: Associated Business Press, 1979.

Mumford, E., and MacDonald, B. *EXEL's Progress: The Continuing Journey of an Expert System*. New York: John Wiley, 1989.

Parker, M. M., and Benson, R. J. *Information Economics*. Englewood Cliffs, NJ: Prentice-Hall, 1987.

PEP. *Three Case Studies in Automation*. London: (Political and Economic Planning, 1957.

Porter, M. E. *Competitive Advantage: Creating and Sustaining Superior Performance*. New York: The Free Press, 1980.

Porter, M. E. "How Competitive Forces Shape Strategy," *Harvard Business Review* March/April, 1979, pp. 137-145.

Strassmann, P. A. *The Business Value of Computers: an Executive's Guide*. New Canaan, CT: The Information Economics Press, 1990.

Strassmann, P.A. *Information Payoff: The Transformation of Work in the Electronic Age*. New Canaan, CT: The Information Economics Press, 1985.

Trist, E. *The Evolution of Socio-Technical Systems,* Occasional Paper No 2, Ontario Ministry of Labour, June 1981.

Trist, E. L., and Bamforth, K. W. "Some Social and Psychological Consequences of the Long Wall Method of Coal-getting," *Human Relations* (4:1), 1951, pp. 6-24; 37-38.

Trist, E., Higgin, G., Murray, H., and Pollock, A. B. *Organisational Choice*. London: Tavistock Publications, 1963.

Willamson, O.E. *Economic Organisation*. London: Wheatsheaf Books, 1986.

Willcocks, L. (ed.). *Information Management: Evaluation of Information Systems Investments*. London: Chapman & Hall, 1994.

Zangemeister, C. *Nutzwertanlyse in der Systemstechnik: Eine Methode zur Multidimensionsien Bewartung und Auswahl von Project Alternativen*. Munich: Wittmannsche Buchhandlung, 1970.

About the Author

Frank Land received his Bsc (Econ) from the London School of Economics in 1950. In 1952, he joined J. Lyons, the UK food and catering company that pioneered the use of computers for business data processing, building its own computer, the LEO (Lyons Electronic Office). He stayed with LEO until 1967 when he was invited to establish teaching and research in information systems at the London School of Economics. In 1982, he was appointed Professor of Systems Analysis. In 1996, he became Professor of Information Management at the London Business School. On his retirement in 1992, he was appointed Visiting Professor of Information Management at the LSE. Frank is a past Chairman of IFIP WG 8.2. He can be reached by e-mail at FLandLSE@aol.com.

9 COLLABORATIVE PRACTICE RESEARCH

Lars Mathiassen
Aalborg University
Denmark

Abstract

This paper reports from a systems development research tradition, which emphasizes relating research activities to practice and establishing fruitful collaboration between groups of researchers and practitioners. The paper describes and evaluates a specific research project in which a large group of researchers and practitioners worked together to understand, support, and improve systems development practice over a period of three years. The case is used to reflect on the research goals, approaches, and results involved in this tradition for researching systems development practice. A combined approach—based on action research, experiments, and conventional practice studies—is suggested as one practical way to strike a useful balance between relevance and rigor in practice research. The paper concludes with a general discussion of the relation between research and practice as well as advice on how to design collaborative research efforts.

Keywords: Systems development, research, practice, collaboration

1. Introduction

The Information Systems discipline has for quite some time been preoccupied with improving the ways in which we do research. This concern for research methodology has played a major role in maturing the discipline and has resulted in a rich discussion of different approaches (Boland and Hirschheim 1987; Cash et al. 1989; Galliers 1991;

Galliers and Land 1987; Lee, Liebenau, and DeGross 1997; McFarlan 1984; Mumford et al. 1985; Nissen, Klein, and Hirschheim 1991). In our efforts to become a respected research discipline, we have also established an impressive portfolio of scientific journals and international conferences that serve as the primary media for publishing research findings. Our discussions of research methodology have, therefore, concentrated on how to support researchers in designing dedicated research activities that lead to good scientific papers.

A number of scholars within our discipline have recently made a strong plea for more relevance without abandoning rigor (Applegate 1999). Benbasat and Zmud (1999) recommend ways to increase the relevance of our research by reconsidering topic selection, the purpose and content of the articles we write, the readability of an article, and the reviewing process. Davenport and Markus (1999) suggest more radical interventions that challenge core academic values around research rigor, publication outlets and audiences, and consulting. Lyytinen (1999) supports such a broader view and encourages us to critically rethink the institutional policies and incentive schemes that govern research, the organization of research groups, the professional image of Information Systems researchers, and, last but not least, the ways in which we study practice. Lee (1999) argues for a need to go beyond a positivistic research tradition.

This paper addresses many of these concerns by reporting from a research tradition that for some years has studied practice in close collaborations between groups of practitioners and researchers. Such a collaborative approach introduces new interpretations of the relation between research and practice, it raises many practical problems and conflicts, and it is not easily implemented into any institutional setting. The purpose of the present argument is, therefore, not to criticize well-established research traditions. Rather, it is to critically rethink key issues in designing practice research based on experiences from a particular tradition in which relevance is emphasized without abandoning rigor.

Starting from research methods, or from the point of view of writing scientific papers, invites us to think in terms of choosing between different research methods (see, for example, Galliers 1991; Galliers and Land 1987). This viewpoint is extremely useful when one wants to understand the variety and the relative strengths and weaknesses of available research methods. But when designing and organizing research projects based on collaboration with practitioners, the challenge is not so much which methods to choose. Rather, it is to find practical ways to combine qualitatively different research approaches to support the diverse, and partly contradictory, goals involved in such an effort. In the following, I am in favor of such a combined approach—based on action research, experiments, and conventional practice studies—as one practical way to strike a useful balance between relevance and rigor in practice research.

Section 2 presents and discusses a particular research project in which a large group of researchers and practitioners worked together to understand, support, and improve systems development practice based on the so-called Software Process Improvement (SPI) paradigm (see, for example, Emam, Drouin, and Melo 1998; Humphrey 1988, 1989; Kuvaja et al. 1994; Paulk et al. 1993). Based on this case and on the related tradition for doing practice research, I review classical issues and state-of-the-art literature related to research goals (section 3), research approaches (section 4), and research results (section 5). Section 6 examines the underlying view of the relation between research and practice and provides advice on how to design useful collaborations between researchers and practitioners. While the issues raised are of interest to Information Systems research in general, they are discussed based on experiences from the particular field of systems development.

2. Research Practice

The SPI project was a research collaboration involving practitioners from four software organizations and researchers from universities and technology institutes (Johansen and Mathiassen 1998). The collaboration lasted over three years (1997-1999) with a US $4 million budget, of which half was financed by the Danish Ministries of Commerce and Research while the other half was sponsored by the participating software organizations. The project involved more than 10 researchers, each spending between 25% and 75% of their time on the project, and involved three to seven practitioners from each organization as active members of the research project.

Establishing such a research effort is in no way easy. We had good contacts with all levels of the four participating organizations, national research programs provided funding to make researchers collaborate more closely with industry, and the particular theme of the project, i.e., SPI, was known to provide attractive opportunities for software organizations. Finally, we had considerable experience in working closely with practitioners and our institutional setting was positive toward collaboration with industry. Our approach to research grew out of the Scandinavian trade union research tradition (Bjerkness and Bratteteig 1995; Nygaard and Bergo 1975), was later inspired by Checkland's approach to action research (Checkland 1981; Checkland and Scholes 1990), and has been adapted and developed since the early 1980s to suit the study of systems development practice (Mathiassen 1998a, 1998b).

The industry related mission of the project was to systematize SPI knowledge in Danish companies, to tailor and further develop the most promising models for SPI so they apply to the Danish software industry, to develop frameworks for managing, organizing, and implementing SPI activities in Danish companies, and, finally, to communicate and publish knowledge about SPI to Danish companies. These missions were addressed through research efforts in which the following research questions were addressed:

- **Evaluation**: How can we interpret or assess an organization's capability to develop systems, identify appropriate improvement areas and strategies, and measure the effect of the implemented improvement?
- **Modeling:** How can we model systems development processes, the conditions under which they are performed, and their capability to develop quality systems?
- **Improvement:** How can we manage, organize, and carry out initiatives that in a sustainable way improve an organizations capability to develop quality systems?

To meet this diverse set of industry and research related objectives, the project was organized as a loosely coupled system of interacting agendas each addressed by a dedicated forum of actors. The overall organization of the project is illustrated in Figure 1.

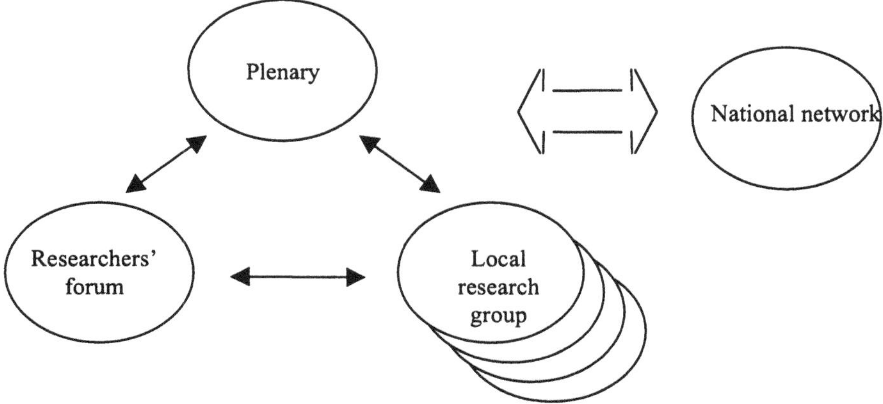

Figure 1. Overall Research Organization

First, a local research group was established to study software process improvement in each of the four software organizations (see Figure 2). This group worked tightly with the local management (both informally and formally through a steering committee), the local SPI group, and the ad hoc projects that were established to implement specific improvement initiatives. Each research group included three to seven practitioners from the software organization (normally the SPI group) and three to four external researchers. The research group met eight to 10 times a year and followed the SPI initiative closely. The group supported the software organization in adapting and using improvement approaches; it participated in some of the organization's dedicated improvement projects; and it continuously evaluated the way in which SPI knowledge was adapted, used, and further developed within the organization.

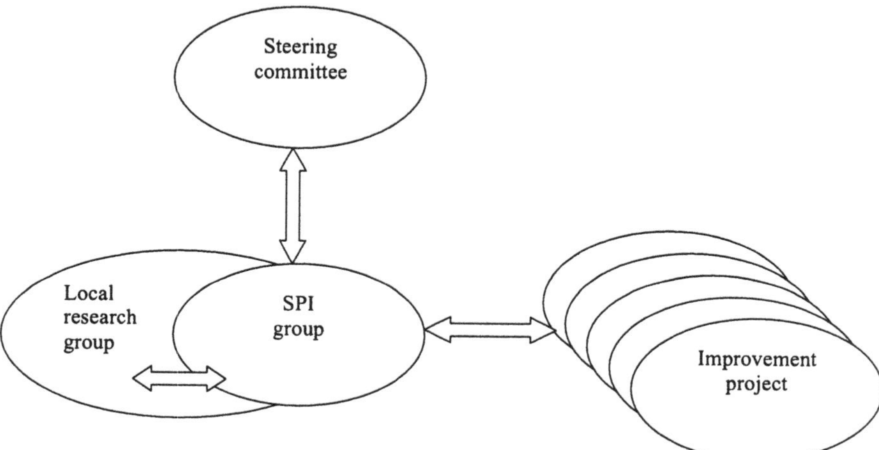

Figure 2. Local Research Organization

Second, a plenary, consisting of the four local research groups, was formed to support interaction between the researchers and practitioners involved in the project. This plenary of 25 to 35 people met twice each year in two-day workshops where experience and knowledge were exchanged across the companies, new knowledge on process improvement was presented, ideas related to specific improvement areas were discussed, and experiences were interpreted and put into perspective through general discussions of software development, software management, and organizational learning and change.

Third, a nation-wide network was formed to support software process improvement initiatives in other organizations. More than 50 companies and 80 individuals participated in this network that met for one-day seminars three to four times a year. The project also organized or participated in a number of conferences on SPI targeting the Danish software industry.

Last, but not least, a researchers' forum was formed to stimulate publication of results and collaboration between the researchers involved in the project. The researchers' forum met for half a day six to eight times a year to identify emerging research themes, plan dedicated research initiatives, form new patterns of collaboration between researchers, discuss research approaches, present and evaluate preliminary results, and discuss relevant theoretical frameworks and related research. The researchers' forum also served as a means to support close collaboration with a number of international research colleagues. Their participation was established to improve the research process by including experiences from similar research projects and knowledge from relevant reference disciplines.

From the point of view of the individual researcher, this organization constituted a collaborative space in which dedicated research initiatives and shifting patterns of collaboration took shape as the process unfolded. Typically, each researcher participated in two local research groups, took part in all plenary meetings, and was actively involved in the researchers' forum. When adding additional ad hoc meetings related to preparation of seminars and workshops, planning of joint publications, and supervision of Ph.D. thesis work, each researcher would typically participate in one to three joint activities each week during the three year course of the project. The number of shared obligations, some of a rather practical nature, that follow from participating as a researcher in such a collaborative effort is, therefore, extremely high. However, I will argue that the opportunities to create relevant research results are at the same time extremely good.

3. Research Goals

Collaborative research involving both researchers and practitioners must serve different interests. In the SPI project, there was an industry-related mission together with a set of research goals; there was also an ambition to add to the body of knowledge within the systems development profession while at the same time advancing practices in each of the participating organizations. Collaborative practice research is, in this way, constantly confronted with dilemmas between practice-driven and research-driven goals and between general and specific knowledge interests. It is well known that researchers in such situations easily turn into consultants (Baskerville and Wood-Harper 1996) and it was, therefore, not surprising that we had to actively promote our research interests in the SPI project. SPI efforts are both demanding and exciting so the researchers were constantly encouraged to engage themselves in the practical struggle to make things happen and succeed in the four organizations. In response to this pressure, we build a strong sub-culture around the researchers' forum to maintain critical reflection, publication, and research methodology as key issues.

Underlying the specific goals of a collaborative practice research effort, we find a deeper level of related research goals and activities. These goals can be expressed in terms of the types of knowledge that the effort intends to develop. Adapting the framework offered by Vidgen and Braa (1997), we can distinguish between different types of knowledge as illustrated in Figure 3 (Mathiassen 1998a, 1998b). The arrows inside the triangle represent *distinct*, and in some respects divergent, research activities through which each type of knowledge is developed. First, to develop our understanding of systems development, we must engage in interpretations of practice. Second, to build new knowledge that can support practice, we must design normative propositions or artefacts, e.g., guidelines, standards, methods, techniques, or tools. Third, to learn what it takes to actually improve practice we must engage in different forms of social and technical intervention. Most Information Systems research restricts itself to understanding and supporting practice. A commitment to improve practice is the distinguishing feature of collaborative practice research and of action research in general (Baskerville and Wood-Harper1996).

The three goals are distinct and can be pursued in isolation, but that would seriously reduce the opportunities to learn about practice. The triangle symbolizes that the involved activities presuppose each other: we reach a deeper understanding of practice as we attempt to change it; we need to understand practice to design useful propositions; and the propositions and our interpretations of practice are ultimately tested through attempts to improve practice. This *unity* of the goals is a simple expression of the elements in organizational learning and change: to appreciate the situation—to invent new options—to change the situation. We find these elements in different forms and relations within theories of the field (e.g., Argyris and Schön 1978; Schein 1985) and in practical approaches as well (e.g., Checkland 1981; Checkland and Scholes 1990; Davenport 1993).

The research goals apply to different levels of practice. First, they apply to systems development. Due to the complex and dynamic nature of the discipline, systems developers need to practice reflection-in-action (Mathiassen 1998a, 1998b; Schön 1983). They must interpret the situations in which they find themselves (understand); they must develop what-if scenarios to reflect on opportunities for action (support); and they must enact some of these to establish and maintain satisfactory working situations (improve). Second, these goals and activities correspond closely to the main ingredients involved in dedicated improvement activities as expressed in the so-called IDEAL model (McFeeley 1996) for SPI (Initiate improvement effort, Diagnose current practices and form strategy for intervention, Establish specific improvement projects, Act to improve, and Learn from the initiatives). The IDEAL process represents a specific way to enact the general learning cycle expressed in the triangle: to appreciate the situation—to invent new options—to change the situation and so on. Third, the triangle expresses the types of knowledge and activities involved in collaborative practice research. Practice, improvement of practice, and practice research are in this way both similar and different in nature. We will further explicate the relation between them in Section 6.

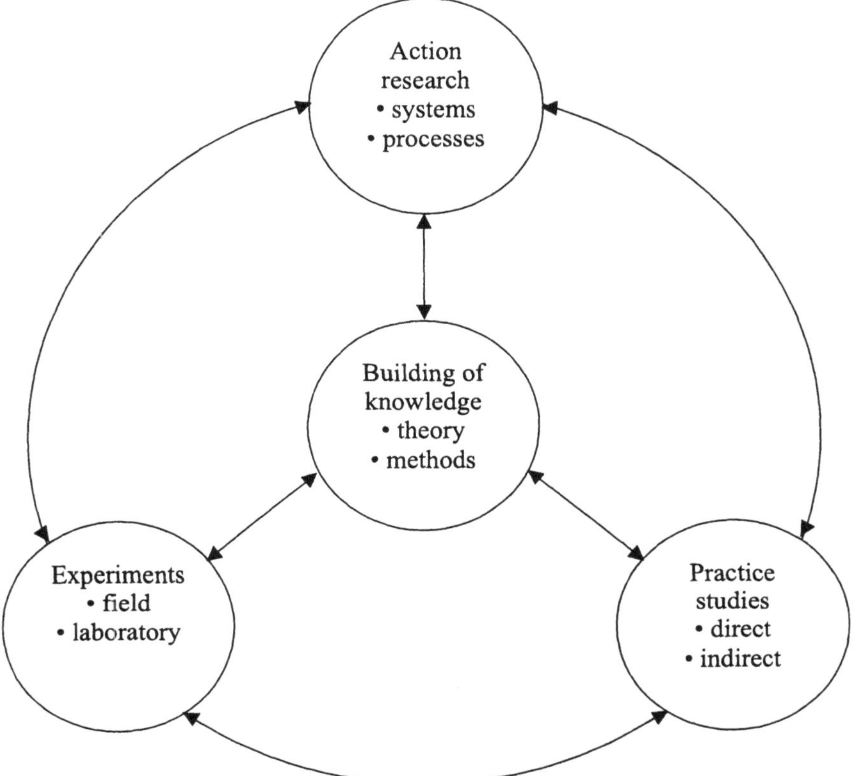

Figure 4. Approaches to Study Systems Development Practice (Adapted from Nunamaker, Chen, and Purdin 1991)

number of contributions have been reported based on this approach (e.g., Baskerville and Stage 1996; Boehm, Gray, and Seewaldt 1984; Boland 1978; Floyd 1986; Guindon, Krasner, and Curtis 1987; Mathiassen 1998b [Chapters 3 and 7]; Selby, Basili, and Baker 1987; Vitalari 1985; Vitalari and Dickson 1983).

Practice studies cover a wide variety of approaches to study systems development without the active involvement of the researchers. Some approaches study practice directly, e.g., field studies and case studies, whereas others are indirect, based on people's opinions and beliefs, e.g., surveys or interviews. The strengths of this approach are that it focuses on practice and that provides the researchers with a vast repertoire of techniques to structure the process and the findings. The weakness is that it separates research from practice. The researchers observe and interpret the actions and beliefs of practitioners and the practitioners do not take active part in the research process. Most of the empirical literature on systems development is based on practice studies (e.g., Aaen et al. 991; Bansler and Bødker 1993; Benbasat, Dexter, and Mantha 1980; Boehm and

Papaccio 1988; Boland and Day 1982; Ciborra and Lanzara 1994; Curtis, Krasner, and Iscoe 1988; Elam et al. 1987; Gould and Lewis 1985; Kozar 1993; Krasner, Curtis, and Iscoe 1987; Madabusyhi, Jones and Price 1993; Markus 1983; McKeen 1983; Necco, Gordon, and Tsai 1987; Stolterman 1992; Tan 1994; Waltz, Elam, and Curtis 1993; White 1984; White and Leifer 1986).

The SPI project was basically organized as an action research effort to gain optimal conditions for interacting closely with practice and to support close collaboration between practitioners and researchers (Baskerville and Wood-Harper 1996). This basic approach was, however, complemented with experiments and with practice studies to establish a more complete and solid foundation for producing rigorous research results. Such a combined strategy supports the variety of research goals discussed above and compensates for the greatest weakness of action research: the limited support that it provides for structuring the research process and findings. The action research approach in the SPI project was implemented through the local research groups (see Figures 1 and 2). The agendas of these groups reflected the local SPI initiatives and the groups served as a forum for evaluating SPI practices, for experimenting with new or modified approaches, and for learning about the impact of SPI approaches on practice. Field experiments were then staged as dedicated research initiatives within this setting and focused practice studies were initiated to learn about selected SPI practices and their impact on the organization. Some examples will illustrate this combined approach.

The major challenge faced by the SPI initiative in one of the participating organizations was to motivate the systems developers, and in particular the project managers, to commit themselves to improvement efforts. Traditional SPI initiatives are based on normative models (e.g., Emam, Drouin, and Melo 1998; Kuvaja 1994; Paulk et al. 1993), but none of these models were considered useful by the SPI group or the developers. The SPI group, therefore, decided to use problem diagnosis techniques rather than assessments based on general models to learn what the developers considered to be key problems. This research initiative is documented in Iversen, Nielsen, and Nørberg (1998) and is primarily based on action research as proposed by Checkland (1981; Checkland and Scholes 1990).

Each improvement project that is initiated as part of an SPI initiative is facing a complex and often quite risky task, for example to develop, implement, and institutionalize processes to support subcontract management in the organization. The SPI group in one of the organizations wanted to develop tools that could support specific improvement projects in managing risks, thereby minimizing the chance of failure. A dedicated risk management tool for improvement projects was, therefore, developed based on the experiences from that organization in combination with insights from the SPI literature. The resulting tool is documented in (Iversen, Mathiassen, and Nielsen 1999) and the underlying research approach is primarily a field experiment.

Some events during the course of the SPI project were considered so interesting that they attracted special attention. One such case was the implementation of a metrics program to help the organization measure the effects of their improvement efforts. A focused study of this particular case was based on the "natural traces" of the SPI program, such as project plans, meeting minutes, and memos. In addition to this, we tape-recorded the monthly meetings in each local research group as well as some of the working sessions and workshops to collect supplementary data to be used in dedicated and focused research initiatives. The relevant segments for the metrics program were transcribed. The case and the lessons learned were published (Iversen and Mathiassen 2000) based on a combination of direct and indirect practice studies.

We see in these examples how a variety of research approaches are used in dedicated research initiatives within the larger project. The SPI project can, from this point of view, be seen as a collaborative space in which specific and dedicated research initiatives are formed to report (1) key lessons from selected parts of the action research activities, (2) proposals based on field experiments that are designed to provide local support, and (3) insights from practice studies of events that emerge as interesting or surprising cases. These three types of research initiatives are exemplified above and they are initiated based on the agendas of the local research groups (see Figure 1). The agenda of the researchers' forum is, however, broader. It includes classical concerns, in which SPI is seen as one instance of technology-related organizational change, and it involves theoretical issues and the use of reference disciplines. Therefore, nsights from practice were also used in the SPI project as inspiration for theoretical studies. A few examples illustrate these theoretical activities.

The SPI literature is extensive but rather practical, with little concern for fundamental research questions. In the SPI project, we had many discussions on the identity and boundaries of the SPI approach as described in the literature. We found no explicit, shared understanding of SPI as a strategy for change in systems development organizations. Based on our practical experiences, we decided to survey the SPI literature to explicate important underlying assumptions and related strategies for change (Aaen et al. 2000).

There are also only a few, rather weak, relations between the SPI literature and relevant reference disciplines. It is, therefore, interesting to interpret SPI experiences using contemporary frameworks from other research areas. One such example is the use of Nonaka's (1994) theory of knowledge creation to understand better how tacit and explicit forms of knowledge can be combined in software organizations and how one can support interaction between individual, group, and organizational knowledge creation processes (Arent and Nørbjerg 2000). Such studies serve to interpret and inform SPI practices and they often provide interesting examples for the reference discipline in question.

Other theoretical contributions from the SPI project are based on subjective/argumentative approaches (Galliers 1991; Galliers and Land 1987). SPI is typically based on a rather narrow view of the systems development process, e.g., from the time a contract is signed until a software system is delivered, and the main concern is to improve processes within the software organization. Bjerknes and Mathiassen (2000) reflect on the nature of well-functioning customer-supplier relations, evaluate those SPI models that are concerned with improved customer relations, and propose specific initiatives to improve the collaboration between customer and supplier organizations. Such initiatives were subsequently implemented in one of the participating organizations. Hence, the research inspired innovative activities that could lead to alternative SPI strategies.

5. Research Results

Collaborative practice studies, as I have discussed them here, are both practice- and research-driven and serve general knowledge interest as well as knowledge interests that are specific for the participating organizations. The results of such efforts are, therefore, of a more diverse nature than those of conventional research projects.

There was a strong inclination in the SPI project to focus on practical results. The rationale to participate was, from the point of view of the software organizations, to engage in collaborative activities that could lead to improved systems development performance and that could stimulate learning within the organization. It goes without saying that actual improvements in processes, infrastructure, and competencies were the key success criteria for the local SPI groups. This constant pressure to focus on practical results implies that research results tend to have secondary priority. The project was, from the very start, based on a shared commitment between the involved organizations, the individual practitioners, and the researchers to build new knowledge that could be published as a contribution to the body of knowledge on systems development. This commitment was maintained by having research issues integrated into the agendas of all groups in the project organization (cf. Figure 1).

Conventional research publications played an important role in the SPI project. Three Ph.D. studies were carried out as part of the project and more than 30 papers, addressing a variety of issues, were published. The important difference in relation to more conventional research projects was that practitioners were included as an important target group for the project. First, more than 10 papers with results from the project were published at conferences or in journals mainly for practitioners (e.g., Jakobson 1998; Johansen and Mathiassen 1998; Vinter 1998). Second, it was decided to publish key lessons from the project in a book titled *Learning to Improve*. Each chapter in the book presents lessons on SPI from the project, it is written with a practitioners orientation, the foundation is academically sound, based on documented research results from the project, and most of the chapters are coauthored by researchers and practitioners. Papers published in *IEEE Software* were used as model examples for each chapter and an editor from that journal was engaged to guide the authors and facilitate the editing process.

Professional Systems Development: Experiences, Ideas, and Action (Andersen et al. 1986, 1990), *Quality Management in Systems Development* (Bang et al. 1991), and *Object Oriented Analysis and Design*" (Mathiassen et al. 1997) are examples from previous projects illustrating that publication for practitioners is given high priority in this research tradition. There is a sound rationale for pursuing such a strategy in practice research. The researchers are constantly challenged to develop and express results they believe to be useful in practice. Practitioners study the contributions, they attempt to use them, and that reveals strengths and weaknesses of the published results. The research contributions are in this way instrumental in establishing and maintaining a dialogue between research and practice that goes beyond specific research projects. Such a dialogue serves to test and further develop new knowledge, it plays a major role in developing higher education within our discipline, and it helps build an image of the researcher as actively contributing to improved professional practices. This, in turn, makes it considerably easier to establish new collaborations with practitioners.

Unfortunately, such research contributions are given quite low, and in some cases no, priority in making career decisions within the established research community. The incentive to publish for practitioners is, therefore, minimal and a simple cost/benefit analysis will lead most researchers to the conclusion that it is not worthwhile to become engaged in collaborative practice studies. In that way, we risk isolating ourselves as an appendix to the Information Systems profession instead of being a major force in strengthening its position in society. We should not, of course, stop our efforts to mature as a respected research discipline with high quality journals and research conferences. But we should take care to combine these ongoing initiatives with modified incentive schemes, intensified collaboration with practitioners, and more publications that address practical concerns. That will help us develop new knowledge that will prove to be relevant and it will strengthen our position within the profession and society in general.

6. Research and Practice

Collaborative practice research is not merely a way to organize and conduct research. Underlying the specifics of the SPI project is a coherent view of practice, of research, and of the relations between the two. This view is part of a research tradition that has been developed over the past 20 years. The tradition has been discussed as "the ordinary work practices approach" (Hirschheim and Klein 1992) or "the professional work practices approach" (Hirschheim, Klein and Lyytinen 1995a, 1995b; Iivari and Lyytinen 1997). I have presented the underlying perspective as "Reflective Systems Development" (Mathiassen 1998a, 1998b) to acknowledge the relation to Schön's (1983) ideas on how professionals think in action and to stress the intrinsic relation between research and practice. The SPI project should be seen as an effort to further develop this tradition by addressing some of its weaknesses, e.g., a too simplistic view of what it takes to improve practice and lack of an explicit notion of what it means to change practices to the better (Hirschheim, Klein and Lyytinen 1995a, 1995b).

Reflective systems development is illustrated in Figure 5 based on Checkland's notion of the experience-action cycle (Checkland and Scholes 1990). The challenges and opportunities involved in systems development practice are considered the starting point for systems development research. Research activities yield experience-based knowledge that leads to new, and hopefully improved, practices. The knowledge that is developed in this process is both interpretive and normative. Part of it remains local, individual, and even tacit, while other parts are explicated and made publicly available as systems development guidelines, professional books, and research contributions. The research activity is primarily informed by systems development practice, but supported by various reference disciplines (e.g., design theory, organization science, management science, and philosophy), and by dialectic reflections that help us understand change and the contradictory nature of our discipline (Robey 1995). The reference disciplines and dialectics encourage the researchers to go beyond the limited perspectives of approaches such as SPI and to frame their thinking and results in ways that contribute to the building of research-based knowledge.

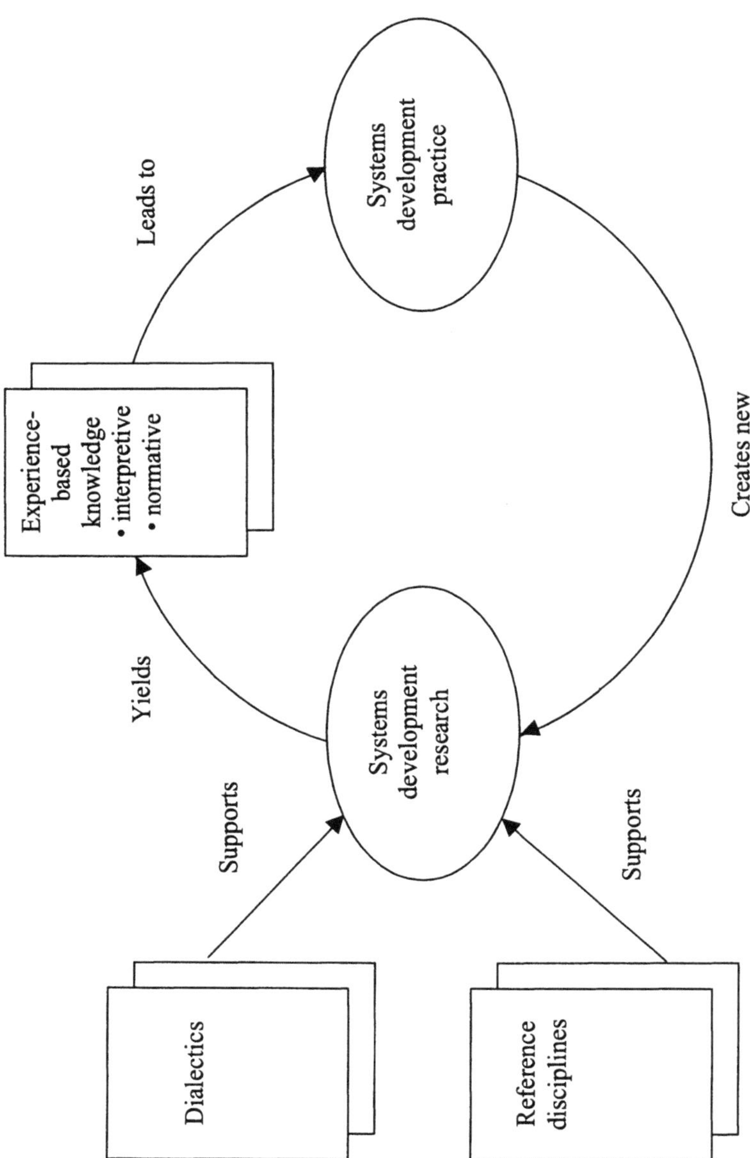

Figure 5. Reflective Systems Development (Adapted from Checkland and Scholes 1990)

Reflective systems development expresses one, coherent view of practice *and* research with two different modes of inquiry, a research and a practice mode. This is expressed in Table 1 based on Checkland's ideas on how intellectual frameworks are used in relation to specific application areas (Checkland and Scholes 1990, p. 283). The table explicates the different but related purposes of research and practice, the underlying intellectual frameworks, the type of process in which they are applied, and the shared arena to which they are applied.

In this view, research becomes an activity in which practitioners (can) participate and collaborate with researchers.

> The practitioner does not function as a mere user of the researcher's product. He reveals to the reflective researcher the ways of thinking that he brings to his practice, and draws on reflective research as an aid to his own reflection-in-action. Moreover, the reflective researcher cannot maintain distance from, much less superiority to, the experience of practice...he must somehow gain an inside view of the experience of practice. [Schön 1983, p. 323]

This view of research implies and builds on a complementary view of practice in which reflection and learning are key elements. Systems developers must, in addition to mastering a repertoire of general methods and tools, know how to cope with the specific environment in which they work. Many situations involve uncertainty, instability, uniqueness, and contradictions and they require an ability to go beyond the relatively safe territory of general professional knowledge. Systems developers must open their minds and engage in reflections and dialogues to generate the necessary insights into the situation at hand.

Table 1. Reflective Systems Development as an Approach to Research and Practice

	Research	**Practice**
Purpose	to develop knowledge to understand, support, and improve practice as part of the ongoing professional development	to develop computer-based information systems as part of the ongoing transformation of organizations and society
Framework	dialectics reference disciplines	systems development theory systems development methods
Process	action research	reflection-in-action
Arena	systems development practice	systems development practice

From this point of view, the SPI project is an attempt to practice reflective systems development as a close collaboration between researchers and practitioners. I have discussed the project from a researcher's perspective and in the process provided concepts and experiences that can be used to organize similar initiatives. The discussion can be summarized as a number of lessons.

- *Lesson 1: Implement a full learning cycle of understanding, supporting, and improving practice.* Understanding, supporting, and improving expresses the basic knowledge interests involved in studying practice. Dedicated research initiatives will have different emphasis, but it is important to organize the overall project so that it includes full learning cycles in which our understanding of present practices are confronted with explorations of possible alternatives to form new, and hopefully improved, practices.

- *Lesson 2: Organize the project as a loosely coupled system of related agendas.* Research collaboration with practitioners involves a multiplicity of partly contradictory goals. Research projects should be organized to support diversity, but at the same time function as a shared space in which dedicated research initiatives can be formed as new opportunities emerge. It is particularly important to have separate, interacting agendas for local involvement and detached research.

- *Lesson 3: Combine action research, experiments, and practice studies.* Action research should be used as the basic form to establish a close relation to practice, but whenever feasible and useful it should be supplemented with experiments and practice studies. Such a combined strategy supports the variety of research goals involved and helps establish a useful balance between rigor and relevance.

- *Lesson 4: Establish a basic documentation system to support longitudinal practice studies.* The danger in action research is to become too involved in the problems of practice. Longitudinal field research has been developed in response to this challenge emphasizing the need to establish systematic documentation efforts involving in-depth interviews, documentary and archive data, and observational and ethnographical material (Nielsen 1999; Pettigrew 1990). Such a documentation system serves as the backbone for organizing dedicated research initiatives that are focused on particular events.

- *Lesson 5: Facilitate collaboration in dedicated research initiatives.* Research projects should offer opportunities for establishing shifting patterns of collaboration between the researchers involved and between researchers and practitioners. Such patterns are established as interesting issues emerge and new, dedicated research initiatives are formed. Each new initiative involves specific actors and is based on its own combination of research methods to suit the task.

- *Lesson 6: Combine scientific publication with publications targeting practitioners.* Traditional research publications should be supplemented with publications that inform practitioners about research results. Such publications challenge the researchers to evaluate the relevance of their efforts and are instrumental in maintaining a dialogue between researchers and practitioners beyond the specific project.

- *Lesson 7: Engage yourself fully, but only for a while.* Collaborative practice research as described here offers good opportunities to develop relevant research results, but it requires a dedicated effort involving both research work and organizational work. During a project, it is necessary to spend a major effort on the collaboration. After the project, it is advisable to return to a more traditional activity pattern, to reflect on the experiences from the project, and to publish more of the insights that were gained during the project.

These lessons are meant as inspiration for those having the opportunity and the motivation to engage in collaborative practice research. Such efforts are, however, difficult to create and demanding to manage. This is partly because of the diverse and contradictory nature of the interests and goals involved. However, in many cases, it is also because our institutional settings and incentive schemes do not encourage researchers to engage in close collaboration with practitioners (Lyytinen 1999).

Acknowledgments

This research has been partially sponsored by the Danish National Centre for IT Research. I wish to thank the software organizations, practitioners, and researchers participating in the Danish SPI project. I also want to thank the associate editor and the anonymous reviewers for many valuable suggestions.

References

Aaen, I., Siltanen, A., Sørensen, C., and Tahvanainen, V.-P. "A Tale of Two Countries: CASE Experience and Expectations," in The Impact of Computer Technologies on Information Systems Development, K. E. Kendall, K. Lyytinen, and J. I. DeGross (eds.). Amsterdam: North-Holland, 1992.

Aaen, I., Arent, J., Mathiassen, L., O. Nwgenyama, O. "A Conceptual MAP of Software Process Improvement," *The Computer Journal*, 2000, forthcoming.

Andersen, C. V., Arent, J., Bang, S., and Iversen, J. "Project Assessments: Supporting Commitment, Participation, and Learning in SPI," in *Proceedings of Thirty-third Hawaii International Conference on System Sciences*. Los Alamitos, CA: IEEE Computer Society, 2000.

Andersen, N. E., Kensing, F., Lassen, M., Lundin, J., Mathiassen, L., Munk-Madsen, A., and Sørgaard, P. *Professional Systems Development. Experiences, Ideas, and Action.* Copenhagen: Teknisk Forlag (in Danish), 1986.

Andersen, N. E., Kensing, F., Lassen, M., Lundin, J., Mathiassen, L., Munk-Madsen, A., and Sørgaard, P. *Professional Systems Development. Experiences, Ideas, and Action.* Englewood Cliffs, NJ: Prentice-Hall, 1990.

Applegate, L. M. "Rigor and Relevance in MIS Research," *MIS Quarterly* (23:1), 1999.

Arent, J., and Nørbjerg, J. "SPI as Organizational Knowledge Creation: A Multiple Case Analysis," in *Proceedings of Thirty-third Hawaii International Conference on System Sciences.* Los Alamitos, CA: IEEE Computer Society, 2000.

Argyris, C., and Schön, D. *Organizational Learning.* Reading, MA: Addison-Wesley, 1978.

Avison, D. E., and Wood-Harper, A. T. "Information Systems Development Research: An Exploration of Ideas in Practice," *The Computer Journal* (34:2), 1991.

Avison, D., Lau, F., Myers, M., and Nielsen, P. A. "Action Research: Making Academic Research Relevant," *Communications of the ACM* (42:1), 1999.

Bang, S., Efsen, S., Hundborg, P., Janum, H., Mathiassen, L., and Schultz, C. *Quality Management in Systems Development.* Copenhagen: Teknisk Forlag. (in Danish), 1991.

Bansler, J., and K. Bødker "A Reappraisal of Structured Analysis. Design in an Organizational Context," *ACM Transactions on Information Systems* (11:2), 1993,

Basili, V. R., and Weiss, D. M. "A Methodology for Collecting Valid Software Engineering Data," *IEEE Transactions on Software Engineering* (10), 1984.

Basili, V. R., Selby, R. W., and Hutchens, D. H. "Experimentation in Software Engineering," *IEEE Transactions on Software Engineering* (12), 1986.

Baskerville, R., and Wood-Harper, A. T. "A Critical Perspective on Action Research as a Method for Information Systems Research," *Journal of Information Technology* (11), 1996.

Baskerville, R. L., and Stage, J. "Controlling Prototype Development Through Risk Management," *MIS Quarterly* (20:4), 1996.

Benbasat, I., Dexter, A. S., and Mantha, R. W. "Impact of Organizational Maturity on Information System Skill Needs," *MIS Quarterly* (4:1), 1980.

Benbasat, I., and Zmud, R. W. "Empirical Research in Information Systems: The Practice of Relevance," *MIS Quarterly* (23:1), 1999.

Bjerknes, G. "Dialectical Reflection in Information Systems Development," *Scandinavian Journal of Information Systems* (3), 1991.

Bjerknes, G., and Bratteteig, T. "User Participation and Democracy: A Discussion of Scandinavian Research on Systems Development," *Scandinavian Journal of Information Systems* (7:1), 1995.

Bjerknes, G., and Mathiassen, L. "Improving the Customer-Supplier Relation in IT Development," in *Proceedings of Thirty-third Hawaii International Conference on System Sciences*. Los Alamitos, CA: IEEE Computer Society, 2000.

Boehm, B. W., Gray, T. E., and Seewaldt, T. "Prototyping versus Specifying: A Multiproject Experiment," *IEEE Transactions on Software Engineering* (10:3), 1984.

Boehm, B. W., and Papaccio, P. N. "Understanding and Controlling Software Costs," *IEEE Transactions on Software Engineering* (10:4), 1988.

Boland, R. J. "The Process and Product of System Design," *Management Science* (24:9), 1978.

Boland, R. J., and Day, W. "The Process of System Design: A Phenomenological Approach," in *Proceedings of the Third International Conference on Information Systems*, M. J. Ginzberg and C. A. Ross (eds.), Ann Arbor, Michigan, 1982.

Boland, R. J., and Hirschheim, R. A. (eds.). *Critical Issues in Information Systems Research.* Chichester: John Wiley, 1987.

Cash, J. I., Benbasat, I., Kraemer, K. L., and Lawrence, P. R. (eds.). *The Information Systems Research Challenge*, Volumes 1–3. Boston, MA: Harvard Business School, 1989.

Checkland, P. *Systems Thinking, Systems Practice.* Chichester: John Wiley, 1989.

Checkland, P., and Scholes, J. *Soft Systems Methodology in Action.* Chichester: John Wiley, 1990.

Ciborra, C., and Lanzara, G. F. "Formative Contexts and Information Technology: Understanding the Dynamics of Innovation in Organizations," *Accounting, Management and Information Technology* (4:2), 1994.

Cotterman, W., and Senn, J. *Challenges and Strategies for Research in Information Systems Development.* Chichester, England: Wiley Series in Information Systems, 1992.

Curtis, B., Krasner, H., and Iscoe, N. "A Field Study of the Software Design Process for Large Systems," *Communications of the ACM* (31:11), 1988.

Davenport, T. H. *Process Innovation—Reengineering Work through Information Technology.* Boston, MA: Harvard Business School Press, 1993.

Davenport, T., and Markus, M. L. "Rigor vs. Relevance Revisited: Response to Benbasat and Zmud," *MIS Quarterly* (23:1), 1999.

Elam, J. J., Waltz, D. B., Krasner, H., and Curtis, B. "A Methodology for Studying Software Design Teams: An Investigation of Conflict Behaviors in the Requirements Definition Phase," *Empirical Studies of Programmers. Second Workshop.* Norwood, NJ: Ablex Publishing, 1987.

Emam, K. E., Drouin, J-N., and Melo, W. *SPICE: The Theory and Practice of Software Process Improvement and Capability Determination.* Los Alamitos, CA: IEEE Computer Society, 1998.

Floyd, C. "A Comparative Evaluation of Systems Development Methods," in *Information Systems Design Methodologies. Improving the Practice*, T. W. Olle et al. (eds.). Amsterdam: North-Holland, 1986.

Galliers, R. D. "Choosing Appropriate Information Systems Research Approaches: A Revised Taxonomy," in *Information Systems Research: Contemporary Approaches and Emergent Traditions*, H-E. Nissen, H. K. Klein, and R. Hirschheim (eds.). Amsterdam: North-Holland, 1991.

Galliers, R. D., and Land, F. F. "Choosing Appropriate Information Systems Research Methodologies," *Communications of the ACM* (30:11), 1987.

Gould, J. D., and Lewis, C. "Designing for Usability: Key Principles and What Designers Think," *Communications of the ACM* (28:3), 1985.

Guindon, R., Krasner, H., and Curtis, B. "Breakdowns and Processes During the Early Activities of Software Design by Professionals," *Empirtcal Studies of Programmers. Second Workshop.* Norwood, NJ: Ablex Publishing, 1987.

Hirschheim, R., and Klein, H. K. "Paradigmatic Influences of Information Systems Development Methodologies: Evolution and Conceptual Advances," in *Advances in Computers (33)*, M. Yovits (ed.). New York: Academic Press, 1992.

Hirschheim, R., Klein, H. K., and Lyytinen, K. *Information Systems Development and Data Modeling. Conceptual and Philosophical Foundations.* Cambridge, England: Cambridge University Press, 1995a.

Hirschheim, R., Klein, H. K., and Lyytinen, K. "Exploring the Intellectual Structures of Information Systems Development: A Social Action Theoretic Analysis," *Accounting, Management and Information Technology* (6:1-2), 1995b.

Humphrey, W. S. "Characterizing the Software Process," *IEEE Software* (5:5, 1988, pp. 73-79.

Humphrey, W. S. *Managing the Software Process.* Reading, MA: Addison-Wesley, 1989.

Iivari, J., and Lyytinen, K. "Information Systems Research in Scandinavia: Unity in Plurality," in *Rethinking Management Information Systems*, W. Currie and R. D. Galliers (eds.). London: Oxford University Press, 1997.

Iversen, J., Johansen , J., Nielsen, P. A., and Pries-Heje, J. "Combining Quantitative and Qualitative Assessment Methods in Software Process Improvement," in *Proceedings of 1998 European Conference on Information Systems*, Aix-en-Provence, France, 1998.

Iversen, J., and Mathiassen, L. "Lessons from Implementing a Metrics Program," in *Proceedings of Thirty-third Hawaii International Conference on System Sciences.* Los Alamitos, CA: IEEE Computer Society, 2000.

Iversen, J., Mathiassen, L., and Nielsen, P. A. "Risk Management in Software Process Improvement," in *Proceedings of the 1999 European Conference on Information Systems*, Copenhagen, Denmark, 1999.

Iversen, J., Nielsen, P. A., and Nørbjerg, J. "Problem Diagnosis in Software Process Improvement," in *Information Systems: Current Issues and Future Changes*, T. J. Larsen, L. Levine, and J. I. DeGross (eds.). Laxenburg, Austria: IFIP, 1998, pp. 111-130.

Jakobsen, A. B. "Tricks of Bottom-Up Improvements," *IEEE Software*, January 1998.

Johansen, J., and Mathiassen, L. "Lessons Learned in a National SPI Effort," *EuroSPI'98,* Gothenburg, Sweden, 1998.

Kaiser, K. M., and Bostrom, R. P. "Personality Characteristics of MIS Projects Teams: An Empirical Study and Action-Research Design," *MIS Quarterly* (6:4), 1982.

Knuth, D. "The Errors of TEX," *Software—Practice and Experience* (19:1), 1989.

Kozar, K. A. "Adopting Systems Development Methods. An Exploratory Study," *Journal of Management Information Systems* (5:4), 1993.

Krasner, H., Curtis, B., and Iscoe, N. "Communication Breakdowns and Boundary Spanning Activities of Software Design by Professionals," *Empirical Studies of Programmers. Second Workshop*. Norwood, NJ: Ablex Publishing, 1987.

Kuvaja, P., Similä, J., Krzanik, L., Bicego, A., Saukkonen, S., and Koch, G. *Software Process Assessment and Improvement—The Bootstrap Approach*. Oxford: Blackwell, 1994.

Lau, F. "A Review on the Use of Action Research in Information Systems Studies," in *Information Systems and Qualitative Research*, A. S. Lee, J. Liebenau, and J. I. DeGross (eds.). London: Chapman & Hall, 1997, pp. 31-68.

Lee, A. "Rigor and Relevance in MIS Research: Beyond the Approach of Positivism Alone," *MIS Quarterly* (23:1), 1999.

Lee, A. S., Liebenau, J., and DeGross, J. I. (eds.). *Information Systems and Qualitative Research*. London: Chapman & Hall, 1997/

Lyytinen, K. "Empirical Research in Information Systems: On the Relevance of Practice in Thinking of IS Research," *MIS Quarterly* (23:1), 1999.

Madabushi, S. V. R., Jones, M. C., and Price, R. L. "Systems Analysis and Design Models Revisited: A Case Study," *Information Resources Management Journal*. Winter 1993.

Markus, M. L. "Power, Politics, and MIS Implementation," *Communications of the ACM* (26:7), 1983.

Mathiassen, L. *Reflective Systems Development*. Unpublished Dr. Techn. Thesis, Aalborg University, 1998a (available at www.cs.auc.dk/~larsm).

Mathiassen, L. "Reflective Systems Development," *Scandinavian Journal of Information Systems* (10:1-2), 1998b.

Mathiassen, L., Munk-Madsen, A., Nielsen, P. A., and Stage, J. *Object Oriented Analysis and Design*. Aalborg: Marko (in Danish; to appear in English), 1997.

McFarlan, F. W. *The Information Systems Research Challenge*. Boston, MA: Harvard Business School Press, 1984.

McFeeley, B. *IDEAL: A User's Guide for Software Process Improvement*. Pittsburgh: SEI Handbook, CMU/SEI-96-HB-001, 1996.

McKeen, J. D. "Successful Development Strategies for Business Application Systems," *MIS Quarterly* (7:3), 1983.

Mumford, E. *Designing Human Systems: The ETHICS Method*. Manchester, England: Manchester Business School, 1983.

Mumford, E., Hirschheim, R. A., Fitzgerald, G., and Wood-Harper, A. T. (eds.). *Research in Information Systems*. Amsterdam: North-Holland, 1985.

Munk-Madsen, A. *Knowledge About Systems Development*. MARS Report No. 13. Aarhus University (in Danish), 1986.

Necco, C. R., Gordon, C. L., and Tsai, N. W. "Systems Analysis and Design: Current Practices," *MIS Quarterly* (11:4), 1987.

Nielsen, P. A. "Action Research and the Study of IT in Organizations: Making Sense of Change," Department of Computer Science, Aalborg University, 1999.

Nissen, H.-E., Klein, H. K., and Hirschheim, R. A. (eds.). *Information Systems Research: Contemporary Approaches and Emergent Traditions*. Amsterdam: North-Holland, 1991.

Nonaka, I. "A Dynamic Theory of Organizational Knowledge Creation," *Organization Science* (5:1), 1994.

Nunamaker, J., Chen, M., and Purdin, T. D. M. "Systems Development in Information Systems Research," *Journal of Management Information Systems* (7:3), 1991.

Nygaard, K., and Bergo, O. T. "The Trade Unions. New Users of Research," *Personnel Review* (4:2), 1975.

Parnas, D. L., and Clements, P. C. "A Rational Design Process: How and Why to Fake It," *IEEE Transactions on Software Engineering* (12:2), 1986.

Paulk, M. C., Weber, C. V., Curtis, B., and Chrissis, M. B. *The Capability Maturity Model: Guidelines for Improving the Software Process*. Reading, MA: Addison Wesley, 1993.

Pettigrew, A. M. "Longitudinal Field Research on Change: Theory and Practice," *Organization Science* (1:3), 1990.

Robey, D. "Theories that Explain Contradiction: Accounting for the Contradictory Organizational Consequences of Information Technology," in *Proceedings of the Sixteenth International Conference on Information Systems*, J. I. DeGross, G. Ariav, C. Beath, R. Hoyer, and C. Kemerer (eds.), Amsterdam, 1995.

Salaway, G. "An Organizational Learning Approach to Information Systems Development," *MIS Quarterly* (11:2), 1987.

Selby, R. W, Basili, V. R., and Baker, F. T. "Cleanroom Software Development: An Empirical Evaluation," *IEEE Transactions on Software Engineering* (13:9), 1987.

Schein, E. K. *Organizational Culture and Leadership: A Dynamic View*. San Francisco: Jossey-Bass, 1985.

Schön, D. A. *The Reflective Practitioner: How Professionals Think in Action*. New York: Basic Books, 1983.

Stolterman, E. "How Systems Designers Think about Design and Methods: Some Reflections Based on an Interview Study," *Scandinavian Journal of Information Systems* (4), 1992.

Stowell, F., West, D., and Stansfield, M. "Action Research as a Framework for IS Research," in *Information Systems: An Emerging Discipline*, J. Mingers and F. Stowell (eds.). New York: McGraw-Hill, 1997.

Tan, M. "Establishing Mutual Understanding in Systems Design: An Empirical Study," *Journal of Management Information Systems* (10:4), 1994.

Vidgen, R., and Braa, K. "Balacing Interpretation and Intervention in Information Systems Research: The Action Case Approach," in *Information Systems and Qualitative Research*, A. S. Lee, J. Liebenau, and J. I. DeGross (eds.). London: Chapman & Hall, 1997, pp. 524-541.

Vinter, O. "Using Defect Analysis to Initiate the Improvement Process," in *EuroSPI'98*, Gothenburg, Sweden, 1998.

Vitalari, N. P. "Knowledge as a Basis for Expertise in Systems Analysis: An Empirical Study," *MIS Quarterly* (9:3), 1985.

Vitalari, N. P., and Dickson, G. W. "Problem Solving for Effective Systems Analysis: An Experimental Exploration," *Communications of the ACM* (26:11), 1983.

Waltz, D. B., Elam, J. J., and Curtis, B. "Inside a Software Design Team: Knowledge Acquisition, Sharing, and Integration," *Communications of the ACM* (36:10), 1993.

White, K. B. "MIS Project Teams: An Investigation of Cognitive Style Implications," *MIS Quarterly* (8:2), 1984.

White, K. B., and Leifer, R. "Information Systems Development Success: Perspectives from Project Team Participants," *MIS Quarterly* (10:3), 1986.

Wynekoop, J. L., and Russo, N. L. "System Development Methodologies: Unanswered Questions and the Research-Practice Gap," in *Proceedings of the Fourteenth International Conference on Information Systems*, J. I. DeGross, R. P. Bostrom, and D. Robey (eds.), Orlando, Florida, 1993.

About the Author

Lars Mathiassen is a professor in the Computer Science Department of Aalborg University (Denmark). His research interests are within Software Engineering and Information Systems. He has published several books and papers on systems development, object-orientation, risk management, and the philosophy of computing. More details are available at www.cs.auc.dk/~larsm. Lars can be contacted by e-mail at larsm@cs.auc.dk.

10 PROCESS AS THEORY IN INFORMATION SYSTEMS RESEARCH

Kevin Crowston
Syracuse University
U.S.A.

Abstract

Many researchers have searched for evidence of organizational improvements from the huge sums invested in ICT. Unfortunately, evidence for such a pay back is spotty at best (e.g., Brynjolfsson 1994; Brynjolfsson and Hitt 1998; Meyer and Gupta 1994). On the other hand, at the individual level, computing and communication technologies are increasingly merging into work in ways that make it impossible to separate the two (e.g., Bridges 1995; Gasser 1986; Zuboff 198). This problem—usually referred to as the productivity paradox—is an example of a more pervasive issue: linking phenomena and theories from different levels of analysis.

Organizational processes provide a bridge between individual, organizational, and even industrial level impacts of information and communication technologies (ICT). Viewing a process as the way organizations accomplish desired goals and transform inputs into outputs makes the link to organizational outcomes. Viewing processes as ordered collections of activities makes the link to individual work, since individual actors perform these activities. As well, process theories can be a useful milieu for theoretical interplay between interpretive and positivist research paradigms. A process-centered research framework is illustrated with an analysis of the process of seating and serving customers in two restaurants. The analysis illustrates how changes in individual work affect the process and thus the organizational outcomes and how processes provide a theoretical bridge between work at different levels of analysis.

1. Introduction

Many researchers have searched for evidence of organizational productivity improvements from investments in information and communication technologies (ICT). Unfortunately, evidence for such payback is spotty at best (e.g., Brynjolfsson and Hitt 1998; Meyer and Gupta 1994). On the other hand, at the individual level, ICT are increasingly merging into work in ways that make it impossible to separate the two (e.g., Bridges 1995; Gasser 1986; Zuboff 1988). The contrast between the apparently substantial impact of ICT use at the individual level and the apparently diffuse impact at the organizational level is but one example of the problem of linking phenomena and theories from different levels of analysis.

The goal of this paper is to show how individual-level research on ICT use might be linked to organization-level research by detailed consideration of the organizational process in which the use is situated. Process as used in this paper means an interrelated sequence of events that occur over time leading to an organizational outcome of interest (Boudreau and Robey 1999). Understanding this linkage is useful for those who study ICT, and especially useful for those who design them (Kaplan 1991).

In the remainder of this section, the problem of cross-level analysis is briefly discussed. The following section discusses the concept of a process to explain how processes link to individual work and ICT use, on the one hand, and to organizational and industrial structures and outcomes, on the other. As well, a brief discussion of the potential use of process theories as a milieu for interplay between research paradigms is presented. In later sections, the application of this framework in a study of the use of an information system in a restaurant is illustrated. The paper concludes by sketching implications of this process perspective for future research.

1.1 The Problem of Multi-level Research

Information systems research has in recent years shifted its attention to organizational issues (Benbasat, Goldstein, and Mead 1987). Organizational research in turn has historically been divided between micro- and macro-level perspectives. Unfortunately, many organizational issues are multi-level and thus incompletely captured by single-level theories. ICT impact is clearly multi-level, as the same ICT has discernable impacts on individuals, groups, and organizations. For such topics, multi-level theories are preferable because they provide a "deeper, richer portrait of organizational life—one that acknowledges the influence of the organizational context on individuals' actions and perceptions *and* the influence of individuals' actions and perceptions on the organizational context" (Klein, Tosi, and Cannella 1999, p. 243). However, multi-level research is difficult, so theorizing at different levels is often disconnected, leading to misleading theoretical conclusions.

Klein, Dansereau, and Hall (1994, p. 196) stress the primacy of theory in dealing with levels issues. However, multi-level work to date has been restricted to a few domains, such as climate or leadership (Klein, Dansereau, and Hall 1994, p. 197). The lack focus of focus on information issues suggests that there is an opportunity and a need for multi-level research and theorizing on ICT use.

2. Processes as Theory

Most theories in organizational and IS research are variance theories. Variance theories comprise constructs or variables and propositions or hypotheses linking them. Such theories predict the levels of dependent or outcome variables from the levels of independent or predictor variables, where the predictors are seen as necessary and sufficient for the outcomes. A multi-level variance theory is one that includes constructs and variables from different levels of analysis. The link between levels takes the form of a series of bridging or linking propositions involving constructs or variables defined at different levels of analysis.

An alternative to a variance theory is a process theory (Markus and Robey 1988). Rather than relating levels of variables, process theories explain how outcomes of interest develop through a sequence of events (Mohr 1982). Typically, process theories are of some transient process leading to exceptional outcomes, e.g., events leading up to an organizational change or to acceptance of a system. However, this paper will focus instead on what might be called "everyday" processes: those performed regularly to create an organization's products or services.

A description of a process has a very different form from the boxes-and-arrows of a variance theory, but it is still a theory, in that it summarizes a set of observations and predictions about the world. In the case of a process theory, the observations and predictions are about the performance of events leading up to organizational outcomes of interest. Such a theory might be very specific, that is, descriptive of only a single performance in a specific organization. More desirably, the theory might describe a general class of performances or even performances in multiple organizations. As Orlikowski (1993) puts it, "Yin (1984) refers to this technique as 'analytic generalization' to distinguish it from the more typical statistical generalization that generalizes from a sample to a population. Here the generalization is of theoretical concepts and patterns."

Kaplan (1991, p. 593) states that process theories can be "valuable aids in understanding issues pertaining to designing and implementing information systems, assessing their impacts, and anticipating and managing the processes of change associated with them." The main advantage of process theories is that they can deal with more complex causal relationships than variance theories, and provide an explanation of how the inputs and outputs are related, rather than simply noting the relationship. As well, it is argued here that process theories provide a link between individual and organizational phenomena and a milieu for interplay between research paradigms. However, to make this point, first the components of a process theory, in contrast to the variables and hypotheses of a variance theory, will be described.

2.1 Components of a Process

This section develops a series of increasingly elaborate process conceptualizations. It begins by discussing processes as wholes and then as compositions of activities with constraints on assembly. The goal of this discussion is to understand the connection between processes and individual work, on the one hand, and processes and organizational outcomes on the other.

2.1.1 Processes as Wholes

A simple view is that processes are ways organizations accomplish desired goals. In fact, as Malone et al. (1999) point out, processes are often named by the goals they accomplish (for example, product development or order fulfillment). The goal identifies the desired result or output of the process, or the set of constraints the process satisfies (Cyert and March 1963; Simon 1964), which is necessary to link to organizational outcomes (i.e., how quickly or efficiently different process options meet the constraints and produce the output). By focusing at the level of a process, the paper tries to avoid the problems outlined by March and Sutton (1997), who noted the instability of organizational performance.

A related view is that a process is a transformation of an input to an output. This view focuses on the resources that flow through the process. The business process concept has strong roots in industrial engineering (IE) and its subfield of process engineering (Sakamoto 1989). Other process concepts borrow heavily from operations research (OR) and operations management (OM), in particular, the design and control of manufacturing and product-producing processes of the firm.

This view of a process is also similar to the root definition (RD) from Soft Systems Methodology (SSM) (Checkland and Scholes 1990). A key point in SSM, to which this paper also adheres, is that there is not a single correct RD for a process. Instead, there can be many RDs reflecting different view of the process. For example, one RD might focus on the official rationale for the process and the concrete items created. Another might focus on the way the organization allocates resources to different processes. Instead of arguing that whichever model chosen is a true representation of the work, the description is viewed as a discursive product, that is, as an artifact, with an author, intended to accomplish some goal. Checkland (1981, p. 81) similarly describes models as "opening up debate about change" rather than "what ought now to be done."

Describing a process as a way to accomplish a goal or as a transformation of an input to an output establishes the link between processes and organizational outcomes. For example, at this level of detail the efficiency of a process can be stated as the process outputs divided by the inputs. However, at this level of detail, the link to individual work or ICT use is not yet apparent.

2.1.2 Processes as Activities and Interdependencies

To progress further, we need a more detailed view of processes that will allow us to say more about differences in how individuals contribute to processes and especially how the use of ICT might make a difference to these contributions. To do so, we start with the definition of a process as a sequence of events, focusing specifically on events as activities performed by individual or groups. Such a description will be a theory of the process in the sense that it summarizes a set of observations about what activities happened when the process was performed in the past and a set of predictions about what will happen when the process is performed in the future.

Representing a process as a sequence of activities provides insight into the linkage between individual work and processes, since individuals perform the various activities that comprise the process. As individuals change what they do, they change how they perform these activities and thus their participation in the process. Conversely, process changes demand different performances from individuals. ICT use might simply make individuals more efficient or effective at the activities they have always performed. However, an interesting class of impacts involves changing which individuals perform which activities. ICT might also be used to automate the performance of certain activities, thus changing the activities that comprise the process. Analysis of these possibilities requires an even more detailed view of the process, which is presented next.

To understand how changes in individual work might affect the process, it is necessary to examine the constraints on assembling activities that limit the possible arrangements and rearrangements of activities into processes. To identify these constraints, we focus in particular on the implications of dependencies for process assembly. In focusing on dependencies, we both follow and diverge from a long tradition in organization theory. Thompson (1967) viewed subunit interdependency as the basic building block of organizational structure and behavior. Following Thompson, two basic conceptualizations of organizational interdependency have evolved: resource interdependency, generated through exchanges between organizational members (e.g., people); and workflow interdependency, generated between organizational units located in the division of labor (Victor and Blackburn 1987).

In both cases, dependencies were seen as arising between individuals or groups. In contrast to these earlier views, the belief expressed here is that conceptualizing dependencies as arising between *activities* provides more insight into processes. This view makes it easier to consider the implications of reassigning work to different actors. In this view, the limits on the orders of activities arise from the flow of resources between them, that is, on resource interdependencies.

Malone and Crowston (1994) proposed two major classes of dependencies: *flow* or *producer/consumer* dependencies and *shared resource* dependencies. *Producer/consumer* dependencies arise when one activity creates a resource that is then used by another activity. *Shared resource* dependencies arise when two or more activities require the same resources (because of space limitations, this class of dependency will not be discussed further in this paper).

Both kinds of dependencies have implications for changes to processes. Since the activities can not be performed without the necessary resources, the existence of the dependencies constrains how the process can be assembled. In particular, *producer/consumer* dependencies restrict the order in which activities can be performed. On the other hand, activities that are not involved in a dependency can be freely rearranged. Therefore, we can limit possible arrangements of the activities in analyzing existing processes or in designing new ones.

As well as constraining the order of activities, interdependencies often require additional activities to manage them. According to Malone and Crowston, the *producer/consumer* interdependency described above not only constrains the order of the activities (a *precedence* dependency), but may also require additional activities to manage the *transfer* of the resource between or to ensure the *usability* of the resource. *Precedence* requires that the producer activity be performed before the consumer activity. This dependency can be managed in one of two ways: either the person performing the first activity can notify the person performing the second that a resource is ready, or the second can monitor the performance of the first. ICT may have an affect by providing a mechanism for cheap monitoring. *Transfer* dependencies are managed by a range of mechanisms for physically moving resources to the actors performing the consuming activities (or vice versa). For example, inventory management systems can be classified here. *Usability* can be managed by having the consumer specify the nature of the resources required or by having the producer create standardized resources expected by the user (among other mechanisms).

In general, there may be numerous different coordination mechanisms that could be used to address a given dependency. Different organizations may use different mechanisms to address similar problems, thus resulting in a different organizational form. Because these coordination mechanisms are primarily information processing, they may be particularly affected by the use of ICT.

2.2 Processes as a Milieu for the Interplay of Research Paradigms

As should be clear from the preceding discussion, developing a model of a process raises numerous problems, such as how activities are identified and determined to be relevant to the process or choosing an appropriate level of decomposition for the process description. These choices can be problematic because processes involve numerous individuals with possibly different interpretations of the process. Resolution of these choices raises questions about the theoretical assumptions underlying the theory.

As a framework for discussing these underlying assumptions, Burrell and Morgan (1979) suggest a 2x2 categorization of social theories: order-conflict and subjective-objective (assumptions about ontology, epistemology, human nature, and methodology). The combination of these two dimensions results in four distinct paradigms for research. Burrell and Morgan present their four paradigms as incommensurable approaches to research. However, Schultz and Hatch (1996) suggest a research project can draw on and contrast multiple paradigms. They identify several ways research might cross paradigms, including sequential (e.g., Lee 1991), parallel, bridging, and interplay. Schultz and Hatch argue that interplay "allows the findings of one paradigm to be recontextualized and reinterpreted in such a way that they inform the research conducted within a different paradigm."

In the Burrell and Morgan framework, theories of processes clearly focus on the ordering of society—stability, integration, functional coordination, and consensus—rather than on conflict. However, they could provide a milieu for interplay between subjective and objective perspectives. A process study might contrast realist and nominalist ontologies to achieve a richer description. Activities performed might be viewed as real (e.g., stamping metal) or nominal (e.g., many information processes). Flows of physical goods have a physical reality, although many interesting processes are largely information processing for which a nominalist position is more appropriate.

A study might contrast positivist and anti-positivist epistemologies. On the one hand, viewing a process as a way to accomplish organizational goals implies a positivist conception of the process. On the other, focusing on individuals and their conceptions of their work implies an anti-positivist view of activities. A possible result of this contrast is to explicitly problematize the question of how individuals come to contribute to the higher-order goals. For example, even though individuals make sense of the world themselves, there must still be some degree of agreement among members of a group, e.g., about the meaning and nature of a shared process, meaning that individual perceptions are subjective but not completely arbitrary. Numerous researchers have investigated the nature of such shared cognitions and the social processes by which they are built (Walsh 1995). For example, Weick and Roberts (1993) show how aircraft carrier flight deck operations are made reliable by the "heedful interrelating" of flight deck personnel.

A study might contrast deterministic and voluntaristic assumptions about human nature. Individuals working in a group do not have total freedom in what they do if they are to contribute to the group, but are not totally constrained either. Again, consideration of interplay between these positions is possible. For example, Simon (1991) raises the question of why individuals adopt organizational goals in the first place.

To summarize, the objective-subjective debate is often presented as a dichotomy and a matter of prior assumption. However, as Schultz and Hatch say, "the assumption of impermeable paradigm boundaries reinforces and is reinforced by 'either-or' thinking. We believe that paradigm boundaries are permeable and claim that when paradigm contrasts are combined with paradigm connections, interplay becomes possible." Process theories provide a milieu for such interplay.

2.3 A Process-centered Research Framework

Crowston and Treacy (1986) noted that linking the use of ICT to any kind of organizational-level impact requires some theory about the inner workings of organizations. Processes provide a possible bridge between individual and organizational (and even industrial) level outcomes of the use of ICT. This framework is shown pictorially in Figure 1. The framework acknowledges that ICT, by themselves, do not change organizations, nor are they merely tools of managerial intent. Rather, ICT use opens up new possibilities for individual work, and these changes in work in turn have implications for the processes and thus the organizations in which these individuals participate.

These work and process changes, in turn, may involve changes in organizational structures and outcomes (and vice versa). In other words, as individual workers incorporate various forms of ICT in their work, they alter both how they conduct their work and how they participate in the organization's structure, and thus indirectly how their organizations participate in the industry-wide value-chain. Conversely, there are organizational and industry-wide forces shaping how work is done. These forces also affect how individuals do their work. The interaction of these forces is what shapes the uses of ICT, new forms of work and new ways of organizing.

In the next section, this framework is used in the study of the use of an information system in a restaurant. It shows how processes can provide a link between individual and organizational level phenomena.

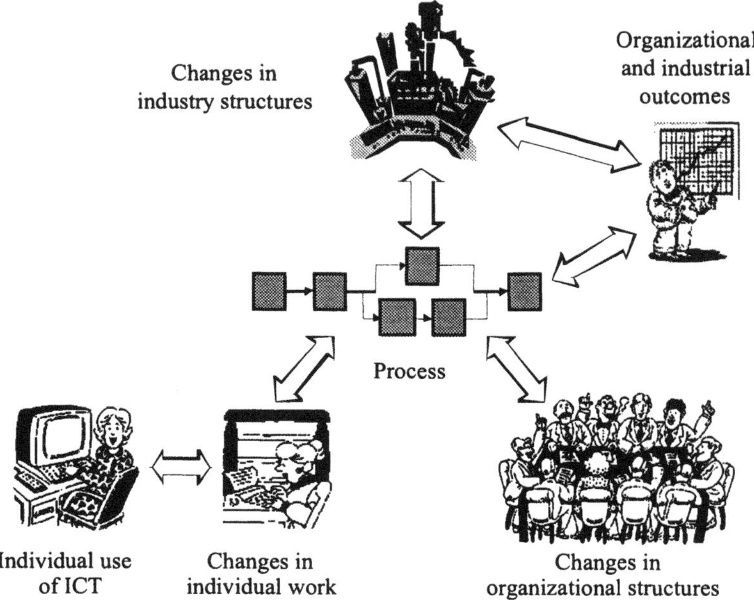

Figure 1. The Relationship between ICT-induced Changes in Individual Work and Changes in Organizational and Industrial Structures and Outcomes

3. Illustrative Example: Service Processes in Two Restaurants

To illustrate the use of this framework, we will analyze and compare the service processes in two restaurants, one with and one without a seating information system (Crowston 1994). This example demonstrates how consideration of the process helps to link phenomena observed at the individual and organizational levels. Restaurants have long been studied as important forums for coordination. The essential characteristics of restaurants—many customers, many orders, frequent deliveries, continuous monitoring of customers and of personnel in accomplishing work, and perishable products—makes them particularly illuminating for studies of logistical flows, information flows, and resultant needs for coordination.

3.1 The Research Setting

The two restaurants compared—one in Lake Buena Vista, Florida, and the other in Southfield, Michigan—both belonged to the same national chain. They differed significantly, however, in their use of information technology. The description and analysis is based on observations of lunch and dinner service at the two restaurants, discussions with staff, and analysis of documentation describing the IT system provided by the software services company that developed and sold the system to the restaurant chain (Karp 1994; Rock Systems 1994).

The Southfield restaurant was a conventional sit-down restaurant, organized for high-volume operations. Seats were allocated by assigning entries in a conventional grease pencil-and-acetate record used by the hostess. Communications were face-to-face. By contrast, the Lake Buena Vista restaurant used an information system to track table status and to automate some communications between restaurant staff.

When I arrived at the Lake Buena Vista restaurant, the hostess consulted a computerized display of tables in the restaurant to select a table for us. The system can balance customers across wait staff or maintain a waiting list if the restaurant is full. As I was seated, my hostess pointed out a button under the table. Pressing the button updated the status of the table in the information system, e.g., from free, to occupied, to waiting-to-be-bused, and finally back to free. The system also included pagers carried by the wait staff. When the table button was pressed indicating I had been seated, the system paged the waitress responsible for the table, indicating there were new customers. When my meal was ready, the kitchen used the pagers to inform the waitress my order was ready to be picked up and served. When the waitress collected the bill after I had left, she could page a buser to clean that table. Similarly, when the buser had finished, he or she could inform the hostess (and the system) that the table is available and the next customer could be seated.

This system apparently had a significant practical impact: it is reported, for example, that "diners spend 15 to 30 minutes less time in the restaurant [after the installation of the system] because of swifter service" (Karp 1994). The question to be answered is, why does the system have such a profound impact on organizational performance? This question can not be answered by a single-level theory. On the one hand, focusing only on individual use of the system can not explain how the system has an effect on the overall performance of the organization, especially considering that the system does not seem to dramatically affect how any individual works. On the other hand, considering only the organization as a whole (e.g., by comparing a number of organizations with and without systems), quantifies but does not illuminate how the system provides benefit.

3.2 Analysis

This section provides an analysis of the process of seating and serving customers in the two restaurants that illustrates how changes in individual work affect the process and thus the organizational outcomes. The changes in individual work have been described above: use of an information system to track table status and to communicate between individual employees. The organizational outcomes have also been described: reduced waiting time and increased table turns and profitability. The question addressed here is how consideration of the process can clarify the link between these phenomena.

The first step in this analysis is to develop a description of the activities involved in the process. A simple description of these steps is shown in Figure 2. This figure shows actors on the left and activities performed by each across the page in time-order. Activities performed jointly are connected by dotted lines. While there may be some dis-

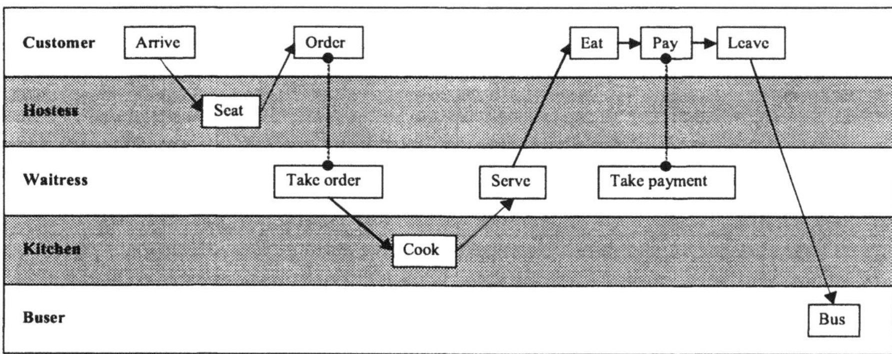

Actors are shown down the left side, activities performed by each are shown in order across the page. Activities performed jointly are connected with dotted lines.

Figure 2. The Restaurant Service Process

agreements about details, the belief is that most people will recognize the sequence of activities as representative of a restaurant. It was argued above that process descriptions should be viewed as resources for action rather than as necessarily valid descriptions of reality. In that spirit and in deference to a limited page count, the paper will bracket discussion of the validity of this model and instead focus on the insights possible from the analysis.

In the case of these restaurants, a particularly important type of dependency is the producer/consumer dependency between activities. These dependencies can be easily identified by noting where one activity produces something that is required by another. These resource flows and the dependencies between activities are shown in Figure 3. For example, the activity of cooking creates food that can then be served and eaten; the customer's departure produces a table ready for busing; and busing and resetting a table produces a table ready for another customer.

This distinction clarifies the role of the information system used. Recall that in Malone and Crowston's (1994) analysis, such a dependency can be managed in one of two ways: either the person performing the first activity can notify the person performing the second that a resource is ready, or the second can monitor the performance of the first. Employees in Southfield can not be easily notified that they can now perform an activity. They must instead spend time monitoring the status of the previous activity. For example, a bused table, ready for a customer, waits until the host or hostess notices it. In Lake Buena Vista, by contrast, the buser can use the system to notify the host or hostess that a table has been bused and is ready. Similarly, the wait staff can monitor the kitchen to notice when an order is ready or, using the system, the kitchen can page the wait staff to notify them that it is. Similar changes can be made throughout the process. The appropriate waiters or waitresses can be paged when customers arrive at their tables; a buser can be paged when the table has been vacated and is waiting to be bused.

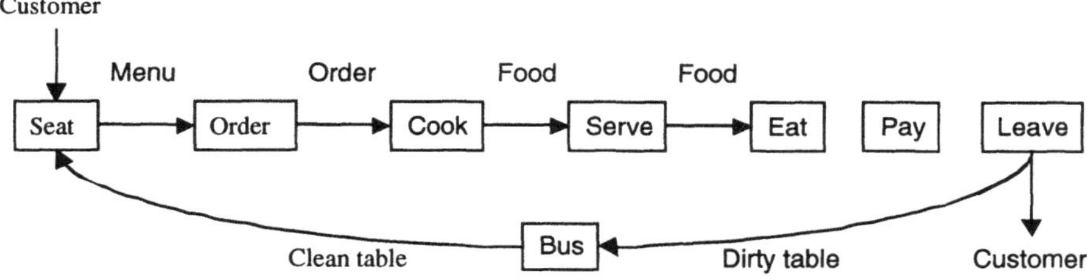

Figure 3. Flow of Resources between Activities and Resulting Dependencies in the Restaurant Service Process

The effect of this change in coordination mechanism is to slightly reduce the interval between successive activities. The change likely comes from increasing the pace at which the restaurant employees work. Since there are many such intervals, the result of the system can be a noticeable decrease in the interval between successive customers or, alternately, a higher number of table turns and increased utilization of the restaurant's tables. (Of course, this analysis assumes that there are a large number of customers waiting to be seated and that these customers are not seeking a leisurely dining experience, both factors that were true of the restaurants studied.)

3.3 Summary

This example demonstrates how examination of the process helps to link phenomena observed at the individual and organizational levels. The changes in individual work include use of an information system to track table status and to communicate between individual employees. The organizational outcomes include reduced waiting time and increased table turns and profitability. The analysis of the process suggests that the system allows individuals to change how they manage precedence dependencies, from noticing to notifying, thus decreasing the interval between activities, and, overall, increasing table turns and profitability for a certain class of restaurant.

4. Recommendations for Process Research and Practice

It was argued above that a focus on processes makes contributions to the study of ICT use and organizations. Overall, it seems reasonable to urge adoption of a process perspective when investigating the many organizational problems that have an ICT component. Five specific recommendations are outlined below for incorporating processes in ICT research and practice.

4.1 Develop Richer Process Analysis and Design Techniques

First, researchers need to develop richer process analysis and design techniques. Analyses of processes must consider the flow of resources, the dependencies created by these flows, and how these dependencies are managed (Crowston and Osborn 1998), not just the sequence of activities. Researchers in these areas might consider how their instruments can be adapted for broader usage.

A more difficult challenge is developing a meta-theory for processes comparable to the well-defined and well-understood set of terms and concepts for variance theories (e.g., construct, variable, proposition, hypothesis, variance, and error) and statistical tools for expressing and testing hypotheses. The framework developed in this paper is but a small first step toward such a meta-theory.

4.2 Use Processes as a Unit of Analysis

Organizational theorists have found it problematic to develop generalizations that hold for entire organizations, reflecting the diversity of activities and micro-climates found in most modern organizations. Mohr (1982) describes organizational structure as "multi-dimensional—too inclusive to have constant meaning and therefore to serve as a good theoretical construct." Processes provide a useful level of analyses to narrow the study of organizational form (Abbott 1992; Mohr 1982). As Crowston (1997, p. 158) states, "to understand how General Motors and Ford are alike or different, researchers might compare their automobile design processes or even more specific subprocesses." Within this finer focus, it may be possible to reach more meaningful conclusions about a range of theoretical concerns (Price and Mueller 1986).

For example, March and Sutton (1997) note the difficulties in studying antecedents of organizational performance due to the instability of this construct. However, it may be meaningful to consider performance at the level of a process. Similarly, it is probably not meaningful to measure the level of centralization or decentralization of an entire organization (Price and Mueller 1986), but such measures may be quite appropriate and meaningful within the context of a single process.

4.3 Develop the Theory of Organizational Processes

More research is necessary to properly establish processes and the various constraints on process assembly as valid theoretical constructs. For example, research methods need to be developed or adapted to operationalize activities, resource flows, and dependencies and to validate models built around these constructs. As well, additional research is needed to characterize the range of possible dependencies and the variety of coordination mechanisms possible and, in general, to document the assembly rules used in organizations. Work already done on work design and agency needs to be adapted to the general process perspective. Most importantly, research is needed to characterize the tradeoffs between different mechanisms. Ultimately, such work may allow some degree of prediction of the performance of a selected configuration of activities.

4.4 Expand to Richer Contexts

Consideration of organizational processes has been used primarily in an applied fashion and, as a result, its use has mostly been restricted to processes in companies, often with the intent of designing a more efficient process, employing fewer workers. Certainly, the belief expressed here is not that this is the only or even most interesting application of these ideas. Therefore, it is recommended that the use of organizational process analysis be expanded to a richer and more complex range of contexts.

4.5 Use Multiple Theories

Cannella and Paetzold (1994) argued that use of multiple theories is a strength of organizational science. Following their argument, the use of a process perspective with complementary theories, resulting in a multi-level and multi-paradigm understanding of the organization, is recommended. One example of this approach is an ongoing study of the use of ICT in the real estate industry (Crowston, Sawyer, and Wigand 1999; Crowston and Wigand 1999; Sawyer, Crowston, and Wigand 1999). To accomplish the objectives of this research, the researchers synthesize several theoretic perspectives to integrate findings from multiple levels of data collection. Specifically, at the individual level, they draw on theories of work redesign and social capital. At the organizational and industrial levels, they apply transaction cost and coordination theory.

6. Conclusion

This paper presented the argument that individual-level research on ICT use can be linked to organization-level research by detailed consideration of the organizational process in which the use is situated. Viewing a process as the way organizations accomplish desired goals and transform inputs into outputs makes the link to organizational outcomes. Viewing processes as ordered collections of activities makes the link to individual work, since individual actors perform these activities. As well, process theories can be a useful milieu for theoretical interplay between interpretive and positivist research paradigms (Schultz and Hatch 1996). An analysis of the process of seating and serving customers in the two restaurants illustrates how changes in individual work affect the process and thus the organizational outcomes.

Acknowledgments

Sections of this paper are derived from work done jointly with Jim Short, Steve Sawyer, and Rolf Wigand. The paper has been greatly improved by comments from the track chair, associate editor and two anonymous reviewers. As usual, the author accepts all responsibility for the current paper. This work has been partially funded by the National Science Foundation, Grant IIS 97-32799, and by a grant from the Office of the Dean of the School of Information Studies, Syracuse University. This version of the paper has been edited to fit the page restrictions of the conference. A complete version is available from the author.

References

Abbott, A. "From Causes to Events: Notes on Narrative Positivism," *Sociological Methods and Research* (20:4), 1992, pp. 428-455.

Benbasat, I., Goldstein, D. K., and Mead, M. "The Case Research Strategy in Studies of Information Systems," *MIS Quarterly* (11:3), 1987, pp. 369-386.

Boudreau, M.-C., and Robey, D. "Organizational Transition to Enterprise Resource Planning Systems: Theoretical Choices for Process Research," in *Proceedings of the Twentieth International Conference on Information Systems*, P. De and J. I. DeGross (eds.), Charlotte, North Carolina, 1999.

Bridges, W. *Job Shift*. Reading MA: Addison-Wesley, 1995.

Brynjolfsson, E. "The Productivity Paradox of Information Technology," *Communications of the ACM* (36:12), 1994, pp. 67-77.

Brynjolfsson, E., and Hitt, L. "Beyond the Productivity Paradox," *Communications of the ACM* (41:8), 1998, pp. 49-55.

Burrell, G. and Morgan, G. *Sociological Paradigms and Organizational Analysis*. London: Heinemann, 1979.

Cannella, A. A., Jr., and Paetzold, R. L. "Pfeffer's Barriers to the Advance of Organizational Science: A Rejoinder," *Academy of Management Review* (19:2), 1994, pp. 331-341.

Checkland, P. *Systems Thinking, Systems Practice*. New York: Wiley, 1981.

Checkland, P., and Scholes, J. *Soft Systems Methodology in Action*. Chichester, UK: Wiley, 1990.

Crowston, K. "Organizational Processes for Coordination. Symposium Presentation," in *Academy of Management Conference*, Dallas, Texas, 1994.

Crowston, K. "A Coordination Theory Approach to Organizational Process Design," *Organization Science* (8:2), 1994, pp. 157-175.

Crowston, K., and Osborn, C. S. *A Coordination Theory Approach to Process Description and Redesign*. Technical Report Number 204, Massachusetts Institute of Technology, Center for Coordination Science, 1998.

Crowston, K., Sawyer, S., and Wigand, R. "Investigating the Interplay Between Structure and Technology in the Real Estate Industry," paper presented at the Organizational Communications and Information Systems Division, Academy of Management Conference, Syracuse University, School of Information Studies, 1999.

Crowston, K., and Treacy, M. E. "Assessing the Impact of Information Technology on Enterprise Level Performance," in *Proceedings of the Sixth International Conference on Information Systems*, L. Maggi, R. Zmud, and J. Wetherbe (eds.), Indianapolis, Indiana, 1986, pp. 299-310.

Crowston, K., and Wigand, R. "Real Estate War in Cyberspace: An Emerging Electronic Market?" *International Journal of Electronic Markets* (9:1-2), 1999, pp. 1-8.

Cyert, R. M., and March, J. G. *A Behavioral Theory of the Firm*. Englewood Cliffs, NJ: Prentice-Hall, 1963.

Gasser, L. "The Integration of Computing and Routine Work," *ACM TOOIS* (4:3), 1986, pp. 205-225.

Kaplan, B. "Models of Change and Information Systems Research," in *Information Systems Research: Contemporary Approaches and Emergent Traditions*, H-E. Nissen, H. K. Klein, and R. Hirschheim (eds.). Amsterdam: North-Holland, 1991, pp. 593-611.

Karp, D. "Programming Lunch from 'Table's Ready' to 'Here's your Check'," *New York Times*, August 24, 1994, p. B1.

Klein, K. J., Dansereau, F., and Hall, R. J. "Levels Issues in Theory Development, Data Collection and Analysis," *Academy of Management Review* (19), 1994, pp. 195-229.

Klein, K. J., Tosi, H., and Cannella Jr., A. A. "Multilevel Theory Building: Benefits, Barriers and New Developments," *Academy of Management Review* (24:2), 1999, pp. 243-249.

Lee, A. "Integrating Positivist and Interpretive Approaches to Organizational Research," *Organization Science* (2:4), 1991, pp. 342-365.

Malone, T. W., and Crowston, K. "The Interdisciplinary Study of Coordination," *Computing Surveys* (26:1), 1994, pp. 87-119.

Malone, T. W., Crowston, K., Lee, J., Pentland, B., Dellarocas, C., Wyner, G., Quimby, J., Osborne, C., Bernstein, A., Herman, G., Klein, M., and O'Donnell, E. "Tools for Inventing Organizations: Toward a Handbook of Organizational Processes," *Management Science* (43:3), 1999, pp. 425-443.

March, J. G., and Sutton, R. I. "Organizational Performance as a Dependent Variable," *Organization Science* (8:6), 1997, pp. 698-706.

Markus, M. L., and Robey, D. "Information Technology and Organizational Change: Causal Structure in Theory and Research," *Management Science* (34:5), 1988, pp. 583-598.

Meyer, M., and Gupta, V. "The Performance Paradox," *Research in Organizational Behavior* (16), 1994, pp. 309-369.

Mohr, L. B. *Explaining Organizational Behavior: The Limits and Possibilities of Theory and Research.* San Francisco: Jossey-Bass, 1982.

Orlikowski, W. "Case Tools as Organizational Change: Investigating Incremental and Radical Changes in Systems Development," *MIS Quarterly* (20:3), 1993, pp. 309-340.

Price, J. L., and Mueller, C. W. *Handbook of Organizational Measurement.* Marshfield, MA: Pitman, 1986.

Rock Systems. *Prohost Promotional Material*, 1994.

Sakamoto, S. "Process Design Concept: A New Approach to IE," *Industrial Engineering*, March 1989, p. 31.

Sawyer, S., Crowston, K., and Wigand, R. "ICT in the Real Estate Industry: Agents and Social Capital," in *Advances in Social Informatics and Information Systems Track, Americas Conference on Information Systems*, Milwaukee, Wisconsin, 1999.

Schultz, M., and Hatch, M. J. "Living with Multiple Paradigms: The Case of Paradigm Interplay in Organizational Culture Studies," *Academy of Management Review* (21:2), 1996, pp. 529-557.

Simon, H. A. "On the Concept of Organizational Goal," *Administrative Sciences Quarterly* (9:1), 1964, pp. 1-22.

Simon, H. A. "Organizations and Markets," *The Journal of Economic Perspectives* (5:2), 1991, pp. 25-44.

Thompson, J. D. *Organizations in Action: Social Science Bases of Administrative Theory.* New York: McGraw-Hill, 1967.

Victor, B., and Blackburn, R. S. "Interdependence: An Alternative Conceptualization," *Academy of Management Review* (12:3), 1987, pp. 486-498.

Walsh, J. P. "Managerial and Organizational Cognition: Notes from a Trip Down Memory Lane," *Organization Science* (6:3), 1995, pp. 280-321.

Weick, K., and Roberts, K. "Collective Mind in Organizations: Heedful Interrelating on Flight Decks," *Administrative Science Quarterly*, 1993, pp. 357-381.

Yin, R. K. *Case Study Research: Design and Methods.* Beverly Hills, CA: Sage, 1984.

Zuboff, S. *In the Age of the Smart Machine.* New York: Basic Books, 1988.

About the Author

Kevin Crowston joined the School of Information Studies at Syracuse University in 1996. He received his Ph.D. in Information Technologies from the Sloan School of Management, Massachusetts Institute of Technology (MIT), in 1991. Before moving to Syracuse, he was a founding member of the Collaboratory for Research on Electronic Work at the University of Michigan and of the Center for Coordination Science at MIT. His current research focuses on new ways of organizing made possible by the extensive use of information technology. Kevin can be reached by e-mail at crowston@syr.edu.

Part 4:

Transforming Toward New Challenges

11 TOWARD AN INTEGRATED THEORY OF IT-RELATED RISK CONTROL

M. Lynne Markus
Claremont Graduate University
U.S.A.

1. Introduction

In business today, awareness of risk is growing, and risk management is increasingly seen as a critical practical discipline (Teach 1997). Further, risk is being defined broadly to include *anything* that could have a significant negative effect on the business.

For example, Microsoft recently began an integrated approach to risk management (Teach 1997). Twelve categories of business risk were identified, including financial, operational, people, and political risks (see Figure 1). Having identified these risks, Microsoft set about mapping them on several dimensions, such as potential frequency of loss producing event, potential severity, and adequacy of insurance. This analysis revealed that less than 50% of Microsoft's total business risk was adequately covered—an insight that led to the development of more effective risk management plans.

In the field of Information Systems, there has long been an interest in the risks associated with information systems and technology (Davis and Olson 1985; McFarlan 1988). However, as a field, we have not taken an integrated approach to the identification, analysis, and management of IT-related risk. By failing to do so, we are missing an important opportunity to make a major contribution in an area of pressing business need.

In this paper, I first show that our field's approach to the topic of IT-related risk has been quite fragmented, and I make the case that business people need a more integrated view of the topic. Next, I discuss some issues that help frame a theoretical perspective on IT-related risk. Finally, I examine some efforts at conceptual integration and show where they need bolstering for an integrated framework of IT-related risk management.

1. Business partners (inderdependency, confidentiality, cultural conflict, contractual risks, etc.)
2. Competitive (market share, pricing wars, industrial espionage, antitrust allegations, etc.)
3. Customer (product liability, credIT-related risk, poor market timing, inadequate customer support, etc.)
4. Distribution (transportation, service availability, cost, dependence on distributors, etc.)
5. Financial (foreign exchange, portfolio, cash, interest rate, etc.)
6. Operations (facilities, contractual risks, natural hazards, internal processes and controls, etc.)
7. People (employees, independent contractors, training, staffing adequacy)
8. Political (civil unrest, war, terrorism, enforcement of intellectual property rights, change in leadership, revised economic policies, etc.)
9. Regulatory and legislative (export licensing, jurisdiction, reporting and compliance, environmental, etc.)
10. Reputational (corporate image, brands, reputations of key employees, etc.)
11. Strategic (mergers and acquisitions, joint ventures and alliances, resource allocation and planning, organizational agility, etc.)
12. Technological (complexity, obsolescence, the year 2000 problem, workforce skill sets, etc.)

Figure 1. Microsoft's 12 Categories of Business Risk (Source: Teach 1997, p. 71)

2. IT-related Risk in the IS Literature and in Business Practice

The IS field has provided many useful insights for IS professionals and business managers in the domain of IT-related risks. Many individual types of IT-related risk have been isolated and studied, although they are not always labeled as risks. For example, there is a sizable IS literature on the topic of IS *project failure* (Keil 1995; Lyytinen and Hirschheim 1987; Sauer 1993). IS project failure can usefully be categorized as an IT-related risk, but it is only one of many. Other IT-related risks include *operational failure* or lack of reliability (Markus and Tanis 2000), *security breaches* (Baskerville 1993; Straub and Nance 1990; Straub and Welke 1998), *reputational damage* to a company owing to its failure to safeguard the privacy of customer data (Smith 1994), and *strategic risk* (Vitale 1986), such as adopting a new IT too soon or too late.

In general, IS research on risks falls into two broad categories: (1) risks related to the *development* of information systems and (2) risks related to the ongoing *operation* of information systems. This grouping of IT-related risks mirrors the way such risks are managed in practice by IS professionals. Most IS organizations structurally separate applications development from operations. In some cases, application development reports directly to business units, while operations reports to the CIO. Thus, the separate frameworks developed for managing project risks (Keil 1995) versus operational risks (Straub and Welke 1998) meet the needs of the different categories of IS professionals engaged in these tasks.

The major exception to the general statement that the academic treatment of IT-related risks is fragmented into discrete investigations of a variety of project and operational risks is Peter Neumann's (1995) treatise on *Computer-Related Risks* compiled from the Internet newsgroup, *The Risks Forum*, that Neumann has moderated for years. Neumann justly refers to his work as integrated treatment of computer-related risks, and I will discuss this work more later, but it is more properly viewed as a Computer Science contribution than a work in the Information Systems tradition.

However well the divergent approach to studying IT-related risks fits current IS practice, it is disadvantageous *from the perspective of the executive leadership of companies making major investments in information technology*. Executives tend to become involved in critical IT decision making only when new IS development or enhancement projects are initiated. They rarely become involved in IT operational issues unless there has been a significant problem, such as a major operational failure or a serious security breach. Thus, executives tend to be distanced from making decisions concerning operational IT-related risks. Integrated frameworks for IT investment decision making that combine benefits, costs, and both development and operational risks could help ensure better decisions from an organization-wide perspective.

From the business point of view, dividing the management of IT-related risk into development risk and a collection of disparate operational risks is counterproductive. Businesses should think of their IT initiatives as investments that are intended to pay off over their entire lifecycles. While it is true that the nature of IT-related risk changes as an IT investment progresses through its lifecycle, an integrated approach to IT-related risk management allows for intelligent tradeoffs between development costs and risks on the one hand and operational costs and risks on the other. Lack of an integrated approach to IT-related risk management makes possible the situation in which decisions designed to *reduce* project cost and schedule risk (e.g., ignore control needs, shortcut training and testing, etc.) may actually *increase* the operational risks of non-use, external threat, and contingencies.

The case of the Fox-Meyer Drug Company provides a useful illustration of how companies can suboptimize total business risk while attempting to manage project metrics. When Fox-Meyer Drug chartered its ERP system implementation, the CIO was aware that it was a "bet your business" proposition (Bulkeley 1996). Yet, the $60 million project was approved at the same time that the company also embarked on a state-of-the art $18 million automated warehouse. During the Project Phase, bad luck intervened: Fox-Meyer Drug lost a large customer, accounting for 15% of its business. To increase revenues, the company aggressively bid on new business: they figured contract pricing on the assumption that the projected annual $40 million savings from the SAP project would be realized immediately on startup, and they decided to advance the SAP rollout by 90 days. That close to the end of the project, little was left to do other than training and testing. So project team members decided not to test modules that had not been customized (thus failing to detect configuration errors). Cutover to the new system resulted in disaster. Meanwhile, the automated warehouse also did not perform as planned. It was estimated that the company sustained an unrecoverable loss of $15 million from erroneous shipments. The company was forced into bankruptcy and shareholders have since sued both the enterprise systems vendor and the integration consultant for $500 million each.

In this example, there are multiple, interacting factors in the failure. Some of them lay within the company's control, others did not. But the example clearly shows that decisions made to address development issues can have much wider consequences. One wonders whether Fox-Meyer's executives would have been so willing to advance the project schedule if they had taken a serious look at the likelihood and consequences of operational failure owing to poor project testing.

Today, in the business world, there is much discussion of "TCO," the total cost of ownership of IT systems. The TCO concern first arose in the context of standalone PCs (Strassmann 1990). But the TCO issue has acquired special significance in the context of packaged enterprise resource planning (ERP) software. Initially, executives focused only on the sizable license costs of this software. Later, they learned that ERP software license costs often pale in comparison with the costs of configuration consulting, technical platform, end-user training, and maintenance. Today, it is generally considered best business practice for executives to consider both the total lifecycle costs and the total lifecycle benefits of an IT investment. Should not the third major component of an executive's IT investment decision making be *total lifecycle risks,* where this concept comprises both project and operational risks?

An integrated approach to the management of IT-related risk is especially important in the current era for two reasons. First, worldwide connectivity through the Internet increases the opportunities for widespread fraud and cascading operational failure. Second, organizations are increasingly relying on outside parties for the development, operation, and management of their information systems. One could argue that IT-related risk management (ensuring investment payoff, while controlling potential negative consequences) is the *only* IT job an organization has left in an environment of total outsourcing.

3. Framing the Discussion of IT-related Risk

An integrated approach to IT-related risk must start with basic definitions. In this section, I define IT-related risk, present a typology of IT-related risks, discuss the issue of stakeholders—whose goals are to be served?—and outline the academic case for an integrated approach to this important topic.

3.1 What is IT-related Risk?

A significant obstacle in the way of an integrated approach to IT-related risk is definitional: what is the appropriate level of analysis of IT-related risk? And what is risk?

Discussions of IT-related risk often take a computer-based information system—a technical artifact—to be the appropriate level of analysis for the study of risk. Certainly, important technical issues, such as the existence of "trap doors" in software or the vulnerability of much modified code, must be addressed in any complete treatment of IT-related risks. However, the perspective I am advocating in this article suggests that there is need for an integrated treatment of IT-related risks at the organizational and interorganizational levels of analysis. At the organizational level, the knowledge and skills of users and their social interactions while using computer-based information systems are as important to an understanding of risk as is the technical system itself. Since so many of today's most interesting IT developments involve "business-to-business" and "business-to-consumer" e-commerce, it is often necessary to extend an analysis of risks beyond the boundaries of a single organization.

Different connotations and definitions of IT-related risk can be found in the literature and in common usage. One common definition holds risk to be uncertainty; alternatively, a risk is a wager (or an attempt at rewards). A second definition considers risk to be the possibility of loss. A third definition views risk as a negative quality of an opportunity that must be effectively managed.

These definitions of risk reflect very different emotional stances toward risk and its management. Someone who views risk as a wager for potential benefits to be maximized is likely to approach risk management quite differently than someone who views risk as the possibility of loss to be minimized. For my purposes here, I choose the third definition, because my focus is on managing the business risks associated with investments in information technology. At the same time, I am aware that risk is an emotionally difficult topic, because some people approach risks avidly, some people avoid thinking about them, and still others consider them emotionally neutral and completely amenable to rational analysis.

> IT-related risk is the likelihood that an organization will experience a significant negative effect (e.g., technical, financial, human, operational, or business loss) in the course of acquiring, deploying, and using (i.e., maintaining, enhancing, etc.) information technology either internally or externally (i.e., facing customers, suppliers, the public, etc.)

3.2 What Kinds of IT-related Risk Are There?

An additional obstacle in the way of an integrated treatment of IT-related risk is that many things one might label as risk have been discussed in the IS literature under other names. For example, the IT project failures literature rarely uses the label of "risk." A few authors (Clemons 1995; Lyytinen and Hirschheim 1987) have proposed typologies of IT-related risk. Building on their efforts and drawing on a wide range of literature, I propose the following 10 categories of IT-related risk:

1. Financial risk (the technology costs more than expected, yields fewer financial benefits, etc.)
2. Technical risk (the technology used is immature, poorly understood, unreliable, obsolete, etc.)

3. Project risk (the project is late, there is turnover of key personnel, the project becomes a "runaway," etc.)
4. Political risk (the project/system/technology is subject to political infighting or resistance)
5. Contingency risk (accidents, disasters, viruses, etc.)
6. Non-use, underuse, misuse risk (the intended users do not use the technology, they do not use it sufficiently or in a manner that would lead to the intended benefits, inappropriate use, etc.)
7. Internal abuse (malicious or felonious destruction, theft, abuse, etc., by company insiders)
8. External risk (hacking, theft of assets, willful destruction, etc., by company outsiders)
9. Competitive risk (negative reactions by customers, competitors, suppliers, etc., to the company's IT initiatives)
10. Reputational risk (negative reactions by the public at large, the media, the government, etc., to a company's IT initiatives)

In short, IT-related risk includes *anything* related to IT that could have significant negative effects on the business or its environment from the perspective of an executive investing in IT.

3.3 Who Are the Stakeholders in IT-related Risk?

It should be clear from the preceding discussion that there are many stakeholders where IT-related risk is concerned. Inside the focal company, stakeholders include executives, IS applications developers, IT infrastructure maintainers, and many different types of users. External to the company are customers, business partners, the public at large, investors, regulators, competitors, and others. These many stakeholders have widely differing interests in the risks of IT systems and their interests are likely to conflict often.

This paper proposes an integrated view of IT-related risks from the perspective of an executive decision maker, not because I think that executives are smarter, more ethical, or more important than other stakeholders. Instead, I am arguing that it is in the best interests of rational, well-informed executive decision makers to manage the total lifecycle risks of IT investments, regardless of who might be most affected by the risks. Given the fragmentation of IS practice into development versus operational concerns, it is not likely that IS professionals are as well placed as organizational executives to manage the full spectrum of IT-related risks. I hasten to add, however, the IS professionals are essential *partners* in the effective management of IT-related risk.

3.4 The Academic Case for an Integrated Approach to IT-related Risk Management

Here and there, academics have called for an integrated treatment of IT-related risk. The most comprehensive argument is that of Neumann, whose focus encompasses security, reliability, safety, privacy, and other operational risks. Writing on the need for an integrated treatment of security and reliability, for example, Neumann notes:

> Considerable commonality exists between reliability and security. Both are weak-link phenomena. For example, certain security measures may be desirable to hinder malicious penetrators, but do relatively little to reduce hardware and software faults. Certain reliability measures may be desirable to provide hardware fault tolerance, but do not increase security. On the other hand, properly chosen system architectural approaches and good software engineering practice can enhance both security and reliability. Thus, *it is highly advantageous to consider both reliability and security within a common framework*, along with other properties such as application survivability and application safety. [Neumann 1995, pp. 129-130, emphasis added]

In the IS system failure literature, there is widespread recognition that development and implementation/use issues must be jointly considered. For example, Lyytinen and Hirschheim discuss both development failures and use failures (i.e., failure in operation) and argue that different failure types must be addressed in a common framework. Markus and Keil (1994) and Markus and Tanis make similar points.

In the area of IS security, Baskerville makes a compelling case for an integrated approach to system development and the management of operational IT security risks. By tracing the evolution of system development and security management methods, he shows that initially there was no integration. Gradually, security risk analysis and management procedures have become built into system development methods. By extension, management of all other IT-related risks listed earlier in this paper should also be incorporated in system development methods.

In short, here and there in the literature, it is possible to find arguments that, when assembled, call for an integrated approach to the management of IT-related risk. Such an approach would encompass both project failure and a range of operational risks, including those related to safety, reliability, security, privacy, non-use, and reputation.

4. Theoretical Perspectives on IT-related Risk and Risk Control

Since Neumann claims to have taken an integrated approach to computer-related risks, what more needs to be done? Cannot the IS field simply adopt Neumann's framework and declare the problem solved?

On the contrary, I argue that, while Neumann's work is an important first step toward an integrated theory of IT-risk and risk management, it is lacking in a number of areas that require theoretical integration. Those areas are the social psychological dimensions of risk perception, the structural conditions of risk management, the dynamics of risk control, and the dynamics of risk.

4.1 Social Psychological Dimensions of Risk Perception

Neumann does a thorough job of treating the various system-related causes, both accidental and intentional, of computer-related disasters. While he includes a chapter on "the human element," this chapter falls far short of capturing the social-psychological processes that go into the human side of the equation. Notably absent from his discussion is the phenomenon of "escalating commitment" to a losing course of action that has proved so useful in analyzing certain IS project failures (Keil 1995; Staw 1993). Also lacking is a treatment of the many cognitive biases that are known to affect people's judgment in making decisions involving risk (Sitkin and Pablo 1992), in dealing with crises (Pearson and Mitroff 1993), and in problem solving in complex situations (Dörner 1989). For a wonderful complement on the human side to Neumann's technology-oriented analysis, see Dörner.

4.2 Structural Conditions of Risk Management

A second area in which Neumann's analysis of computer-related risks needs augmentation for the IS domain is that of the structural conditions in which IT-related risk is created and materializes into problems. Structural conditions are the social and economic arrangements (e.g., reporting relationships and policies) that influence the processes and outcomes of IS work (Orlikowski 1992). Examples of relevant structural conditions include the separation of development from operations work in many IS departments and the outsourcing of selected IT-related tasks to consultants and vendors. Neumann discusses large programming projects as a human source of risk, but he says nothing about how variations in the organization and management of such projects might contribute to the incidence or control of risks.

4.3 Dynamics of Control

A third area in which Neumann's analysis of computer-related risks needs enhancement is that of the dynamics of risk *control* strategies. Neumann devotes an entire chapter to strategies for controlling risk, such as modeling and simulation, complexity management, reliability improvement approaches, and so forth. Interestingly, while his analysis of computer-related risks is quite holistic, his approach to risk control is not: it is in essence a laundry list of techniques and rules of thumb. He does not attempt an integrated methodology of system engineering for the prevention of risk nor tackle the issue of how to recover from failure. He also does not tackle the difficult problem of ensuring that people follow acceptable methodologies (hence the importance of understanding the structural conditions under which IS work gets done).

An integrated theory of risk control needs to address what is known about the different types of strategies for gaining and maintaining control over people and organizational processes. Review of the control literature suggests that the types of control potentially useful in managing IT-related risk are as varied as the risks themselves (Handy 1995; Simons 1995a, 1995b). For example, Straub and Welke identify four distinct, sequential activities involved in the management of systems security risks: deterrence, prevention, detection, and recovery. Combining these sources, a list of risk control strategies would surely include the following:

1. Plans (e.g., backup, disaster recovery, etc.)
2. Policies (e.g., regarding unauthorized use of a company's computer resources, etc.)
3. Operational controls (e.g., budgets, performance evaluations, etc.)
4. Automated controls (e.g., passwords, access monitoring, etc.)
5. Physical controls (e.g., cardkeys, etc.)
6. Audit and detection (e.g., post project audits, system penetration detection, etc.)
7. Risk awareness building (e.g., training, bulletins, etc.)
8. Belief systems (e.g., beliefs about value of customer privacy, etc.)
9. Social systems (e.g., behavioral norms and reminders about confidentiality, etc.)

Further, the control literature suggests that control attempts are not invariably successful. While it is generally recognized that too little control is bad, a sizable body of literature suggests that too much control is also bad. Too much control has been associated with three types of negative consequences. First, too much control is expensive. In fact, the high cost of control is a major reason given for reengineering business processes with looser control (Sia and Neo 1997). Second, too much control can interfere with business operation and flexibility and can damage the relationship between controllers and controllees (Block 1993). Third, too much control is associated with unintended human and social consequences (Handy 1995; Sitkin and Roth 1993). These negative consequences include low morale and circumventing the rules; ironically, excessive control can also promote fraud. There is a science of control, just as there is a science of technology failure, and an integrated theory must incorporate both.

4.4 Dynamics of Risk

As we move toward an integrated theory of IT-related risk and risk management, it is important to keep in mind the empirical evidence about how problems, crises, and disasters materialize from risk. Studies of nuclear power plant accidents (Perrow 1984) and IT-related accidents (Neumann 1995) show that crises, disasters, and failures often have multiple independent or correlated causes. These "weak link" phenomena remind us that we should not take a static view of risk but should recognize that risk is a dynamic function of technology developments and human interventions.

The literature on IT project risk often appears to assume that risk is greatest at the start of the project when the unknowns are greatest, then decreases over the life of the project as work progresses toward completion. This view suggests that risk does not remain static, but changes as a function of prior decisions and behavior. Therefore, one can posit the concept of "residual risk" that varies throughout a project (Nidumolu 1995) and by extension throughout the lifecycle of a system.

Residual risk is often assumed to decrease monotonically over the life of IT projects; but, in the IT domain, one must consider the possibility that residual risk will actually *increase* over a system's lifecycle. In the first place, as systems age, they are maintained and enhanced; over time, this process increases their fragility or failure-proneness (Lientz and Swanson 1980). (This is why there is such a strong emphasis on development for maintainability, another example of the need for an integrated risk management approach.) Further, there is an increasing trend toward the integration of formerly discrete systems. As systems are integrated with other systems, complexity and tight-coupling increase the chances of failure (Neumann 1995; Perrow 1984). For example, in a recent project involving the implementation of SAP R/3 financials, Microsoft loaded financial data into a data warehouse and provided access to the data and preformatted reports via the corporate intranet. Integration of SAP R/3 with data warehousing and intranet technology vastly increased the number of people who had access to financial data and vastly increased the risks of non-use, internal abuse, external risk, etc. (Bashein, Markus, and Finley 1997).

An additional consideration is the actions people take to remedy problems that arise as projects and systems pass through their lifecycles. In situations involving system development and operation (as in the progression of a nuclear power plant incident), people may misdiagnose the causes of problems and apply attempted solutions that actually make the situation worse (Markus and Tanis 2000). They thus create new situations that call forth additional actions and changes (Orlikowski 1996). Therefore, an integrated theory of IT-related risks must also take into account the second-order consequences of human problem-solving behavior.

5. Conclusion

Much of the research in the IS field deals directly or indirectly with issues of IT-related risk, although that term is seldom used. The business world is beginning to see the value of an integrated approach to identifying and managing business risk; the time is right for the IS field to begin developing an integrated approach to identifying and managing IT-related risk. Not only will such an approach be useful to businesses in their attempts to obtain maximum value from their IT investments, it will also help bring together a large part of the IS literature under a common conceptual umbrella. By viewing system development and maintenance along with package acquisition and outsourcing as part of the business's IT investment process, risk management becomes the center of attention. By viewing system development failure, security breaches, and competitive threats as different types of the unitary phenomenon of IT-related risk, it becomes possible to make intelligent end-to-end tradeoff decisions throughout the lifecycles of systems in organizations.

References

Bashein, B. J., Markus, M. L., and Finley, J. B. *Safety Nets: Secrets of Effective Information Technology Controls*. Morristown, NJ: Financial Executives Research Foundation, Inc., 1997.

Baskerville, R. "Information Systems Security Design Methods: Implications for Information Systems Development," *ACM Computing Surveys* (25:4), 1993, pp. 375-414.

Block, P. *Stewardship: Choosing Service Over Self-Interest*. San Francisco: Berrett-Koehler Publishers, 1993.

Bulkeley, W. M. "A Cautionary Network Tale: Fox-Meyer's High-Tech Gamble," *Wall Street Journal Interactive Edition*, November 18, 1996.

Clemons, E. K. "Using Scenario Analysis to Manage the Strategic Risks of Reengineering," *Sloan Management Review* (36:4), 1995, pp. 61-71.

Davis, G., and Olson, M. *Management Information Systems: Conceptual Foundations, Structure and Development*. New York: McGraw-Hill, 1985.

Dörner, D. *The Logic of Failure: Recognizing and Avoiding Error in Complex Situations*. Reading, MA: Addison-Wesley, 1989.

Handy, C. "Trust and the Virtual Organization," *Harvard Business Review* (73:3), 1995, pp. 40-50.

Keil, M. "Identifying and Preventing Runaway Systems Projects," *American Programmer* (8:3), 1995, pp. 16-22.

Lientz, B. P., and Swanson, E. B. *Software Maintenance Management: A Study of the Maintenance of Computer Applications in 487 Data Processing Organizations*. Reading, MA: Addison-Wesley, 1980.

Lyytinen, K., and Hirschheim, R. "Information Systems Failures: A Survey and Classification of the Empirical Literature," in *Oxford Surveys in Information Technology*, P. I. Zorkoczy (ed.), 4. Oxford: Oxford University Press, 1987, pp. 257-309.

Markus, M. L., and Keil, M. "If We Build It They Will Come: Designing Information Systems That Users Want To Use," *Sloan Management Review*, Summer, 1994, pp. 11-25.

Markus, M. L., and Tanis, C. "The Enterprise Systems Experience-From Adoption to Success," in *Framing the Domains of IT Research: Glimpsing the Future Through the Past*, R. W. Zmud (ed.). Cincinnati, OH: Pinnaflex, 2000.

McFarlan, F. W. "Portfolio Approach to Information Systems," *Harvard Business Review* (59:5), 1988, pp. 142-150.

Neumann, P. G. *Computer Related Risks*. New York: The ACM Press, 1995.

Nidumolu, S. "The Effect of Coordination Uncertainty on Software Project Performance: Residual Performance Risk as an Intervening Variable," *Information Systems Research* (6:3), 1995, pp. 191-219.

Orlikowski, W. J. "The Duality of Technology: Rethinking the Concept of Technology in Organizations," *Organization Science* (3:3), 1992, pp. 398-427.

Orlikowski, W. J. "Improvising Organizational Transformation Over Time: A Situated Change Perspective," *Information Systems Research* (7:1), 1996, pp. 63-92.

Pearson, C. M., and Mitroff, I. I. "From Crisis Prone to Crisis Prepared: A Framework for Crisis Management," *The Academy of Management Executive* (7:1), 1993, pp. 48-59.

Perrow, C. *Normal Accidents: Living With High-Risk Technologies*. New York: Basic Books, 1984.

Sauer, C. *Why Information Systems Fail: A Case Study Approach*. London: McGraw-Hill, 1993.

Sia, S. K., and Neo, B. S. "Reengineering Effectiveness and the Redesign of Organizational Control: A Case Study of the Inland Revenue Authority of Singapore," *Journal of Management Information Systems* (14:1), 1997, pp. 69-92.

Simons, R. "Control in an Age of Empowerment," *Harvard Business Review* (73:2), 1995a, pp. 80-88.

Simons, R. *Levers of Control: How Managers Use Innovative Control Systems to Drive Strategic Renewal*. Boston: Harvard Business School Press, 1995b.

Sitkin, S. B., and Pablo, A. L. "Reconceptualizing the Determinants of Risk Behavior," *Academy of Management Review* (17:1), 1992, pp. 9-38.

Sitkin, S. B., and Roth, N. L., "Explaining the Limited Effectiveness of Legalistic 'Remedies' for Trust/Distrust," *Organization Science* (4:3), 1993, pp. 367-392.

Smith, H. J. *Managing Privacy: Information Technology and Corporate America.* Chapel Hill, NC: The University of North Carolina Press, 1994.

Staw, B. M. "Organizational Escalation and Exit: Lessons From the Shoreham Nuclear Power Plant," *Academy of Management Journal* (36:4), 1993, pp. 701-732.

Strassmann, P. A. *The Business Value of Computers: An Executive's Guide.* New Cannan, CT: The Information Economics Press, 1990.

Straub, D. W., and Welke, R. J. "Coping with Systems Risk: Security Planning Models for Management Decision Making," *MIS Quarterly* (22:4), 1998, pp. 441-469.

Straub, D. W., and Nance, W. D. "Discovering and Disciplining Computer Abuse in Organizations: A Field Study," *MIS Quarterly* (14:1), 1990, pp. 45-60.

Teach, E. "Microsoft's Universe of Risk," *CFO*, March 1997, pp. 69-72.

Vitale, M. R. "The Growing Risks of Information Systems Success," *MIS Quarterly* (10:4), 1986, pp. 327-334.

About the Author

M. Lynne Markus is Professor of Management at the Peter F. Drucker Graduate School of Management and Professor of Information Science at the School of Information Science, Claremont Graduate University. She has also taught at the Sloan School of Management (MIT), the Anderson Graduate School of Management (UCLA), the Nanyang Business School, Singapore (as Shaw Foundation Professor), and Universidade Tecnica de Lisboa, Portugal (as Fulbright/FLAD Chair in Information Systems). She began thinking about IT-related risks while conducting research for the Financial Executives Research Foundation. Lynne can be reached by e-mail at m.lynne.markus@ cgu.edu.

12 INDIVIDUAL, ORGANIZATIONAL, AND SOCIETAL PERSPECTIVES ON INFORMATION DELIVERY SYSTEMS: BRIGHT AND DARK SIDES TO PUSH AND PULL TECHNOLOGIES

Julie E. Kendall
Rutgers University
U.S.A.

Kenneth E. Kendall
Rutgers University
U.S.A.

Abstract

Two competing visions of humans and whether they eschew or embrace the use of pull and push technologies can help us to envision the future. It is said that George Orwell feared that what we hate *will ruin us. On the other hand, Aldous Huxley feared that what we* love *will ruin us (Postman 1985). Information delivery systems, also called pull and push technologies, are ways to obtain and deliver information to users. This article briefly reviews the types of pull and push technologies, then goes on to explore the major benefits of both pull and push. But as the benefits of information delivery systems are elaborated, the darker side of push and pull are also exposed. Some alarming possibilities for future use of these technologies are identified, but this article goes on to discover and create both remedies and resolution mechanisms to counter the threats to individuals, organizations, and society. When remedies have no effect, the alternative may be unplugging.*

1. Introduction

Information delivery systems (IDS) is used as a collective term to include both pull and push technologies for obtaining materials over the Internet (and its successors). In the first section of the paper, we explore the world of a person making use of pull technology, or seeking out information on the Web. The term pull technology can mean simply surfing the Net to allowing an ever changing, independent evolutionary agent explore the Web for you. The term push technology can be used to describe anything from broadcasting to selective content delivery with sophisticated evolutionary filtering using data mining techniques.

Decision makers need no longer surf the Net or use simple search engines with little precision or unsatisfactory recall. The advent of advanced push technologies means that content is delivered to users on a periodically scheduled basis, without the need for a user request (Richardson 1997). Push technology is an efficient way to feed content to millions of consumers simultaneously (if desired). However, push technology, which takes advantage of several new developments that enable Web-based material to be located, downloaded, and delivered by intelligent agents, is being greeted with a variety of responses that range from enthusiasm to disparagement and even depression. Much of the criticism focuses on demands for an examination of how the push delivery systems are changing decision makers.

Postman states

> Our conversations about nature and about ourselves are conducted in whatever "languages" we find possible and convenient to employ, we do not see nature or intelligence or human motivation or ideology as "it" but only as our languages are. And our languages are our media. Our media are our metaphors. Our metaphors create the content of our culture. [Postman 1985, p. 15]

But there is a danger of backlash as people dislike the content that is pushed and, therefore, give up on new push technologies before they have truly emerged, and well before their benefits are realized. There is the possibility that people will become disenchanted and unplug, before the best applications are developed. There is also a chance that people will suffer information overload and that in turn the stress on their information processing capabilities will cause them to have great difficulty in sorting through what is a useful pull or push technology and what is not. Corporations could conceivably ban certain services from the office as being off-limits since on the surface they do little if anything to further corporate objectives.

Our approach to this material is what Graber (1976) terms an intuitive method of verbal analysis. (Verbal here refers to both oral and written material.) The steps taken in our analysis include setting a goal for the investigation, sampling the verbal output, analyzing it for clues, and finally putting it together and interpreting it. The approach has additional layers of complexity since it requires a simultaneous analysis of the context provided by the society as well as the interactions of the various writers and their opinions, and their short and long term objectives. In this article, we define various types or levels of pull and push technologies and discuss their value. Next, using the intuitive method of verbal analysis, we expose some alarming features of pull and push and their implications for individuals, organizations, and societies. We then pose remedies and resolution mechanisms, and if these remedies do not work, we describe the phenomenon of unplugging. Finally, we discuss what may happen in the future.

2. Information Delivery Systems

Individuals get the information they need in a couple of ways. They can seek information, hoping that their search process, tools, and training are sufficient to find the information they need. They can also subscribe to delivery systems that bring information to the individual without asking for it. In this case, the individual hopes that the delivery mechanism and content delivered adequately provides the information they need.

Pull technologies involve the seeking process and push technologies consist of the subscription and delivery process. Most individuals depend on a combination of these two methods. We describe each of these processes in the next section.

2.1 Pull Technologies

When a person goes to the library, they pull a book off the shelf; similarly, they pull a piece of information from the Web. The word *pull* connotes grabbing and yanking something from the Internet.

Pull technologies can be simple or complex. Kendall and Kendall (1999a, 1999b) identify four types of pull technologies, beginning with surfing the Net. In this instance, users can browse by clicking on links. Tools available to the surfer include bookmark managers and software and hardware that helps speed up downloading of Webpages. This is called alpha-pull.

The second type of pull, beta-pull, involves using a search engine in order to locate information more effectively. Search engines do not actually search the Web on request. They are merely databases in which information is collected during off-peak times using spiders or bot, short for robot. Users measure effectiveness in terms of precision and recall (Stohr and Viswanathan 1999).

Gamma-pull, the third type of pull, uses a personal assistant, a spider, or bot to search the Web. Spiders or bots do what they are told to do. Therefore they gather information on what the user really wants. This is based on information provided by the individual in a user profile.

The fourth type of pull is very advanced, because an evolutionary agent is used to observe the user's behavior and then modify the searches to more accurately locate the information. By observing the behavior of a user the evolutionary agent forms an opinion about what the user really ***needs***, rather than what the user ***wants***. This last type of pull technology is called delta-pull.

The revolution created by the adoption and use of pull technologies will be fueled by how heartily individual users, companies, and the societal culture at large embrace them. Burdensome searching may be partially replaced with the full deployment of delta-pull technology, because the evolutionary agent will observe and understand a user's behavior, seeking out information the user ***needs***. Web searches will also be more efficient with full use of delta-pull technology. More satisfying results from initial searches will translate into more effective searchers.

Pull technologies are not the only way for an individual to obtain the information needed for decision making. The other side of information systems delivery is push technology. It mirrors pull technology, but from the user's perspective push technology provides information without having to ask for it. We now discuss push technology and what it means to individuals, organizations, and society.

2.2 Push Technologies

Similar to pull technologies, there are four types of push technologies (Kendall and Kendall 1999a, 1999b). The simplest form of push technology can be classified as Webcasting. It is one-to-many, unidirectional, and there is little personalization in this most basic form of push technology. Webcasting is analogous to television broadcasting. Several of the most popular push technologies take the form of screen savers that flash news headlines, stock quotes, and the like.

Webcasting, or alpha-push, can also serve as event reproduction where, according to Apple computer, "The goal is to provide an interactive experience, allowing the user to dig deeper and go behind the scenes of the event itself" (Johnson 1997)—providing real-time event coverage, making live Webcasts not just like being there, but "better than being there."

The second level of push, called beta-push, allows the most flexibility. As the amount of information increases, users will want to filter out unnecessary information that is being pushed on them and corporations will consider some information to be too sensitive to broadcast to every employee. At this level, users can choose channels to include or exclude certain types of content. For example, if seeking information from a news provider, an individual may want to include news and sports, but exclude weather.

A third level, called gamma-push, uses intelligent agents to personalize and filter out messages even further. The agent will adapt user profiles to screen out specific articles based on a set of predetermined criteria. For example, the agent will only display scores from the preselected football teams about which the user expressed interest.

Finally, delta-push, the fourth level, takes this one step further. Services will be custom-designed for the individual, based not only on demographics and data mining (Codd 1995; Gray and Watson 1998; Watson and Haley 1997), but on the behavior of the user. For example, if an evolutionary agent observes the behavior of an individual over time and realizes that the individual is a devoted sports enthusiast, the agent will take this into consideration and push sports news, scores, and information toward the user. In addition, the agent will push commercial advertisements in that direction as well. The important point here is that, based on observation of a user's behavior, the evolutionary agent delivers what the user ***needs***, not merely what the user ***wants***.

Table 1. The Four Levels of Pull and Push Technologies Illustrate that Users Obtain Information in a Variety of Ways, Each with its Own Set of Consequences

Level	Pull Technologies	Push Technologies
Alpha	Surfing the Net and using simple tools like bookmark managers	Viewing Webcasting, which is one-to-many, impersonal, and uni-directional
Beta	Using search engines to narrow the search to relevant information	Selecting channels to choose delivery of relevant categories of information only
Gamma	Using a personal assistant or bot that finds what the user wants, based on a set of well-defined user criteria (user profile)	Using intelligent agents, based on a user's profile, to include information the user wants and to screen out particular information with a channel that the user does not want
Delta	Relying on an evolutionary agent that changes over time to find what the user needs	Relying on an evolutionary agent to observe a user's behavior and then choose to deliver only the information a user really needs

Table 1 summarizes the four levels of pull and push technologies. Researchers need to reflect on how users obtain information on each of these levels. As one moves from the alpha technologies to the delta technologies, the searching or delivery appears to be much easier, but the consequences need to be considered as well. For an extended discussion about the evolutionary agent see Kendall (1996).

Push technologies will change the orientation and vision of organizations. We are in the early phases of a transforming process. Some of the early corporate leaders have been National Semiconductor, Wheat First Securities, MCI, and Church & Dwight (the maker of Arm and Hammer baking soda products) (Sliwa and Stedman 1998). Many of these first and early efforts involve the use of corporate intranets to feed information to managers rapidly so that decisions can be made in a timely manner when competition is keen. Additionally, IDS can assist in situations that are unusual, whether highly dynamic or so extreme as to be termed an emergency. For example, MCI uses PointCast to send outage information to 7,000 operations employees. Astound Webcast and Intermind Communicator are both push platforms that can be used with corporate intranets (Strom 1997). Push will make its presence felt in what were often mundane corporate systems such as e-mail. Users in corporations can now create a multimedia experience for the receiver. Where e-mail was originally the province of scientific researchers, Downes and Mui (1998) note its adaptation for advertising and information delivery. Even more startling, they believe that push technology can wind up entirely revamping human communication.

Delta push technology uses memes (Brodie 1996) to replicate and spread content. The advantage of delta push technology is in the successful reproduction of messages that will alter behavior in a way that the provider desires. The copying of a meme in this situation can be referred to as a push-virus. For example, advertisers that sponsor Webcasts use gamma push technology to formulate "catchy" songs, ideas, and slogans that stay with users of the media long enough for them to purchase a product or service that has been crystallized in the meme.

The advantage of delta push technology is that push-viruses are so strong that they can result in "catching" an idea, a song, a way of problem solving, and so on that may be beneficial only if large groups of people share in it quickly. Preparing large segments of a population for an impending emergency such as a forecast hurricane is an example of an advantageous use of delta push. Mounting a concentrated effort for solving a large scale transportation strike or unifying negotiators in looking for a solution to a peace process within a short time frame might also be worthwhile uses of delta push technology.

The concept of push technology itself is subject to constant (and not always kind) revision. Office workers may view push as a way to obtain news, weather, and stock quotes without having to click on or call up anything. Librarians may look at push as a data retrieval system. Executives may experience push as a way to gather information needed to make decisions and, perhaps, push will find itself one day at the center of many executive information systems.

As we will see later in this article, use of pull and push technologies harbor major changes for the way that decision makers think, act, and communicate. Businesses using pull and push technologies are competing to allow access or deliver content to users in ways that will make it abundantly apparent that, as Postman writes, "The medium is the metaphor."

Content is determined by structure. Therefore, in the future, pull and push technologies will evolve. Pull technologies will reflect pull content. Push technologies, therefore, are all about push content.

We have just presented the positive aspects of pull and push technologies. The next section examines the darker side of push and pull.

3. Alarming Aspects of Information Delivery Systems

All four push technologies and all four pull technologies change the way we think. The counter point to the advantages is provided by the existence of very real threats that can have alarming consequences not just for the individual but for organizations and society as a whole as they work with the pull and push technologies. Just as the alphabet forever changed the written culture, so information technology transforms our culture in ways we are just beginning to contemplate. Postman compares and contrasts the view of the future put forward by Aldous Huxley with that of George Orwell's ubiquitous and, by now, well-known Big Brother:

> But in Huxley's vision, no Big Brother is required to deprive people of their autonomy, maturity and history. As he saw it, people will come to love their oppression, to adore the technologies that undo their capacities to think. What Orwell feared were those who would ban books. What Huxley feared was that there would be no reason to ban a book, for there would be no one who wanted to read one. Orwell

> feared those who would deprive us of information. Huxley feared those who would give us so much that we would be reduced to passivity and egoism. Orwell feared the truth would be concealed from us. Huxley feared the truth would be drowned in a sea of irrelevance....In 1984, Huxley added, people are controlled by inflicting pain. In the Brave New World, they are controlled by inflicting pleasure. In short, Orwell feared that what we hate will ruin us. Huxley feared that what we love will ruin us. [Postman 1985, pp. vii-viii]

3.1 Implications for Individuals

We know that information does affect human beings; we cannot predict exactly how it will be manifested. For individuals, the potential problems of pull and push technologies include information overload, anxiety, addiction, and disorientation.

We do know that those with addictions to the relationships they strike up on the Internet mirror patterns found in people addicted to other behaviors such as substance abuse, gambling, or compulsive sex. However, in the instance of those addicted to surfing the Web for days on end, normal endeavors and life styles are displaced by Web-based activities, resulting eventually in the loss of jobs, break up of homes, and other socially malignant events.

McArthur (in Kelley 1998, p. G1) states,

> There's a great line from John Le Carre, "They're fed up and asking for more." Which is another way of stating that users are experiencing an overload, which they decry, but ironically they are also eager to get ever-increasing amounts of information; more of what they love.

Other thinkers share similar concerns. The contemporary philosopher Heim (1993) wrote, "With a mind-set fixed on information, our attention span shortens. We collect fragments. We become mentally poorer in overall meaning."

How does possessing the capability for 24-hour connectedness influence our behavior with push and pull technologies? Friedman (in Kelley 1998) states that, "It's not clear that (connectedness) enhances the quality of our lives or the productivity of our work lives." Others are amazed at the potential for electronically induced "soul sickness"(Crabb as quoted in Kelley 1998) of exhibitionists and voyeurs alike as people procure Web cameras to exhibit every manner of bodily function to viewers on the Internet.

The stress of togetherness causes illness and dysfunction or lack of functioning in any human culture (Kelley 1998, p. G8). The positive aspects of having a private time and place to oneself become evident.

A steady onslaught of broadcast information with no customization has been termed a "wasted expression" by Reid (1996, p. 10), who reminds us that

> All pushed information inevitably reaches many people who just aren't interested in it. Consider the Sunday newspaper...The many misses of

> pushed communication mean that every spot on *hit* is expensive to reach, and this expense prices a tremendous amount of would-be content out of the distribution channel.

Years ago there was a comedic play, written by the comedian and playwright Ben Elton, produced in London, called *Gasping*. In it, big business interests had created a commodity of the very air we breathe. Citizens were literally gasping for air, because the air market had been cornered by ruthless business moguls.

When we see individuals disconnecting, we might be able to recognize that those individuals are themselves "gasping." Gasping for the air of the old order, the old culture.

3.2 Implications for Organizations

The use of alpha push on Webcasts such as PointCast and others may result in a junk culture where everything is pushed to the user with little discrimination and, because the users have already discounted the value of the content (since what is conveyed through the medium of alpha push is so often junk), content is largely ignored. Alpha push technologies may be passed over by users as being too mass oriented; as being both "high tech and low touch" in an era when personalization and customization (whether real or counterfeit) are prized. Vast sums of money are spent to push content that is irrelevant.

Another disadvantage of delta push is lack of access or denied access because of too small of a bandwidth. A message such as "Server too busy" means that the information cannot be accessed.

3.3 Implications for Society

Wise researchers and commentators have not let the advent of mass media pass unremarked. Because push and pull technologies alter our discourse and thereby the way that we think, we must acknowledge that the media with which we engage each other are not empty vessels waiting to be filled with good or bad content. Rather they are the revolution in and of themselves. Postman writes credibly in arguing for the belief that the change has already taken place:

> We have reached...a critical mass in that electronic media have decisively and irreversibly changed the character of our symbolic environment. We are now a culture whose information, ideas and epistemology are given form by television, but not by the printed word...print is now a residual epistemology, and it will remain so, aided to some extent by the computer, and newspapers and magazines made to look like television screens. [Postman 1985, pp. 27-28]

One way to discuss what happens with delta push technology is to use the metaphor of viruses that mutate and change, infecting those susceptible rapidly, with no known cure. For instance, delta push technology may feature evolving memes with an evil message of ethnic hate, that eventually stiffen a nation's resolve to go to war against a neighboring country, or even its own citizens. With the use of delta push technology, the evolving memes may be too overwhelmingly memorable to permit other forms of thought.

4. Remedies and Resolution of IDS Threats

What are the remedies that can be taken so that the darker side of IDS do not prevail? Our analysis and interpretation reveal that there are several steps that can be taken and that many of these remedies are already being employed on a small, unsystematic basis, but still with good results. What is needed is an adoption and application of these coping strategies in a more systematic and all-encompassing way.

One possible remedy is to focus the attention of the users on understanding the value of the IDS technology. Problems often disappear when people begin to appreciate what technology has to offer. Another effective remedy is the education of all people; not just superficially so that the tangible uses of information technology are harnessed, but education in a deeper fashion so that the actual cultural departure represented by changing to an information society can be grasped. A third remedy is to use humor to satirize the changing activities, priorities, and livelihoods of users. Since humor is a relentlessly human exercise, it provides a strong countervailing force in response to new technologies.

4.1 Striving Toward Understanding the Value of IDS

One of the key remedies that can be taken to quell the frightening aspects of pull and push technologies is that of understanding, and then valuing, what we are up against. Postman writes:

> Public consciousness has not yet assimilated the point that technology is ideology...this, in spite of the fact that before our very eyes technology has altered every aspect of life in America during the past eighty years....To be unaware that a technology comes equipped with a program for social change, to maintain that technology is neutral, to make the assumption that technology is always a friend to culture, is, at this late hour, stupidity plain and simple. [Postman 1985, p. 157]

Johnson states that "One of the central problems with push media is that it takes away much of the users' control without providing many of the benefits that traditional broadcasting does in the form of solid linear narrative." In other words, push technologies are designed to be in front of you but otherwise unobtrusive (placed on pagers or screen savers, etc.); they are "there" without causing disruption to a typical task. However, because of this, push media does not do a good job of telling stories. This sounds trivial but is actually quite important. Strength and presence of narrative alters the passive nature of TV and radio, and makes them "work" (Johnson 1997). Others would agree, noting that current television news in the U.S. has an insatiable appetite for what *The Economist* (July 1998) has dubbed "water-cooler stories." The key criteria for qualifying as a water-cooler story is that it "must have recognizable characters and a developing drama." Thus the point is made that editors, who are normally quite competitive, fulfill their competitive assignment by falling all over each other spending large budgets on getting "the same old story."

4.2 Educating Users and Providers about Designing IDS

One of the remedies for the alarming aspects of pull and push technologies is the education of all people who are impacted by them. We need to educate people about how media changes the structure of our discourse. We can begin by educating decision makers and other users of pull media about the need for balance.

We can also educate push providers about the need for balance. Perhaps pushing a coherent story is more important than displaying a fragment. For example, although CNN Headline News has added a one-sentence headline to the bottom of its screen, which is continuously changing, stories are often incomprehensible because of the small amount of display space permitted. An additional self-imposed constraint seems to be that the headline must fit at the bottom of the screen without benefit of expansion via scrawling. Balance in stories presented like this is utterly lacking, and this lack can be addressed.

4.3 Using Humor to Alleviate the Stress in IDS Implementation

Edward de Bono (1992) suggests that humor is the heart of creativity. Postman asserts that we should laugh at ourselves and even parody ourselves. With tongue in cheek, he also notes that this last assumes that we can distinguish between parody and some other event that might be picked up and played out through the media. Kendall (1997) found that information systems designers could use the posting of cartoon humor by users to assess the gap between the actual versus intended addressing of critical success factors important to systems implementation.

The ability to regard the eight pull and push technologies as often producing humorous results, along with their capability to pose serious threats, puts the designer in a superior position to understand the complexity of what they are doing and the possible influence of their push and pull designs on the human spirit.

5. Unplugging

Unplugging is a reaction to overload. An example of unplugging is when an individual feels inordinate pressure from being connected to technology on a 24-hour, seven day per week basis and subsequently takes some sort of action to disconnect from the artificial word of technology. The individual resumes life in the physical world, the one populated with people, not just machines.

Unplugging can take on different meanings at the individual, organizational, and societal levels. We will discuss each of these in turn.

5.1 Meaning for Individuals

Individuals may have to seek real life (high touch, low tech) experiences instead of being "plugged in." Heim has written about information fixation and has stated that *what* people did while unplugged was more important than the length of time they spent away from technology. He notes that technology retrains our nervous system, creating a new pace and tempo to which we adapt, but with which we could ultimately be uncomfortable. Many spiritual groups ask their participants to unplug for the duration of their retreats (Yuen 1998).

We envision entire technology-free sectors (e.g., public parks) where people can go and be unplugged without worrying that they will be accosted by others being able to "reach" them electronically in the wilderness. These parks could include places for retreats, concerts, and hideaways. It is incumbent on the new generation to create cultural norms that encourage thinkers to think, actors to act, and so on where the structure of thought is not dictated by the incessant and ubiquitous interactions with Web push and pull technologies.

This suggestion is not in any way to be construed as Luddite in origin. Quite the contrary. If we are to continue being creative, we need time away from push and pull technologies to develop new paradigms and innovative approaches to solving our problems (some of which paradoxically arise due to our interactions with new technologies). We need time to reflect. Quite possibly, we even need time alone. Once we can safely unplug, our creative and innovative solutions to new and old problems should be promising enough for the society to want to continue to inculcate this practice. The fruits should be plentiful, and they should be obvious.

5.2 Meaning for Organizations

Picarille (1997) acknowledges increasing information overload among users of push technology on the Internet, intranets, and extranets. Although one remedy has been characterized as "pushing back" by simply getting rid of their multiple channels, Picarille notes that "corporate users are looking for better ways to manage the critical information they receive."

In place of electronic connectivity, the organization can legitimize formal or informal face-to-face time when certain thinking forms can be introduced. These might include brainstorming or other ways to stimulate creativity or evaluate new ideas (see, for example, the negative assessment step suggested by Abu Hamdieh and Kendall [1999] for evaluation of systems projects.) Somewhat surprisingly, planned face-to-face activities that take managers completely out of the reach of push and pull systems are already the norm for executives in some companies such as Microsoft.

Other researchers and philosophers have suggested that the *topics* that occupy managers' minds during times of being unconnected are every bit as important as the *length of time* spent away (Heim 1993). Another example of a corporate remedy is that supplied by National Semiconductor, who added its own channel to PointCast, which they have called National Advisor. Three streams of product-related data are featured: (1) daily traditional sales and order information recorded by field sales staff; (2) data captured on their Website, which specifically captures the frequency of visits, keeps track of what product information customers are looking for, and also which items are requested more often as samples; and (3) analysis of e-mail questions that are fielded on the Web which are sorted according to recognizable product categories (Cronin 1997).

The values of the corporation should be such that (most) employees are not expected to exist with 24 hour connectivity. Just as connectivity can be part of the strategic IS plan, so can sanctioned time away from e-mail and the incessant interaction of collaborative work systems.

5.3 Meaning for Society

How will push and pull technologies shape and impact the existence and persistence of particular countries, their boundaries, their world views? The world will undoubtedly experience problems and face decisions resulting from information overload and a diversity of perspectives because of push and pull technologies. Neither do we know what the effects are of continuous connection, instant access, or even bombardment with information. In some ways, we know more about what happens to people totally deprived of sensory input, for example, than what happens to a society deluged with information.

6. Conclusions

After the necessities of love and loyalty, what is one quality we would seek in a friend? Perhaps it is their ability to know what it is we like, in other words, the information and activities that we uniquely prefer. We disclose these preferences to our friend, and they remember and oblige. They observe us further, watch our behavior, and then guide us as we seek our objectives. This is what we can expect of delta-pull technology in the future.

Sometimes, if we have been friends for a long while, our friends intuit what we want, without requiring us to utter a word. Then, rather than guiding us, they bring us what we need when we need it. In the future, delta-push technology will do just this.

Computers using software based in part on intelligent, evolutionary agents can simultaneously deliver information to decision makers on a variety of topics. One of our most valued aspects of friendship can be addressed by the right system. What happens to our expectations when this occurs? What do pull and push technology mean for the way that we seek information and the way decision makers perceive it?

Similarly, for pull technology, the evolutionary agent helps users to seek out exactly the information they need on the World Wide Web or the Internet, and to download it to their computers, to use it at their convenience. Pull technology allows users to have the world at their fingertips, whenever they need it.

The story of the sorcerer's apprentice is illustrative of what happens when push or pull technologies run amok. The sorcerer has power to command an intelligent agent to perform a certain behavior. Without feedback, the apprentice continues unchecked until it has carried enough water to fill an entire room. Then, not only does the apprentice accomplish a task, but continues executing a task to the point where the results of engaging with the technology actually becomes harmful.

What does the future hold for researchers of push and pull technologies? What will corporations and the cultures within which they exist do when confronted with emerging technologies? Will our accepted models of adoption and use of technologies hold true? Perhaps instead we will experience a revolution of adoption processes as well as the revolutionary technologies. Push and pull technologies are being greeted with a wide spectrum of reactions. There are users and organizations that are fascinated with IDS. There are small forays into technological innovation and adoption (almost too tentative to be called experiments) by some organizations. Some companies have encouraged toying with IDS, but it is almost as if invoking the game precludes them from imagining wider implications. Indeed, their view is that employees will likely tire with the new toy of push and pull technologies, abandoning it for whatever the next fad is. In this scenario, IDS are not so much greeted with alarm as with eventual boredom, lapsing into disuse.

Our thought is that, since information delivery systems (IDS) will continue to emerge and evolve, we need not assume that the evolution and spread of this new technology will unfold in the same old (predictable) patterns. One way that IDS can gain acceptance on the grass roots level is by making all of the personal information devices such as palmtops, digital watches, pagers, mobile phones (and other intriguing combinations) deliver a greater quantity of customized information in a more personalized manner. This is a departure from older models of the adoption of new information technologies, which dictated that because of the cost of the technology, it would be a mass medium, broadcasting to the masses, impersonally targeting non-segmented groups. Portable devices and hand held devices will be able to communicate to each other in an *ad hoc* way. Old barriers to mass media, including the clumsy footprint left by required cables, may not be viewed as barriers any longer.

What we learn from the birth and growth of push and pull technologies is both distressing and invigorating. We recommend that we understand, educate, and use humor in order to really come to grips with what is happening. We are undergoing a revolution of dramatic proportions. Push and pull technologies will enable the transformation of the print culture. Very few people understand the proportions of the revolution. As Postman writes, "Only through a deep and unfailing awareness of the structure and effects of information, through a demystification of media, is there any hope of our gaining some measure of control over television, or the computer, or any other medium." We heartily concur. By considering pull and push technologies at their most alarming and then examining the possible remedies, they can be demystified and, in many cases, put to good use for individuals, organizations, and society as a whole.

References

Abu Hamdieh, and Kendall, J. E. "Assessing the Negative Effects of Proposed Information Systems: A Method Based on de Bono's Thinking Hats," in *Proceedings of the Fifth International Meeting of the Decision Sciences Institute*, K. Zopounidis (ed.), Athens, Greece, 1999.

Brodie, R. *Virus of the Mind.* Seattle, WA: Integral Press, 1996.

Codd, E. F. "Twelve Rules for On-Line Analytic Processing," *Computerworld*, April 13, 1995, pp. 84-87.

Cronin, M. "Using The Web To Push Key Data To Decision Makers," *Fortune* (136:6), September 29, 1997, p. 254.

De Bono, E. *Serious Creativity: Using the Power of Lateral Thinking to Create New Ideas*. New York: Harper Business, 1992.

Downes, L., and Mui, C. *Unleashing The Killer App: Digital Strategies For Market Dominance*. Boston: Harvard Business School Press, 1998.

Graber, D. A. *Verbal Behavior and Politics*. Chicago: University of Illinois Press, 1976.

Gray, P., and Watson, H. J. *Decision Support in the Data Warehouse*. Upper Saddle River, NJ: Prentice Hall, 1998.

Heim, M. *The Metaphysics of Birtual Reality*. New York: Oxford University Press, 1993.

Johnson, M. R. "Webcasting," 1997 (http//www.mindspring.com/cityzoo/mjohnson/papers/webcasting/introduction.html; accessed on July 1, 1998).

Kelley, T. "Only Disconnect (for a While, Anyway): A Few of the Well-connected Who Take Time Off from E-mail, and Survive," *The New York Times*, Thursday, June 28, 1998, pp. G1, G8.

Kendall, J. E. "Examining the Relationship Between Computer Cartoons and Factors in Information Systems Use, Success, and Failure: Visual Evidence of Met and Unmet Expectations," *The DATA BASE for Advances in Information Systems* (28:2), 1997, pp. 113-126.

Kendall, J. E., and Kendall, K. E. (1999) "Information Delivery Systems: An Exploration of Web Push and Pull Technologies," *Communications of AIS* (1:14), April 23, 1999.

Kendall, J. E., and Kendall, K. E. "Web Pull and Push Technologies: The Emergence and Future of Information Delivery Systems," in *Emerging Information Technologies: Improving Decisions, Cooperation, and Infrastructure*, K. E. Kendall (ed.), Thousand Oaks, CA: SAGE Publications, Inc., 1999, pp. 265-288.

Kendall, K. "Artificial Intelligence and Götterdämerung: The Evolutionary Paradigm of the Future," *The DATA BASE for Advances in Information Systems* (27:4), Fall 1996, pp. 99-115.

Picarille, L. "Push Comes to Shove on the Web," *Computer Reseller News* (749) August 11, 1997, pp. 113-114.

Postman, N. *Amusing Ourselves To Death: Public Discourse in the Age of Show Business*. New York: Penguin Books, 1985.

Reid, R. H. *Architects of the Web: 1,000 Days that Built the Future of Business*. New York: John Wiley & Sons, Inc., 1996.

Richardson, D. "Data Broadcasting- The Ultimate Push Technology?" *IEE Conference Publication*, 447, Stevenage, UK: IEE, 1997, pp. 36-42.

Sliwa, C., and Stedman, C. "'Push' Gets Pulled Onto Intranets," *Computerworld* (32:12), March 23, 1998, p. 6.

Stohr, E. A., and Viswanathan, S. "Recommendation Systems: Decision Support for the Information Economy" in *Emerging Information Technologies: Improving Decisions, Cooperation, and Infrastructure*, K. E. Kendall (ed.). Thousand Oaks, CA: SAGE Publications, Inc., 1999.

Strom, D. "Tune in to the Company Channel," *Windows Sources*, (5:9), 1997, pp. 149-158 passim.

The Economist. "The News Business," July 4, 1998, pp. 17-19.

Watson, H. J., and Haley, B. "A Framework for Data Warehousing," *Data Warehousing Journal* (2:1), 1997, pp. 10-17.

Yuen, E. "An Ideal Buddhist Vacation: Travel Combined with Mindfulness Meditation," *Philadelphia Inquirer*, Sunday, July 12, 1998, p. H7.

About the Authors

Julie E. Kendall, Ph. D., is an associate professor of MIS in the School of Business-Camden, Rutgers University and is currently the Vice-Chair of IFIP WG 8.2. Julie has published in *MIS Quarterly*, *Decision Sciences*, *Information & Management*, *Organization Studies*, and many other journals. Additionally, Julie has recently co-authored a college textbook with Kenneth E. Kendall, *Systems Analysis and Design*, fourth edition, published by Prentice Hall. She has served as a functional editor of MIS for *Interfaces* and as an associate editor for *MIS Quarterly*. She is on the editorial boards of *Journal of AIS*, *Journal of Management Systems*, and *Journal of Database Management*, and is on the editorial review board of *Information Resource Management Journal*. She was recently elected to serve as Vice President for Decision Sciences Institute. Julie's research interests include developing innovative qualitative approaches for information systems researchers interested in systems analysis and design. She is researching societal implications of push and pull technologies. Julie can be reached by e-mail at julie@thekendalls.org; her home page can be accessed at www.thekendalls.org.

Kenneth E. Kendall, Ph. D., is a professor of Information Systems in the School of Business-Camden, Rutgers University. He recently co-authored a text, *Systems Analysis and Design*, fourth edition, published by Prentice Hall and edited *Emerging Information Technologies: Improving Decisions, Cooperation, and Infrastructure* for Sage Publications, Inc. Ken has had his research published in *MIS Quarterly*, *Management Science*, *Operations Research*, *Decision Sciences*, *Information & Management*, and many other journals. He is one of the founders of the International Conference on Information Systems (ICIS). Ken is the past Chair of IFIP Working Group 8.2 and served as a Vice President for the Decision Sciences Institute. He is the MIS editor for the *Journal of Management Systems* and an Associate Editor for *Decision Sciences*, the *Information Systems Journal*, and the *Information Resources Management Journal*. He has also served as a functional editor of MIS for *Interfaces*. Ken's research focuses on studying push and pull technologies and developing new tools for systems analysis and design. He can be reached by e-mail at ken@thekendalls.org; his home page can be accessed at www.thekendalls.org.

13 GLOBALIZATION AND IT: AGENDA FOR RESEARCH

Geoff Walsham
University of Cambridge
United Kingdom

Abstract

A precise definition of globalization is elusive, but it is widely accepted that the world is becoming increasingly interconnected in terms of its economic, political, and cultural life, and that IT is deeply implicated in the change processes that are taking place. However, these processes are not uniform in their effects. Individuals, groups, organizations, and societies remain distinct and differentiated, and the challenge is to design information systems that enable increased connectivity but also support this inherent diversity. This paper considers the actual and potential contribution of IS research to this challenge, using five levels of analysis ranging from the individual to the societal. Conclusions are drawn on the need for in-depth studies, a broad and evolving research agenda, and an anti-ethnocentric approach.

Keywords: Globalization, IS research agenda, levels of analysis, culture

1. Introduction

It is widely acknowledged that major social transformations have taken place in organizations and societies over the last few decades, and most commentators believe that the pace of change is unlikely to slow in the foreseeable future. For example, in the last decade, many large organizations have undergone transformation processes, such as de-layering and job redefinition, associated with business process reengineering. The detailed work life in these organizations, and in small and medium sized enterprises, is often significantly different in style and content to what was the case only a few years earlier. In addition to changes in work and work-life, the norms and values of society outside the sphere of paid employment have also shifted, often dramatically. Attitudes to gender, the environment, race, sex, family life, and religion have been transformed in the latter part of the 20th century in many countries of the world. The term globalization reflects one linking thread for the changes that have taken place. Although hard to define in any precise way, the concept refers to the increasing interconnection of individuals, organizations, and societies in terms of their economic, political, and cultural life.

It is relatively easy to catalogue changes as outlined above, but it is important to remember that there has also been continuity and stability. Societies remain clearly distinct, despite increasing interconnection, and their citizens normally pursue a life-style that would be immediately recognizable to their ancestors of 50 years earlier. Organizations have transformed themselves in many ways, but the human processes involved, for example, in leadership, team working, and the pursuit of personal aspirations remain much the same in essence. Individuals may work or socialize in a different way, and some social norms and values may have shifted, but we all need self-respect, a community or communities to which we belong, and ways of giving meaning to our lives.

The specific role of information and communication technologies in the interweaving of stability and change that has taken place, and that might take place in the future, is the subject of much debate. Some commentators argue that these technologies are the driving forces for change, while others would see them in a more supporting role. It is, however, generally accepted that they are a fundamental element in the changed nature of work processes, in organizational restructuring and in societal transformation. Terms such as the "information society," or even the "information revolution," are relatively common in the academic and business world and are increasingly recognized by the ordinary citizen. However, they often reflect a wide variety of different views on the precise nature of these phenomena, and thus obscure more than they reveal.

In order to investigate the role of information technology in more detail, it is necessary to go down from the level of broad generalizations to the more specific level of organizations, groups, and individuals. However, in doing this, we are confronted with a bewildering array of ever-changing technologies, work processes and organizational forms. Much current debate in business organizations centers around technologies such as groupware, the internet, and enterprise systems, and around topics such as e-commerce and knowledge management, although we can be sure that these will be supplanted in due course by the as-yet-unknown technologies and topics that will be considered central a few years hence.

In addition to the shifting multiplicity of technologies and topics, there are multiple levels at which we can analyze the role of information technology. For example, one can focus at the individual level of personal or professional identity, at the level of teams and group working, at the level of the organization, on interorganizational networks, on a given society, or at the inter-societal level. All these levels can be analyzed separately, but they are, of course, inextricably interconnected.

Much of the literature in the English language that addresses the wide range of issues outlined above has a strong "Western country" bias, both in terms of the subjects of study normally being located in these countries and in terms of the nationalities of the authors.

However, major economic activity is increasingly taking place in a wider global arena, and what is happening in the countries of Asia, Africa, or Latin America is a subject of concern for the whole world. The cultures of these countries are often radically different to those of the Western countries and insights from work in the latter do not necessarily apply to the former. Japan presents one obvious example of such a difference, and one where significant amounts of published work exist in English. However, much less has been written about, for example, the nature of work and organizations in India or China, that together account for over two billion people.

So, how should we try to make sense of the role played by information technology in this multi-technology, multi-level, multi-cultural arena? Most research in this domain has been carried out by selecting one part of the mosaic and exploring issues in depth in one country, one organization, one group, or even with respect to one individual. This work has provided us with many insights, but the resultant picture of the role of information technology in the contemporary world is highly fragmented, patchy in its coverage, and lacking in overall synthesis. The purpose of this paper is to discuss future research agendas that might help to improve our understanding of IT in the contemporary world. This discussion takes place in the third section. However, prior to that, in the next section, an overall backdrop for the discussion is provided by a brief analysis of the concept of globalization.

2. What is Globalization?

The term globalization has achieved the unusual status, in a relatively short time, of becoming fashionable in academic debates in the social sciences, in the business world, and to some extent in the popular media. However, even a cursory examination of these sources demonstrates that the term is highly ambiguous and that it masks a wide variety of opinions on what is happening in the world. The purpose of this section is to summarize and critique some of these views.

Robertson (1992) wrote an influential book on globalization in which he said that "Globalization as a concept refers both to the compression of the world and the intensification of consciousness of the world as a whole" (p. 8).

The first of these two points relates directly to time-space compression, largely mediated by information and communications technologies. The second point is less obvious, and certainly refers to the world as a whole rather than just Western society. The widespread accessibility of communications media such as the television, even in remote rural villages in the Third World or underprivileged urban communities anywhere, means that news of happenings in the world as a whole are available to the great majority of the world's population. This does not necessarily imply a well-informed world, since the "news" are chosen, condensed, filtered, and manipulated by a host of complex mechanisms. However, it does mean that remoteness and isolation are not the same in the contemporary age and that most people are more aware than they were of a wider global arena within which their own community forms only a small part.

Changes in perception and access to wider information sources have significant cultural and political implications, but globalization also refers to important economic phenomena. Giddens has joined the globalization theorists in recent years, and he argues that global financial flows are a key element of the globalization process:

> Geared as it is to electronic money—money that exists only as digits in computers—the current world economy has no parallels in earlier times. In the new global electronic economy, fund managers, banks, corporations, as well as millions of individual investors, can transfer vast amounts of capital from one side of the world to another at the click of a mouse. As they do so, they can destabilize what might have seemed rock-solid economies—as happened in East Asia. [Giddens 1999]

Global financial flows are one element in trends toward more global business as a whole, although we need to be wary of simplistic generalizations here. Although there is much talk in the business world, and the management schools, of global businesses, global markets, and global supply chains, the degree to which this has occurred to date, and the degree to which it might occur in the future, remains in dispute. For example, Doremus et al. (1998) investigated a range of multinational corporations, mainly in Germany, Japan, and the USA, and argued that such companies, who after all should surely be at the forefront of the move toward globally-minded enterprises, remained tied to approaches derived from their unique national identities:

> However lustily they sing from the same hymn book when they gather together in Davos or Aspen, the leaders of the world's great business enterprises continue to differ in their most fundamental strategic behavior and objectives. [Quoted in Kogut 1999]

2.1 Globalization and Diversity

This leads on to one of the most controversial issues in the globalization debate, namely the issue of homogenization and diversity. The broad question is whether the globalization phenomena that we have outlined above, such as time-space compression, an increased awareness of the world as a whole, and movements toward global business, will inevitably lead to a decrease in differences among nations, companies, and/or individuals. There is a school of thought, prevalent among the Western business community, for example, which takes this "end of history" idea for granted. The argument runs that there is only one economic system now, capitalism, and that enterprises need to compete globally under this one set of rules. Therefore, all companies that wish to survive will need to adopt the practices of the winners, leading toward more homogeneous ways of doing things and, by extension to the wider society, to a less-diverse cultural world.

There is some force in this argument, but a range of writers have taken exception to the conclusion of the inevitability of homogenization. For example, Robertson discusses the way in which imported themes are "indigenized" in particular societies, with local culture constraining the receptivity to some ideas rather than others, and adapting them all in specific ways. He cites Japan as a good example of this blending of the "native" and the "foreign" as an ongoing process. While accepting the idea of time-space compression facilitated by information technology, Robertson argues that one of its main consequences is an exacerbation of collisions between global, societal, and communal attitudes.

A second example of the counter-argument to the homogenization thesis is provided by Gopal (1997), writing specifically about the developing countries. He argues that the vision of an IT-driven world of progress, efficiency, unlimited markets, individualism, and the superiority of the Western developmental trajectory needs to be challenged, not least because the effects of the use of IT in the West itself have not always been benign. He cites concerns about unemployment, de-skilling, privacy, and surveillance as examples. However, he also argues that each developing country must forge its own path in the future, and not try to imitate inappropriate Western models:

> They [developing countries] have different pasts; their historical trajectories have led to different configurations of valences, patterns of trust, responsibilities, and allegiances. They have, as a result, different presents: priorities, voices, capabilities, and capacities are arranged in patterns quite unlike those of the societies from which the technologies originate. And, in spite of the attempts of a few to sediment in the popular imagination a singular vision of prosperous IT-driven existence, they have different imagined futures (Appadurai 1997); their varieties of aspirations and expectations bear little resemblance to the visions embedded in the technology. [p. 140]

Castells (1996) is an influential writer on contemporary society, and his views on globalization echo a number of the above themes, but within a broader conceptualization of the new society that he believes is emerging from current processes of change. He argues that this society is both capitalist and informational, and he defines the latter as follows:

> The term informational indicates the attribute of a specific form of social organization in which information generation, processing and transmission become the fundamental sources of productivity and power, because of new technological conditions emerging in this historical period. [p. 21]

Castells believes that globalization is real, in the sense that the markets for goods and services are becoming increasingly globalized. This does not mean that all firms sell worldwide, but that the strategic aim of all firms, large and small, is to sell wherever they can throughout the world, either directly or via their linkage with networks that operate in the world market. However, in the informational economy, there is complex interaction between historically rooted political and social institutions and increasingly globalized economic agents. There is, thus, wide variety in the way in which individual countries and regions act as part of the globalized world. Castells argues that the global economy is characterized by its interdependence, but also by its asymmetry, and the increasing diversification within each region.

2.2 Globalization and Self-identity

The debate about the extent to which various processes linked to globalization may lead to homogenization in terms of business processes or national cultures rests to a significant extent on the effect of the processes on the individual member of society, whether in the Western or developing world. The argument will be made here that there is no strong evidence of any simple standardization of humanity, and indeed that global forces may, somewhat paradoxically perhaps, have some effects that tend toward the opposite.

Giddens (1990, 1991) and Beck (1992) are influential writers about self-identity in contemporary Western society. They both discuss the need for individual life projects, specific to a person's own past history and context, and to his or her future trajectory and aspirations. They argue that the world of relatively set rules, traditions, social classes, and job roles has been undermined. The new, uncertain world requires active navigation. Authors such as these have sometimes been criticized as emphasizing individual freedom of action in societies where the underprivileged can be thought to have little choice. It is certainly true that some people's choices are more constrained than others, and that life chances are very different at birth dependent on one's parentage and background. Nevertheless, any individual growing up and working in the 21st century will need to chart their own course with some vigor, within the range of possibilities available to them, or risk being carried away by the waves of change which will undoubtedly continue to roll.

Robertson talks about this issue related specifically to globalization and he adds a subtle distinction regarding concepts such as life projects themselves, which can be seen to be Western individualistic constructs. More group-oriented societies, such as many of those in Asia, and religions, such as Islam, which place great emphasis on community, would tend not to see the world as composed of distinct individuals with discrete life goals, opportunities, and problems. However, unless one feels that the cultural imperialism of Western individualism will sweep away all this potential resistance, and even then we are back to individualistic life projects, the heterogeneity of these cultural phenomena does not lead us in the direction of individual sameness.

This academic theorizing could be seen to be fiddling while Rome burns. When a large part of the world watches Hollywood movies, it is possible to buy Coca Cola virtually everywhere, and the internet is spreading like wildfire, are we not seeing the Americanization of the world and the end of cultural diversity? It would certainly seem to be the case that more people in the world are exposed to particular influences, such as aspects of American culture as portrayed in films and other media, than has been the case at any time in world history. But there is little evidence that this produces a higher proportion of cultural dopes than in previous eras, and the processes of selective indigenization, due to the collision of external ideas with internal norms and values, may proliferate the production of new hybrids, at the level of the individual, group or society.

3. Designing for Diversity: Research Agenda

The brief discussion in the previous section cannot be used to produce a definitive statement about globalization, nor about the particular role of IT. Nevertheless, three points of connection can be identified that broadly unite the authors cited. First, the contemporary world is undergoing major processes of change that affect all countries of the world and IT is deeply implicated in the changes that are taking place. The change processes will continue to have profound effects on self-identity, the nature of work and employment, organizational structure and networking, and the nature and governance of the nation-state. Second, the change processes are not uniform in their effects and individuals, groups, organizations, and societies will remain distinct and differentiated, although increasingly interconnected. Third, there is an increased need for reflection and action on the part of individuals, groups and societies in order to address issues of global change, including the role that IT plays in this.

If we accept these broad themes regarding globalization and IT, what agendas are implied for information systems designers and academics studying such systems? Is it possible to design systems that enable increased connectivity between individuals, groups, organizations, and societies, but also support the inherent diversity within these categories? The focus of this section of the paper will be on defining research agendas in these various domains, divided into five levels of analysis, ranging from the individual to the societal. At each level, examples of relevant published work are given, the current depth and coverage of work in this domain is discussed, a brief linkage is made to globalization themes, and future IS research agendas are identified. A summary of key points in the section is given in Table 1.

3.1 Individual

An example of work in this category is the description by Schultze and Boland (1997) of the work of three systems administrators, Ilana, Dan, and Jon, who were employed by a consulting company called Consultco. They were hired on a contract basis to work for US Co, a large manufacturing firm in a Midwestern U.S. state. The particular work for which they had been hired involved a knowledge management system called KnowMor based on a platform of the groupware Lotus Notes. The responsibility of the system administrators was to set up and operate the system environment to ensure the smooth running of the application in US Co.

Schultze and Boland describe and discuss the perceptions, feelings and actions of the three system administrators, and their search for identity:

> We depict these contractors as people struggling to "go on" in a world in which space and place, mind and body, logical and illogical explanations, clean and dirty work, physical and mental activity, and specialized and transcendent knowledge are posed as dualisms. [p. 556]

Table 1. Summary of IS Research and Links to Globalization Themes

Level of Analysis	Example	Current Depth of Coverage	Linkage to Globalization Themes	Future Research Agenda
Individual	Schultze and Boland (1997)	Very limited	Major shifts in personal and professional identity in contemporary world	• Explore shifting identity linked to IT in wide variety of contexts
Group	Ciborra (1996)	Good work on groupware	Groups need to communicate across time/space	• Groupware in multi-cultural contexts • Intranets
Organization	Knights and McCabe (1998a, 1998b)	Much interesting work here	Global initiatives and approaches such as BPR, ERP	• Research on current themes such as enterprise systems
Inter-organization	Barrett and Walsham (1999)	Growing literature	Networking for global reach	• Critical in-depth case studies, e.g., of IOS, e-commerce
Society	Lind (1991)	Under-researched, particularly in Third World contexts	Ironic under-representation in age of globalization	• Role of IS in different cultures: anti-ethnocentrism

The argument here is that the system administrators felt the need to present their work as involving logical, clean, mental activity based on knowledge that transcends particular places. Such work is contrasted with physical, local, specific work, in which "illogical" things happen and fixes and work-arounds are the norm. However, in practice, the work of Ilana, Dan and Jon involved both types of activity, inextricably intermingled. In terms of personal identity, the system administrators struggled between the self-disciplining needed to present their work as the former, while often identifying closely with the latter.

It is possible to connect the particular case here with more general globalization themes. The widespread emphasis on core competencies, at least in Western organizations, and the related outsourcing of many tasks previously carried out by permanent employees, has led to a legion of contract workers carrying out activities on a commodity basis. The representation of the work carried out by these people is a key element in determining the price of the commodity. For example, Consultco wished to reinforce the high-level status of their workers by presenting them as being the possessors of transcendent knowledge that is not locally specific and can be seamlessly transferred across space. However, this created pressures in terms of the workers themselves, who had to confront the realities of their work as well as its representation.

Other work on IT and shifting identity includes a case study of health care workers in the UK by Bloomfield and McLean (1996). Walsham (1998) explored the changing identities of particular professional groups, related to new IT, in the banking, insurance, and pharmaceutical industries. Despite such studies, and earlier pioneering work by Zuboff (1988), the topic of shifting identity linked to information technology remains under-researched. The discussion on globalization in this paper emphasized the importance of shifting identity in the contemporary world, but the social theorists on globalization do not go into any detail on individual cases. There are clear agenda here for IS researchers, namely to explore identity linked to new information technology in a wide variety of situations, organizations, and cultures.

3.2 Group

The study of the work of groups linked to information technology includes interesting research in the 1990s on groupware. A good example is the book edited by Ciborra (1996). Seven different cases are described in the book, in a variety of U.S. and Western European contexts, and involving a range of organizations and sectors. Ciborra, in the first summary chapter of the book, argues that the development and use of groupware technology in large, complex organizations is variable, context-specific and drifts over time. The book is entitled *Groupware and Teamwork: Invisible Aid or Technical Hindrance?* Ciborra argues that the question mark remained at the end of the study, since there is no *general* answer to the question. The very definition of groupware depends on how an organization molds it to the specific context: "'What groupware is' can only be ascertained *in situ*, when the matching between plasticity of the artefact and the multiform practices of the actors involved takes place" (p. 9).

In addition to the Ciborra book, the literature contains a relatively wide range of other in-depth studies of groupware technology, with Lotus Notes being used in the majority of the reported cases, reflecting its relatively dominant market position in groupware technology in the 1990s. For example, Karsten (1999) summarized the results of 17 different applications of Notes reported in the literature. The pressure on organizations to use groupware technology can be seen as a response to globalization trends, requiring individuals, groups, and organizations to communicate across time/space. The use of Lotus Notes by a wide range of organizations could be viewed as a form of homogeneity and standardization. However, the outcomes of the groupware cases were enormously varied, resulting from the complex matching of the plasticity of the technology and the practices of the actors in the specific context. This supports our earlier arguments on the persistence of diversity despite the increased interconnectivity of the world.

What of future agendas for IS research at the group level of analysis? First, a number of the cases reported in the literature cited above mentioned problems with the utilization of groupware where the systems were not restricted to a single country. For example, Orlikowski (1996) described the use of Notes to support incident tracking in the customer support department of a software company called Zeta, with headquarters in the Midwest of the U.S. There is an intriguing reference in the case to difficulties in trying to extend the technology to the three main overseas support offices in the UK, Europe (country not specified), and Australia. However, this aspect of the case was not explored further. In a similar vein, Ciborra and Patriotta (1996) describe a Notes-based application in Unilever designed to support innovation centers that were responsible for the global co-ordination of specific product categories. The authors note that there were "misunder-standings" due to the introduction of an international environment linking different cultures, but no further analysis is provided. There are clear agendas here for IS researchers to investigate in more detail the role of groupware in multi-cultural contexts.

A second area for future IS research concerns the use of intranets. At the present time, few in-depth case studies of intranets have been published, although research studies are currently taking place. Groupware and intranets are not synonymous, of course, but lessons from the use of groupware can inform our thinking about intranets. For example, the theme of the plasticity of the artefact and its matching to the context is immediately relevant. More specifically, effective intra- or inter-community collaboration, using intranets or groupware, needs a context and climate in which people and groups feel able to share information relatively freely, and where improvisation and change are a permanent and natural part of evolving work process.

3.3 Organization

In-depth IS research conducted at the level of the organization is perhaps the best represented of all levels of analysis in the current literature, with many published studies being available (e.g., Orlikowski 1991; Walsham 1993). An example of an interesting recent study, which also addresses wider global agendas, is that reported by Knights and McCabe (1998a, 1998b), concerning a business process reengineering (BPR) case in a medium-sized UK bank, which they called Probank. The bank underwent a series of restructuring initiatives from the late 1980s onward, using telecommunications and on-line customer database technology, designed to shift customers calls away from local branches to centralized back offices. Many customers resented having to contact the back office instead of their local branches. In addition, the "armchair banking" staff were often unable to cope with the volume of calls. In an attempt to address this latter problem, multi-skilled teams were created in the back office, but reactions of staff to this were mixed depending on their earlier job role, with staff whose earlier role involved only the telephone being generally more favorable. Relatively high stress levels remained the norm, however, and operation staff were said to begin "screaming and running about" when "the phone goes wild."

BPR can be seen as one example of global approaches that have affected many organizations across the world. Authors such as Willmott (1994) see BPR as a front for job losses, intensification of work, and tighter management control. What does the Probank case have to say about these issues? Well, certainly there were widespread losses, particularly in the branches. In terms of intensification of work, there appears to have been a more nuanced position, with at least some back office staff feeling that multi-skilled teams were a less stressful environment than their previous seven-hour telephone day. Management attempted to tighten control, but there was some evidence of "spaces for resistance" in various forms, some of them quite subtle in terms of the tone and style of staff response to customers. Knights and McCabe (1998a) argued, interestingly, that resistance to new ways of working in change programs such as BPR in Probank is not solely concerned with organizational politics centered on workers' resistance to management or one group trying to achieve advantage over another. It also needs to be thought of in terms of identity relations, or how individuals seek, through political maneuvering, to further or secure their individual careers and identities in an uncertain world. This ties back to issues at the individual level of analysis discussed above.

In terms of future IS research agendas, organization-level studies such as the one reported above are an essential means to get beyond the simple "good" or "bad" label for a new technology or change initiative, technological utopianism or anti-utopianism (Kling 1994), toward a more grounded position based on a range of in-depth studies in particular contexts. A particular example of a current rich vein for IS research is that of enterprise systems. These have largely taken over from BPR as the fashionable large-organization approach to integrated information systems and associated organizational change. There has been an enormous growth in the sales of such systems in the 1990s, with the market leader being SAP at the time of writing. Davenport (1998) quoted estimates of US $10 billion per year being spent on enterprise systems in 1997/98, which he argued could probably be doubled if consulting expenditures were added. Davenport described "success stories" at various companies such as Elf Atochem, but he also outlined "horror stories" at Mobil Europe, Dell Computers, and Dow Chemicals among others. Davenport provides no detail on these stories, however, and IS researchers could have much to offer in providing such detail. An example of a study which moves further toward this level of detail is that by Hanseth and Braa (1998).

3.4 Interorganization

There has been much emphasis in recent years on the partnering of organizations and their interconnection in networks. These initiatives are designed, for example, to exploit synergies between partnering organizations, or to improve links with suppliers and customers. They are related to issues of global reach and the striving for increased efficiency and speed of response in rapidly changing markets. Interorganizational information systems (IOS) are often a fundamental part of partnering processes, or the connection of organizations in networks. These systems are increasingly carried over the medium of the Internet, as the scope, familiarity and ease of use of the Net increases.

An example of a specific IOS case study is that by Barrett and Walsham (1999) on the London Insurance Market, which is an important part of the UK general insurance industry, built up around Lloyd's of London. It is a network of hundreds of semi-autonomous players, including underwriting groups, brokerage firms and Market managers. Responding to the tough global competitive environment of the insurance industry in the 1990s, the Market has sought to develop and use IT to lower costs by streamlining business processes and to increase service quality and interorganizational efficiency in the Market. Some success was achieved in claims management and settlement systems, and in accounting systems. However, Barrett and Walsham describe how a key system, called the electronic placing system (EPS), encountered many problems and achieved only low levels of adoption and use.

The new EPS system was designed to replace the face-to-face and paper approach to the negotiation and agreement of insurance risk between underwriters and brokers with an electronic system for working at-a-distance. The resistance to the system in the Market can be considered as largely arising from concerns that the new system would undermine the established system of personal trust and relationships and result in a worsening of the competitive position of the London Market rather than the explicit goal of the reverse. Barrett and Walsham conclude that, while electronic trading offers opportunities for speed, efficiency, and the bridging of time and space, complex insurance risks need delicate and sophisticated negotiation, and asynchronous electronic media are not necessarily well-suited to many aspects of this. They also argue that the shift to an electronic placing system places a high value on the explicit representation of knowledge in electronic documents on the network and past information stored in computers. However, these imply uncertain shifts in power-knowledge relations between brokers and underwriters, and resistance to the EPS system could also be seen as reflecting participants' concern regarding their future autonomy, prestige, and control.

There is an extensive and growing literature on IOS (see, for example, Choudhury 1997; Holland 1995) and this can be seen to reflect increased efforts at global reach in the current period. However, there remains a shortage of critical in-depth case studies in this area and the provision of these can be added to our IS research agendas. A crucial area for these future studies is e-commerce or e-business. Despite the enormous hype and interest in the use of the Internet for business-to-business and business-to-consumer applications, it is hard to find in-depth case study material that adopts a critical stance, in the sense of examining the benefits and weaknesses of electronic trading over the Net, viewed from a variety of stakeholder perspectives. The IS research community has a major potential contribution to make in this area.

3.5 Society

Most of the developments in information technology have taken place in Western countries, which themselves display a heterogeneity of culture that is relatively unexplored in the IS literature. But the so-called developing countries are normally even further removed from the contexts in which a particular technology was developed. The mainstream IS community is notable for its almost total disregard of issues of technology transfer and implementation in Third World contexts. However, there is a small body of literature that addresses these issues.

Lind (1991) explored the subject in some depth, using a detailed example of the introduction of a computerized production control system, Copics, into an Egyptian car manufacturing company, Nasco. His basic conclusion was that the model of a production system implicit in Copics was a poor match with the actual functioning of the production system in Nasco, where the rationality was based on Egyptian norms and values. Specific examples of the mismatch included the assumption that users of the computer system will react instantly to problems which arise in their area and will take action to correct things. However, middle-level supervisors and managers were not given the responsibility to do this. Lind argues that systems like Copics are designed around a Western approach to decision making with respect to delegation of authority, whereas in Egypt, decision making tends to be highly centralized. A second example concerns the more uncertain external environment of an Egyptian company, factors that were not present in Copics. These included the availability of imported material, great fluctuations in lead times, and changes imposed by government. Lind summarizes his conclusions as follows:

> This book is an attempt to point out how computer programs, developed in the more advanced industrialized countries and based on models and conceptions of reality that are prevailing in these countries, tend to be inappropriate under different conditions in developing countries. The reason, so it is argued in the book, is that models do not have the same explanation value in different cultures. [p. xiii]

Similar themes related to the use of IS in different cultural contexts have been developed by other researchers (see, for example, Malling 1998; Sahay and Walsham 1997). However, this area remains under-researched, doubtless because it is not perceived to affect the richer countries of the world in a direct way. However, there is some irony in this state of affairs, as noted by Castells (1997) in contrasting the age of globalization with the narrow ethnocentricity of much social science:

> There is in this book a deliberate obsession with multiculturalism.... This approach stems from my view that the process of techno-economic globalization shaping our world is being challenged, and will eventually be transformed, from a multiplicity of sources, according to different cultures, histories and geographies....I would like also...to break the ethnocentric approach still dominating much social science at the very moment when our societies have become globally interconnected and culturally intertwined. [p. 3]

Tricker (1999) supported this anti-ethnocentric approach in the specific domain of IS research and practice. He argued that, as the effects of information management become global, the cultural dimension is not just an interesting attribute of IS development, but is of fundamental significance to its effectiveness.

4. Conclusions

This paper has major limitations. The description of the literature on globalization is very selective and brief, as is the attempt to link globalization themes to the five levels of analysis. The paper has not done full justice to the exemplary papers cited, having reduced them to a few paragraphs, and not all streams of IS research are represented. In addition, splitting the IS literature into levels of analysis ignores the fact that the levels are inextricably inter-linked. For example, it is not possible to study shifting personal identity without taking some account of the groups, organizations and society to which the individual belongs. Despite these clear limitations, it is hoped that the main purpose of defining some future research agendas has been achieved. Three general points about these agendas will now be made in conclusion.

First, in order to investigate IS that enable connectivity but support diversity, it is necessary to study *particular* individuals, groups, organizations, or societies in detail, and in context. Large scale surveys or laboratory experiments may yield interesting data on occasions, but any attempt to study inherent variety, as outlined in the proposed research agendas in this paper, requires methods that go into depth in particular situations. There are a range of ways of approaching this, such as interpretive case studies, ethnographies, or action research projects, but we need to see more of these as a proportion of the totality of IS research in the future.

A second general point is that the IS academic community should focus future research agendas on what is important at the time, and not on some static definition of the "discipline." For example, IS researchers should engage fully with e-commerce at the present time, even though this area cuts across other disciplines such as marketing and strategy. This is not an argument based on simple expediency or short-term fashion. Rather, it is saying that we should engage with the important organizational issues of the current period, bringing to bear what we have learned from earlier topics, but also looking for what is new and different. For example, we learned a lot about groupware in the 1990s that is relevant to the study of intranets at the current time, but the two research domains are not synonymous.

A final point to end the paper concerns the most striking omission in published IS research in the context of globalization, namely the shortage of serious work on IT in non-Western cultures. In addition, some of the sparse literature in this domain takes culture as an impediment to Western-style "progress," rather than trying to see merit in other cultures' ways of seeing and acting in the world. An understanding of IT in cultures other than one's own can be justified on economic grounds alone, in terms of increased understanding of foreign markets or supply chains, for example. However, a deeper ethical rationale is that we should be concerned with the world as a whole, and not just some subset within which we happen to be located through accident of birth or background.

References

Appadurai, A. *Modernity at Large: Cultural Dimensions of Globalization.* New Delhi: Oxford University Press, 1997.

Barrett, M., and Walsham, G. "Electronic Trading and Work Transformation in the London Insurance Market," *Information Systems Research* (10:1), 1999, pp. 1-22.

Beck, U. *Risk Society: Towards a New Modernity.* London: Sage Publications, 1992.

Bloomfield, B. P., and McLean, C. "Madness and Organization: Informed Management and Empowerment," in *Information Technology and Changes in Organizational Work*, W. J. Orlikowski, G. Walsham, M. R. Jones, and J. I. DeGross (eds.). London: Chapman & Hall, 1996, pp. 371-393.

Castells, M. *The Rise of the Network Society.* Oxford: Blackwell Publishers, 1996.

Castells, M. *The Power of Identity.* Oxford: Blackwell Publishers, 1997.

Ciborra, C. (ed.). *Groupware and Teamwork: Invisible Aid or Technical Hindrance?* Chichester, UK: Wiley, 1996.

Ciborra, C., and Patriotta, G. "Groupware and Teamwork in New Product Development: The Case of a Consumer Goods Multinational," in *Groupware and Teamwork: Invisible Aid or Technical Hindrance?,* C. Ciborra (ed.). Chichester, UK: Wiley, 1996, pp. 121-142.

Choudhury, V. "Strategic Choices in the Development of Interorganizational Information Systems," *Information Systems Research* (8:1), 1997, pp. 1-24.

Davenport, T. H. "Putting the Enterprise into the Enterprise System," *Harvard Business Review,* January-February 1998, pp. 121-131.

Doremus, P. N., Keller, W. W., Pauly, L. W., and Reich, S. *The Myth of the Global Corporation.* Princeton, NJ: Princeton University Press, 1998.

Giddens, A. *The Consequences of Modernity.* Cambridge, UK: Polity Press, 1990.

Giddens, A. *Modernity and Self-Identity.* Cambridge, UK: Polity Press, 1991.

Giddens, A. Reith Lectures, 1999 (http://news.bbc.co.uk/hi/english/static/events/ reith_99).

Gopal, A. "Information Technology and Globalization: Exploring the Underbelly," in *Proceedings of Workshop on Understanding Information Technology, Globalization, and Changes in the Nature of Work*, University of Alberta, Edmonton, 1997, pp. 135-144.

Hanseth, O., and Braa, K. "Technology as Traitor: Emergent SAP Infrastructure in a Global Organization," in *Proceedings of the Nineteenth International Conference on Information Systems*, R. Hirschheim, M. Newman, and J. I. DeGross (eds.), Helsinki, Finland, 1998, pp. 188-196.

Holland, C. P. "Cooperative Supply Chain Management: The Impact of Interorganizational Systems," *Journal of Strategic Information Systems* (4:2), 1995, pp. 117-133.

Karsten, H. "Collaboration and Collaborative Information Technology: What is the Nature of Their Relationship?" in *Information Systems: Current Issues and Future Changes*, T. J. Larsen, L. Levine, and J. I. DeGross (eds.). Laxenburg, Austria: IFIP, 1999, pp. 231-254.

Kling, R. "Reading 'All About' Computerization: How Genre Conventions Shape Nonfiction Social Analysis," *The Information Society* (10:3), 1994, pp. 147-172.

Knights, D., and McCabe, D. "When 'Life is but a Dream': Obliterating Politics through Business Process Reengineering," *Human Relations* (51:6), 1998a, pp. 761-798.

Knights, D., and McCabe, D. "'What Happens When the Phone Goes Wild?': Staff, Stress and Spaces for Escape in a BPR Telephone Banking Regime," *Journal of Management Studies* (35:2), 1998b, pp. 163-194.

Kogut, B. "What Makes a Company Global," *Harvard Business Review*, January-February 1999, pp. 165-170.

Lind, P. *Computerization in Developing Countries: Model and Reality.* London: Routledge, 1991.

Malling, P. "Information Systems and Human Activity in Nepal," in *Implementation and Evaluation of Information Systems in Developing Countries*, C. Avgerou (ed.). London: London School of Economics, 1998, pp. 120-128.

Orlikowski, W. J. "Integrated Information Environment or Matrix of Control? The Contradictory Implications of Information Technology," *Accounting, Management and Information Technologies* (1:1), 1991, pp. 9-42.

Orlikowski, W. J. "Evolving with Notes: Organizational Change Around Groupware Technology," in *Groupware and Teamwork: Invisible Aid or Technical Hindrance?*, C. Ciborra (ed.). Chichester, UK: Wiley, 1996, pp. 23-59.

Robertson, R. *Globalization: Social Theory and Global Culture*. London: Sage Publications, 1992.

Sahay, S., and Walsham, G. "Social Structure and Managerial Agency in India," *Organization Studies* (18:3), 1997, pp. 415-444.

Schultze, U., and Boland, R. J. "Constructing High Tech Space: Mind, Body and Place in Knowledge Work," in *Proceedings of the AOS Conference on Accounting, Time and Space*, H. K. Rasmussen (ed.). Copenhagen: Copenhagen Business School, 1997, pp. 539-558.

Tricker, R. I. "The Cultural Context of Information Management," in *Rethinking Management Information Systems*, W. L. Currie and R. D. Galliers (eds). Oxford: Oxford University Press, 1999, pp. 393-416.

Walsham, G. *Interpreting Information Systems in Organizations*. Chichester, UK: Wiley, 1993.

Walsham, G. "IT and Changing Professional Identity: Micro-Studies and Macro-Theory," *Journal of the American Society for Information Science* (49:12), 1998, pp. 1081-1089.

Willmott, H. "Business Process Re-engineering and Human Resource Management," *Personnel Review* (23:3), 1994, pp. 34-46.

Zuboff, S. *In the Age of the Smart Machine*. New York: Basic Books, 1988.

About the Author

Geoff Walsham is a Research Professor of Management Studies at the Judge Institute of Management Studies, Cambridge University, UK. His teaching and research is centered on the development, management, and use of computer-based information systems, and the relationship of information and communication technologies to stability and change in organizations and societies. He is particularly interested in the human consequences of computerization in a global context, including both industrialized and developing countries. His publications include *Interpreting Information Systems in Organizations* (Wiley 1993) and *Information Technology and Changes in Organizational Work* (edited with Orlikowski, Jones, and DeGross, Chapman & Hall, 1996). Geoff can be reached by e-mail at g.walsham@jims.cam.ac.uk.

Part 5:

Reformation of Conceptualizations

14 STUDYING ORGANIZATIONAL COMPUTING INFRASTRUCTURES: MULTI-METHOD APPROACHES

Steve Sawyer
Pennsylvania State University
U.S.A.

Abstract

This paper provides guidelines for developing multi-method research approaches, provides several examples of their use, and discusses experiences with conducting a multi-method study of one organization's computing infrastructure changes. The focus on organizational computing infrastructures is due to the contemporary belief that these are increasingly critical to organizational success. However, understanding the value of an organization's computing infrastructure is difficult. This is because of their uniqueness, pervasiveness, context-driven nature, temporality, the constant changes in underlying technologies, and the variety of their effects at multiple levels in the organization. These difficulties are especially pronounced in organizations with distributed computing environments because the dispersion of computing accentuates these effects.

Keywords: Organizational informatics, computing infrastructure, multi-method research, ERP, implementation, organizational change, fieldwork

1. Introduction

Computing infrastructures, and the collections of information technologies (IT) that form their fabric, are seen as increasingly critical to organizational success (Goodman and

Sproull 1989; Kling 1992; Yates and Van Maanen 1996). However, the contributions of a computing infrastructure to organizational success are difficult to assess, in part, because of the methodological challenges. This paper addresses two of those challenges. First, what are the methodological issues in researching organizational computing infrastructures? Second, how can we best address these issues?

An organization's computing infrastructure includes the hardware and software, the administrative roles and rules that support that collection, and the informal norms and behaviors that grow up around these IT and the rules governing their use[1] (Barley 1990; Kling 1987; Kling and Iacono 1984; Kling and Scacchi 1982). A computing infrastructure helps to connect the disparate parts of its host organization. This forms the basis of what Kling and Scacchi call the "web of computing." At a more abstract level, a computing infrastructure embodies the organization's information processing capabilities (Galbraith 1974; March and Simon 1958; Yates 1988).

There are at least two reasons why an organization's computing infrastructure is critical to its success. The first is that this infrastructure is increasingly embedded into the conduct of work (Bridges 1994; Sproull and Kiesler 1991; Star and Ruhleder 1996; Wigand, Picot, and Reichwald 1997). For example, help-desk workers require that the phone system, databases, and problem tracking programs are "up" in order for them to perform their work (Pentland 1992). The second reason an organization's computing infrastructure is seen as critical is the growing trend to view IT as levers of organizational strategy (i.e., Henderson and Venkatraman 1991).

There are at least two reasons why it is difficult to understand the roles a computing infrastructure plays in an organization. The first is our limited understanding of the effects of computing infrastructures on work (e.g., Johnson and Rice 1987; Kling 1995; Truex, Baskerville, and Klein 1999). This stems, in turn, from their context-dependent nature, their pervasiveness, and the unpredictable and varied nature of their effects at multiple levels in the organization (i.e., Burkhardt 1994; Manning 1996). The distribution of computing seems to accentuate these effects.

A second reason that makes it difficult to understand the role of a computing infrastructure is the rate at which core IT are changing. Examples include client/server computing, Internet-based technologies, the use of information appliances, and the pervasiveness of personal computing. The constant change in these IT mean that, at any given time, several components of a computing infrastructure may be in flux. Moreover, most computing infrastructures, and especially those that are in the process of distributing, are constructed using products provided by a changing array of hardware and software vendors. The issues with implementing enterprise resource systems (ERPs) are a contemporary example of this trend (Davenport 1998).

This paper continues in four parts. In the first, we address the first question by setting out methodological concerns with studying organizational computing infra-

---

[1]Viewing an organization's computing infrastructure as comprising technical, administrative, and social aspects broadens the view provided by alternative conceptualizations. For example, the socio-technical (Bostrom and Heinen 1978a, 1978b; Fairhurst, Gren, and Courtright 1995; Holsapple and Luo 1995; Shulman, Penman, and Sless 1990); and soft-systems approach (Checkland and Scholes 1990) also highlight that computing infrastructures are more than the information technologies that form their technical core. However, aggregating both the formal rules and procedures and the norms and behaviors of use limits the interplay of the two non-technology-centered aspects of infrastructure and obscures the roles these forces play in enmeshing computing infrastructures into their host organization (Burkhardt 1994; Kling 1995).

structures. The second presents the first steps to addressing the second question by presenting a discussion of issues regarding multiple method research design. The third presents three examples of multiple method approaches to studying organizational computing infrastructures. In the fourth, a summary comparison of the three case studies is presented and the methodological issues that arise from this comparison are discussed.

2. Organizational Computing Infrastructures

As outlined below, understanding an organization's computing infrastructure demands a research approach that can accommodate the unique, pervasive, multi-level/multi-effect, and time-variant nature of these systems.

2.1 Uniqueness

Sproull and Goodman (1989, p. 255) argue that each organization's computing infrastructure (which they call a technical system) is unique since common technical components are woven into an organization's administrative and social fabric. This uniqueness helps to account for some of the variations across organizations using identical IT. Two examples are the variations in the use of Lotus Notes across departments of a large consulting firm (Orlikowski 1993) and the use of the same technology at Zeta Corp. (Orlikowski 1996). Other examples include the variations in use of computer-aided software engineering (CASE) tools among software development teams (Guinan, Cooprider and Sawyer 1997) and in the use of computer-aided design (CAD) (Collins and King 1988; Kelley 1990; Salzman 1989).

2.2 Pervasiveness

By pervasive we mean that the organization's computing infrastructure directly or indirectly affects each member of that organization. With the dispersion of computing (via networks), this reach is more expansive and direct than at any other time in the history of computing. This proliferation of computing increases both the number and distribution of people who support this infrastructure. Indeed the distinction between other workers and technologists (those who exist to serve the computing needs of the organization) is blurring (Wigand, Picot, and Reichwald 1997). For example, Gasser (1986) discusses the way in which new computing technologies were incorporated into work at two firms. Zuboff (1988) described the increasing integration of computing into work as "informating." Salzman (1989) discusses how CAD system use helped to reshape engineering design work by blurring the distinction between technician and engineer. Laudon and Marr (1995) found that increased IT use has helped reshape occupational structures in the U.S. federal government.

2.3 Multi-level/Multi-effect

Because an organization's computing infrastructure is both pervasive and unique, its uses differ based on purpose. For example, lower-level workers interact with IT in their jobs as a matter of course, thus shaping its functions through use (Clement 1994; Kraut, Dumais, and Koch 1987; Manning 1996; Zuboff 1988). Managers plan and act around and with its assistance (Orlikowski 1993). Senior managers actively try to shape and direct its alignment with their strategy (Henderson and Venkatraman 1991).

Variations in IT use across the levels of an organization imply that the effects will also vary. For example, Zuboff describes "panoptic power," where senior managers can exert more control over work even as the workers have more flexibility to make decisions (see also Wilson 1995). Koppell (1994), in his study of a hospital's use of a new integrated computing system, finds that doctors, nurses, and orderlies each used the system in different ways.

2.4 Temporality

Organizations vary over time and the time variance of an organization's computing infrastructure is reflected in every aspect of its design. There are ongoing changes in base IT, steady adaptation of the rules and the roles played by people who shape its use, and constant evolution of social norms surrounding its use. This temporality is a form of organizational homeostasis—the effort to maintain the status quo over time (Lee 1999). Note that this conceptualization of an organization reflects only a dynamic equilibrium and not a statement of success or failure (March and Simon 1958).[2] Over time, the computing infrastructure's form emerges through its uses (Goodman and Sproull 1989; Markus and Robey 1988; Truex, Baskerville, and Klein 1999). Current literature advocates IT as levers to alter an organization's homeostasis (Benjamin and Levinson 1993; Davenport 1998).

2.5 Methodologic Ramifications

This set of forces suggests that the most useful research methods for understanding computing infrastructures should span levels of analysis and explicitly address the ways in which computing infrastructures are entwined within their host organization. Not surprisingly, research into computing infrastructures has generally employed multiple methods of data collection (e.g., Adler 1995; Barley 1986; Burkhardt 1994; Gasser 1986; Kelley 1990; Kling and Scacchi 1982; Koppel 1994; Ruhleder 1995; Salzman 1989). /These authors represent a range of disciplines and conceptual perspectives. Each

[2]Concepts such as organizational homeostasis and dynamic equilibrium reflect the ongoing debate between structuralism and functionalism. In this paper, the focus is not on the debate. Rather, it serves as a background and the author's position regarding this debate is implied by the use of the homeostasis/dynamic equilibrium.

provides a rich set of descriptions regarding the roles and issues with various aspects and elements of the computing infrastructures in the organizations that they study. However, discussions of how the authors developed and used their multiple methods is not as extensive. While this is understandable, as these papers report empirical findings and do not specifically focus on methods, it leaves the researcher interested in studying organizational computing infrastructures and/or conducting multi-method research with little explicit guidance.[3]

3. Defining Multi-method Research

A multi-method approach to social science research involves the use of several data collection techniques in an organized manner to provide multiple data sets regarding the same phenomena. This is typically done by drawing on a set of data collection methods that accommodate each other's limitations (Gallivan 1997; Jick 1979). Since both the conceptual bases and data collection techniques help to shape the phenomena of interest, there are many ways to conduct multi-method research (Brewer and Hunter 1989). This paper focuses on "multi-method fieldwork": blending fieldwork with surveys.

Fieldwork includes participant observation, interviewing, and the collection of archival records—characteristics of both intensive and prolonged involvement with the social units being studied. Surveys involve data collection instruments (often self-administered) to collect responses to *a priori* formalized questions on predetermined topics from a valid sample of members of identified social categories. Surveys are a mainstay of the quasi-experimental field research tradition on which most IS research is based; survey-based studies comprise more than 49% of research done on the use of IT in organizations (Orlikowski and Baroudi 1991). Explicit multi-method studies represent about 3% of the same research base (Gallivan 1997).

A multi-method fieldwork approach uses surveys in a manner that differs from traditional quasi-experimental field research. In the latter, survey data are extracted from the field and quantified. Non-survey data are used to support or enrich findings from survey data. A multi-method approach sees the two forms of data collection as intertwined. That is, each data collection method must both stand on its own and also be combinable (typically called triangulation) (Brewer and Hunter 1989; Gallivan 1997; Jick 1979; Kaplan and Duchon 1988).

There are few common conventions to describe the process of triangulating data (Howe and Eisenhardt 1990; Jick 1979; Williams 1986). (One exception is the multi-trait/multi-method matrix described by Campbell and Stanley 1966.) This often limits the

[3]Fieldwork is one of the many qualitative or intensive methods being used to study information technologies in organizations (e.g., Avison and Myers 1995; Kling 1980; Lee 1989; Lee and Markus 1995; Markus 1983; Markus and Robey 1988; Walsham 1995), although this approach is not without criticism (e.g., Sandstrom and Sandstrom 1995). There are also special issues of leading journals (e.g., *Administrative Science Quarterly* in 1979 and *Organization Science* in 1990) describing approaches to fieldwork methods for organizational research in general. Edited volumes by Mumford et al. (1984), Nissen, Klein, and Hirschheim (1991), and Lee, Liebenau, and DeGross (1997) provide additional direction for researchers conducting fieldwork on the use of IT in organizations. Electronic sources are also avaialble (i.e., http://comu2.auckland.nc.nz/~isworld/quality.htm).

value of this type of research to the broader community because describing the methods used in triangulating is both space-consuming and important for establishing credibility (Lincoln 1995; Sieber 1973; Sutton and Staw 1995). The need to write extensively about methods typically comes at the cost of reduced space devoted to discussing findings.

Both fieldwork and the more common survey-based approaches have strong ideological bases. A multi-method approach that combines surveys with fieldwork seeks to integrate these perspectives. The result is a new method, not just an aggregation of styles (Brewer and Hunter 1989, p. 17). For example, Kaplan and Duchon, (1988), Trauth and O'Connor (1991), and Wynekoop, (1992) collected data on various IT uses in organizations using both observation/ interviews and surveys. In this way, the studies draw on the strengths of the combination of data collection methods (Gallivan 1997; Jick 1979).

4. Conducting Multi-method Research

Zuboff (1988, p. 423) began the discussion of her methodology by writing: "behind every method is a belief." The belief expressed here is that multi-method fieldwork is well-suited to research on organizational computing infrastructures. This belief stems from the fact that the multi-method approach is typically longitudinal and builds on interpreting data collected from observations, interviews, archival records, and surveys. Analysis draws on different combinations of data sets that provide different perspectives on the same phenomena. In a multi-method fieldwork approach, new events demand re-analysis over time. This implies that an understanding of the data develops from a constant reinterpretation of existing data/observations in the new light of unfolding events (Jackson 1987; Van Maanen 1988).

Multi-method research is guided by a number of factors such as validity and generalizability. These factors are well documented in other work and not discussed here (see Brinberg and McGrath 1984; Creswell 1994; Danzin 1970). There are additional factors specific to multi-method research: the role of theory, method independence, insulation, data interdependence, analytic integration, and data comparability versus contrast (Brewer and Hunter 1989). These are defined in Table 1 and discussed below.

4.1 The Role of Theory

Central to most research is the development and/or testing of theory (Blalock 1971; Popper 1968). Contributions toward theory can be seen along a continuum from development to testing (Bagozzi 1979; Blalock 1971; Glaser and Strauss 1967; Hoyle 1995; Sutton and Staw 1995; Yin 1989). However, theory development and testing are often not done well in organizational research, a subset of which is research on organizational computing infrastructures (Merton 1967; Sutton and Staw 1995; Weick 1995). For example, Grunow (1995) reported on the methodological and theoretic approaches to 303 papers concerning organizational research. He found that 78% of these papers did not align theory to their research questions and 82% of the papers could not contribute meaningful results to theory development. As Sutton and Staw (1995, p. 371) state: "references, data, variables, diagrams and hypotheses are not theory."

Table 1. Six Issues with Multi-method Data Collection (from Brewer and Hunter 1989)

Role of theory	Used to guide the study and establish relationships between multiple data sets
Independence	The effect of one data collection method on another
Insulation	Exposing subjects to effects of multiple waves of data collection
Interdependence	Providing for intentional links between data
Integration	Combined analysis of multiple data sets
Comparability v. Contrast	Analysis highlights differences caused by different type of data and can lead to incongruities in analysis

Even if Sutton and Staw are correct and Grunow's analysis is accurate, some form of theoretical rationale still forms the basis of the analysis and discussion sections of most scholarly papers on IT use in organizations. For instance, papers reporting theory development provide extensive rationales. Typically, this rationale is based on several sources of evidence (see Eisenhardt 1989). For example, Zuboff approaches her fieldwork using a grounded theory approach across her series of cases (see Glaser and Strauss 1967). In his fieldwork, Barley (1986) draws on structuration and compares findings from two sites over time. Markus (1983) uses three theoretic perspectives to reflect on the data drawn from a case study of one implementation, arguing that one perspective best fits the data. Orlikowski (1993), however, uses *a priori* theory as a basis for her work.

One benefit, and the primary differentiator, of multi-method research is multiple data sets. This suggests that theory is both the source of guidance and a means for uniting the various data collection approaches. While some authors protest against using formal theory (see Van Maanen 1995a; 1995b), theory serves as a stabilizing force for multi-method-based research. A theory-based approach helps sort through the blur of reality, providing a way to characterize observation and interpretation (Weizenbaum 1976). Vaughan (1992) calls this "theory elaboration" and Weick (1995, p. 385) calls it "theorizing." Hence, using theory does not preclude description, nor does it demand prescription.

4.2 Dependence and Insulation: Collecting Multiple Data Sets

Multiple collection methods draw data from the same subjects. These subjects interact with the researcher(s) in several ways over the course of data collection (as participants in interviews, as subjects of observation, and as respondents to surveys). This means that the survey effort is also related to the interview effort in that they are typically done by the same researcher(s). Hence, researchers must consider how their *total* presence will affect subjects prior to entering the field.

A second issue concerns the manner in which the data will be collected when the observer is one of the instruments of data collection. Inherent to multi-method fieldwork is the disturbance caused by the (multiple) presence of the researcher(s). Approaches to reducing this disturbance lie on a continuum from unobtrusive observation to direct participation (i.e., action research) (Argyris, Putnam, and Smith,1985). Researchers can also strive to be unobtrusive and focus on constantly mitigating their role in the surroundings (Barley 1990). Barley writes: "with time, I believe I came to be seen as a harmless, but perhaps eccentric, individual without factional loyalties" (p. 243). Still, as he writes, his acceptance may have depended on a madcap drive through Boston with a technician from one of the hospitals to help them cover for a critical supply shortage.

Pettigrew (1990) takes a more participant-oriented approach. He writes of "an open and reciprocal relationship between the researchers and their host organizations," stating that "running a research in action workshop is not only a sound instrumental act ... it is also a clear sign of respect for your new partner" (p. 286). He goes on to detail the manner and substance of the reciprocity in activities such as status meetings, workshops, project reports, and deliverables.

Active participant/observers must also attend to ethical issues. Schein (1990, p. 204) says of intervention that "we should not do this lightly, and we should have a clear picture of what our motivation is when we do it." Regarding the role(s) of the researcher as a participant/observer, Schein (1990, p. 207) writes simply: "Be careful." Being involved in the outcomes of an organizational process is a tremendous responsibility for the researcher (e.g., Schön 1983).

Mintzberg (1979) left open the role of the observer, contending that each field researcher is forced to tune their use of participation and observation to sites in which they are involved. Given this advice, some contextual information helps to explain the variance in approaches. For example, Barley studied work-level changes brought about by a new imaging system while Pettigrew's emphasis was to understand organizational change from a managerial perspective. Schein's focus is on describing leadership's role in affecting change. Still, the role of the participant observer remains a research decision open to interpretation. The fieldworker is an interpreter both of the events seen and roles played. Each interaction between the researcher and the host site requires answering questions about the level of involvement such as: What is the appropriate role of the researcher? For what purpose? To what effect? How does this relate to the observations? The various levels of an organization can serve as insulation, allowing the researcher to have different roles at different levels.

Another concern with dependence in multi-method fieldwork involves determining when data collection begins and ends. Case research is often post hoc, making this a relevant but not critical question (Benbassat, Goldstein, and Mead 1987; Glick et al. 1990; Lee 1989; Yin 1989). However, this is a major consideration for research conducted in "real time." Since most organizational changes have many antecedents, no matter when data collection begins there is a history. This history may be relevant to the decision about when to begin data collection. Equally difficult is the decision to end data collection: when is something "over"? Dubinskas (1988) argues against work where "it is assumed that we really can peg all events conveniently to that same time line" (p. 8). Barley (1988) suggests the researcher define these dates within the context of the study. Pacanowsky (1988) asserts the starting and ending points are arbitrary. If the researcher seeks to understand issues or events in an organizational setting, then the research is likely to start after the issue "began" and finish before the issue "ends."

4.3 Integration as Comparison or Contrast: Conducting Multi-method Analysis

Analyzing multiple data sets that are focused on a common phenomenon often leads to paradoxical results. That is, findings drawn from these different data sets may be contradictory (Jick 1979). These contradictions represent the potential for new learning or for exposing methodological flaws (Robey 1995). Developing such findings suggests the value of interim analyses to help define differences among data sets (Miles 1979).

In this integration, interviews and observations are often seen as a rich source of interesting data. However, the volume of data collected can make analysis seem overwhelming. This raises the question of how to characterize the interpretations that arise from the data. Mintzberg (1979) argues for an inductive approach. Miles (1979), drawing on his own work, suggests four points: intertwining analysis and data collection, formulating classes of phenomena, identifying themes, and provisional testing of hypotheses. This implies the importance of iteration as the basis to both formulating the classes of phenomena and identifying themes. This sets up the ongoing testing of interim hypotheses.

Miles' four points are often modified to fit the particular needs of the researcher/ effort. For example, both Zuboff (1988) and Leonard-Barton (1990) do the interim analysis between cases, using an early case to establish the classes and themes that are modified by data obtained in later cases. Pacanowsky used ongoing reflection during his data collection. Then, following his stay at Gore-Tex, he rebuilt his analysis. Bogdan (1972, p. 59) argues for balancing participant observation with breaks from the field. Barley (1986, 1990) used a seven-week break from fieldwork to help make sense of what he was seeing. However, Glick et al. developed their research almost completely on post hoc analysis.

These examples illustrate the variations in the conduct of multi-method fieldwork. Being close to the data, as field work demands, often prevents the researcher from stepping back to see the, often quiet, evolutions of contemporary organizations and their computing infrastructures. Too much distance and the researcher is no longer able to view the flow of life—a central aspect of fieldwork (Geertz 1973).

This tension reflects a second issue: the multi-method researcher must be both close to and distant from the data. To make sense of mixed forms of field data, ongoing analysis is critical and deeply reflective analysis is demanded. Prescriptive analytic techniques—available in more traditional experimental and quasi-experimental data analysis (e.g., Pedhauzer and Schmelkin 1991)—are not as well developed for qualitative analysis (e.g., Miles and Huberman 1994). Still, there is some guidance. For example, using explanatory matrices—where issues form one axis, sources form the other, and supporting data fill the intersecting cells—is one flexible technique (Miles and Huberman

1994). Another technique is to build evidence chains, where an issue is stated and then the supporting evidence is laid out. Both of these imply immersion in the data to develop ways to categorize the corpus of data and to extract the relevant segments.

5. Assessing Multi-method Research Examples

The discussion in this part centers on three examples of multi-method studies on computing in organizations (see Table 2 for a summary).[4]

5.1 The Kaplan and Duchon Study of a Clinical Information System

Kaplan and Duchon (1988) report on the implementation of a clinical information system in nine laboratories of a large, urban, teaching hospital. The study team was comprised of information systems and organizational behavioral researchers. They pursued a set of questions concerning what happens when a new system is implemented and the effects of this implementation on work. The research team relied on interviews, participant observation, and surveys to collect data. However, the original theories guiding the research were so disparate that the first rounds of analysis revealed both a lack of clarity and the difficulty in triangulating among the sets of data. Theory generation emerged as the conceptual bond since the lack of coherence in the original approach led members of the team to build theory out of the data.

During data collection, the undiscovered differences with the *a priori* theoretic bases meant that the dependence among the data collection approaches varied. For example, some of the researchers did cursory interviewing with the goal of using the data solely to build surveys. Others supported their interviews with extensive note-taking and descriptive field notes. Furthermore, while the survey was developed by the team, drawing from both standard scales and context-specific questions generated from the data collected by interview and observation, all participant observation was done by one person

Two independent groups analyzed the data sets. The survey data were analyzed by part of the team, the rest by another part of the team. Findings from the survey were considered uninteresting relative to the *a priori* theory and this led to part of the team leaving the project. However, the qualitative data revealed several issues for the remaining two team members. After much discussion, Kaplan convinced Duchon to re-analyze the survey data in light of the findings from the interviews and observation. This re-analysis led to significant findings and the combination of analyses across several sets of data led to theory generation—the job orientation model. Thus, theory generation turned out to be the "conceptual glue" that provided the unifying tie between the disparate data sets. And the multi-method approach led to unexpected findings.

---

[4]An author from each of these studies spoke during a panel on the use of multiple methods at the 1997 International Conference on Information Systems. That presentation and subsequent extended informal discussions with the authors of these studies form the basis of the presentation of their studies in this section.

Table 2. Contemporary Multi-method Studies

	Kaplan and Duchon (1988)	Guinan, Cooprider, and Sawyer (1997)	The MSU Study[a]
Role of theory	Originally explicit, ended up as a grounded theory approach.	Extensive *a priori* integrative model.	Three interrelated *a priori* theories as interpretive framework.
Independence	Based on who conducted the data collection and by time.	Both theoretically-based chronology and using multiple researchers.	Multiple phases combined with separation of data collection by both levels of analysis and specific roles for each researcher.
Insulation	Time, the use of multiple researchers.	Time sequencing. The use of multiple researchers.	Time sequencing, the use of multiple researchers, and disparate data collection methods.
Interdependence	Theory-connected.	*A priori* links specified and incorporated into data collection instruments. This constrained adaptation over the course of the project.	Theory-connected. Seen as secondary to interdependence.
Integration	Ad hoc use based on role of grounded theory. Driven by qualitative data.	Theory driven. Problematic with the *a priori* theory did not hold up under the rigors of longitudinal research.	Both theory and data-driven approaches. Reliance on interim analyses. Current work suggests that this is difficult.
Comparability	Driven by use of grounded theory.	Driven by *a priori* linkages between theories.	Driven by on-going analysis of data. Using multiple analysis techniques.

[a]See Sawyer (forthcoming) and Sawyer and Southwick (1996, 1997).

5.2 The Guinan, Cooprider, and Sawyer Study of CASE Tool Use

Guinan, Cooprider, and Sawyer (1997) report on a four year study on the use of CASE tools by 100 software development teams at 22 sites of 15 companies. The study was conducted by 15 faculty and/or doctoral students. Like Kaplan and Duchon, they had multiple research objectives. Moreover, they built an extensive, normative, and multi-faceted *a priori* theoretical model that crossed levels of analysis and incorporated multiple theories. This conceptual structure framed both the survey designs and the interviews.

Data collection involved extensive surveying, done four times for each project team as it moved from initiation through implementation. This required using 28 instruments to gather data from more than 2,000 respondents. Survey data collection was supported by interviews of participants and other data collection efforts such as independently assessing development metrics and collecting documentation regarding the development effort. This led to amassing more than 500,000 data points. Because of the scope and level of effort, both academic and industrial advisory boards provided oversight and met frequently.

The *a priori* theoretical model served both as a means to organize data collection and as a time line that highlighted both the intended populations and the links between the data collection methods. The research team was both trained and managed to assure that this plan would be followed. However, the ability to adapt to issues that arose in the course of the data collection (where the time with each team spanned about 18 months) was constrained by the prescriptive nature of the *a priori* theory and the size of the data collection effort.

The extensive data sets, the large number of participants, the difficulty with incorporating adaptations along the data collection process and the multiplicity of interests, perspectives, and needs, made analysis difficult. That is, the large scope of the effort made it hard to provide concise explanations while a partial focus on sub-sets of the data often limited the discussion regarding important inter-relationships. For example, Guinan, Cooprider, and Faraj (1998) focus on the survey data and do not explicitly draw on the field work data.

5.3 The ERP Implementation at Medium Sized University

The third example draws from an ongoing study of the computing changes at one organization. From 1994 to 1999 a medium-sized university (MSU) installed both a client/server-based computing infrastructure and an enterprise resource package (ERP) to replace their mainframe computing infrastructure and proprietary, standalone administrative information systems. The goal of the research has been to identify how changes to MSU's computing infrastructure are manifested in the technical, social, and administrative structures of the organization. (See Sawyer forthcoming; Sawyer and Southwick 1996, 1997).

The research was designed to build on the factors listed in Table 1. For example three *a priori* theories formed the interpretive frame of the research (Sutton and Staw 1995). These theories overlap at various levels of analysis and provide a means both to integrate various data sets and to analyze multiple data sets. To address independence and insulation issues, the research was planned in four phases. This allows for data collection to be separated by level of analysis and provides for some control over selection of participants. Further, this phased approach allowed for multiple researchers to have distinct roles in each phase. And, since the work is longitudinal, both levels of analysis and temporal distance insulate data collection. The research design also emphasized data interdependence over independence and insulation for two reasons. Thus, data collection methods were sequenced to help gather data first, then adjusted to deal with method independence and insulation issues.

6. Observations from Comparing the Examples

Several observations arise from comparing these examples. First, in all three examples, the research team included multiple researchers. Typically, each researcher had a particular method expertise (and perspective). This suggests that integrating the data sets collected using the various methods also meant integrating the various researcher's perspectives. Kaplan and Duchon intimate the intensity of the struggle to negotiate and integrate the various research perspectives and interests of the research team. In the MSU study, one researcher leads the effort and works with various collaborators within the research framework. Thus, each research team member has a predefined role.

A second observation is that all three studies use *a priori* theory—albeit each in a different way. The Kaplan and Duchon study struggled with *a priori* theory because it made it difficult for researchers to interpret the data. The empirical basis of the Guinan, Cooprider, and Sawyer study reflects the difficulty with adjusting the theoretical basis to accommodate emerging issues and findings. The MSU study was explicitly developed as an interpretive study so the *a priori* theory serves as a guide.

The third observation concerns the struggle with conducting the analysis that each study team faced. For instance, regarding the first example, this struggle reached the point where continuation relied on one author convincing the other to keep working (see Kaplan and Duchon 1988, p 580). The second example suggests an ongoing struggle to find a jointly agreed-upon view of the large set of findings as is suggested by the broad coverage of the summary paper (see Guinan, Cooprider, and Faraj 1998; Guinan, Cooprider, and Sawyer 1997). The emerging issues from the MSU study suggest that the interpretive approach allows for flexibility (see Sawyer forthcoming; Sawyer and Southwick 1997).

6.1 Additional Observations from the MSU Research Experience

The ongoing experiences with conducting multi-method research at MSU raise three additional issues. The first issue is establishing and maintaining the research. For instance, the research team took nearly 11 months to learn more about the prestudy period and to gain the confidence of the organization's members (its senior management, the technologists, the line workers, and their managers). This confidence-building was done using both a series of one-page outlines, each directed to specific audiences, and multiple meetings with key stakeholders of each group. For example, the outline for senior managers included a promise to provide periodic feedback—where this feedback meant observation and not prescription. Line workers were promised that the research would document their issues and concerns in an open way—including sharing data with them on demand (as suggested by Barley 1990). Throughout the study, this relationship support has been a constant aspect of the fieldwork (see Pettigrew 1990).

A second observation from the MSU experience is the constant mixing of participation and observation In the field-based, participant/observer portions of data collection, the researcher becomes the central instrument of data collection. This is one factor in what Barley calls "becoming a research tool" (1990, p. 237). That means that conducting fieldwork makes the researcher inseparable from the data collection and analysis. This demands constant reflection and redefinition of the research team member's roles. For example, in the first phase of the MSU study, each research team member's role was to be unobtrusive and to focus on establishing the trust that is so important and difficult to define (see Barley 1990, p. 239; Bogdan 1972, p. 24). In the follow-on phases, research team member's roles differ. For the researcher observing the senior managers, a stance similar to Pettigrew is used. That is, the researcher is more active: interacting more with the participants. The researcher(s) observing the technologists approach the research more as Barley (1990) encourages: being distant but collegial, not taking sides, and being involved but not "going native." In the third phase, the researchers immersed themselves in several projects for the extent of that project. Finally, in the post-implementation phase, the members of the team doing the survey will have very little contact with the host site and the study's participants.

A third experience from the MSU study is the need to follow the implementation. For instance, the original plan for the MSU research was developed to focus on the levels of analysis across the period of implementation. It became clear almost immediately that this needed to be modified. The primary reason is that the computing infrastructure change at MSU is actually a series of projects (numbering nearly 40). Thus this is becoming a study of these projects, each of which can be viewed in the macro/micro way advocated by Wynekoop.

6.2 Multi-method Research on Organizational Computing Infrastructures

Since "what we know is always shaped by how we came to know it" (Brewer and Hunter 1989, p. 66), the purpose of using multiple data collection methods is to draw on the strengths of some to make up for the (often well-known) weaknesses of others (e.g., Campbell and Stanley 1966; Creswell 1994; Jackson 1987; Miller 1991; Miles and Huberman 1994).

Perhaps the limited discussion on the use of multi-method research for studying the roles of information technologies in organizations is due to the different sets of skills involved in presenting descriptive research (Van Maanen 1995a, 1995b). Perhaps it is due to the publication format, in which journals limit the length and, thus, the descriptive approach, of multi-method research (Yanow 1995). Perhaps this is also due to present expectations about the use of theory and the role of empirical data (Sutton and Staw 1995). Still, given the potential value of multi-method fieldwork-based research, it is important for such discussions to continue.

References

Adler, P. "Interdepartmental Interdependence and Coordination: The Case of Design/ Manufacturing Interface," *Organization Science* (62), 1995, pp. 147-167.

Argyris, C., Putnam, R., and Smith, D. *Action Science: Concepts, Methods, and Skills for Research and Intervention*. San Francisco: Jossey-Bass, 1985.

Avison, D., and Meyers, M. "Information Systems and Anthropology: An Anthropological Perspective on IT and Organizational Culture," *Information Technology & People* (8:3), 1995, pp.; 43-56.

Bagozzi, R. "The Role of Measurement in Theory Construction and Hypothesis Testing: Toward a Holistic Model," in *Conceptual and Theoretical Developments in Marketing*, O. Ferrell, S. Brown, and C. Lamb (eds.). Chicago: American Marketing Association, 1979.

Barley, S. "Images of Imaging: Notes on Doing Longitudinal Field Work," *Organization Science* (1:3), 1990, pp. 220-249.

Barley, S. "Technology as an Occasion for Structuring: Evidence from Observations of CT Scanners and the Social Order of Radiology Departments," *Administrative Science Quarterly* (31), 1986, pp. 78-108.

Barley, S. "On Technology, Time and the Social Order: Technically Induced Change in the Temporal Order of Radiological Work," in *Making Time: Ethnographies of High-Tech Organizations,* F. Dubinskas (ed.). Philadelphia: Temple University Press, 1988.

Benbasat, I., Goldstein, D., and Mead, M. "The Case Research Strategy in Studies of Information Systems," *MIS Quarterly* (11:3), 1987, pp. 369-386.

Benjamin, R., and Levinson, E. "A Framework for Managing IT-Enabled Change," *Sloan Management Review* (34:4), 1993, pp. 8-18.

Blalock, H. *Causal Models in the Social Sciences*. Chicago: Aldine, 1971.

Bogdan, R. *Participant Observation in Organizational Settings*. Syracuse: Syracuse Press, 1972.

Bostrom, R., and Heinen, S. "MIS Problems and Failures: A Socio-Technical Approach: Part 1: Causes." *MIS Quarterly* (1:4), 1978a, pp. 11-27.

Bostrom, R., and Heinen, S. "MIS Problems and Failures: A Socio-Technical Approach: Part 2: The Application of Socio-Technical Theory," *MIS Quarterly* (1:4), 1978b, pp. 17-32.

Brewer, J., and Hunter, A. *Multi-Method Research A Synthesis of Styles*. Newbury Park, CA: Sage, 1989.

Bridges, W. *JobShift: How to Prosper in a Workplace without Jobs*. Reading, MA: Addison Wesley, 1994.

Brinberg, D., and McGrath, J. *Validity and Research Process*. Beverly Hills, CA: Sage, 1984.

Burkhardt, M. "Social Interaction Effects Following a Technological Change: A Longitudinal Investigation," *Academy of Management Journal* (37:4), 1994, pp. 869-898.

Campbell, D., and Stanley, J. *Experimental and Quasi-Experimental Designs for Research*. Chicago: Rand-McNally, 1966.

Checkland, P., and Scholes, J. *Soft Systems Methodology in Action*. New York: John Wiley & Sons, 1990.

Clement, A. "Computing at Work: Empowering Action By 'Low-level Users,'" *Communications of the ACM* (37:1), 1994, pp. 52-63.

Collins, P., and King, D. "Implications of Computer-Aided Design for Work and Performance," *The Journal of Applied Behavioral Science* (12:2), 1988, pp. 173-190.

Creswell, J. *Research Design: Qualitative and Quantitative Approaches*. Thousand Oaks, CA: Sage, 1994.

Danzin, N. *The Research Act: A Theoretical Introduction to Sociological Methods*. Chicago: Aldine Publishing Company, 1970, pp. 297-313.

Davenport, T. "Putting the Enterprise intro the Enterprise System," *Harvard Business Review*, 1998, pp. 121-134.

Dubinskas, F. *Making Time: Ethnographies of High-Tech Organizations*. Philadelphia: Temple University Press, 1988.

Eisenhardt, K. "Building Theories from Case Study Research," *Academy of Management Review* (14:4), 1989, pp. 532-550.

Fairhurst, G., Green, S., and Courtright, J. "Inertial Forces and the Implementation of a Socio-Technical Systems Approach: A Communication Study," *Organization Science* (6:2), 1995, pp. 168-185.

Galbraith, J. "Organization Design: An Information Processing View," *Interfaces* (43), 1974, pp. 28-36.

Gallivan, M. "Value in Triangulation: A Comparison of Two Approaches for Combining Quantitative and Qualitative Methods," in *Qualitative Method in Information Systems*, A. Lee, J. Liebenau, and J. DeGross (eds). London: Chapman & Hall, 1997, pp. 83-107.

Gasser, L. "The Integration of Computing and Routine Work," *ACM Transactions on Office Information Systems* (43), 1986, pp. 205-225.

Geertz, C. *The Interpretation of Cultures*. New York: Basic Books, 1973.

Giddens, A., and Turner, J. *Social Theory Today*. Stanford, CA: Stanford University Press, 1987.

Glaser, B., and Strauss, A. *The Discovery of Grounded Theory*. New York: Aldine de Gruyter, 1967.

Glick, W., Huber, G., Miller, C., Doty, D., and Sutcliffe, K. "Studying Changes in Organizational Design and Effectiveness: Retrospective Event Histories and Periodic Assessments," *Organization Science* (1:3), 1990, pp. 293-312.

Goodman, P., and Sproull, L. *Technology and Organizations*. San Francisco: Jossey-Bass, 1989.

Grunow, D. "The Research Design in Organization Studies," *Organization Science* (6:1), 1995, pp. 93-103.

Guinan, P., Cooprider, J., and Sawyer, S. "The Effective Use of Automated Application Development Tools: A Four-Year Longitudinal Study of CASE," *IBM Systems Journal* (38), 1997, pp. 124-141.

Guinan, P., Cooprider, J., and Faraj, S. "Enabling Software Development Team Performance During Requirements Gathering: A Behavioral Versus Technical Approach," *Information Systems Research* (9:2), 1998, pp. 101-125.

Henderson, J., and Venkatraman, N. "Strategic Alignment: A Model for Organizational Transformation via Information Technology," in *Transforming Organizations*, T. Kockan and M. Useem (eds.). New York: Oxford Press, 1991.

Holsapple, C., and Lou, W. "Organizational Computing Frameworks: Progress and Needs," *The Information Society* (11:1), 1995, pp. 59-74.

Howe, K., and Eisenhardt, M. "Standards for Qualitative and Quantitative Research: A Prolegomena," *Educational Researcher* (19:4), 1990, pp. 2-9.

Hoyle, R. *Structural Equation Modeling: Concepts, Issues Applications*. San Francisco: Sage, 1995.

Jackson, B. *Field Work*. Urbana, IL: University of Illinois Press, 1987.

Johnson, B., and Rice, R. *Managing Organizational Innovation: The Evolution from Word Processing to Office Information Systems*. New York: Columbia University Press, 1987.

Jick, T. "Mixing Qualitative and Quantitative Methods: Triangulation in Action," *Administrative Science Quarterly* (24), 1979, pp. 602-611.

Kaplan, B., and Duchon, D. "Combining Qualitative and Quantitative Methods in Information Systems Research: A Case Study," *MIS Quarterly* (12:4), 1988, pp. 571-586.

Kelley, M. "New Process Technology, Job Design, and Work Organization: A Contingency Model," *American Sociological Review* (55), 1990, pp. 209-223.

Klein, K., Danserou, F., and Hall, R. "Level Issues in Theory Development, Data Collection, and Analysis," *Academy of Management Review* (19:2), 1994, pp. 195-229.

Kling, R. "Behind the Terminal: The Critical Role of Computing Infrastructure in Effective Information Systems' Development and Use," in *Challenges and Strategies for Research in Systems Development*, W. Cotterman and J. Senn (eds.). London: John Wiley & Sons, 1992.

Kling, R. *Computerization and Controversy*. San Diego: Academic Press, 1995.

Kling, R. "Defining Boundaries of Computing Across Complex Organizations," in *Critical Issues in Information Systems*, R. Boland and R. Hirschheim (eds.). New York: John-Wiley & Sons, 1987.

Kling, R. "Social Analyses of Computing: Theoretical Perspectives in Recent Empirical Research," *Computing Surveys* (12:1), 1980, pp. 61-110.

Kling, R., and Iacono, S. "The Control of Information Systems Development after Implementation," *Communications of the ACM* (27:12), 1984, pp. 1218-1226.

Kling, R., and Scacchi, W. "The Web of Computing: Computing Technology as Social Organization," *Advances in Computers* (21). New York: Academic Press, 1982.

Koppel, R. "The Computer System and the Hospital: Organizational Power and the Control of Information," in *Software By Design: Shaping Technology and the Workplace*, H. Salzman and S. Rosenthal (eds.). New York: Oxford University Press, 1994, pp. 143-170.

Kraut, R., Dumais, S., and Koch, S. "Computerization, Productivity, and the Quality of Work-Life," *Communications of the ACM* (32:2), 1989, pp. 220-228.

Laudon, K., and Marr, K. "Information Technology and Occupational Structure," *Proceedings of the 1995 AIS Americas Conference*, Pittsburgh, PA: Association for Information Systems, 1995, pp. 166-168.

Lee, A. "A Scientific Methodology for MIS Case Studies," *MIS Quarterly* (13:1), 1989, pp. 32-50.

Lee, H. "Time and Information Technology: Monochronicity, Polychronicity and Temporal Symmetry," *European Journal of Information Systems* (8:1), 1999, pp. 16-26.

Lee, A., Liebenau, J., and De Gross, J. (eds.). *Information Systems and Qualitative Research*. London: Chapman & Hall, 1997.

Lee, A., and Markus, L. "Special Call for Longitudinal Research," *MIS Quarterly*, 1995.

Leonard-Barton, D. "A Dual Methodology for Case Studies: Synergistic Use of a Longitudinal Single Site with Replicated Multiple Sites," *Organization Science* (1:3), 1990, pp. 248-266.

Lincoln, Y. "Emerging Criteria for Quality in Qualitative and Evaluative Research," *Qualitative Inquiry* (1:3), 1995, pp. 275-289.

Manning, P. "Information Technology in the Police Context: The 'Sailor' Phone," *Information Systems Research* (7:1), 1996, pp. 52-62.

March, J., and Simon, H. *Organizations*. New York: Wiley, 1958.

Markus, M. "Power, Politics and MIS Implementation," *Communications of the ACM* (26:6), 1983, pp. 430-444.

Markus, M., and Robey, D. "Information Technology and Organizational Change: Causal Structure in Theory and Research," *Management Science* (34:5), 1988, pp. 583-598.

Merton, R. *On Theoretical Sociology*. New York: The Free Press, 1967.

Miles, M. "Qualitative Data as an Attractive Nuisance: The Problem of Analysis," *Administrative Science Quarterly* (24), 1979, pp. 590-610.

Miles, M., and Huberman, M. *Qualitative Data Analysis*, 2nd ed. Thousand Oaks, CA: Sage, 1994.

Miller, D. *Handbook of Research Design and Social Measurement*. Newbury Park, CA: Sage, 1991.

Mintzberg, H. "An Emerging Strategy of 'Direct' Research," *Administrative Science Quarterly* (24), 1979, pp. 582-589.

Mumford, E., Hirschheim, R., Fitzgerald, G., and Wood-Harper, T. (eds.). *Research Methods in Information Systems*. Amsterdam: North-Holland, 1984.

Nissen, H-E., Klein, H., and Hirschheim, R. (eds.). *Information Systems Research: Contemporary Approaches and Emergent Traditions*. Amsterdam: North-Holland, 1991.

Orlikowski, W. "Learning from Notes: Organizational Issues in Groupware Implementation," *The Information Society* (9), 1991, pp. 237-250.

Orlikowski, W. "Improvising Organizational Transformation Over Time: A Situated Change Perspective," *Information Systems Research* (7:1), 1996, pp. 63-92.

Orlikowski, W., and Baroudi, J. "Studying Information Technology in Organizations: Research Approaches and Assumptions," *Information Systems Research* (2:1), 1991, pp. 1-28.

Pacanowsky, M. "Communication in the Empowering Organization," in *Communication Yearbook* (11), J. Anderson (ed.), 1988, pp. 356-379.

Pedhauzer, E., and Schmelkin, L. *Measurement, Design and Analysis*. Hillsdale, NJ: Lawrence Erlbaum Associates, 1991.

Pentland, B. "Organizing Moves in Software Support Hot Lines," *Administrative Science Quarterly* (31), 1992, pp. 527-548.

Pettigrew, A. "Longitudinal Field Research on Change: Theory and Practice," *Organization Science* (1:3), 1990, pp. 267-292.

Popper, N. *The Logic of Scientific Discovery*. New York: Harper Torchbooks, 1968.

Robey, D. "Theories that Explain Contradiction," *Proceedings of the Sixteenth International Conference on Information Systems*, J. I. DeGross, G. Ariav, C. Beath, R. Hoyer, and C. Kemerer (eds.), Amsterdam, The Netherlands, December 1995.

Ruhleder, K. "Computerization and Changes to Infrastructures for Knowledge Work," *The Information Society* (11), 1995, pp. 131-144.

Salzman, H. "Computer-Aided Design: Limitations in Automating Design and Drafting," *IEEE Transactions on Engineering Management* (36:4), 1989, pp. 252-261.

Sandstrom, A., and Sandstrom, P. "The Use and Misuse of Anthropological Methods in Library and Information Science Research," *The Library Quarterly* (65:2), 1995, pp. 161-199.

Sawyer, S. "A Market Based Perspective on Software Development," *Communications of the ACM*, forthcoming.

Sawyer, S., and Southwick, R. "Implementing Client-Server: Issues from the Field," in *The International Office of the Future*, B. Glasson, D. Vogel, P. Bots, and J. Nunamaker (eds). London: Chapman & Hall, 1996, pp. 287-298.

Sawyer, S., and Southwick, R. "Transitioning to Client/Server: Using a Temporal Framework to Study Organizational Change," in *Information Systems and Qualitative Research*, A. Lee, J. Liebenau, and J. De Gross (eds.). London: Chapman & Hall, 1997, pp. 343-361.

Schein, E. *Organizational Psychology*, 3rd ed. Englewood Cliffs, NJ: Prentice-Hall, 1980.

Schön, D. *The Reflective Practitioner: How Professionals Think in Action*. London: Temple Smith, 1983.

Shulman, A., Penman, R., and Sless, D. "Putting Information Technology in its Place: Organizational Communication and the Human Infrastructure," in *Applied Social Psychology in Organizational Settings*, J. Caroll (ed.). Hillsdale, NJ: Lawrence Erlbaum, 1992, pp. 155-192.

Sieber, S. "The Integration of Fieldwork and Survey Methods," *American Journal of Sociology* (78:6), 1973, pp. 1335-1359.

Sproull, L., and Goodman, P. "Technology and Organizations: Integration and Opportunities," in *Technology and Organizations*, P. Goodman and L. Sproull, (eds.). San Francisco: Jossey-Bass, 1989, 254-266.

Sproull, L., and Kiesler, S. *Connections: New Ways of Working in the Networked Organization*. Cambridge, MA: MIT Press, 1991.

Star, S., and Ruhleder, K. "Steps Toward and Ecology of Infrastructure: Design and Access for Large Information Spaces," *Information Systems Research* (7:1), 1996, pp. 111-134.

Sutton, R., and Staw, B. "What Theory is Not," *Administrative Science Quarterly* (40), 1995, pp. 371-384.

Trauth, E., and O'Conner, B. "A Study of the Interaction Between Information, Technology and Society: An Illustration of Combined Qualitative Research Methods," in *Information Systems Research: Contemporary Approaches and Emergent Traditions*, H-E. Nissen, H. Klein, and R. Hirschheim, (eds.). Amsterdam: North-Holland, 1991, pp. 131-144.

Truex, D., Baskerville, R., and Klein, H. "Growing Systems in Emergent Organizations," *Communications of The ACM* (42:8), 1999, pp. 117-124.
Van Maanen, J. "Crossroads: Style as Theory," *Organization Science* (6:1), 1995a, pp. 132-143.
Van Maanen, J. "Fear and Loathing in Organizational Studies," *Organization Science* (6:6), 1995b, pp. 687-692.
Van Maanen, J. *Tales of the Field*. Chicago: University of Chicago Press, 1988.
Vaughan, D. "Theory Elaboration: The Heuristics of Case Analysis," in *What is A Case: Exploring the Foundations of Social Inquiry*, C. Ragin and H. Becker (eds.). Cambridge, MA: Cambridge University Press. 1992.
Walsham, G. "The Emergence of Interpretivism in IS Research," *Information Systems Research* (6:4), 1995, pp. 376-394.
Weick, K. "What Theory is Not: Theorizing Is," *Administrative Science Quarterly* (40), 1995, pp. 385-390.
Weizenbaum, J. *Computer Power and Human Reason*. New York: W. H. Freeman and Company, 1976.
Wigand, R., Picot, A., and Reichwald, R. *Information, Organization and Management*. London: Wiley Interscience. 1997.
Williams, D. "Naturalistic Evaluation: Potential Conflicts Between Evaluation Standards and Criteria for Conducting Naturalistic Inquiry," *Educational Evaluation and Policy Analysis* (8:1), 1986, pp. 87-99.
Wilson, F. "Managerial Control Strategies within the Networked Organization," *Information Technology and People* (83:3), 1995, pp. 57-72.
Wynekoop, J. "Strategies for Implementation Research: Combining Research Methods," in *Proceedings of the Thirteenth International Conference on Information Systems*, J. I. DeGross, J. D. Becker, and J. J. Elam, Dallas, Texas, December 1992, pp. 185-194.
Yates, J. *Control Through Communication: The Rise of System in American Management*. Baltimore: Johns Hopkins Press, 1988.
Yates, J., and Van Maanen, J. "Editorial Notes for the Special Issue," *Information Systems Research* (7:4), 1996, pp. 1-4.
Yanow, D. "Crossroads: Writing Organizational Tales: Four Authors and Their Stories About Culture," *Organization Science* (6:2), 1995, pp. 224-237.
Yin, R. *Case Study Research*, 2nd ed. Beverly Hills, CA: Sage Publications, 1989.
Zuboff, S. *In the Age of the Smart Machine*. New York: Basic Books, 1988.

About the Author

Steve Sawyer is an associate professor at the Pennsylvania State University's School of Information Sciences and Technology where he conducts social informatics research. Current projects focus on the social processes of software development, systems implementation, and related organizational changes. Steve earned his doctorate at Boston University and has also had the privilege of serving on the faculty of Syracuse University's School of Information Studies. To date, he has published in journals such as *Computer Personnel, Communications of the ACM, IBM Systems Journal, Information Technology & People,* and the *International Journal of Information Management*. With co-authors, Rob Kling, Holly Crawford, Howard Rosenbaum and Suzie Weisband, his first book, *Information Technologies in Human Contexts: Learning form Social and Organizational Informatics*, is due out in 2000. Steve can be reached by e-mail at sawyer@ist.ps.edu.

15 INFORMATION SYSTEMS RESEARCH AT THE CROSSROADS: EXTERNAL VERSUS INTERNAL VIEWS

Rudy Hirschheim
University of Houston
U.S.A.

Heinz K. Klein
State University of New York,
Binghamton
U.S.A.

Abstract

The advent of the third millennium provides the backdrop for exploring the current intellectual stage of IS research in both substance and matters of research methods. The paper first identifies and reflects upon two key aspects that have shaped our discipline for the past three decades and are critical for us as a community to get right if the discipline is to flourish. These aspects surround: (1) the external view of the state of the community and (2) the internal state of the community. It is our contention that, in both areas, the discipline faces significant problems. In the second part of the paper, we identify some promising steps to be taken next in IS research and its institutionalization as we cross the millennium threshold.

Keywords: Fragmentation of information systems discipline, internal view, external view, influences, external expectations, IS theory, paradigms, generality, pluralism, rigor, relevancy, values and ethics of IS research, axiology of relevancy.

1. Introduction

We are thankful to the organizers of this conference for this rare venue and opportunity to offer a broad picture of what we consider central issues of IS as an academic discipline and practical profession. This opportunity allows us to step back and reflect upon what has occurred in our discipline over the past 30 years that has led to the current state of affairs and anticipate what might be the focus of attention in the future. What is of particular concern to us—and hopefully everyone else in the field—is the realization that the IS discipline has reached a state of fragmentation that carries with it the threat that its members will become emasculated through dispersal into other disciplines or business functions. The institutionalized IS discipline, as we know it now, may cease to exist. Neither practitioners nor academics can escape this peril, because without a thriving practitioner community, there is little need for an academic one.

The principal purpose of this paper is to present a bird's eye view of the intellectual landscape, in which academics—in particular—and practitioners currently find themselves. As is typical for *terra incognita*, new opportunities and perils abound and it is often difficult to tell one from the other. In order to reach some understanding of the terrain through which we must pick a path leading to a promising future, we need to look back and analyze from where we have come. In particular, we focus on how the path through the terrain that the field has chosen—either implicitly or explicitly—has led to where we are now. Naturally such a venture runs the risk of resembling more a personal reflection than an objective account of history. Nevertheless, such a reflection, biased and incomplete as it may be, is important because we believe the field seriously needs to understand where it stands. To this end, we identify and discuss what we see as the two key dilemmas facing the field in its current situation. These dilemmas revolve around the two different views on the state of the IS community. One is the external view of the community; the other is the internal view. These should help us to see where we are now in the evolution of our field. We explore this evolution under the heading "where are we now?" In the second part of the paper, we conclude with some observations about "where do we go from here?" Our intention is to identify some promising steps to be taken next in IS research if we are to be successful in the next millennium.

2. Where Are We Now?

It has been, and continues to be, common place to bemoan the lack of relevancy of IS research for professional practice. However, we contend that this is only part of the problem. Let us focus on the dependence of the IS community (consisting of both practitioners and researchers) on external actors, i.e., senior management and other business functions. We contend that the current situation is characterized by four "disconnects." First, there is a disconnect between expectations as formulated by senior management and the practice of IS departments in the way they interpret their mission. Second, there is insufficient relevance of current IS research to IS practice. Third and *a fortiori*, there is a disconnect of current IS research from senior management expectations.[1] Fourth, there is a disconnect within the IS research community in that there are numerous research sub-communities that do not communicate with each other (internal fragmentation). In the following discussion, we lump together the disconnect of IS as a whole (including its researchers and practitioners) from senior management as "the external view of the state of the community" in that it revolves around outside (non-IS community) perceptions and expectations of IS. The second problem, which we take up below, involves the internal disconnect of the field encompassing the fragmentation of IS research within the field. We refer to this as "the internal view of the state of the research community."

2.1 The External View of the State of the Community

Our interpretation of the external state of the community is based on formal and informal interviews of hundreds of IS managers on three continents over a 10 year period which have culminated in a variety of publications (Bhattacherjee and Hirschheim 1997; Hirschheim and Lacity 2000; Hirschheim and Miller 1993; Lacity and Hirschheim 1993, 1995; Sabherwal, Hirschheim, and Goles 2000). They point to five expectations that IS managers are confronted with from their board and peers in the other business functions.[2] These five expectations stand in contrast to the regularly published "top issues facing the CIO," published in outlets such as *MIS Quarterly*, because they have more persistence about them. As our data have been accumulated over a decade, we reached the conclusion that they have continually shaped the quality of the relationships of IS practitioners with the organizational environment in two ways: horizontally with the other business functions and vertically with senior corporate management. These expectations are:

1. Lower costs of the IS function
2. Increasing speed of delivery of IS products and services
2. Comprehensive, cross-functional data availability
4. Demonstrable value add
5. Leadership in shaping corporate strategic direction

Even a cursory examination makes it apparent that the IS community has not been able to meet these "external" expectations in the past nor is likely to do so in the foreseeable future.[3]

Lower IS Costs. True, hardware functionality in terms of processor speed, bandwidth and storage capacity has been increasing while prices decreased and this trend

---

[1]We do not wish to imply the identification of such disconnects that IS research must necessarily cater to industry ideology. As we will suggest below, IS research can and should help shape such ideology.

[2]In part, these expectations and where they come from, how they are formed, and what they lead to are taken up in Poora and Hirschheim (1999).

[3]Of course there have been times, such as in the early 1990s, when many IS departments did lower their budgets, typically through downsizing, but this often led to a concomitant rise in hidden IT spending in the business units that didn't show up in the corporate IS spend figure. Hence, the perception that IS costs had actually gone down during this period is somewhat illusory. And where it wasn't illusory, organizations typically suffered.

is likely to continue, but this misses the point. Lower IT costs typically do not translate into lower overall costs for IS functionality, but the IS practitioner community has historically had difficulty providing persuasive grounds as to why this is so. Simply put, IT costs only make up a fraction of overall IS costs. As the IS function delivers more and more products and services to the business units, its overall costs go up inevitably. Consider the analogy with the car industry: Consumers expect greater functionality from cars, yet don't necessarily expect them to go down in price, even given a decrease in price of some car components. So does it make sense for senior business executives to expect lower costs from their IS units?

Speed of delivery. One of the persistent problems confronting IS is the speed with which it can deliver products and services. As the number of products and services demanded from it increase, the function struggles to meet expectations. Just as senior management use the decline in hardware costs to buttress their belief that IS costs should go down, they also believe that new technology should aid IS to deliver products and services more expeditiously. And it does, to some extent. But just as senior management's belief that IS costs should decrease is misguided, so too is this. The IS function is increasingly attempting to balance a greater and greater number of system demands. Whether it is implementing an ERP across the organization, maintaining legacy applications, putting in a new telecommunications infrastructure, and/or preparing the organization for electronic commerce, these demands simply place an enormous burden on IS. And the burden grows daily as more and more business units request new, complex services.

Comprehensive, cross-functional data availability. Although organizations for some time have wished for and—to some extent—have been promised information systems that could deliver comprehensive data spanning the entire organization, such a desire was more pipe-dream than reality. Nor did IS organizations help themselves with the overselling of database technology, and decision support and executive information systems, promising senior executives exciting integrative possibilities now that data were accessible from across the entire organization. In fact, these systems, while potentially beneficial to management, were not the panacea they were often sold as. While they did provide richer data that was easier to access and understand, they were neither comprehensive nor truly cross-functional. Senior management needs this kind of information but perhaps until recently—with the advent of client/server based ERP= applications—such wishes could not be met. Again, in many ways this mismatch of expectations is in part IS's fault. Delivering cross-functional data and applications is not only a technological problem but a political one as well. The IS function has simply not done a good job in making this situation visible to all.

Demonstrable value-add. Perhaps the most intractable problem that IS faces is the issue of how to get management to see the value-add that the IS function provides. This issue is connected to the evaluation of IS, i.e., the evaluation of IS products and services, as discussed in the previous section of the paper. While IS evaluation, indeed, has received a fair amount of attention in research, the results have been disappointing in that no reliable method has been found to measure the value of an IS before it has been built (and even after it is built, *cf.* Smithson and Hirschheim 1998). *A fortiori* it follows that the evaluation of the IS function as a whole is even more intractable. According to Porra and Hirschheim (1999), the IS function is often perceived as "overhead"; that is, a cost of doing business but one to be minimized. As such, business unit managers want IS costs to decrease using the rallying cry "these IS overhead costs are killing us." In such an environment, it is hard for IS to demonstrate its strategic contribution to the organization as it has had to focus its attention on justifying why it charges what it does to the business units. Simply put, if IS continues to be seen as overhead, with all the negative connotations this brings with it, it will be increasingly difficult for the IS executive to focus on its strategic role and, more specifically, get senior management to focus on IS's strategic potential.

Leadership in shaping corporate strategic direction. Somewhat paradoxical to the last point where IS is typically perceived as overhead, organizations often wish IS would take an active role in shaping a corporation's strategic direction. The belief seems to be that the IS function is uniquely suited for this role for two reasons. (1) Because IS develops systems for all of the business units in the organization, it has to understand how the different systems fit together. Therefore IS leaders are well suited to having a good overview of all the functions and systems of the organization. (2) IS is perhaps the most knowledgeable group of individuals in the organization on new technologies. Given these new technologies may provide opportunities for the organization to get into new businesses, new markets, etc., such technology expertise could prove invaluable for setting strategic direction. Hence IS is uniquely placed to provide leadership in shaping a company's strategic direction. Yet, here too we note a serious inconsistency. On the one hand, IS has been perceived as being an overhead, as too expensive, as failing to provide comprehensive data across the organization and insufficient value for money; while on the other hand, it is supposed to provide leadership in strategic direction. The first set of problems with IS has led the function to being ignored for corporate senior management positions. How often does one find a former IS director in the capacity of chief officer of a corporation? Answer: not very often. To us, it is totally inconsistent to expect IS to take a lead role in shaping strategic vision and direction without giving IS the necessary access to senior management positions and the knowledge and motivation that come with it.

So where does this leave us? Clearly there is a problem with the non-IS practitioners' view of IS. They have an unrealistic image of IS and, concomitantly, unrealistic expectations about what IS can and cannot accomplish. In placing these expectations into a historical perspective, we have reached the conclusion that the theoretical concept and image of the nature of information systems to which the IS research community has subscribed since its beginnings has been incongruous with that held by the external "consumers" of IS research. As a result, they do not look for enlightenment through IS research, because they have given up on our research a long time ago. Maybe they are right in that IS theories are truly irrelevant for practice *(horribile dictu)*. If we truly believe that at least some IS theories are, indeed, relevant for practitioners, we must have done a very poor job of communicating in a convincing way which theories are relevant for practice. (We don't want to imply that every bit of theoretical exploration has to be immediately relevant for practitioners.) There are two sides to this incongruency issue. First, the IS concept held by the IS practitioners is at best partially supported by some of the theories that guide IS research. Second, the IS concept held by non-IS practitioners, i.e., senior management or business units, is even more at odds with the academic notions of information systems and also quite different from the IS practitioners' beliefs about the nature of IS. This contributes to a credibility crises of IS as a whole that engulfs both academia and practice. In the following, we shall concentrate on what we perceive to be the academic root cause of this problem: the IS theories underlying the "schools of thought"[4] that have produced the most influential publications since the late 1960s.

In the field of IS, most theories take the form of conceptual frameworks. They provide a coherent way for reasoning about the nature and roles of information systems and their preferred ways of development; i.e., what kinds of methods and tools in principle are needed for the analysis, design, and implementation of IS? In fact, for some schools of thought, the concern for the right systems development approach came first and assumptions about the nature of information systems were implied by the preferred systems development approach with its specific methods and tools for ISD. The theoretical framework about the nature of information systems and the preferred way of IS development are an essential part of the fundamental assumptions that researchers within a certain school of thought make. These assumptions concern issues such as the nature of reality (ontology) and the preferred methods of IS research (epistemology). They also guide the formation of specific research programs, allowing certain IS problems to come into focus while at the same time obscuring other problems. For the sake of brevity, we shall refer to this complex of ideas as "IS theory." IS theory affects the selection of research problems and the formulation of research strategy more then anything else.

But what are the IS theories that have emerged from the various schools of thought that make up the history of our discipline since the 1960s? We propose that Table 1, in brief, captures the visions of the most influential founding fathers of our discipline and, moreover, their visions are still the defining ideas for current research schools. (See Klein and Hirschheim 1999 for more details.)

Issues lost. We claim that the fundamental direction of current research agendas and strategies is shaped more by the ramifications of their founding fathers' ideas and their interplay with other research schools than by senior management expectations directed at the IS function executives. This could be substantiated by tracing how the ideas from the 10 IS theories in Table 1 have proliferated through the publications of the founding fathers' students and their students' students and how much of the literature can be accounted for in this way.[5] Hence it is not too surprising that IS research is disconnected from the expectations of managers, because for the most part it is driven by its internal traditions. In this way it is like most other disciplines, its short history notwithstanding. The issues caused by these internal dynamics ultimately define what counts as knowledge and we term the issues that this raises as "lost," because they seem to have been largely ignored in the IS literature.

---

[4]When we speak of "schools of thought," we are referring to key, historical research programs that have shaped the course of the IS discipline. Giving a particular research program such a designation invariably raises some controversy, but we are prepared to defend our position on the basis that history provides a reasonable metric by which to "score" the impact of particular research programs.

[5]In fact, for the Scandinavian schools of IS theory listed in Table 1, this type of analysis has already been done for the most part in Iivari and Lyytinen (1999, *cf.* especially p. 139, Figure 1). It should be uncontroversial to extend this kind of analysis to the remainder of the Western world, where most of the IS research has been published. For example, our own research has mostly been influenced by the STS, inquiring systems, and social action theories of IS. It is beyond the scope of this paper and even unnecessary to trace the principal influences for the other schools in detail, but see Klein and Hirschheim (1999).

Table 1. The Principal Schools of Thought of IS Theory

Original Source of IS Theory	Proposed theoretical concept of information systems as:
Langefors (1966, 1973)	datalogical and infological systems
Blumenthal (1969)	reporting and control systems[6]
Teichroew and Hershey (1977); Yourdon (1978)	formally specified technical systems[7]
Churchman (1971)	inquiry systems
Dickson (1968, 1981); Davis (1974)	behavioral systems
Nygaard (1975)	weapons of class struggle
Mumford and Henshall (1978); Bostrom and Heinen (1977)	socio-technical systems[8]
Checkland (1981)	human activity systems[9]
Goldkuhl and Lyytinen (1982); Winograd and Flores (1986, 1987)	linguistic systems in social action
McFarlan (1984)	competitive weapons

[6]It would be more accurate to characterize Blumenthal's ideas primarily as a modular, incremental design strategy of building reporting and control information systems on top of transaction-based systems. The implied IS theory is that of a parametric feedback loop hierarchy of the type envisioned by Ashby and by Forrester's *Individual Dynamics*.

[7]The earlier observation that "for some of the IS theories, the concern for the right design approach came first and assumptions about the nature of the IS were implied by the preferred design approach" especially applies to the IS theories that conceive of IS as a mechanistic, technical artifact. In the case of Teichroew, his original concern was the automation of IS design including software generation from an exact "problem statement" (expressed in PSL) using AI techniques. When this proved impossible, the research moved to supporting the specification process, i.e., IS modeling. Basically, for Teichroew, an IS was whatever could be specified consistently in a repository. De facto, this was very similar to what Yourdon tried to teach with manual specifications, even though Teichroew claimed that the ISDOS project was independent from any specific methodology.

[8]As was noted in Hirschheim and Klein (1994), there are some underlying consistencies between the original STS theory of IS (*cf.* Mumford 1983), the inquiring systems notion, and the social action theory of IS. Basically, the original STS concept can be made theoretically more rigorous by interpreting it as a type of social action system.

[9]Checkland more than anyone else gave equal thought to a theoretical concept of an IS and a development strategy consistent with this concept in his SSM (soft systems methodology).

IS emasculation and diffusion. From the external viewpoint, the disturbing outcome of neither researchers nor practitioners having been able to meet senior management expectations—unrealistic as they may seem to us—has been the emasculation and diffusion of the IS function. This diffusion takes the form of outsourcing and dispersing IS personnel to the various business units. The accompanying withdrawal of resources has begun to emasculate many IS departments to a skeleton of what they once were. If this trend continues, it must lead to dire consequences for the IS academic community in terms of changing curricula, fewer students, reduced grants, and last, but not least, research opportunities. Therefore, it is not surprising that Markus (1999) questions the very existence of the IS discipline in a recent provocative article. She poses the question: "What happens if the IS field as we know it goes away?" For her, the field is at a crossroads. On the one hand, it could grow to become one of the most important areas for business, since no organization can ignore the inexorable development and application of new information technology and expect to survive. On the other hand, there is a move to emasculate and devolve the field, moving IT tasks and skills into the business functions. She writes:

> With dollars and people moving from corporate IS units to business units (and even further out to IT industry firms), central IS units (not just hardware) have been downsized. Corporate IS groups that once hired hundreds of people are today mere shadows of their former selves—sometimes only five or six people. [Markus 1999, p. 184]

Lucas (1999) supports Markus' concern noting that the migration of IS skills to other disciplines is occurring and that many U.S. business school deans have adopted this "disturbing belief." Indeed, we note that some universities no longer support a vigorous and expanding IS group. But this analysis needs further scrutiny, and we offer the following two comments. The first is simply the observation that moving IS personnel to business units and closer to users need not necessarily reduce demand for IS professionals. True, but what we have in mind is the issue of a center for professional identity and institutional support. If IS tasks are widely absorbed by user departments, then the core competency of IS will be lost and with it the institutional support structures. The second weakness of the above analysis is more difficult to overcome: it leaves out the recent rise of Internet-driven demand, in particular from e-commerce development. Is it true, as one anonymous reviewer of our paper contended, that net-centric applications move (or will move) information technology

> to be the core enabler of the enterprise....The IT network is moving into a societal artifact and most organizations are beginning to believe that survival is impossible without IT. Rather than viewing the IS field as diminishing or dispersed, the field has moved to the core of the enterprise.

In response, we note that the last sentence need not follow from the first two, because of the immense potential for standardization, outsourcing, and licensing that Internet application offers. Right now, we see much experimentation and creativity and hence high demand for Internet web site development professionals. This was similar with ISD before the arrival of software packages as a commodity. (How many DSS programmers have been employed since spreadsheet software was standardized?) Because of the public nature of Internet web sites, we believe that within the next few years we will see a limited number of standardized and parameterized "web site packages." It is quite likely that, using such packages, skilled users can customize most of the common web applications they need. And for the few that require new development, they will be outsourced. In fact, we believe that e-commerce raises more organizational issues than technical ones: professionals with a good understanding of more than one business function and the ability to realign and integrate their processes in a way that exploits the new technologies will be in high demand. Currently most academic IS programs presume business knowledge, focusing more on IT rather than teaching how to exploit it. Maybe this is our opportunity if we are able to redefine the core of the IS discipline "to get with it." Adding a new e-commerce elective course is, however, not what we have in mind here. It is going to take more than that.

Hence, in conclusion, we as researchers have to ask ourselves whether the trend noted by Markus is the beginning of a dramatic change for the field or simply a passing fad. Will companies be forced to "backsource" as they become more and more aware of their lack of IS expertise? Is there anything that we as academics can do to illuminate the issues at stake from a longer term perspective, thereby contributing to the avoidance of costly, maybe even dangerous, mistakes?

Currently the state of IS research appears to be ill equipped to recognize let alone address this issue. In large part this is because of the continuing relevancy gap between IS research and senior management expectations. An important factor explaining the relevancy gap is the images of IS that have been developed in 30 years of IS research and which simply do not resonate with the five expectations of senior management. Therefore, the research stimulated by those images does not connect to the expectations either. A further implication of the anchoring of current research agendas to these IS images is reinforcement of the fragmentation that characterizes the internal state of the IS research community, as will be explored next.

2.2 The Internal View of the State of the IS Research Community

From an analysis of influential treatments of the foundations of IS, it becomes clear that the images of the nature of information systems (*cf.* Table 1) are not the only primary influence of the state of the IS research community. The discussion about preferred reference disciplines and paradigms has also shaped the current ways of thinking and agendas in the IS research community more than anything else.[10] Together, the effects of

[10]We do not overlook the importance of reference discipline focus as a source of differentiation, i.e., training in preferred references disciplines and professional experience as an engineer, accountant, economist, etc. This source of influence was of particular importance during the early era of IS when there were no internally trained IS faculty. This influence works through the personality of influential researchers. It affects their vision of an IS and their paradigmatic assumptions. Hence it is indirectly acknowledged. A more detailed treatment of this source is beyond the scope of this paper.

conflicting paradigms and commitment to incompatible visions of the nature of IS have fragmented the IS research community along several dimensions to the point that it is now often called a "fragmented adhocracy."[11] Hence the greatest issue that the IS community faces internally is its fragmentation into numerous specializations (or what we might call "sects"). They are in want of intellectual synthesis that could emerge from a fruitful discourse. However, we lack a set of shared assumptions and language. Hence large conferences like ICIS, AIS, or HICSS are reincarnations of the Tower of Babel. Fruitful cross-sectional debate almost never occurs. The situation can be dated back to at least 1985 and possibly before (*cf.* Hirschheim 1986).

Traditionally the relationship between alternative paradigms was conceived as being one of the following: Dominance, synthesis, incommensurability, eclecticism, or pluralism (Morgan 1983). For example, positivism—through the centuries—has enjoyed great success. Its position was one of dominance. More recently, however, critics have surfaced calling into question positivism's dominance. A call has gone out for pluralism rather than dominance in research (*cf.* Lincoln and Guba 1985). From an historical perspective, one can distinctly see the uneasy tension that has existed in the application of positivism in the social sciences. This has given rise to what Tashakkori and Teddlie (1998) have termed "the paradigm wars": battles fought by the adherents of positivism against those from other paradigms.

For Landry and Banville (1992), the paradigm wars can be recast into three types of researchers, each with its own outlook on paradigm appropriateness. They have characterized these types or groups as mainstream navigators, unity advocates, and knights of change. The first group, *mainstream navigators*, is composed of supporters of the dominant orthodoxy. Their epistemological roots are in logical positivism, which cements them in the functionalist paradigm. The second group, *unity advocates*, is more concerned with the acceptance of information systems as a scientific discipline than with a specific paradigm. In the unity advocates' view of the world, an immature or pre-science discipline is characterized by the existence of several competing paradigms. A more desirable state, that of a full-fledged scientific discipline, is characterized by the reign of a single dominant paradigm. They would be agreeable to using any paradigm as long as it granted them scientific respectability. Since the current state of information systems research is dominated by positivism, unity advocates tend to cluster toward this end of the paradigm dimension. The third group, *knights of change*, is of the opinion that

---

[11]By applying Whitley's (1984a, 1984b) model of cognitive and social institutionalization of scientific fields (or academic disciplines), Banville and Landry (1989) conclude that the field of IS is a "fragmented adhocracy." This is so because in order to work in IS one does not need a strong consensus with one's colleagues on the significance and importance of the research problem as long as there exists some outside community for support. Nor are there widely accepted, legitimized results or procedures on which one must build "in order to construct knowledge claims which are regarded as competent and useful contributions" (Whitley 1984a, pp. 88 -123, as quoted by Banville and Landry 1989, p. 54). In addition, research involves high task uncertainty, because problem formulations are unstable, priorities vary among different research communities, and there is little control over the goals by a professional leadership establishment (such as bars or licensing boards for physicians and engineers). For example, some IS research groups may choose to define and cherish projects that do not follow the familiar patterns of engineering or empirical social science, although such groups are generally in the minority. There appears—to some extent at least —to be local autonomy to formulate research problems and standards for conducting and evaluating research results (*cf.* Goles and Hirschheim 2000).

reality is multifaceted and forged from the interpretations and interactions of individual actors. They also give credence to the belief that no single research approach can fully capture the richness and complexity of what we experience as reality. Thus they champion a collection of assorted research approaches arising from multiple paradigms.

Yet even the knights of change, with their clarion call for methodological pluralism, argue for change *within* Burrell and Morgan's (1979) four paradigms. Others argue that Burrell and Morgan's framework, by virtue of its widespread acceptance and impact, has normalized and rationalized emerging streams of research, constraining alternative perspectives.

> In time, influential frameworks can become as restraining and restrictive as those they originally challenged....we are sometimes presented through responses to a conceptual framework ...with a new, rich set of alternative perspectives through which we can continue our study and talk about our subject matter. [Frost 1996; p. 190]

In addition to the three types of paradigm warriors (mainstream navigators, unity advocates, and knights of change) identified by Landry and Banville, a new group is emerging which is calling for an end to the paradigm wars: the pacifists. These theorists and researchers argue that there are strengths and weaknesses in both the positivist and non-positivist positions and point out that the conflicting paradigms have, in spite of the best efforts of their most ardent supporters, achieved a state of coexistence (Tashakkori and Teddlie 1998). Datta (1994) has presented five compelling arguments in support of this assertion: (1) Both paradigms have been in use for a number of years; (2) There are a considerable (and growing) number of scholars arguing for the use of multiple paradigms and methods; (3) Funding agencies support research in both paradigms; (4) Both paradigms have had an influence on various policies; and (5) Much has been learned via each paradigm.

In the IS research arena, the existence of such paradigm pluralism can be found but it is not as wide spread as either Datta or Tashakkori and Teddlie suggest. Worse, the supposed interplay between researchers of different paradigms does not occur because of the communication gap that exists between the alternative paradigms.

Although the "internal and external views of the community" paint a somewhat worrying picture of the field, all may not be lost. In the next section, we offer two avenues for future work that might help to mitigate the deleterious effects of IS fragmentation, and the disconnect of expectations of—and the demands for—IS. We offer some suggestions for improving the generality of IS research and this primarily concerns the internal view of IS. From the external view, more generality should help to improve the transparency and interest of IS research to the external communities. This is part and parcel of a research strategy to address the relevancy disconnect of IS research.

3. Where Do We Go from Here?

If, as we claim, paradigmatic pluralism is more fruitful than harmful and that the notion of paradigm wars (*cf.* Datta 1994; Tashakkori and Teddlie 1998) is unhelpful, what should be the next steps in the evolution of IS research? It seems to us that there are two arenas where research in the future needs to head. One reflects the internal view of the field and revolves around the issue of the generality, transparency and applicability of our work, and the embracing of pluralism. The second focuses on the need to involve the external community (i.e., non-IS) in what we do in a new way. This we recast in terms of the age-old "relevance vs. rigor" debate.

3.1 Generality and Pluralism

IS as a field has had difficulties with generalization from its beginnings. It started with story telling of experiences that were generalized into insights that should apply to many situations (e.g., the five lessons from Ackoff [1967] or the "myths" of Dearden [1966]). Such reasoning was later debunked as "unscientific" and replaced with "rigorous" hypothesis testing. It now appears as if IS research has come full circle by returning to new forms of story telling (the politically correct term is, of course, "narratives"). Whereas the new story telling movement can point to much better and more explicit philosophical grounding than "the great wise men" had for their stories, this does not necessarily make them "more general." In fact, they lag in generality behind the insights offered by the earliest authors. Moreover, they share this weakness with the failure of positivist research to offer a few broad theories that contribute to general orientation and bring some measure of order to the perpetual confusion in our field. Hence interpretivists have little reason to be gleeful about the failures of positivism. They are no stronger in theory formulation than the positivists. Should it turn out that they are equally unable to deliver results of general interest, i.e., that reduce the complexity of coping with reality by applying to a large number of instances, they may fall into disrepute quicker than the attraction of positivism is waning. So how does one deal with this dilemma?

It appears that the generalization deficit is a concern that affects interpretivists and positivists alike, yet is ignored by both. We propose that it could be addressed by a change in paper reviewing practices in the direction of giving generalization the same weight as methodological rigor. Often authors are discouraged from generalization, because they cannot support it with the same degree of plausible evidence as narrowly conceived hypotheses. This practice discourages prospective authors from connecting specific hypotheses or ethnographic findings to broader theoretical constructs that might qualify as some form of "general theory," at least within a specific sub-community. If papers are rejected for lack of rigor, then the same should apply for lack of generalization. The degree of rigor required should be tempered in relation to the degree of generalization attempted. The more generalization, the less rigor would be expected. Unfortunately, the current theory part of most papers consists of minor building blocks for more general theories that are at best implied and at worst *ad hoc* and hence do not even exist. This leads to a multitude of hypotheses with associated significant tests or ethnographic insights with associated thick descriptions, which as a whole go nowhere. They go nowhere because their interconnections do not exist or are at best transparent to the insiders who know the literature of a specific sub-community.

We believe that a new genre of papers should be encouraged, which take a major block of specific studies and mold them into a larger theoretical framework. There are examples of this kind of work (e.g., Ives, Hamilton, and Davis 1980; Zmud 1979), but they are far too few. Of course to engage in such work not only requires considerable effort, but typically leads to paper that are longer and conceptual rather than empirical in nature. Most journals have page limitations that specifically militate against such efforts. Thus in encouraging these new papers, journal editors would have to revise their editorial policies in the following ways: (1) engage sympathetic associate editors and reviewers to broaden their view about what types of papers are acceptable to their journals; (2) set different and realistic new page limitations for such papers; and (3) revise the scope and aims of their journals to reflect the broader focus. Additionally, there are a number of books such as Checkland and Scholes (1990) and Walsham (1993) that offer good generalizations drawn from action research and detailed field studies respectively.

3.2 Relevancy and its Relationship to Rigor

The call for better generality in our research is important for the IS community as a whole, but there is also a need to better connect IS to the outside community. One way for this to occur is through research that is more relevant: more relevant for IS practitioners and more relevant for other groups who depend on IS, such as senior management and business unit managers and personnel.

One pointer on how this might be accomplished can be found in Lynne Markus' address to the 1997 IFIP8.2 conference in Philadelphia. She argued that one of the directions the field should now take is "the appreciation of practicality in IS research" (Markus 1997, p. 18). The intent of what she terms practical research is not to replace or overshadow research that builds or tests academic theory, but rather to complement theoretical research with "rigorous research that describes and evaluates what is going on in practice" (Markus 1997, p. 18). This is underscored by the conference theme of ICIS'97 with its emphasis on "the issue of relevance and relationship of IS research to practice" (Kumar 1997, p. xvii). More recently, *MIS Quarterly* announced a renewed thrust aimed "at better imbuing rigorous research with the element of relevance to managers, consultants, and other practitioners" (Lee 1999a, p. viii). The discussions presented in Benbasat and Zmud (1999), Applegate and King (1999), Lyytinen (1999), and Lee (1999b) support this thrust.

Of course, this call for relevance in research is neither particularly new nor confined to the IS field. Ormerod (1996), for example, called for "the synergistic combination of consulting and academic research in IS." Davenport and Markus (1999), in like fashion, note the value of consulting and academic research learning from each other. Similarly, Avison et al. (1999) advocate a greater use of action research to make IS academic research more relevant to practitioners.

But this call for more relevancy in IS academic research embodies a number of underlying questions; questions that do not necessarily have simple yes/no answers. We see three basic questions that need to be addressed in the area of IS relevancy: (1) relevancy for whom; (2) the relationship between relevancy and rigor; and (3) axiology of relevancy.

(1) Relevancy for whom: When we speak of "relevancy," to whom, exactly, are we referring? One constituency for whom IS research might want to be relevant is the external community. But who is this external constituency: is it senior management, business unit managers, society, customers, the government? Is it some of these, all of these, none of these? Each of these constituencies is likely to have a very different notion of what is relevant for them. Moreover, not only will what is relevant for each of them differ, they may in fact be in conflict. Society might wish research on IS to lead to more democratic societal forms embodying empowered citizens. Corporate executives, on the other hand, might wish IS research to focus on systems that offer new market opportunities rather than opportunities for empowered workers. Another constituency likely to have a different view of relevancy is the IS practitioner community. Here, relevant research would consist of knowledge leading to, for example, better tools and approaches for developing information systems. So clearly, when we speak of making IS research more relevant, we need to define which group we are attempting to reach. Most calls for relevancy seem to mean relevant for the IS practitioner. As we have suggested above with the set of senior management expectations of IS, clearly this is an area where research attention needs to be placed. Another external community that is the recipient of our research is other academic disciplines. In relevancy vs. rigor discussions, the focus tends to be on practitioners. The fact is, much IS research is done with an eye to other academics—academics from other departments in business schools and other faculties on campus. These communities, often at odds with each other, may have entirely different sets of goals and objectives for IS. Even if they share the applied focus of IS (like the other disciplines), that does not mean that marketing, finance, organization, or operations management have consistent expectations for their IS colleagues. Hence there is continuing pull on IS academics to make their research and teaching useful for others, i.e., to be servants of too many masters. These pressures can be especially strong for junior IS faculty if they know that their tenure committees are dominated by non-IS faculty. They need to strike alliances lest they perish even if they publish.

Beyond this, many of the disciplines on campus are not applied, and see applied research as "unacademic" and hence not valued. And if such work is not valued, this poses a problem for the IS academic whose rewards (tenure and promotion) and punishments (failed tenure and promotion cases, rejected research proposals) heavily depend on other academic groupings. IS academics can ill afford to ignore what these groups consider relevant. Many IS researchers have succumbed to this pressure by undertaking highly theoretical work, which is relevant not only for the broader university community but also one other important community: IS academics themselves. While such a strategy has helped make IS an arguably accepted discipline,[12] it has done so at the cost of practitioner relevancy.

In fact, the strategy has led to the rather dubious condition of what might be termed "the vicious cycle" of academic research. This is a particularly unflattering condition of which many academics feel we are part. n the traditional model of research, the purpose of research is to generate knowledge. The model starts out with a problem, which leads to research, which in turn leads to knowledge, which in turn informs practice, which in turn encounters new problems, which start the whole cycle over again. In the "vicious" version of this cycle model, the purpose gets distorted to one where a research problem

---

[12]We are, of course, aware that not everyone would agree with this view, but we'll leave that discussion for another time.

leads to research, which leads to new research problems, which leads to research, which leads to more research problems and on without end. The feedback control loop to practice is lost; research remains entirely in the ivory tower.

To summarize, there are many recipients of IS research, each with their own particular view of what relevance means. Currently, IS research has pursued relevance in the context of relevance for academic communities. But such singularity of focus has led to a possible crisis in IS. We in the discipline need to broaden our notion of relevancy to include other groups who use or could use the knowledge generated by IS research. We need research on how all the various external groups come to understand IS and how they form their perceptions about the proper role of IS. We need to research the interaction patterns between business unit management among themselves and with senior executives of the company. And if there is a mismatch between what business unit managers and senior corporate executives see as the role of IS, then hopefully IS research can lead to persuasive arguments to solve the mismatch. Fundamentally, we clearly must extend our notion of relevancy to include stakeholders other than just IS practitioners.

(2) Relationship between relevancy and rigor: A dictionary definition of rigor typically uses terms such as severity, sternness, strictness, stringency, and harshness to describe its nature.[13] In academic research, the term rigor has become the touchstone for quality and scholarship. If it isn't rigorous, it isn't scholarly. If it isn't rigorous, it shouldn't be published. Rigor seems to have taken on a life of its own in academic research. Rigor is usually manifested in research through Greek symbols, mathematical formulas, number of experimental controls, and conforming to the standard of the best research the community of scholars interprets it has done so far. Typically, this means applying the hypothetico-deductive method—the accepted methods of science. Such a view of rigor, however, excludes other forms of scholarly research, which do not subscribe to such positivist standards.

We contend that there are many scholarly vehicles for knowledge creation and they need to be recognized as rigorous as long as they employ sound forms of reasoning and giving evidence. Indeed, we believe that any knowledge claim emanating from research should be scrutinized using a sound reasoning process. This might embrace Habermas' (1984) notion of the "force of the better argument," where competing knowledge claims are evaluated and the knowledge claim based on the better reasoning, arguments, and evidence being judged as "accepted." Such a process can be used to evaluate interpretive as well as positivist research, even though interpretive research is considered inherently more difficult to be "evaluated objectively" because the community consensus about its standards have not yet solidified. Nevertheless, it can be done. An objective evaluation would typically involve considering three aspects of the research: intelligibility, novelty, and believability. *Intelligibility* relates to the question of how well the research approach and results are comprehended, i.e., how closely others can follow them with similar qualifications. *Novelty* can be judged in at least three ways: (1) by the amount of new insight added; (2) by the significance of the research reported in terms of the implications it has for seeing important matters in a new light and/or providing a new way of thinking about the phenomenon under study; (3) by the completeness and coherence of the

[13]For some interpretivist researchers, such notions of rigor seem totally understandable as that is the way their research often seems to be treated by reviewers, i.e., harshly!

research report(s). Can the author provide an overall picture so that its components link up to each other without major holes in the picture that is being painted? *Believability*, on the other hand, relates to how well the research arguments make sense in light of our total knowledge. The key question for believability is how well the research in method and results fits with other ideas and arguments that are taken for granted within the current state of knowledge. A first measure of this is the number of references with which it is consistent (or which it challenges). New research inevitably challenges some part of the web of beliefs that makes up the current state of knowledge that is best characterized as a "web of beliefs" that are only sparsely connected to "hard" evidence (Quine 1970). Another useful image of the state of knowledge is the fact net and its proliferation in the sense of Churchman (1971). The more references are challenged the less believable, but potentially the more significant, is the research. For believability, each author must demonstrate how his research "fixes" the net of our knowledge. If the research "takes out" certain parts of the knowledge net, the author must reconnect the loose parts. Often this requires relating to some forgotten or remote parts of the web of belief (which may be the domain of some other research community). By bringing in more references from other areas, authors can often successfully challenge major parts of a local (within one discipline) web of belief.

Hence intelligibility and believability are inextricably linked (and often inversely related to the degree of novelty). Ultimately they are a measure of what the research community terms "the validity of the research." Although all research projects must produce results that are intelligible, novel, and believable for the community of scholars to bestow the label "contribution to the state of knowledge" upon them, the criteria are perhaps more subjective in interpretive research projects. To this end, Klein and Myers (1999) offer assistance. They present a set of criteria on how to judge knowledge claims generated from interpretivist research. We feel this is critically important for the IS discipline because interpretivist research can often offer better insights for practice than its counterpart. Practitioners can usually better relate to interpretive research as the research is closer to practice, involves actual case studies, involves real people in real situations, and is undertaken in real world settings. They talk about the results offering new insights, and the results are more translatable into the ways people actually work in organizations. However, interpretive research is typically weak in implying clear advice on what to do or how to improve matters in practice. This leads us to the values and ethics of relevancy, which is the issue we turn to next.

(3) Axiology of relevancy: Practice necessarily involves decisions and actions that are intrinsically value laden. If one speaks of improving practice, or providing insights (from research) which improve practice, there is no way to escape the issue of what values guide the research and for what purposes the results of the research are to be used. Fay's (1975) cogent treatment of values shows how they invariably find their way into political practice and, hence, why it undesirable to ignore this important domain. Yet, most IS research has shied away from value laden research issues or simply taken ideological values for granted (like more effective management control is desirable). However, as the debate in Accounting shows,[14] industry executives are often quite open

---

[14]There has been a recent call from the big accounting firms for research and teaching of ethics in accounting departments in universities.

to critical treatment of value questions that have ethical implications. Could the same not also hold in the domain of IS? And if so, how could such a program fertilize new and relevant IS research?

Knowing about values and their behavioral implications is a type of knowledge—normative knowledge—and that an important form of normative knowledge is that of the 'rule'. Rules are important in the practice of any profession: e.g., management, engineering, medicine, law, and information systems. To characterize the kinds of rules typical for normative knowledge, it is useful to bear in mind Kant's (*cf.* 1785, 1788 – for translations see Kant 1964, 1929) distinction between rules of skill, rules of prudence, and categorical rules (to which we shall refer as ethical values). *Rules of skill* are concerned with the physical propensity and dexterity to carry out certain operations to achieve specified ends. *Rules of prudence* are concerned with judgments to achieve ends that informed and reasonable people would not question as being worthwhile, such as designing systems that are acceptable to their intended users. *Categorical rules* are concerned with choices where the ends themselves are in question, such as a choice between developing information systems that are preferable to one set of stakeholders (e.g., workers on the shop floor) or another (e.g., a company's shareholders). Categorical rules are particularly important when our discipline has to deal with social value conflicts in practice. They are concerned with ultimate values or ethical standards. Elsewhere, we (Klein and Hirschheim 1998) provide a fuller treatment of the practical relevance of research into ultimate value standards (or design ideals). This should be of interest to highest levels of policy formation. Elsewhere (Hirschheim and Klein 1994; Iivari, Hirshheim, and Klein 1998) we have also dealt with rules of prudence—what are more or less prudent principles of ISD in broadly characterized situations. Rules of prudence are best researched through comparative analysis of classes of similar methodologies—approaches as they are called in Iivari, Hirschheim, and Klein. This level of value discourse is of practical significance to project managers or IS executives who are concerned with finding the proper set of methods and tools for their organization. However, many more research contributions are needed in this area as "maybe the biggest challenge for the future of IS research is to come to grips with the need for normative knowledge in our field that goes beyond rules of skill" (Klein, 1999, p. 23).

4. Conclusions

In the introduction, we noted the perils of emasculation and dispersal of the IS function, which to some extent has already happened. However, in addressing this issue through increased relevance and improved generalization, we have become more confident that the challenges we all face could be successfully met by reforming certain institutional practices. We see the need for increasing the amount of research directed at understanding non-IS practitioners and engaging them in a discourse about a realistic set of expectations for what the IS function can and cannot deliver. We also need to provide well-articulated arguments to the IS practitioners by which they can state their case to senior management that a thriving IS department is needed along with the other functional units.

Another important change in institutional publication practices concerns redefining the concept of rigor in research. It needs to be augmented to include a wide range of scholarly inference and evidence giving, on the one hand, and tightened, on the other, to include the linking of detailed models or hypotheses to broader theory to arrive at expanded categories of knowledge. Finally, we note that while pluralism is here to stay, we need to worry more about communicating across the narrow boundaries of our preferred academic sub-communities. For that purpose, all publication venues, in particular the large conferences such as AIS, ICIS, HICSS, and ECIS and all first tier journals need to provide some visible vehicles (e.g., special sessions, special subsections or issues) for broad syntheses that are interesting and comprehensible to all members of the IS community.[15] We need more historical analyses of the various areas that make up the IS domain. And we need to get the IS community to truly value such contributions. In the current situation, we seem to have an overabundance of specialty papers for in-group members with the result that the IS community as a whole suffers from a serious communication gap. The current publication culture of narrowly focused, highly specialized papers is one of the major impediments to making our research more relevant to practitioners. We simply must attempt the difficult, but invaluable, syntheses that pull together research results from the various sub-communities into broader analyses of potential interest to practitioner communities.

References

Ackoff, R. "Management Misinformation Systems," *Management Science*, December, 1967, pp. B147-156.

Applegate, L. and King, J. "Rigor and Relevance: Careers on the Line," *MIS Quarterly* (23:1), 1999, pp. 17-18.

Avison, D., Lau, F., Myers, M., and Nielsen, P. A. "Action Research," *Communications of the ACM* (42:1), 1999, pp. 94-97.

Banville, C., and Landry, M. "Can the Field of MIS be Disciplined?" *Communications of the ACM* (32:1), 1989, pp. 48-60.

Benbasat, I., and Zmud, R. "Empirical Research in Information Systems: The Practice of Relevance," *MIS Quarterly* (23:1), 1999, pp. 3-16.

Bhattacherjee, A., and Hirschheim, R. "IT and Organizational Change: Lessons from Client/ Server Technology Implementation," *Journal of General Management* (23:2), Winter, 1997, pp. 1-16.

Blumenthal, S. C. *Management Information Systems: A Framework for Planning and Development*. Englewood Cliffs, NJ: Prentice-Hall, 1969.

Bostrom, R., and Heinen, S. "MIS Problems and Failures: A Sociotechnical Perspective - Part I: The Causes, *MIS Quarterly* (1:3), 1977, pp. 17-32.

Burrell, G., and Morgan, G. *Sociological Paradigms and Organizational Analysis*. London: Heinemann Books, 1979.

Checkland, P. *Systems Thinking, Systems Practice*. Chichester: John Wiley & Sons, 1981.

Checkland, P., and Scholes, J. *Soft Systems Methodology in Action*. Chichester: John Wiley & Sons, 1990.

Churchman, C. W. *The Design of Inquiring Systems*. New York: Basis Books, 1971.

[15]In computer science, the journal *ACM Computing Surveys* serves such a purpose. There is no equivalent in our field, although the new e-journal *MIS Review* should come close. IS needs more outlets like this.

Datta, L. "Paradigm Wars: A Basis for Peaceful Coexistence and Beyond," in *The Qualitative-Quantitative Debate: New Perspectives*, C. Reichardt and S. Rallis (eds.). San Francisco: Jossey-Bass, 1994, pp. 53-70.

Davenport, T., and Markus, M. L. "Rigor vs. Relevance Revisited: Response to Benbasat and Zmud," *MIS Quarterly* (23:1), 1999, pp. 19-23.

Davis, G. B. *Management Information Systems.* New York: McGraw-Hill, 1974.

Dearden, J. "Myth of Real-time Management Information," *Harvard Business Review*, May-June, 1966, pp. 123-132.

Dickson, G. "Management Information-decision Systems," *Business Horizons* (11), 1968, December, pp.17-26.

Dickson, G. "Management Information Systems: Evolution and Status," in *Advances in Computers*, M. Yovits (ed.). New York: Academic Press, 1981.

Fay, B. *Social Theory and Political Practice.* London: George Allen & Unwin, 1975.

Frost, P. "Crossroads," *Organization Science*, 7(2), March-April, 1996, p. 190.

Goldkuhl, G., and Lyytinen, K. "A Language Action View of Information Systems," in *Proceedings of the Third International Conference on Information Systems*, M. Ginzberg and C. Ross (eds.), Ann Arbor, MI, 1982, pp. 13-30.

Goles, T., and Hirschheim, R. "The Paradigm is Dead, the Paradigm is Dead.... Long Live the Paradigm: The Legacy of Burrell and Morgan," *OMEGA,* 2000 forthcoming..

Habermas, J. *The Theory of Communicative Action: Reason and the Rationalization of Society*, Volume I (transl. T. McCarthy). Boston: Beacon Press, 1984.

Hirschheim, R. "Office Systems: Themes and Reflections," in *Office Systems*, A. Verrijn-Stuart and R. Hirschheim (eds.). Amsterdam: North-Holland, 1986, pp. 193-199.

Hirschheim, R., and Klein, H. K. "Realizing Emancipatory Principles in Information Systems Development: The Case for ETHICS," *MIS Quarterly* (18:1), March, 1994, pp. 83-109.

Hirschheim, R., and Lacity, M. "Information Technology Insourcing: Myths and Realities," *Communications of the ACM* (43:2), February, 2000.

Hirschheim, R., and Miller, J. "Implementing Empowerment Through Teams: The Case of Texaco's Information Technology Division," *Proceedings of the 1993 ACM SIGCPR Conference Managing Information Technology: Organizational and Individual Perspectives*, M. Tanniru (ed.), St. Louis, April 1-3, 1993, pp.255-264.

Iivari, J., Hirschheim, R., and Klein, H. K. "A Paradigmatic Analysis Contrasting Information Systems Development Approaches and Methodologies," *Information Systems Research* (9:2), June, 1998, pp.164-193.

Iivari, J,, and Lyytinen, K. "Research on ISD in Scandinavia," *Scandinavian Journal of Information Systems* (10:1/2), 1999, pp. 135-185.

Ives, B., Hamilton, S., and Davis, G. B. "A Framework for Research in Computer-based Management Information Systems," *Management Science* (26:9), 1980, pp. 910-934.

Kant, I. *The Critique of Practical Reason.* New York: Harper Torch Books, 1929.

Kant, I. *Groundwork of the Metaphysics of Morals*, (translated and analyzed by H. J. Patton). New York: Harper Torch Books, 1964.

Klein, H. K. "Knowledge and Methods in IS Research: From Beginnings to the Future," in *New Information Technologies in Organizational Processes, Field Studies and Theoretical Reflections on The Future of Work*, O. Ngwenyama, L. Introna, M. Myers, and J. I. DeGross (eds.). Boston: Kluwer Academic Publishers, 1999, pp. 13-25.

Klein, H. K., and Hirschheim, R. "The Rationality of Value Choices in Information Systems Development," *Foundations of Information Systems*, 1998 (http://www.cba.uh.edu/~parks/fis/kantpap.htm).

Klein, H. K., and Hirschheim, R. "The Evolution of the Information Systems Discipline: *Quo vadis et cui bono*?" 1999, submitted for publication.

Klein, H. K., and Myers, M. "A Set of Principles for Conducting and Evaluating Interpretive Field Studies in Information Systems," *MIS Quarterly* (23:1), 1999, March, pp. 67-94

Kumar, K. "Program Chair's Statement," in *Proceedings of the Eighteenth International Conference on Information Systems*, K. Kumar and J. DeGross (eds.), Atlanta, GA. December 15-17, 1997, xvii-xix.

Lacity, M., and Hirschheim, R. "The Information Systems Outsourcing Bandwagon," *Sloan Management Review* (35:1), Fall, 1993, pp.73-86.

Lacity, M., and Hirschheim, R. "Benchmarking as a Strategy for Managing Conflicting Stakeholder Perceptions of Information Systems," *Journal of Strategic Information Systems* (4:2), 1995, pp.165-185.

Landry, M., and Banville, C. "A Disciplined Methodological Pluralism for MIS Research," *Accounting, Management, and Information Technology* (2:2), 1992, pp. 77-97.

Langefors, B. *Theoretical Analysis of Information Systems*. Philadelphia: Auerbach, 1973 (originally published in Lund, Sweden: Studentlitterature, 1966)..

Lee, A. "The MIS Field, the Publication Process, and the Future Course of MIS Quarterly," *MIS Quarterly* (23:1), 1999a, pp. v-xi.

Lee, A. "Rigor and Relevance in MIS Research: Beyond the Approach of Positivism Alone," *MIS Quarterly* (23:1), 1999b, pp. 29-33.

Lincoln, Y., and Guba, E. *Naturalistic Inquiry*, Beverly Hills, CA: Sage, 1985.

Lucas, H. "The State of the Information Systems Field," *Communications of the AIS* (5:1), January 1999.

Lyytinen, K. "Empirical Research in Information Systems: On the Relevance of Practice in Thinking of IS Research," *MIS Quarterly* (23:1), 1999, pp. 25-27.

Markus, M. L. "The Qualitative Difference in Information Systems Research and Practice," in *Information Systems and Qualitative Research*, A. S. Lee, J. Liebenau, and J. I. DeGross (eds.). London: Chapman & Hall, 1997, pp. 11-27

Markus, M. L. "Thinking the Unthinkable: What Happens If the IS Field as We Know it Goes Away?," in *Rethinking MIS*, R. Galliers and W. Currie (eds.). Oxford: Oxford University Press, 1999, pp. 175-203

McFarlan, F. W. "Information Technology Changes the Way You Compete," *Harvard Business Review*, May-June, 1984, pp. 98-103.

Morgan, G. (ed.). *Beyond Method: Strategies for Social Research*, Beverly Hills, CA: Sage Publications, 1983.

Mumford, E. *Designing Human Systems—The ETHICS Method*. Manchester: Manchester Business School, 1983.

Mumford, E., and Henshall, D. *A Participative Approach to the Design of Computer Systems*. London: Associated Business Press, 1978.

Nygaard, K. "The Trade Unions New Users of Research," *Personnel Review* (4:2), 1975, as referenced on page 94 in K. Nygaard, "The Iron and Metal Project: Trade Union Participation," in *Computers Dividing Man and Work*, A. Sandberg (ed.), Stockholm: Arbetslivcentrum, 1975.

Ormerod, R J. "Combining Management Consultancy and Research," *Omega* (24:1), 1996, pp. 1-12.

Poora, J., and Hirschheim, R. "Perceptions and Realities of the IT Function: Lessons from Forty Years of Change at Texaco," 1999, submitted for publication.

Quine, W. V. O. *The Web of Belief*. New York: Random House, 1970.

Sabherwal, R., Hirschheim, R., and T. Goles "The Dynamics of Alignment: A Punctuated Equilibrium Model," *Organization Science*, 2000, forthcoming.

Smithson, S., and Hirschheim, R. "Analyzing Information Systems Evaluation: Another Look at an Old Problem," *European Journal of Information Systems* (7:3), September, 1998, pp. 158-174.

Tashakkori, A., and Teddlie, C. *Mixed Methodology: Combining Qualitative and Quantitative Approaches*. London: Sage Publishers, 1998.

Teichroew, D., and Hershey, E. "PSL/PSA: A Computer-aided Technique for Structured Documentation and Analysis of Information Processing Systems," *IEEE Transactions on Software Engineering* (SE-3:1), January, 1977, pp. 41-48.

Walsham, G. *Interpreting Information Systems in Organizations*. Chichester: John Wiley & Sons, 1993.

Whitley, R. *The Intellectual and Social Organization of the Sciences*. Oxford: Claredon Press, 1984a.

Whitley, R. "The Development of Management Studies as a Fragmented Adhocracy," *Social Science Information* (23:4/5), 1984b, pp. 775-818.

Winograd, T., and Flores, K. F. *Understanding Computers and Cognition*. Norwood, NJ: Ablex, 1986.

Winograd, T., and Flores, K. "A Language/Action Perspective on the Design of Cooperative Work," *Human-Computer Interaction* (3), 1987.

Yourdon, E. *Modern Structured Analysis*. Englewood Cliffs, NJ: Prentice-Hall, 1978.

Zmud, R. "Individual Difference and MIS Success: A Review of the Empirical Literature," *Management Science* (25:10), 1979, pp. 966-979.

About the Authors

Rudy Hirschheim holds the Tenneco/Chase International Chair of Information Systems in the College of Business Administration, University of Houston, and is past Director of the Information Systems Research Center. He was the Sir Walter Scott Distinguishing Visiting Professor, Fujitsu Centre for Managing Information Technology in Organizations, Australian Graduate School of Management, University of New South Wales. He has previously been on the faculties of Templeton College - Oxford and the London School of Economics. He has also worked as a Senior Consultant with the National Computing Centre in Manchester, England. His Ph.D. is from the University of London. He and Richard Boland are the Consulting Editors of the John Wiley Series in Information Systems. He is on the editorial boards of the journals *Accounting, Management and Information Technologies; Information Systems Journal; Journal of Information Technology;* and *Journal of the Association for Information Systems*. He is VP of Publications for AIS. Rudy can be reached by e-mail at rudy@uh.edu.

Heinz K. Klein received his Ph.D. at the University of Munich in business administration. He currently is Associate Professor of Information Systems leading the IS group at the School of Management at the State University of New York at Binghamton, where he has been since 1984. Well known for his interests in foundations and methodologies of information systems development, Heinz has published articles on decision support systems, office information systems, information systems research methodology and information systems development in a variety of journals including *The Computer Journal, Communications of the ACM, the Information Systems Journal, Information Technology and People, Computer-Supported Cooperative Work,* and *MIS Quarterly*. He is a member of the editorial board of the *Information Systems Journal* and the Wiley Series in Information Systems. He has co-authored several books, most recently a monograph on *Information Systems Development and Data Modeling:*

Conceptual and Philosophical Foundations (with R. Hirschheim and K. Lyytinen) published by Cambridge University Press. He is co-editor of conference proceedings such as *Systems Development for Human Progress* (North Holland 1989) and *The Information Systems Research Arena of the 90's: Challenges, Perceptions, and Alternative Approaches* (with H.-E. Nissen and R. Hirschheim, North Holland, 1991). His recent research is informed by Critical Social Theory and other social theories. His applied research interests include language-based information systems modeling, information engineering, and social implications of information technology. Heinz can be reached by e-mail at hkklein@binghamton.edu.

16 THE NEW COMPUTING ARCHIPELAGO: INTRANET ISLANDS OF PRACTICE

Roberta Lamb
University of Hawaii, Manoa
U.S.A.

Elizabeth Davidson
University of Hawaii, Manoa
U.S.A.

Abstract

This paper examines the growth of grass roots intranets as an extension of end-user computing. This perspective helps to characterize the nature of intranet development and use as "islands of practice" and provides a background against which the rapid proliferation of organizational intranets in the 1990s can be compared and contrasted with the explosion of personal computers and "islands of end-user computing" in the 1980s. This retrospective analysis of end-user computing is based on academic and business journal literature. The contemporary analysis of intranet development and use is based upon preliminary results from an ongoing qualitative study of midwest U.S. firms in various industries. These analyses highlight two phenomena that are likely to define the shape of intranets and future computing movements: (1) the integration of "intranet islands," not only within the firm but also across organizational boundaries, and (2) the role changes among IS and business area professionals as they work with intranet technologies. The discussion of these phenomena examines the ways in which intranets present unique opportunities for understanding the initiation and widespread uses of a new technology, while at the same time illuminating the changes in organizational roles that so often attend technological interventions.

1. Introduction and Motivation

Use of Internet technologies for internal organizational communications, or intranets, has grown rapidly since the mid-1990s. The scope and range of intranet applications extend beyond simply posting information in web pages to include document flow management, workgroup collaboration, and database access and update. According to an International Data Corporation study (quoted in Machlis 1999), U.S. companies spent over $10.9 billion, or one fourth of web-related project spending, on intranet projects in 1998. More than half of large U.S. organizations had intranets in place, and more than 30 million people used some form of intranet. Some researchers contend that the use of Internet technologies in applications such as corporate intranets represents a radical shift in the nature of information systems (IS) development, IS services, their delivery and associated organizational processes (Lyytinen, Rose, and Welke 1998). We agree that such changes are likely to be dramatic and substantial. However, we also suggest that the rapid adoption and development of intranets is, to some degree, a case of "dejà vu all over again." There are striking similarities between the introduction and growth of organizational intranets in the second half of the 1990s and the advent of end-user computing (EUC) in the early 1980s. Intranet development is in many ways an extension, albeit in new and exciting technological and content areas, of the end-user computing movement. There are, of course, substantial differences in these two episodes of computerization that warrant careful consideration.

From this perspective, we propose there are lessons to be gleaned from comparing and contrasting the first decades of end-user computing with intranet development occurring in the mid to late 1990s. Consideration of similarities and differences in these two computerization movements may help us to forecast how intranets will develop and mature in organizations and to suggest appropriate IS support strategies. This paper begins such a discussion. We first take a retrospective look at the early stages of the end-user computing movement and review key findings and lessons from this period. We then consider how intranets are being introduced, developed, and used in organizations. This assessment is drawn from a research study of intranets and supplemented by business press reports on organizational experiences with intranets. In the discussion, we compare and contrast these two computerization movements to highlight issues that may be reoccurring in the intranet movement and to identify new questions to be addressed. In particular, we are interested in the roles and responsibilities of IS services and business area professionals in the management and control, integration, operation, and support of intranets. Finally, we consider implications for IS research and suggest potential areas of study.

2. A Retrospective Analysis of the Advent of End-User Computing

End-user computing (EUC) began in the mid-1970s, as a growing number of non-IS personnel started using mainframe timesharing facilities to develop business applications.

It expanded rapidly in 1982, when knowledge workers of all types initiated a wide-scale adoption of personal computers (PCs) and software packages, such as spreadsheet programs and word processors. This end-user computing movement arose from users' frustration with long backlogs for IS services, their impatience with formal systems development methods, the emergence of a more computer-literate end-user community, development of more user-friendly software, and, ultimately, the introduction of low-cost microcomputers (Carr 1987; Dotson 1982; Kling and Iacono 1989). It resulted in a shift from a systems development paradigm, in which IS professionals managed and controlled virtually all aspects of computerized systems development and use, to a new paradigm in which business area staff played a substantial role in the development and operation of such applications.

In many instances, the organizational introduction of EUC technologies was a grass-roots effort led, not by the IS department, but by early adopters from business functional areas (Brancheau and Wetherbe 1990). Some information services departments actively resisted early organizational forays into EUC, particularly those involving personal computers (Benson 1983). IS managers feared that users would fail to coordinate development activities and thus waste corporate resources by "reinventing the wheel," fail to document systems and programs, fail to adhere to standards and policies enacted for the corporation, or fail to exercise adequate controls over data integrity and security, all of which would increase the possibility of major computer-related disasters (Benson 1983; Davis 1982; Guimaraes 1984; Guimaeres and Ramanujam 1986; Mayo 1986; White and Christy 1987). Both academic researchers and writers in the business press warned of inherent issues with non-IS professionals doing their own computing and suggested that the IS department should manage and control their efforts. For example, an early research article commented, "The organization's hardware, software, and data are valuable resources which can be lost or diminished if not properly developed and protected. End users, acting independently, cannot always be expected to use these resources in ways that are optimum for the whole firm. Since end-user computing bypasses the monitoring and control mechanisms built into MIS department computing, there is no formal check on user behavior" (Leistheiser and Wetherbe 1986, p. 338). While acknowledging that the EUC movement would continue to flourish, writers used such terms as "catastrophe" (Kirkley 1989, p. 96) and "disastrous" (Essex, Magal, and Masteller 1998) and outlined disaster scenarios (Davis 1982) that might occur without the IS department's intervention, guidance, and control over EUC. Not surprisingly, by the early 1980s, MIS managers ranked "facilitation and management" of EUC as the second most critical success factor for their organization (Dickson et al. 1984).

IS managers, while doubting end-users' qualifications to manage and control computing activities, faced increased expectations for providing services to this constituency. Early predictions that EUC would reduce the backlog of application development requests and maintenance for the IS department (Leitheiser and Wetherbe 1986; McLean 1980) proved to be unfounded. Instead, EUC addressed the "invisible backlog" for computerized applications (Carr 1987; Garcia 1987) and at times increased the backlog with requests to the IS department for PC-based applications. There was a lack of consensus on whether the IS department's EUC strategy should focus primarily on providing the right information technology tools and training, on actively intervening in EUC activities to enhance organizational productivity, or simply on keeping end users "out of trouble" (Munro and Huff 1988). Strategy options ranged from a *laissez faire* approach to establishing policies and procedures for EUC and strictly enforcing them (Goldberg 1986; Kirkley 1989; Leitheiser and Wetherbe 1986; Munro and Huff 1988).

By the mid-1980s, IS departments had begun to act on their perceived dual responsibilities to support and encourage EUC as well as to control these activities to ensure the corporate good (Leitheiser and Wetherbe 1986). This commonly involved integrating EUC efforts within the organization by standardizing desktop applications and by networking the separate "islands" of computing that were distributed throughout the firm (McFarlan, McKenney, and Pyburn 1982, 1983). As PCs became workstations on the network versus stand-alone "mini-DP shops," the IS department took a stronger role in selecting the standard PC hardware and software platform which they would support and enforcing operating systems standards compatible with networking technology and devices (Freeland 1987; Guimaraes and Ramanujam 1986). Control tactics included having approval over purchases, negotiating approved vendor lists, and restricting access to mainframe computers (Munro and Huff 1988). As controls increased, so did support for end-users, as general consulting, end-user training, help desks, product and technical support, micro-mainframe communications, and software libraries were added to IS services (Guimaraes and Ramanujam 1986; Leitheiser and Wetherbe 1986).

The organizational structures that emerged as the focus of support and control for EUC were the Information Center (IC) and the "user group." The number of ICs grew rapidly between 1982 and 1984, with the growth rate leveling off by 1987 (Garcia 1987). Initially, ICs took on the task of increasing computer literacy by focusing on end-user training (Dotson 1982). Although IC's were typically perceived to be under-funded (Garcia 1987; Goldberg 1986), their organizational mission grew to include promotion, support, control, and management of EUC (White and Christy 1987). By late 1988, a survey comparing services offered by ICs to services desired by organization members indicated that ICs were giving too much emphasis to training and hardware/software evaluation, installation and support, and not enough to developing PC-based applications and helping people understand data sources and data transfer (Ranier and Carr 1992). User surveys indicated that people were most satisfied when IC staff were located close to end-users and offered a variety of support services, but less so when ICs were large and focused on mainframes, possibly because they exerted too much control over EUC (Bergerson, Rivard, and DeSerre 1990). More sophisticated business personnel, defined as end-users who supported others in their functional department by developing computerized applications, had greater perceived needs for IC services, such as standards and guidelines, support staff, and post-development support, and when such support was provided, they were more satisfied (Mirani and King 1994). Such findings indicated that ICs needed to tailor the type and level of support to each end-user community. A key implication was that when IC staff did not have the requisite skills, they should facilitate the formation of user groups, because end-user developers sometimes had more in-depth knowledge of specific development contexts (Mirani and King 1994). In fact, IC growth and expansion was achieved by decentralizing some support activities to these groups (Magal, Carr, Watson 1988). However, this led to role ambiguity for IC staff who were positioned between the IS department and end-users, resulting in dissatisfaction and turnover within the IC (Gupta, Guimaraes, and Raghunathan 1994). In addition, achieving meaningful participation in user groups proved difficult, and the value that the

IC staff contributed to the process tended to be limited to administrative duties (Alexander 1989).

In less than a decade, the end-user computing movement had transformed organizational IS. Beginning as an insurgent, grass roots action, resisted by the IS department, EUC had become an integral part of the corporate-wide computing environment. The degree of institutionalization and routinization of EUC is suggested by the declining importance of EUC issues in surveys of IS managers' most pressing concerns from second position in 1983 to sixth in 1986 and twelfth by 1989-90 (Niederman, Brancheau, and Wetherbe 1991). By the mid-1990s, "end-users" were computer-literate business area professionals who utilized a variety of information technologies to fulfill their functional responsibilities. They commonly controlled the content and function of their EUC activities. The IS department, or its outsourced counterpart, provided the information technology infrastructure and support services that facilitated EUC activities while controlling standards for hardware, software, and the network.

3. A Contemporary Analysis of Intranet Adoption and Use

The rapid growth of intranet applications during the 1990s was in part an outgrowth of the EUC movement of the 1980s. With wide-scale EUC, there existed a computer-literate base of knowledge workers, a distributed network of desktop computers, and a plethora of PC-based software applications. This provided a fertile organizational setting in which intranets, fueled by the development of low cost Internetworking and browser software, and the glamour and success of the Internet's World Wide Web (WWW), could take root and flourish. Internet technology vendors and IS practitioners expected the growth of intranets to follow the innovation diffusion curve of PCs, spreadsheets, and other commercially successful EUC technologies (LaPlante 1997).

Unfortunately, although the EUC movement has been widely observed and richly commented on in practitioner and academic literatures, few studies have examined the theoretical implications of EUC, and even fewer have used theoretical guidelines to inform empirical research (Robey and Zmud 1992). Brancheau and Wetherbe's (1990) study of spreadsheet adoption is one notable exception. Their application of innovation diffusion theory to EUC research suggests that, contrary to popular expectations, this theoretical approach does not provide a complete explanation of technology diffusion in organizations. Interestingly, it did not predict the failure of IS/IC departments to serve as change agents. In their study, only three of 21 departments were involved in introducing personal computers and spreadsheets into their organizations. Brancheau and Wetherbe's quantitative analysis does not offer an explanation for this phenomenon, but they have recommended that future EUC research should examine the organizational contexts of technology diffusion, interpersonal communication channels among adopters, actual use of the technologies and observation of the processes that reshape the organization as well as the technology.

3.1 Theoretically Guided Qualitative Research

In this section, we discuss results from an ongoing intranet study by one of the authors on intranet initiation, development, and use. The study has been designed to focus on the gaps in diffusion theory explanations by using institutional and constructionist approaches to guide qualitative research. In particular, this research seeks to build a more robust explanation of the ways in which new organizational technologies come to be shaped the way the are, and how differential benefits accrue to different user organizations, by examining the discrepancies between theoretically projected and actual uses of intranets.

The findings presented here are based on preliminary analyses of data collected on intranet use in midwest U.S. companies from various industries. To date, over 100 manufacturing firms and law firms have been queried about their intranet development and use; over 20 organizations have been visited to view their intranets; and in-depth onsite studies have been conducted at a Fortune 500 manufacturing firm and at a prominent international law firm. Data has been collected through interviews, direct examination of intranets (Lofland and Lofland 1995), and examination of intranet logs, usage and development guidelines, intranet page samples, and other relevant documentation (Miles and Huberman 1994). One key objective of this study is to determine the influences that shape intranet development and use, particularly the influences of interorganizational networks (Latour 1987). As the study proceeds from one industry to the next, constant comparative methods are being used to formulate data categories, identify cross-case patterns, and develop theoretical leads for further investigation (Strauss and Corbin 1990). The analyses presented in this paper focus on how intranets come to be developed and used the way they are and the roles firm members take on in development and use. To corroborate these preliminary observations, we have supplemented the findings with anecdotes from relevant business press reports of other organizational intranet experiences.

One of the most striking observations made during data collection and preliminary analysis is the way in which intranet development has mirrored EUC experiences (see Table 1). This finding motivated the preceding retrospective analysis and has lent further insight to the ongoing study. We now examine observed intranet phenomena by adopting this new perspective of viewing intranets as an extension of EUC.

3.2 Initiating Intranets Through Grass Roots Efforts

A number of studies and informal reports confirm that the initial spread of intranets throughout organizations, like the early spread of PCs and EUC in the 1980s, has been largely the result of grass roots introductions (Lamb 1999; Rooney 1997; Scheepers 1999). Individuals or small, ad-hoc teams have crafted a set of web pages that link together documents of local interest or Internet sites for common use and have provided access through the existing organizational networks. People are able to do this, even though their primary organizational role may not involve information systems development, for any combination of the following reasons:

Table 1. Similarities between Key Phenomena in the EUC and Intranet Movements

Key Phenomenon	Early EUC Movement	Intranet EUC Movement
Organizational adoption of technology: grass roots efforts	Motivated business users adopted PCs and relevant software, resulting in multiple, diffuse adoptions.	Savvy business personnel develop intranets for specific needs, resulting in multiple, diffuse adoptions.
Development and support roles: conflated roles, initial lack of MIS support	Business area personnel developed application content, operated systems, and used system outputs. IS departments initially resisted EUC, then jumped on bandwagon to provide training and support.	Business area personnel develop content, sometimes operate servers and manage intranet, and use the system. IS departments frequently decline to support grass roots efforts, while simultaneously planning enterprise-wide intranets.
Technology ownership and control: enterprise-wide applications, standards and integration	Concern about integration led to IS departments taking a strong role in setting and enforcing standards, integrating PCs in networks.	Concerns about wasteful duplication of effort, lack of intranet standards, and enterprise-wide integration motivates IS departments to favor top-down approaches to intranet planning and to curtail grass roots efforts.
IS organization: emergence and decline of formal structures	Information centers emerged from IS group to support EUC. As EUC matured, need for IC training and support roles decreased.	Some organizations are establishing small support groups, or Web services, staffed by consultants or in-house Internet technology experts, to help end-users establish intranet sites.

- Internetworking is already in place. Internet e-mail access is available and is used routinely.
- The need for additional hardware is minimal. Most high-end PCs can act as intranet servers.
- Intranet server and browser software is inexpensive, or free, when servers are bundled with operating systems; browsers are freely downloadable.
- Most people working in organizations are already familiar with an array of desktop applications, from e-mail and word processing to spreadsheets and desk-top databases. A number of these applications make it very easy to convert data and documents into web pages.
- Interest in Internet technologies is high and people are willing to use their own time to learn to build an intranet.

These factors alone, however, do not determine whether or not grass roots intranets will develop. External influences are the critical motivators of grass roots intranet development. Findings from this study show that many well-used intranets are in fact technologies-of-interaction between firm members and the outside world. As we discuss later, this raises questions about what to consider when planning intranet integrations and suggests that, in some instances, we may need to carefully examine what "integration" means and who benefits from it.

3.3 Growth and Spread of Intranets Through Islands of Practice

Despite promises from technology vendors that intranets would seamlessly link organizational units into one holistic knowledge network, many organizational intranets consist of a number of loosely connected sites developed through grass roots efforts (Lamb 1999). Like other innovative technologies, intranets have been adopted into the firm multiple times by different groups. These groups are prompted to develop their local intranets by a combination of internal and external influences. Common internal influences are the need to reduce communication delays and to eliminate printing and distribution costs. Some particularly strong external influences come from customers, regulators, and industry standards bodies. At Fortune Manufacturing,[1] for example, corporate research scientists have built an intranet to help them work with customers to develop new product applications. Quality control managers have constructed a quality documentation system to help meet ISO 9000 certification requirements. The communications department has hired an outside vendor to create an intranet platform for corporate communications. The research and development library has established its own intranet that links research project documents and relevant Internet sites, and these librarians are helping Fortune marketers create new customer-focused databases using Lotus Notes intranetworking tools. In large firms, disparate groups like these may know about each other's efforts, but sometimes they don't. This results in sets of computer-based application and documentation resources that are used by only one group within the firm—a configuration that resembles the "islands of computing" that proliferated in the 1980's with the onset of EUC and personal computing. But even when organizational intranets are linked to one another and easily accessed by all firm members, cross-use is uncommon, resulting in what we have termed "islands of practice."

3.4 Developing Corporate Intranets

In organizations where intranets have been in place for more than one or two years, grass roots developments have been followed by a coordinated, IS-led effort to make useful content more widely available, to eliminate duplication of content and development effort, and to standardize, to some degree, the presentation of content and the appearance of intranet sites within the organization. This is the point at which intranets frequently

---

[1]Fortune Manufacturing is an alias for one of the firms studied.

have caught the attention of upper management, who may be struggling to develop a strategy for leveraging Internet technologies, and who then give the go-ahead for further development of enterprise-wide intranet applications. IS department members are enthusiastic about creating these corporate intranets and are encouraged by the show of executive support, which often allows IS to implement some key applications, like enterprise-wide phone directories, and to form Web services groups that will support grass-roots intranet developers.

The expectation that is often set by this sequence of events is that sharing information throughout the organization is mandatory and, therefore, all intranets within the firm should be integrated, even though these corporate intranets often have very different, top-down implementations and communication purposes than the local intranets they seek to integrate. When the organization is comprised of a large number of semi-autonomous entities in various locations, intranet integration becomes problematic. Again, consider Fortune Manufacturing, a parts manufacturer that has grown over the last decade through a series of acquisitions and mergers. At least 15 intranets are currently used in Fortune. Most are linked together and protected by a common firewall. However, a few of the more recently merged firms are still operating on different networks and their intranets are not accessible by others in the larger Fortune organization. A corporate-wide committee was convened in May 1998, to develop an intranet integration plan, but soon afterward, yet another merger was announced. Since then, a few groups have proceeded with limited integration of a small number of closely related intranet projects, but the committee has suspended its corporate-wide effort, waiting for a more opportune and stable time to try again—which may not be anytime soon, given Fortune's current rate and form of growth.

3.5 Enterprise-wide Intranetting and "Killer Apps"

The "killer apps" of end-user computing were word processors, spreadsheets, and games. People worked with these applications on a personal basis as productivity tools, often blurring the line between work and play, bringing work home and personal projects into the office. They shared the documents and files from these applications, first within local work groups and later with remote collaborators through PC networks and email.

The extension of end-user computing into the intranet realm also spans work and play, as well as personal and organizational use. And, despite the prevalence of local use models, a few intranet applications have begun to emerge as the killer apps of corporate intranets: custom portals, compliance systems, and enterprise-wide directories. The use of these applications becomes central to strategic daily operations and generates widespread and heavy daily use: when they come online, overall intranet use skyrockets.

Custom portals, like some of the earliest intranet navigation pages, are filled with links to Internet sites and other firm intranets. These links are specifically relevant to the daily concerns of the individual, or group, that uses them; they typically access sites for customers, news, industry research, baseball scores, local restaurants, travel, stock quotes, and other sites of personal and professional interest. Custom portals constitute a highly specialized "view of the world" and people usually configure their browsers to use these pages as the "home" or startup page, rather than using the one-size-fits-all corporate intranet home page. A few firms have followed Yahoo!'s model for allowing people to configure their own pages through the addition of an unlimited number of links and by adjusting a set of parameters that customize the page layout. At firms that have not provided configurable portals, people often create a personal page on Yahoo!'s site to achieve the same result.

Compliance systems intranets are highly strategic to the firm. Among manufacturing firms, the most common are quality management document systems that adhere to ISO 9000 and QS 9000 standards, Y2K databases that track the status of Y2K compliance throughout the firm, and health, safety, and environment systems that report statistics that government agencies require or industry associations monitor. Quality management intranets, for example, are becoming more prevalent in the process for achieving ISO certification. ISO auditors regard these intranet documentation systems as highly effective ways to ensure that plant operations follow the stated processes, that process changes are quickly disseminated to people that need to know about them, and that the confusion of multiple or outdated paper documents is avoided. Although ISO auditors do not make recommendations about how compliance should be achieved, key customers, like Boeing, Ford, and GM, do; these companies have pushed their suppliers to both achieve compliance and to use intranets in the process.

Portals clearly reflect the influence of interorganizational relationships, as do compliance systems, which specifically focus on meeting the requirements of customers and outside agencies and on making that compliance visible. Enterprise-wide directories, on the other hand, connect people *inside* the firm. People interviewed in a variety of organizations frequently mentioned the online phone directory as the application they use most, particularly when a search engine is integrated into the directory application or when people are working in remote sites of the organization. Some firms have linked these phone directories to human resource and work project databases in a way that profiles the expertise of people in different departments and locations. These enterprise-wide directories help people find experts within their organization, providing a way to navigate between the islands of practice.

3.6 Combining Development, Operational, and Use Roles

Although one might guess that the most intensively used applications would be enterprise-wide applications, local applications are just as likely to receive heavy use. The greatest use seems to happen when intranet developers, in the role of sophisticated end-users, are also content developers, as well as content owners and content users. Scheepers (1999) has identified several key roles that firm members take on when implementing intranets. Some are the familiar "sponsor" and "champion" roles, but a few, like the intranet coordinator, intranet developer, and content provider are new and research confirms their importance to intranet success. However, the findings of this study show that intense use of intranets occurs most regularly when these roles are conflated—that is, effectively combined into one role. Conflated roles emerge when two or more intranet-related roles are all assumed by *one* person, who also uses the content, and also when the responsibilities of one or more roles are shared *among* those who use the content. Such role changes emphasize the local and somewhat isolated nature of intranet development and use.

Conflated roles can be critical to grass roots intranet success, but a merging of roles is not always feasible. Related studies of online service use have shown that innovative uses of online technologies are often tied to knowledge workers' ability to assume dual roles, and that professional hierarchies, like those found in law firms, discourage this type of role conflation (Lamb 1997). Law librarians and other firm staff frequently launch their own intranet sites, and even cater their intranet offerings to firm attorneys. But few attorneys use the sites regularly, and even fewer contribute content or launch their own sites. Because large law firms are actually designed as islands of practice, one might expect each of these practice areas to be fertile ground for grass roots intranets, although it is rare for associate attorneys to be given the time or the encouragement to develop new information systems. However, results from a recent study of Norwegian law firms (Gottschalk 2000) suggest that attorneys are slow to adopt and use intranets, even in firms that encourage cooperation and provide time for knowledge sharing.

3.7 The Role of the IS Department in Supporting and Managing Intranet Development

While grass roots intranets were spreading, most IS departments were preoccupied with upgrading and maintaining desktop computers and the corporate network, dealing with the Y2K bug, and implementing enterprise resource planning packages such as SAP. As a result, they had little attention to devote to local intranet developments. As noted earlier, analogous circumstances prevented IS departments from responding to the computing needs of business professionals during the advent of end-user computing. Beyond these issues of time and resources, there are questions about strategies for supporting or controlling EUC-type activities and appropriate roles and responsibilities.

Preliminary results of the intranet study indicate that IS specialists are often conflicted about their role in intranet development. In some firms, they spearhead the corporate intranet effort. At the same time (and in some of the same organizations), IS managers have tried to squelch or delay the grass roots intranet projects, either by refusing to provide hardware, networking, and software support, or by trying to homogenize all the intranets within a firm. The reasons, expressed in interviews and reiterated in business publications, are varied, multiple, and often easily understandable (Lamb 1999; McCrory 1997; Rooney 1997). Some MIS groups already have an intranet project planned that is more comprehensive and useful to more groups within the organization. They, therefore, view the existing ad-hoc intranets as poorly designed, repetitious, or counter-productive. They express fears that servers not under MIS control could allow for a breach of security and a loss of proprietary information. They also worry that new intranet servers and applications will add to the already heavy MIS burden of computing support. Unfortunately, few of these IS developers have a deep understanding of the content of grass roots intranets to build useful, specialized intranets. In contrast, the grass roots developers are savvy domain experts, with a robust set of technical skills resulting from years of "end-user computing."

4. Discussion: Lessons to Learn and Unlearn from the Past

The end-user computing movement has developed and matured over the last two decades. The intranet movement, and indeed the whole paradigm switch to Internet technologies as the basis for the design and delivery of information services and products, has just begun (Lyytinen, Rose, and Welke 1998). We have tried to illustrate in the preceding sections (and summarized in Table 1) substantive similarities between these two movements to support our argument that lessons learned from the early days of EUC may provide insights into the trajectory of intranet adoption and assimilation processes within organizations. Also important are contextual differences between these two movements (see Table 2) that suggest there may also be lessons to unlearn as we apply the experience gained during the 1980s EUC movement to this new technological development. Our discussion in this section centers on two primary topics: first, the feasibility and the desirability of intranet integration, and second, the roles and responsibilities of various organizational groups, particularly the IS department in fostering and managing the use of intranet technology.

Table 2. Differences between Key Phenomena in the EUC and Intranet Movements

Key Phenomenon	Early EUC Movement	Intranet EUC Movement
Integration	Stand-alone PCs inhibited sharing of data files and communication technologies like e-mail; network integration was highly desirable.	Intranets are already built on corporate networks and standard file transfer and networking technologies facilitate information exchange.
Standardization	Multiple hardware and software platforms limited file exchange and complicated desktop support; standardizing PC platforms reduced costs and improved support.	Intranet technologies already adhere to a range of technical standards.
Role Changes	Few end-users were computer literate; a large-scale training program was needed in the first years of EUC.	Many users are computer literate and are able to utilize intranet development software without formal training or support.
Organizational Changes	The need for IS departments to manage and control security and integrity of computer operations was widely acknowledged.	IS departments have often been outsourced , diminishing their organizational power and influence.

4.1 To Integrate or Not to Integrate?

Grass roots efforts are typical during the initial spread of a new technology (Iacono and Kling 1995), but as the technology matures, a different, often more centrally controlled, approach prevails. This was the case with EUC and the introduction of personal computing. However, with intranets, we may see a different pattern emerge, as the grass roots approach is being refueled by vendors packaging "instant intranet" products expressly geared for deployment by organizational "renegades" (LaPlante 1997). If this grass roots pattern of intranet development is sustained, then perhaps we should pay more attention to its dynamics and motivating factors than we might if we considered it to be just a passing phase in the maturation of this technology.

Soon after islands of computing began to proliferate in the 1980s, IS departments stepped in to standardize applications and to integrate PCs and EUC via local area networks. The islands thus became connected, allowing for more controlled growth. Disconnected use and uncontrolled growth is a problem that currently raises IS managers' concern. Of greater concern is the fact that most grass roots intranets are not integrated and are often designed in a way that makes future integration with other organizational intranets difficult. Although the perceived need for full integration across intranets is common across firms, in fact it is hard to achieve, even in slow-growth organizations, beyond the first or second intranet web page level, where the use of common graphics, logos, and layouts may convey a common theme. Furthermore, this thin veneer of commonality can actually mask the highly varied, unique uses that a single firm may make of intranets. "Integration" can have a unique meaning for the various organizational stakeholders: management, the IS department, content providers, and end-users may each hold very different ideas about what makes a useful intranet "better."

Swanson and Ramiller (1997) have examined the ways in which new technologies, such as intranets, are applied and diffused among organizations. They suggest that an "organizing vision" guides technology use within a community and that each firm is influenced by this vision. Some organizing visions *do* appear to motivate intranet applications and uses, but such visions do not guide actions at the firm level. There are multiple sources and targets of influence within an organization. These spheres of influence are often the source of competing visions held by members at different levels of the firm and we suspect that they may be synonymous with communities-of-practice. With intranets, the visions that matter aren't restricted to the IS/IT professionals and *their* interorganizational contacts, as Swanson and Ramiller imply. Corporate scientists have a vision for intranet use, quality control managers have a vision, communications officers have a vision, industry marketers have a vision, and corporate librarians have a vision. Within a firm, these visions may all compete, but they don't easily merge into one vision, and the intranets that are created to serve each vision don't easily converge to form one system. As Iacono and Kling have observed, the influences that spur technology adoption may be widespread movements, but their effects are distinctly local. Thus, we may have to re-examine whether or not integration is the universal "good" that it often is thought to be. Alternatively, we may need to consider the need for new types of integration: those that integrate intranets across organizational boundaries, as well as those that remain behind the firewall.

4.2 Who Can Best Develop, Operate, and Manage Intranets?

IS departments are engaged in the same type of debate between their perceived need to support and encourage EUC in intranets and their responsibility to control and manage computer and network-related activities as they were fifteen years ago (Keen 1997; Ouellette 1999; Sliva 1999). The IS role in the first decades of the EUC movement was based on the assumption that IS professionals could manage and control some aspects of computerization more effectively than business area personnel, for example, ensuring the security and integrity of data and networks and taking into consideration the needs of the whole organization, particularly for future integration of EUC applications. As the perceived need to connect PC islands through a corporate network and to corporate data sources arose, and as the cost and difficulty of supporting a varied technological infrastructure increased, IS departments were able to set and enforce standards for hardware, operating systems, and some application software. Today, IS professionals are voicing similar fears about the ability of non-IS personnel to develop, operate, and control intranets, citing the need to impose "order from chaos," secure corporate data, monitor network traffic, set limits on what employees can do on an intranet, ensure documentation and continuity of user-developed intranet sites, reduce duplications of effort, and so on (Rooney 1997; Sliva 1999). Although some IS departments are attracted by the glamour and success of intranets and would like to lead their organization's efforts, many IS managers are wary of the demands intranet development could place on their departments, as they were with EUC. They do not want to "be in the content-creation business" (LaPlante 1997). Furthermore, they are aware of the delicate balance between necessary controls and stifling entrepreneurial spirit, which feeds intranet growth (Sliva 1999).

Despite these similarities, careful thought should be given to applying lessons learned in the early days of EUC to EUC in the era of intranets. Intranets are already built on standardized, corporate-wide networks and employ industry-wide standards for network protocol and multi-media data display and manipulation. Furthermore, if local intranets are not integrated with other organizational intranets, the need for the IS function to manage and control intranet development and operations may be even less compelling. One might argue that IS departments are still responsible for ensuring that computing resources are used in ways that contribute to the overall corporate good. However, their ability to take on such a role has been weakened since the early days of EUC. Many MIS departments have now been outsourced, limiting the ability of IS professionals to intervene or even influence EUC activities (Markus 1999; Markus and Benjamin 1996).

The question of how IS departments can add value to EUC efforts in the intranet era must also be carefully considered. The mission of information centers established to support end-users in the early days of EUC was to raise computer literacy through training and to usher in the best hardware and software. ICs were slow to give up this mission and to customize their services to accommodate more sophisticated application developers in the EUC community. In the intranet era, "end users" cannot be treated by IS professionals as low-level, computer-fearful, data-entry staffers who don't know what they want or need in computerized applications. Instead, many are technologically savvy and have far superior knowledge of the content needed in intranet applications. Furthermore, as the study cited in this paper suggests, conflated roles of developer/user may be a critical success factor in intranet use. Such a role is not unique to the intranet movement. This was true in the era of EUC: business personnel developed the content of applications and, even after standardization of hardware and software platforms, they continued to act as end-users **and** developers. Given these contextual factors, the mission, tasks, and activities of an IS department in support of intranet development may be substantially more limited than those of the information centers of the past.

5. Opportunities for Further Research

The foregoing discussion raises some interesting questions about the nature of intranet development and use, its support for interorganizational coordination, and the role challenges it presents to IS professionals and business area professionals alike. Based on this analysis and the differences we see between early EUC and current intranet phenomena (summarized in Table 2), we have identified two key areas where we believe a program of research could sharpen our understanding of how end-users and their intranets alter the dynamics of knowledge work.

5.1 Can Intranet "Integration" Cross Organizational Boundaries?

Intranets can be effectively designed to support communities of practice as well as interfunctional corporate communications. Although study data provides ample evidence for both types of intranets, we have chosen to emphasize the less often reported findings that highlight the role of intranets in support of interorganizational interactions. This focus has prompted us to re-examine what "integration" means within intranet contexts. We have suggested that many intranets may be more effectively integrated *across* organizational boundaries rather than just *within* them. In organizations where intranet use is intense, multiple intranets commonly coexist—each one created and used by a separate group in the firm. Although management may want to facilitate integrated intranets, deeply integrating these islands of practice, beyond the infrastructural components needed for interoperability, may not make sense within the firm.

The fundamental question becomes one of defining what we mean by integration and what it means to "integrate intranets across organizations." This concept could be articulated further, based on an examination of the interactions among members of communities of practice that use intranets. A closer look at intranet implementations could help identify how they are shaped by interorganizational interactions and where opportunities for integration might be found. As the current intranet study proceeds, these interactions and their influences on intranet development will be examined. We expect, based on the institutional and constructionist perspectives that guide this research, to find that different firms will find different types of integrations to be more effective, and that the degree of integration will vary with the degree of intranet institutionalization.

Research can also examine interfirm intranets that already exist. Such studies could indicate how collaborations are currently extending islands of practice into their communities of practice. Data from this study indicate there are at least two ways in which integration with other communities is beginning to happen outside the firm. Both of these involve highly selective access. One example involves direct extension of the intranet infrastructure: the office of a corporate attorney is physically wired to the intranet of their client, a large metropolitan newspaper. Another example uses a new technology called "tunneling": members of a government agency can view limited portions of a research firm's intranet.

It might also be advisable to extend the scope of typical IS research projects to capture the characteristics of intranet collaborations that take place in 24-by-7 "Internet time" as well as the characteristics of the communities of practice these collaborations support. Social constructionists have suggested that the locus of technological innovation often lies within communities of practice (Brown and Duguid 1991; Constant 1987), but the kinds of studies that can effectively investigate the whole community are often beyond the means of individual researchers. We agree with Lyytinen, Rose and Welke (1998), who have suggested that, in order to study networking technologies, IS research studies need to change. Researchers must be prepared to work on long term, large scale, distributed research projects. The one site, in-depth organizational studies that most of our networking knowledge derives from have been insightful in many ways, but will be inadequate in the ever-expanding universe of Internetworking. Even within one nominal organization, an intranet research project more and more often becomes global in scope.

5.2 How Do Intranet-related Roles of IS and Business Area Professionals Need to Change?

We have argued here that the intranet movement of the mid-1990s is an outgrowth of and similar in many ways to the EUC movement of the 1980s, but there are important contextual differences that should give IS departments cause to pause and rethink their role in managing and supporting intranet activities. Prior research on EUC and information has not yet been updated to include the intranet phenomenon. Organizational experiences with intranet development are just now developing (approximately four years from earliest use) to the point where different models of support are emerging and patterns of failure and success may now be evident. A research program to help identify and evaluate potential strategies for providing effective support for intranet development and use could address questions such as these:

- What are the support services most desired and needed by intranet developers? How can these services be structured to accommodate varying levels of business developer sophistication?

- How do IS departments and non-IS professionals view their roles and responsibilities for intranet development? What organizational structures will best fit support needs? Where in the organization should intranet support be located and how should it be organized?

- What are the strategies and tactics that organizations could use to achieve a balance between encouraging and controlling intranet development? Do IS departments still have sufficient credibility to take an active role in this area, or will they be limited to providing and managing the network utility on which non-IS professionals develop intranets?

As previously noted, much of the early research on EUC was a-theoretical, reducing its value in predicting outcomes in this new context. As these topics are addressed, a suitable theoretical lens should be employed, both to re-examine the past and to investigate evolving trends (Robey and Zmud 1992).

Beyond the specific empirical questions posed, questions about IS support in the intranet era point to the larger question of the role the IS department may play in organizations in the future (Markus 1999). Markus and Benjamin (1996) have suggested that IS departments as well as individual IS professionals should expand their understanding of their organizational role from agents of technology change to include roles as facilitators and advocates of technology-enabled organizational change. They recommend a research approach that would address descriptive, explanatory, and prescriptive questions such as those we have posed above. We concur that such an approach would be beneficial when focused on the phenomenon of intranet development and that answering these questions in this context will help us to understand the larger questions about the future of the IS department in organizations.

In addition, we suggest that research attention should be given to the role of end-user developers and the organizational contexts in which they can be most productive and effective in their conflated roles of content owners, developers, and users. In a study by Yates, Orlikowski, and Okamura (1999), the successful use of a networked newsgroup application was attributed to a few members of the software development team who introduced, supported, and used the application themselves. In that study, role conflation provided some basis of credibility for the new application (i.e., the credibility of the developers introducing it), and helped encourage use by others. Yates, Orlikowski, and Okamura concluded that "the use of a new electronic medium within a community is strongly influenced not just by users but also by those individuals who implement the technology, provide training, propose usage guidelines, and alter the technology to adapt it to changing conditions of use" (1999, p. 83). This may help to explain why the most-used intranets are those implemented by content-owners, and why these intranets tend to remain distinct islands-of-practice. There are indications, however, that institutionalized constraints may hinder widespread and sustained development of such roles in some organizational contexts. Conflated roles may also heighten stresses between specialists in the IS department and non-IS employees over access to and proficiency in this glamorous new technology and limit the ability of these groups to collaborate.

These proposed areas of study—intranet integration and the co-evolution of the roles of IS and business area professionals—touch on just a small set of the research questions that the organizational use of Internet technologies presents. But we believe that a focus on intranets could reward researchers with some unique opportunities for understanding the initiation and widespread uses of a new technology, while at the same time illuminating the changes in organizational roles that so often attend technological interventions.

References

Alexander, M. "End Users in Charge," *Computerworld* (23:5), 1989, pp. 43-44.

Benson, D. "A Field Study of End User Computing: Findings and Issues," *MIS Quarterly* (7:4), 1983, pp. 35-45.

Bergeron, R., Rivard, S., and DeSerre, L. "Investigating the Support Role of the Information Center," *MIS Quarterly* (14:3), 1990, pp. 247-260.

Brancheau, J., and Wetherbe, J. "The Adoption of Spreadsheet Software: Testing Innovation Diffusion Theory in the Context of End-User Computing," *Information Systems Research* (1:2), 1990, pp. 115-143.

Brown, J. S., and Duguid, P. "Organizational Learning and Communities-of-Practice: Toward a Unified View of Working, Learning and Innovation," *Organization Science* (2:1), 1991, pp. 40-57.

Carr, H. "Information Centers: The IBM Model vs. Practice," *MIS Quarterly* (11:3), 1987, pp. 325-338.

Constant, E. W., II. "The Social Locus of Technological Practice: Community, System, or Organization?" in *The Social Construction of Technological Systems*, W. E. Bijker, T. P. Hughes, and T. J. Pinch (eds.). Cambridge, MA: MIT Press, 1987, pp. 221-242.

Davis, G. B. "Caution: User Developed Systems Can Be Dangerous to Your Organization," MISRC-WP-82-04, MIS Research Center, University of Minnesota, Minneapolis, Minnesota, 1982.

Dickson, G., Leitheiser, R., Wetherbe, J., and Nechis, M. "Key Information Systems Issues for the 1980s," *MIS Quarterly* (8:3), 1984, pp. 135-154.

Dotson, T. "The Information Center: Fast Relief from Programming Backlog," *Computerworld* (16:19, 1982, pp. 21-33 (in-depth section).

Essex, P., Magal, S., and Masteller, D. "Determinants of Information Center Success," *Journal of Management Information Systems* (15:2), 1998, pp. 95-117.

Freeland, D. "Your Personal Computer Isn't Yours Anymore," *Computerworld* (21:34), 1987, pp. 69-71.

Garcia, B. "Info Center Identity Crisis," *Computerworld* (21:40), 1987, pp. 15-16.

Goldberg, E. "Rise of End-User Computing Brings New Challenges to MIS/DP,' *Computerworld* (20:28), 1986, pp. 77-78.

Gottschalk, P. "Knowledge Management in the Professions: The Case of IT Support in Law Firms," in *Proceedings of the Thirty-third Annual Hawaii International Conference on System Sciences*. Los Alamitos, CA: IEEE Computer Society, 2000.

Guimaraes, T. "The Evolution of the Information Center," *Datamation* (30:11), 1984, pp. 127-130.

Guimaraes, T., and Ramanujam, V. "Personal Computing Trends and Problems: An Empirical Study," *MIS Quarterly* (10:2), 1986, pp. 179-187.

Gupta, Y., Guimaraes, T., and Raghunathan, T. "Attitudes and Intentions of Information Center Personnel," *Information and Management* (22:3), 1994, pp. 151-160.

Iacono, S., and Kling, R. "Computerization Movements: The Rise of the Internet and Distant Forms of Work," in *Ecologies of Knowledge*, S. L. Star (ed.). New York: State University of New York Press, 1995.

Keen, P. "Pendulum Swings between Rock and Hard Place," *Computerworld*, January 2, 1997, p. 76.

Kirkley, J. "What ever happened to the PC revolution?" *Computerworld* (23:11), 1989, pp. pp. 95-100.

Kling, R., and Iacono, S. "The Institutional Character of Computerized Information Systems," *Office: Technology and People* (5:1), 1989, pp. 7-28.

Lamb, R. "Interorganizational Relationships and Information Services: How Technical and Institutional Environments Influence Data Gathering Practices," unpublished Ph.D. Dissertation, University of California, Irvine, 1997.

Lamb, R. "Using Intranets: Preliminary Results from a Socio-Technical Field Study," in *Proceedings of the Thirty-second Annual Hawaii International Conference on Systems Sciences*. Los Alamitos, CA: IEEE Computer Society, 1999.

LaPlante, A. Alice "'Instant' Intranets," *Computerworld*, December 22, 1997 (URL: http://www.computerworld.com/home/online9697.nsf/All/971222main).

Latour, B. *Science in Action: How to Follow Scientists and Engineers Through Society*. Cambridge, MA: Harvard University Press, 1987.

Leitheiser, R., and Wetherbe, J. "Service Support Levels: An Organized Approach to End-User Computing," *MIS Quarterly* (10:4), 1986, pp. 337-349.

Lofland, J., and Lofland, L. H. *Analyzing Social Settings: A Guide to Qualitative Observation and Analysis* (3rd ed.). Belmont, CA: Wadsworth Publishing, 1995.

Lyytinen, K., Rose, G., and Welke, R. "The Brave New World of Development in the InterNetwork Computing Architecture (InterNCA) or How Distributed Computing Platforms Will Change Systems Development," *Information Systems Journal* (8), 1998, pp. 241-253.

Machlis, S. "25% of all Web Spending Goes for Intranets," *Computerworld*, July 19, 1999 (http://computerworld.com/home/news.nsf/all/9907191intra; accessed October 15, 1999).

Magal, S., Carr, H., and Watson, H. "Critical Success Factors for Information Center Managers," *MIS Quarterly* (12:3), 1998, pp. 413-425.

Markus, M. L. "Thinking the Unthinkable: What Happens if the IS Field as We Know it Goes Away?" in *Rethinking Management Information Systems*, Wendy L. Currie and Bob Galliers (eds.). New York: Oxford University Press, 1999.

Markus, M. L., and Benjamin, R. J. "Change Agentry: The Next IS Frontier," *MIS Quarterly*, December 1996, pp. 385-407.

Mayo, D. "Can End-User computing be controlled," *The Internal Auditor* (43:4), 1986, p. 24-28.

McCrory, A. "A Needed Shot in the Arm," *Computerworld*, July 28, 1997 (URL: ttp://www.computerworld.com/home/online9697.nsf/All/970728intra_health).

McFarlan, F. W., McKenney, J. L., and Pyburn, P. "The Information Archipelago: Maps and Bridges," *Harvard Business Review*, September-October, 1982.

McFarlan, F. W., McKenney, J. L., and Pyburn, P. "The Information Archipelago: Plotting a Course," *Harvard Business Review*, January-February, 1983.

McLean, E. "End Users as Application Developers," *MIS Quarterly* (3:4), 1980, pp. 37-46.

Miles, M. B., and Huberman, A. M. *An Expanded Sourcebook: Qualitative Data Analysis, Second Edition*, Thousand Oaks, CA: Sage Publications, 1994.

Mirani, R., and King, W. "Impacts of End-user and Information Center Characteristics on End-user Computer Support," *Journal of Management Information Systems* (11:1), 1994, pp. 141-160.

Munro, M. and Huff, S., (1988.) "Managing End-User Computing," *Journal of Systems Management*, 39:12, pp. 13-19.

Neiderman, F., Brancheau, F., and Wetherbe, J. "Information Systems Management for the 1990s," *MIS Quarterly* (15:4), 1991, pp. 475-500.

Ouellette, T. "Giving Users the Keys to Their Web Content," *Computerworld*, 1999 (http://www.computerworld.com/home/print.nsf/all990726b692; accessed October 15, 1999).

Ranier, R., Jr. and Carr, H. "Are Information Centers Responsive to End User Needs?" *Information and Management* (22:2), 1992, pp. 113-121.

Robey, D., and Zmud, R. "Research on the Organization of End-user Computing: Theoretical Perspectives from Organization Science," *Information Technology & People* (6:1), 1992, pp. 11-27.

Rooney, P. "Imposing Order from Chaos," *Computerworld*, September 22, 1997 (http://www.computerworld.com/home/online9697.nsf/all/970922intra_main).

Rooney, P. "A Grassroots Intranet Built with Foresight," *Computerworld*, September 22, 1997 (http://www.computerworld.com/home/online9697.nsf/all/970922intra_pratt; accessed October 15, 1999).

Scheepers, R. "Key Role Players in the Initiation and Implementation of Intranet Technology," in *New Information Technologies in Organizational Processes: Field Studies and Theoretical Reflections on the Future of Work*, O. Ngwenyama, L. Introna, M. Myers, and J. I. DeGross (eds.). Boston: Kluwer Academic Publishers, 1999.

Sliva, C. "Maverick Intranets a Challenge for IT," Computerworld, 1999 (http://www.computerworld.com/home/print.nsf/all/9903159696).

Strauss, A., and Corbin, J. *Basics of Qualitative Research: Grounded Theory Procedures and Techniques*. Newbury Park, CA: Sage Publications, 1990.

Swanson, E. B., and Ramiller, N. C. "The Organizing Vision in Information Systems Innovation," *Organization Science* (8:5), 1997, pp. 458-474.

White, C., and Christy, D. "The Information Center Concept: A Normative Model and a Study of Six Installations," *MIS Quarterly* (11:4), 1987, pp. 451-458.

Yates, J., Orlikowski, W. J., and Okamura, K. "Explicit and Implicit Structuring of Genres in Electronic Communication: Reinforcement and Change of Social Interaction," *Organization Science* (10:1), 1999, pp. 83-103.

About the Authors

Roberta Lamb has recently joined the Decision Sciences faculty of the University of Hawaii, Manoa, as an assistant professor. She has been researching online technology use for the past eight years, initially in California with the University of California, Irvine, and the Center for Research on Information Technology and Organizations (CRITO), and more recently in the Midwest, where she served as professor of Information Systems at Case Western Reserve University in Cleveland, Ohio. She is presently engaged in a three-year, NSF-funded study of intranet use in commercial organizations. Roberta received her Ph.D. in Information and Computer Science at the University of California, Irvine, in 1997. Previously, she designed software systems and managed the development of software products at two southern California firms for over 10 years. She can be reached by e-mail at lamb@cba.hawaii.edu.

Elizabeth J. Davidson is an assistant Professor in the Department of Decision Sciences at the College of Business Administration of the University of Hawaii, Manoa. Her research and teaching interests include social cognitive aspects of information systems development in organizations, the implementation of clinical information systems, and organizational implications of technology adoption and assimilation. In recent publications, she has examined the role of narratives and metaphors as sense-making devices in requirements definition activities. Elizabeth received her Ph.D. in Information Technologies from the Sloan School of Business at MIT in 1996. She can be reached by e-mail at davidson@cba.hawaii.edu.

Part 6:

Transformation of Conceptualizations

17 INFORMATION TECHNOLOGY AND THE CULTURAL REPRODUCTION OF SOCIAL ORDER: A RESEARCH PARADIGM

Lynette Kvasny
Georgia State University
U.S.A.

Duane Truex III
Georgia State University
U.S.A.

Abstract

This paper introduces the critical social theory of French sociologist Pierre Bourdieu. The objective of Bourdieu's theoretical framework is to uncover the buried organizational structures and mechanisms that are used to ensure the reproduction of social order. This theoretical framework will be used in a research program that examines the structural processes by which information technology may be constrained from emancipating humankind, and may actually be disempowering and abandoning significant numbers of societal members.

Keywords: Bourdieu, critical social theory, cultural capital, field, habitus, organizational change, social reproduction

1. Introduction

Technology may be used for good and for ill. In general, organizational studies on information technology implementation and use assume the neutrality of technology and that the intention of those in power to bring the technology to a given social setting have honorable and nonmalicious intentions. But typically someone, be it management or some class of users, is empowered while others lose power as new technology is deployed. There is a growing body of evidence (McMaster 1999; McMaster, Vidgen, and Wastell 1997; Truex and Ngwenyama 1998) that technology adoption has elements of a zero-sum game.

Information technology may be disempowering significant numbers of societal members. This paper examines how power relations and controls are implicit in the technology and in society's attitudes toward that technology. Hence the deployment of new technology tends to reify the dominant relations in the existing social order. In this paper, we examine the fact that the deployment of information technology is viewed as a mark of cultural distinction, a *cultural good.* Only people with sufficient levels of cultural awareness tend to consume certain technologies. That is, power relations drive consumption patterns. Thus, one tends not to consume those technologies for which one does not perceive a need or benefit. For instance, people may not conceive of the need for an e-mail account if no one with whom they might communicate has or uses this technology. In this situation, people whose social positions privilege them to believe they should have access to information technology can be expected to gain because they will be motivated to expend symbolic and material resources to acquire these technologies. People whose position in society leads to them to other conclusions can be expected to lose out.

Moreover, those who are culturally privileged sustain their advantage through control of the prevailing attitudes toward cultural goods such as information technology. By defining the cultural value of a technology and its meaning within a social group, those in control of the discourse about technology are in a position to reproduce, sustain, and solidify existing economic and social hierarchies. They may do this in undetectable ways because those controlled may not have sufficient ability to understand the ramifications of information technology. Hence the elite tend to go unchallenged as they maintain the existing social order.

In organizational politics, a fundamental question is: Who wins and who loses from technology change? To date, opinions are mixed. There are those who argue technology enables restructuring of power relationships and those who argue that technology simply reinforces existing power relationships. For example, Bell (1976) and Feidson (1986) have predicted that computer and communication technologies will shift power to technocrats. Romm and Pliskin (1997) have suggested that the use of these technologies will strengthen democratic features of the organizations by providing different interest groups with the tools to respond to their opposition. Ross (1997) argues that software has the power to make knowledge work routine and replace professions in insurance, law, travel, finance, and medicine. Still other researchers suggest that organizational elites typically use their control over resources to shape the acquisition and application of technologies in ways that perpetuate their power (Kirsch 1997; Klein and Kraft 1994; Kraft and Truex 1994; Orlikowski 1991). These appear to be contradictory findings when one assumes that structural change implies social change. Bourdieu suggests an alternative way to explain these apparent contradictions.

Bourdieu's theoretical framework helps us understand how changes arising from information technology may actually reinforce existing power structures and help perpetuate the social order. For Bourdieu, change is a self-regenerative mechanism required for the maintenance of stratified organizational hierarchies. Static structures can be figured out and conquered over time. However, changing structures keeps actors off balance, thus leading them to apply familiar strategies in unfamiliar contexts. It is this reuse of learned dispositions in new settings that make existing class positions self-sustaining. According to Swartz (1997), the game is rigged, yet everyone inside the organization must play it out. Those who are culturally (or technologically) disenfranchised have little choice but to participate in the competitive struggle over technological resources despite the fact that there is little chance they will reap any of the benefits.

Until we better understand the reproductive processes enabled and extended through the use of information technology, we may not break the cycle; we will not be able use information technology as a force for universal empowerment. This paper speaks to one effort that is currently trying to understand these processes. It presents a research program designed to explore the nature and degree of the exclusionary aspects of information technology. It is based on a composite critical social theory and poststructuralist framework. It arises from an ideal that technology should be emancipatory rather than enslaving. It is heavily influenced by Bourdieu's theories and empirical studies.

This paper proceeds as follows. The second section explains how computing technologies have been raised to the status of cultural icons. The related myths have self-reproducing and evaluative attributes that characterize technology as a type of capital or power resource. This capital may be husbanded, or misdirected, thus enforcing social roles and stereotypes. This section is designed to convince the reader that these self-reproducing hierarchies should become a focal point of further study. The third section introduces Bourdieu's theory of practice and cultural reproduction. The final sections describe how this theoretical framework may guide an information systems research program.

2. Motivation: Technology as Cultural Icon and Myth

Kling (1998) argues that a popular cultural representation of computing should be taken as a serious influence on the ways that people will use systems as well as on their likely social impacts. Accepting this view necessarily means that the adoption of computers involves much more than installing and using software, hardware, and networks. Organizations are also adopting the cultural value assigned to the technology by "the new tastemakers"—vendors, consultants, and the media. These cultural values shape the manner in which people learn about potential social roles that will be attached to new forms of computing.

Over time, the original limiting context that accompanied the adoption of computers and defined their use may be extended into ways that further reify social inequality within the organization. For instance, Marshall McLuhan (Postman 1985) states that the clearest way to see through a culture is to look at its tools for conversation. Every media favors a particular kind of content and wields influence on the wider culture. Information technology makes possible a new mode of dominant discourse by providing a new orientation for thought and expression. Arguably, the most dramatic change fueled by technology adoption is that it helps define and regulate society's conception of ideas of truth. There is a certain degree of bias in the form that any attempt at truth telling can take. The weight assigned to any particular form of truth telling is influenced by the choice of the communication medium (Postman 1985). Therefore, what the culture claims as significant is subject to change with the introduction of a new communication medium and its attributed values. The cultural importance of technology also grows as the organization's dependency on technology increases. As a result, technology can achieve mythical status wherein its historicity and human construction are no longer evident to people (Turkle 1995).

Myths are ways of understanding the world that are so deeply embedded in an actor's conscious that they become invisible (Barthes 1957). When a concept or icon achieves the status of myth, it becomes part of the "taken-for-granted" assumptions about the order of things. The culture adjusts and embraces those myths. Embracing those myths allows for apparently irrational behaviors such as the continued investment in technology firms that have yet to make a profit; the continued proliferation of computers into schools even though educators do not yet know how to effectively integrate technology into the curriculum (PCAST 1997); and the continued reference to the "information society" even when large segments of the population are systematically excluded from participation (Goslee and Conte 1998; NTIA 1995).

As information technology takes on increasing importance within the culture, the computer is afforded the status of icon with "the computer" being rather like a force of nature that shapes our knowledge about the world as well as our ways of knowing the world. Hence, it is important to broaden our understanding of the computer as a cultural medium, not merely a technology. Moreover, we need to view the socio-political aspects of computing as societal in scope and not simply limited to individual organizations. As a cultural artifact, the computer contains different content, uses, meanings, and power. It becomes a credible form of truth telling, and consequently virtual environments may become replacements for certain aspects of reality. This reliance on the computer is part of something much larger: the spread of a civilization so dominated by technology that those who master these tools are positioned to wield an enormous influence over our social, technical, cultural, and economic policies (Swerdlow 1997).

What theoretical framework can one apply to questions involving practice, power, culture, structures, and issues of agency? Critical social theory explicitly deals with these issues (c.f., Ngwenyama 1991). Yet even in the realm of critical social theories, there are those particularly amenable to this mix of issues. The following section introduces the critical social theory of Pierre Bourdieu as a framework capable of helping us better understand how information technology may at one and the same time be both enabling and disempowering.

3. Bourdieu's Theory of Practice and Cultural Reproduction

The work of French sociologist Pierre Bourdieu is a type of neostructuralist critical sociology of a class Morrow (1994, p. 132) calls genetic or critical structuralism.

Bourdieu focuses on the visible social world and develops a model of social practice. He terms his approach *socioanalysis,* where the role of the researcher is to unveil the "social unconscious" of the social group under investigation (Bourdieu 1984). The social unconscious consists of those unacknowledged interests that actors follow as they participate in a hierarchical social order. According to Bourdieu, the failure to recognize these embedded interests is a necessary condition for the exercise of power by one group upon another. For Bourdieu, the public exposure of embedded interests can help undermine their apparent legitimacy and open up the possibility for altering existing social arrangements. In this respect, Bourdieu's sociology is complementary to other forms of critical social theory (Morrow 1994). However, Bourdieu differs from the Frankfurt school of critical social theory (notably the work of Jürgen Habermas) and from English critical social theory (notably the work of Anthony Giddens).

Where Habermas assumes an ideal state wherein parties share a common goal that is the basis for communicative action, Bourdieu sees symbolic structures (including language) and material structures as formed to serve strategic purposes; typically, to empower or disempower. Thus, in this sense, Bourdieu's theory is one of conflict whereas Habermas' theory, ideally, is one of compromise.

For Bourdieu, as with Habermas, a central issue of concern is *practice*: that is, the outcome of the dialectical relationship between how social actors construct social reality and how structure constrains or enables them. The exercise and reproduction of class-based power informs all practice. In Bourdieu's case, the strong focus on the constraining and enabling features of structure on human action has earned his work the label of "genetic structuralism." Whereas structures for Habermas are largely the result of technical language, Bourdieu posits that structures not only exist in language, but in the social world as well. These "mental structures" take place in an arena that is both animated and constrained by "social structures" (Ritzer 1996).

In the realm of the agency-structure debate, Bourdieu is closer to Giddens as both attempt to bridge the divide between agency and structure. But unlike Giddens, Bourdieu gives more credence to the limiting or conversely enabling aspects of social structures. These structures may be vicious and negative, entrapping structure for the cultural "have nots," or they may positively enable and reinforce privilege for the cultural "haves."

In Giddens' conceptualization of social systems, they do not have structures; rather, they exhibit structural properties. In this view, structure is manifested in social systems in the form of reproduced practices. These social systems are often the unanticipated consequences of human action. On occasion, they are the product of intentional action. Social systems are also manifest in the memory of agents and, as a result, these rules manifest themselves at both a macro and micro level of human consciousness. Thus, for Giddens, the central issue of social order depends on how well social systems are integrated over time and space (Ritzer 1996). Bourdieu posits that structures change over time and space as actors pursue cultural distinctions. However, while structures evolve, positions in the hierarchy remain largely unchanged. Practices serve to reify social structure (Berger and Luckman 1966; Swartz 1997).

Bourdieu takes a critical view of the intentionality of human actions and their consequences, whereas Giddens is more concerned with the consequences than the intentions of social actors. (For example, I have hit you with my car and you have died. The fact that I didn't intend to hit you with my automobile doesn't change the consequences of the act.) Although both theorists see social systems as being integrated over time and space, for Bourdieu, the intent of social practices is to hold actors in self-generating and self-perpetuating social hierarchies. Therefore, practice is tacit and motivated largely by the actor's position within the class structure.

3.1 Bourdieu's Epistemology

In *The Logic of Practice* (1990), Bourdieu provides a critique of theoretical reason. Like Habermas and Giddens, Bourdieu's social theory attempts to overcome the apparent duality between subjectivism and objectivism. Bourdieu argues that these epistemologies, when employed in isolation, provide an impoverished view of social life. To gain a richer understanding of social phenomenon, Bourdieu suggests a two-step reflexivity process, which forces one to consider both sides of the coin.[1]

Subjectivism focuses on the lived experience of the actor, which takes the social world as self-evident. However, subjectivism is limited because it excludes questions about the internalization of social structures that make lived experiences possible. These subjective perceptions can only reinforce the system of domination that is under investigation. Since all human action is situated within determining structures that are not readily apparent in everyday consciousness, the researcher must focus on theoretical practices. This requires critical reflection on the research practice itself, and is essential for establishing the "validity" of the accounts of social phenomena.[2] This is especially important for critical researchers concerned with charges that their work is biased by their own ideologies (Schwandt 1997).

Conversely, objectivism sets out to grasp objective regularities, such as structures, laws, and systems of relationship, that take place irrespective of individual consciousness and will. It assumes the existence of social structures well outside the realm of the individual that exert force over actors. However, objectivism ignores the experiences that are both the condition and the result of structuring operations. Therefore, this epistemology cannot account for the generative features of practice. Bourdieu argues that the researcher must focus on the generative as well as situated character of practice and provides several key concepts to enable the investigation of these dual characteristics of practice.

3.2 Bourdieu's Key Concepts of Practice and of Social Control

Bourdieu makes reference to a number of concepts requiring introduction and explanation. He uses the concepts of the *field* and of the *habitus* to model practice.

---

[1]In this sense, he shares an approach suggested by Latour (1988). We intent to examine the intersection of Bourdieu and Latour in further work, but this analysis lies outside the scope of this paper.

[2]We are quite aware of the host of objections this may raise in the IFIP WG 8.2 research community. Suffice it to say that ours is a hybrid notion that is far from the positivist notions of validity. It is also very different from the "any old interpretation counts" approach. See Baskerville and Lee (1999) for an excellent discussion of this point.

Fields are Bourdieu's notion of social arenas within which individuals struggle to maximize social standing. Fields may be defined as competitive systems of social relations functioning according to rules specified by dominant classes. Hence, in the composite, fields are a social arena within which struggles take place over the accumulation, investment, and conversion of power resources. The outcomes of these struggles influence cultural distinctions. Thus, a field defines the structure of the social setting in which the habitus operates (Swartz 1997). One may, therefore, view society itself is a system of social fields (Morrow, 1994).

The habitus guides practice and behavior in daily life. It is a cognitive construct that arises or "is generative" from personal experience and history. Elements of the habitus are acquired from the social class and status into which one is born. It is, therefore, **both** an individual and a shared concept. That is, one has one's own habitus reflecting one's place in a social structure. But the habitus is not fixed. Rather, while it is durable it is yet malleable, and is in constant negotiation with the world. So it may be seen as a kind of shield in the field of social battle. For instance, if one's habitus has been one of relative social privilege, it serves as a template and provides strategies for continued success. But if one's habitus has been one of relative powerlessness, it provides coping strategies. On the negative side, the habitus may limit social progress in that it defines expectations of the possible and, worse yet, may limit one's aspirations.

Bourdieu employs three additional concepts to model a political economy of symbolic power: *cultural arbitrary*, *capital*, and *symbolic violence*. According to Bourdieu, behind all culture is power in the form of the cultural arbitrary: standards put forth and managed by privileged or ruling classes. These standards explicitly and implicitly determine which capital or stakes are of value and are to be acquired in continuing struggles within fields. Thus capital extends the concept of power to include material and symbolic resources.

Cultural capital is a particularly important in that without it one cannot make cultural distinctions and value judgements. Without the proper cultural capital, one can neither consume nor produce cultural goods. This holds for music, art, scientific formulas, information systems research methods, and computing technology. For instance, if presented with a famous painting by Monet, one might see it as a pretty representation of lily pads on a pond. But with additional cultural capital, one would be better able to understand the significant. One may also understand how it fits into the history of art and particular historical events and, therefore, is deemed to have cultural value.

Symbolic violence is power employed within the field to legitimate and reproduce the class hierarchy. This is pure power exercised through hegemony of norms and political techniques for shaping the mind and the body without the use of physical force or laws (Best and Kellner 1991). Through ideology and discourse, people are trained via the cultural arbitrary as to their place in society and what they may expect from society. In this process, symbolic violence inculcates the habitus to produce durable, but changeable, dispositions. Thus the interplay of symbolic violence and the counter strategies employed by the actor provides the habitus with an ongoing history of continually adjusting and self-perpetuating experiences. This generative habitus coupled with the constraints, demands, and opportunities present in the field determines practice. Through practice, actors struggle and pursue strategies to achieve self-interests within a field. However, in doing so, actors reproduce the class and social structure without conscious recognition and without resistance (Swartz 1997). Capital, and symbolic violence are related as follows: if one doesn't have the right kind or the sufficient amount of capital it is more likely that he or she will be the victim of symbolic violence.

4. Bourdieu in the Domain of Information Technology

Table 1 gives a high level summary of the most salient elements of Bourdieu's theory applied to the domain of information technology. The table and accompanying text in this section are intended to illustrate how Bourdieu's ideas better inform our understanding of technology empowerment and disempowerment. More importantly, this theory suggests how the application of technologies, even when conceived egalitarian motives and a desire to see the technology made available universally, may exclude entire classes of people.

Table 1. Key Theoretical Concepts

Key Term	Application to Information Technology
Field	Information technology organization or a given technology system under investigation
Habitus	Expectations, aspirations, and attitude towards technology; informs practices, e.g., how and if one engages technology
Symbolic Violence	Use of technology to enforce decisions; decision to limit or deny access to technology or use of technology; technology used for surveillance or control
Cultural Capital	Exposure, previous experience and familiarity with information technologies; information technology credentials
Symbolic Capital	Use of highly technical language, sharing of technical expertise, removing manual process in favor of technological processes, denying training in new technology
Social Capital	Access to relationships with others knowledgeable about technology
Economic Capital	Ability to acquire technology and training; choices to allocate resources to the procurement of technology

4.1 Field

A field is the social arena under investigation. This could be an entire industry, an organization dedicated to the development of information systems, or even a virtual community. The researcher may also choose to narrow the focus and investigate a particular information system or business process employed within an organization.

Autonomy is a particularly important measure of the reproductive tendencies of a field. Autonomous fields are those that are more or less free from political intervention and economic constraints from external institutions and, therefore, have greater control over the creation and management of their culture. Autonomy may be enacted through formal methods, procedures, and practices such as hiring, promotion, business planning, and procurement of information technology resources.

In terms of the information systems development organization, autonomy may signal a move to an organization of specialists who are able to progressively develop, transmit, and control their own particular culture. Thus, this group is able to promote its own organizational and professional interests, which may deviate significantly from external interests. With growing autonomy comes the capacity to retranslate and reinterpret external demands to serve the interests of the dominant organization members. Stated more theoretically, a highly autonomous field has greater ability to control the reproduction of the existing social order because it can independently establish criteria for legitimizing the cultural arbitrary (Swartz 1997). Stated more plainly, those with power can use technology to concentrate power. Capital breeds capital.

4.2 Habitus

The habitus is a product of the internalization of the cultural arbitrary. In the information system development organization, for example, the cultural arbitrary confers privileged status to those with expert technical knowledge. The habitus of the software developers has internalized this cultural valuation, which drives their practice to accumulate and convert expert knowledge. Expert knowledge, therefore, becomes an object of cultural distinction used to differentiate and classify actors within the organizational hierarchy.

For instance, Reed (1996) found that software developers not only build information systems to further increase monitoring capabilities of management; developers also monopolize scarce knowledge so that it can't be stolen or duplicated. Reed also describes the politics of expertise where trends such as telecommuting are fragmenting technologists, thereby decreasing their collective power base. However, those who are able to "black box" their expertise—compartmentalize, simplify, and standardize their knowledge in a more portable form—are able to retain their privileged status within the organization. The growing use of credentials by information systems professionals provides an example of how actors attempt to black box expert knowledge (Knights 1992). This demonstrates how the software developers' habitus provides self-preservation strategies that enable them to institute practices that maintain their social position as the structure of the field changes.

4.3 Symbolic Violence

Bourdieu describes the three functions of symbolic violence as (1) knowledge integration, (2) communication, and (3) political domination. The information systems literature on power and politics provides examples of these functions. Typically these studies employ a framework that divides organizational control mechanisms into two categories: formal controls and informal controls (Eisenhardt 1985; Henderson and Lee 1992; Kirsch 1996, 1997; Ouchi 1979, 1980). Formal controls drive behaviors and outcomes through bureaucratic measures; they relate to the knowledge integration function of symbolic violence where written policies, procedures, and methodologies are employed to provide a universal ordering and understanding of the social arena.

Informal controls are derived from clan norms or self-regulation within social classes. Informal controls correspond to the communication function of symbolic violence where controls are embodied in codes that channel deep structural meanings shared by all members of a social class. These deep structural meanings are the result of the collective habitus of actors within a social group.

The final function of symbolic violence, and the function that Bourdieu stresses, is political. The political function of symbolic violence is the process whereby order and social restraint are provided by indirect, cultural mechanisms rather than by direct, coercive social control. This form of symbolic violence is often inflicted with the imposition of technology and the associated dominant-class culture on subgroups (Swartz 1997). For example, Kling and Iacono (1984) found that managers gain their authority and resources through a project champion or steering committee. Together they form a cohort that builds ideology and legitimacy in order to increase the acceptance of information systems. The champion performs "symbolic labor," that is, the masking of the interested practices of managers. Symbolic violence is most powerful when it hides the interested nature of managerial actions and the acceptance of the information systems by the users serves to further perpetuate their domination.

The political function of symbolic violence is also inflicted through the use of surveillance which enables an outsider to observe and penetrate behavior. At the same time, the actor internalizes *the gaze* and regulates her behavior because she is never certain when she is under the gaze (Foucalt 1979). Monitoring telephone calls, examining computer usage logs, and maintaining statistics on the entire workforce place all workers under the gaze. Information systems extend and enhance the exercise of power by an organization, thereby enabling the actions of workers to be supervised by entire managerial groups, not just the immediate supervisor. Supervision with information technology can also overcome time and space distances. Whether the worker sits directly across from the manager or telecommutes from her home, the manager is able to monitor the work effort.

In addition to the use of technology, symbolic violence is enacted via organizational processes such as performance feedback and training programs. Performance evaluations are used to rank and rate workers. These ratings serve to individualize workers by making the individual more identifiable *vis-à-vis* other individuals, while rankings serve to individuate workers by converting individuals into numerical equivalents. Performance evaluations also serve as a confessional where employees are given performance feedback that becomes part of the individual's self-knowledge. The organization strives to control what employees know through the use of formal and informal training programs and examinations. Through these acts of symbolic violence, the organization trains the habitus and, therefore, controls the practices, cultural preferences, aspirations and expectations of actors.

In all cases, the symbolic violence must be "misrecognized" if it is to be successfully imposed on a subculture. That is, the symbolic violence must be seen as a legitimate call for deference to authority, otherwise actors may resist the imposition. For instance, a study by Orlikowski concluded that management introduced CASE technology to formalize and tighten their control over the work practices of consultants. However, the consultants found passive ways to undermine the oppressive software and were, therefore, able to resist the imposition of the technology. An abundance of cultural capital and a habitus generated from a history of privilege enabled the consultants to recognize that symbolic violence was being inflicted upon them and led the consultants to adopt self-preservation practices to maintain their interests.

4.4 Cultural Capital

Ownership of and familiarity with information technology is a type of cultural capital in the same manner as in the art example in section three. It is simply expected that one is confident with technology in business settings. It is a mark of the well educated and erudite. Unless one is socialized in a culture of information technology acceptance and history, one is rather like the indigenous aboriginal people with minimal exposure to the outside world in the film *The Gods Must Be Crazy*. When technology, in the form of a discarded Coke bottle, is encountered, their frames of reference cannot incorporate it into their lives.

Similarly, formal language is a type of cultural capital. More so than any other technology, computing has developed its own extensive and unique lexicon. While the ability to speak intelligently about technology in a business setting immediately signals one's position as upper class, it forces those without a command of the language of technology to a position of deference. Bell (1976) argues that formal language is the central resource in society. However, formal language by definition is not part of everyday vocabulary but the importance of possessing this type of knowledge has grown enormously in advanced post-industrial societies. Thus knowing about technology is an elite form of knowledge, and prestige and respect are given to the technocrats that possess this form of cultural capital (Feidson 1986).

The HomeNet field trials (Kraut et al. 1997) provide empirical support for this notion of cultural capital. This study found that even when technology and helpdesk support was provided, participants were still reluctant to utilize these resources for technical assistance. Winter, Gutek, and Chudoba (1997) noted similar results in their study of the computer literacy of user groups. Respondents to that survey indicated that they rely on peers rather than access a helpdesk in solving their computer problems. In both studies, the participants did not have sufficient cultural capital to self manage the information technology, which forced them to rely on the expertise of others. Rather than defer problem resolution to external groups with greater expertise, the participants preferred to either abandon the technology altogether or find the expertise they needed within their social group. This indicates that, without sufficient cultural capital, the habitus of these actors led them to focus internally to develop coping strategies.

4.5 Symbolic Capital

Symbolic capital is gained when the exercise of power goes unnoticed. It is the use of power in disguised form because it hides underlying interests in order to give them legitimacy. The potency of symbolic capital lies in the fact that it is not perceived as power but as legitimate demands for recognition, deference, obedience, or the service of others (Swartz 1997). By simply conforming to the control mechanisms instituted by managers, workers unwittingly provide management with knowledge that may be used to further manipulate the workforce.

For example, Townley (1993) examined how managerial practices such as performance evaluations, time reporting, and the computerized information systems for monitoring job performance are used to order the population. In all cases, emphasis is on the development of regulatory systems and written policies that record, calculate, observe, and audit employees. Once captured in a database, this information becomes a domain of truth that can be easily translated to other decision-making bodies. Townley concludes that the knowledge obtained through these control systems engenders power because management exercised their privileged status to accumulate the information.

Knights (1992) makes a similar observation when he states that management knowledge is never independent of the power that managers and corporations exercise. Whereas Townley is concerned with the proliferation of information into areas beyond its intended uses, Knights argues that computer records can be used to individuate the worker, constituting the individual as a describable, analyzable object. Technology becomes a tool for normalizing procedures of exclusion where surveillance can be used to physically separate deviants and the incompetent. Knowledge is, therefore, a form of symbolic capital that is firmly rooted in power and authority. Deviance is not merely discovered; knowledge creates entities such as the deviant and the incompetent (Foucalt 1979).

4.6 Social Capital

Information technology also embodies social capital as it opens doors to new relationships, both physical and virtual. With a desktop computer, employees can "network on the network" to search for employment opportunities, seek career advice, and retrieve industry reports that will help them to improve their job performance. White-collar workers can also use the intranet as a political lobbying tool to benefit not only their professional career, but also to support their personal interest (Romm and Pliskin 1997). With access to e-mail, listserves, newsgroups, and web sites, white-collar workers can also use technology to form business related or purely recreational virtual friendships.

4.7 Economic Capital

Economic capital enables one to convert economic resources into information technology resources. Because of its universal applicability and liquidity, economic capital can be defined as the root of all other capitals. In fact, all other forms of capital can be thought of as disguised forms of economic capital (Swartz 1997). For instance, with sufficient economic capital, one can leave the workforce to acquire additional cultural capital through advanced degree programs. In a society where hierarchies are often based on education, additional education buys legitimacy and access to social circles in which one may develop relationships with influential people within different communities of practice.

5. Relevance of Theory

Bourdieu's theoretical framework helps us understand how change arising from information technology may serve to both disrupt social order and still perpetuate existing power structures. While the introduction of technology may provide opportunities for groups and individuals to gain autonomy, the dominant can ensure their privileged position by enacting mechanisms to monopolize their control over scarce and valuable organizational resources. Information technology extends the power resources from which actors may draw upon in their struggle to maximize social position. For instance, new domains of knowledge and mechanisms for control become available with the emergence of technologies such as data mining, knowledge management, and data warehousing. These practices are motivated, consciously or unconsciously, by pursuits for material and symbolic profits.

The contribution of this framework to information systems research lies in the ability to explain and measure the reproductive tendencies of technology in diverse settings such as community, organizational, society, or international studies. This suggests that social order may be an unseen contingency in numerous information systems research areas. For example, changes in job classifications and the rise of new occupations may be explained by the "broken trajectory" effect. This occurs when an agent's aspirations soar above their objective chances. Broken trajectories often occur for those who have not obtained the rewards which their qualifications would have guaranteed in an earlier state of the market. The disparity between aspirations that the educational system produces and the structural reality of the marketplace takes forms that are objectively and subjectively different in the various social classes (Bourdieu and Passeron 1979). The researcher would seek to uncover strategies employed by actors of various social backgrounds to avoid structural downclassing. For instance, one may see the production of new occupations more closely matching the actor's pretensions, or the refurbishment of occupations to which the actor's qualifications do give access. Historical studies of the rise of software consulting firms and computer-related occupations and survey research that examines trends in job placements for college graduates may be informed by this type of social stratification analysis.

Social class analysis may also be relevant in information systems development studies. Because of their highly specialized skills, developers and analysts are granted authority to interpret and translate business requirements into computer systems. However, these functional requirements are sometimes reinterpreted during system design and the resulting information system doesn't perform as expected. This theory would suggest that information systems development is a form of symbolic violence with developers and analysts assuming the role of cultural producers capable of imposing their worldview on the unsuspecting users of their technology products and service.

This theory would also suggest that users and developers are predisposed to develop new information systems that reflect the socialization learned by their habitus. Therefore, calls for participatory design and end user involvement may be questioned because all parties will tend to build systems that reflect their social position. This hypothesis could be tested with a laboratory experiment where participants would be socialized under an initial social order. The participants would then be given specifications that would force them to design an information system that required a different social arrangement. The researcher would be able to measure the reproductive tendencies of an information development effort and determine if one is even able to develop a system outside of one's habitus.

6. Research Approach

Bourdieu's theory of practice and cultural reproduction is useful for examining the relationship between structure, culture, and social class. Bourdieu is a cultural anthropologist, a social theorist, and a prolific empirical researcher. Based on the findings of his empirical studies, Bourdieu's theory is under constant refinement and extension. Bourdieu uses intensive field-based research methodologies and focuses on statistical patterns of relationships gleaned largely from participant observation, document analysis, interviews, and multivariate statistical techniques.

What tools of discovery and analysis may be best brought to bear when using Bourdieu as a theoretical guiding force to examine the role of information technology in organizational settings? What are the units of analysis? How may one frame proper hypotheses and research questions? The answer to each of these questions is, of course, driven by the research question itself. We envision a three phase research agenda for ourselves. It is one that (1) explores specific organizational structures and procedures that serve to either maintain or change the various capital structures described by Bourdieu; (2) identifies cultural features within study organizations that maintain, or on occasion enable, adjustments to the social order (this will allow a further examination of interorganizational cultural and technological hierarchies); and (3) identifies and studies groups and individuals that have "broken the mold," escaped the structural limitations and have moved to new positions in the hierarchy. This agenda will require learning about the struggles and types of symbolic violence employed in organizations. It also requires uncovering and understanding underlying assumptions, beliefs and values about technology within social/cultural strata in organizations.

This research agenda requires a mixed methods approach. We expect that intensive ethnographic fieldwork will be required during the data collection phase. The data will be participant observation field notes and transcripts of conversations with organizational actors as well as organizational documents. The data will be subjected to interpretative discourse analysis. Structured interviews will be mapped using cognitive mapping techniques. All texts will be iteratively examined and will subsequently provide the basis for a deconstructive reading of the organizational setting. Identifying specific organizational practices coupled with Bourdieu's previous work will enable us to develop appropriate instruments for further work. We would like to contribute to the development of a kind of index to measure and "predict" Bourdieu's notion of *life's chances*.

Why? Frankly, we want to change the world. We believe that the promise and benefits of information technology should be more equitably distributed. A first step requires knowing more about how various types of organizational life contribute to limiting or supporting access to the promises of these technologies. Armed with that knowledge, ethical designers and managers of information technology may be able to counter forces that are arrayed to enslave rather than empower.

7. Summary

This paper introduces the social theory of Pierre Bourdieu and describes an information systems research program guided by this theoretical framework. This research program assumes a view of technology as a cultural good that serves as a marker of class distinction. Actors with sufficient cultural capital have acquired an appreciation for technology that enables them to readily consume this cultural good. However, actors with inadequate levels of cultural capital are unable to partake in the potential benefits of technology because they are systematically excluded from the discourse surrounding technology. This research agenda seeks to examine how and why technology is used to reproduce a social order that perpetuates the privileged status of some actors while simultaneously disenfranchising others. Until we understand the processes better, we may not break out of the cycle of social reproduction; we will never be able to use information technology as a more universal force for human empowerment.

References

Barthes, R. *Mythologies*. New York: Hill and Wang, 1957.

Baskerville, R., and Lee, A. "Distinctions Among Different Types of Generalizing in Information Systems Research," in *New Information Technologies in Organizational Processes: Field Studies and Theoretical Reflections on the Future of Work*, O. Ngwenyama, L. Introna, M. Myers, and J. I. DeGross (eds.). Boston: Kluwer Academic Publishers, 1999, pp. 49-65.

Bell, D. *The Coming of Post-Industrial Society: A Venture in Social Forecasting*. New York: Basic Books, 1976.

Berger, P., and Luckman, T. *The Social Construction of Reality: A Treatise in the Sociology of Knowledge*. New York: Anchor Books, 1966.

Best, S., and Kellner, D. *Postmodern Theory: Critical Interrogation*. New York: The Guilford Press, 1991.

Bourdieu, P. *Distinction: A Social Critique of the Judgement of Taste*. Boston: Harvard University Press, 1984.

Bourdieu, P. *The Logic of Practice*. Stanford, CA: Stanford University Press, 1990.

Bourdieu, P. *In Other Words: Essays Toward a Reflexive Sociology*. Stanford, CA: Stanford University Press, 1990.

Bourdieu, P., and Passeron, J. *The Inheritors*. Chicago: University of Chicago Press, 1979.

Eisenhardt, K. "Control: Organizational and Economic Approaches," *Management Science*, (2:31), 1985, pp. 134-149.

Feidson, E. *Professional Powers: A Study of the Institutionalization of Formal Knowledge*. Chicago: University of Chicago Press, 1986.

Foucalt, M. *Discipline and Punish: The Births of the Prison*. New York: Vintage Books, 1979.

Goslee, S., and Conte, C. "Losing Ground Bit By Bit: Low-Income Communities in the Information Age," 1998 (available online from http://www.benton.org/Library/Low-Income; accessed December, 1999).

Henderson, J., and Lee, S. "Managing I/S Design Teams: A Control Theory Perspective," *Management Science* (6:38), 1992, pp. 757-777.

Kirsch, L. "The Management of Complex Tasks in Organizations: Controlling the Systems Development Process," *Organization Science* (7:1), 1996, pp. 1-21.

Kirsch, L. "Portfolios of Control Modes and IS Project Management," *Information Systems Research* (8:3), 1997, pp. 215-239.

Klein, H., and Kraft, P. "Social Control and Social Contract in NetWORKing: Total Quality Management and the Control of Knowledge Work in the United States," *Computer Supported Cooperative Work (CSCW)*, 1994, pp. 89-108.

Kling, R. "Cultural Influences on the Process and Impacts of Computerization," in *Fostering Research on the Economic and Social Impacts of Information Technology*, H. Varian (ed.). Washington, DC: National Academy of Science, 1998, pp. 150-151.

Kling, R.,and Iacono, S. "The Control of Information Systems Developments After Implementation," *Communications of the ACM* (27:12), 1984, pp. 1218-1226.

Knights, D. "Changing Spaces: The Disruptive Impact of a New Epistemological Location for the Study of Management," *Academy of Management Review* (17:3), 1992, pp. 514-536.

Kraft, P., and Truex, D. "Postmodern Management and Information Technology in the Modern Industrial Corporation," in *Information Technology and New Emergent Forms of Organizations*, R. Baskerville, S. Smithson, O. Ngwenyama, and J. I. DeGross (eds.). Amsterdam: Elsevier Science Publications, 1994, pp. 113-127.

Kraut, R., Scherlis, W., Mukhopadhyay, T., Manning, J., and Kiesler, S., "The HomeNet Field Trial of Residential Internet Service," *Communications of the ACM* (38:12), 1997, pp.55-63.

Latour, B. *Science in Action: How to Follow Scientists and Engineers Through Society,* Boston: Harvard University Press, 1988.

McMaster, T., Vidgen, R., and Wastell, D. "Technology Transfer—Diffusion or Translation?" in *Facilitating Technology Transfer Through Partnership: Learning from Practice and Research*, T. McMaster, E. Mumford, E. B. Swanson, B. Warboys, and D. Wastell (eds.). London: Chapman & Hall, 1997, pp. 64-75.

McMaster, T., Vidgen, R., and Wastell, D. "Networks of Association and Due Process in IS Developmentm," in *Information Systems: Current Issues and Future Changes*, T. J. Larsen, L. Levine, and J. I. DeGross (eds.). Laxenburg, Austria: IFIP, 1999, pp. 341-357.

Morrow, R. *Critical Theory and Methodology*. Thousand Oaks, CA: Sage Publication, 1994.

Ngwenyama, O. "The Critical Social Theory Approach to Information Systems: Problems and Challenges," in *Information Systems Research*, H-E. Nissen, H. Klein, and R. Hirschheim (eds.). Amsterdam: Elsevier Science Publications, 1991, pp. 267-294.

NTIA. "Falling Through the Net: Defining the Digital Divide," National Telecommunications and Information Administration, Department of Commerce, Washington, DC, 1999 (available online from http://www.ntia.doc.gov/ntiahome/ fttn99/contents.html; accessed December, 1999).

Orlikowski, W. "Integrated Information Environment or Matrix of Control? The Contradictory Implications of Information Technology," *Accounting, Management, and Information Technology* (1:1), 1991, pp. 9-42.

Ouchi, W. "A Conceptual Framework for the Design of Organizational Control Mechanisms," *Management Science* (9:25), 1979, pp. 833-848.

Ouchi, W. "Markets, Bureaucracies, and Clans," *Administrative Science Quarterly* (3:25), 1980, pp. 129-141.

Postman, N. Amusing Ourselves to Death. New York: Penguin Books, 1985.

PCAST. "Report to the President on the Use of Technology to Strengthen K-12 Education in the United States," President's Committee of Advisors on Science and Technology, Panel on Educational Technology, 1997 (available online from http://www.whitehouse.gov/WH/EOP/OSTP/NSTC/PCAST/k-12ed.html; accessed December, 1999).

Reed, M. "Expert Power and Control in Late Modernity: An Empirical Review and Theoretical Synthesis," *Organizational Studies* (17:4), 1996, pp. 573-597.

Ritzer, G. Modern Sociological Theory. New York: McGraw-Hill, 1996.

Romm, C., and Pliskin, N. "Toward a Virtual Politicking Model," *Communications of the ACM* (40:11), 1997, pp. 95-100.

Ross, P. "Software as Career Threat," in *Computers and Society*, P. Winters (ed.). San Diego: Greenhaven Press, 1997, pp. 44-48.

Schwandt, T. *Qualitative Inquiry: A Dictionary of Terms.* Thousand Oaks, CA: Sage Publications, 1997.

Swartz, D. *Culture and Power: The Sociology of Pierre Bourdieu.* Chicago: University of Chicago Press, 1997.

Swerdlow, J. "Computers and Society: An Overview," in *Computers and Society*, P. Winters (ed.). San Diego: Greenhaven Press, 1997, pp. 16.

Townley, B. "Foucault, Power/knowledge, and Its Relevance for Human Resource Management," *Academy of Management Review* (18:3), 1993, pp. 518-545.

Truex, D., and Ngwenyama, O. "Unpacking the Ideology of Postindustrial Team-Based Management: Self-governing Teams as Structures of Control of IT Workers," paper presented at the Work, Difference and Social Change: Two Decades After Bravermans's Labor and Monopoly Capital, Binghamton, New York, 1998.

Turkle, S. *Life on the Screen: Identity in the Age of the Internet.* New York: Simon & Schuster, 1995.

Winter, S., Gutek, B., and Chudoba, K. "Misplaced Resources? Factors Associated with Computer Literacy Among End Users," *Information & Management* (32:1), 1997, pp. 29-42.

About the Authors

Lynette Kvasny is a doctoral candidate specializing in Computer Information Systems at Georgia State University. Her teaching experience includes Introduction to Business Information Systems, Database Management Systems, and Object-Oriented Programming in C++. Her research interests focus on various social aspects of computerization including the roles of information technology in societal and organizational change, the uses of information technologies in socio-political contexts, and the ways that technology utilization are influenced by social forces and practices. She also has several years of software development and consulting experience at AT&T, Bell Laboratories, and Lucent Technologies. Lynette can be reached at lkvasny@gsu.edu.

Duane Truex is an assistant professor of Computer Information Systems at Georgia State University and a Leverhulme Fellow at the University of Salford, England. His first degrees were in the arts and he spent his early professional life in symphony orchestra and museum management. As a critical social theorist, his work in information systems research has been generally concerned with emancipatory issues of information technologies. He also writes about information systems development and IS research methods. He has published in *Communications of the ACM, Accounting Management and Information Technologies, Information Systems Journal, IEEE Transactions on Engineering Management*, the IFIP transactions series, and in various other proceedings. Duane can be reached at dtruex@gus.edu.

18 THE SCREEN AND THE WORLD: A PHENOMENOLOGICAL INVESTIGATION INTO SCREENS AND OUR ENGAGEMENT IN THE WORLD

Lucas D. Introna
London School of Economics and Political Science
United Kingdom

Fernando M. Ilharco
London School of Economics and Political Science
United Kingdom

Abstract

In this paper, we attempt to show how phenomenology can provide an interesting and novel basis for thinking about screens in a world where screens now pervade all aspects of our daily existence. We first provide a discussion of the key phenomenological concepts. This is followed by its application to the phenomenon of a screen. In our phenomenology of the screen, we aim to give an essential account of a screen, as a screen, in its very screen-ness. We follow Heidegger's argument that the screen will only show itself as a screen in its functioning as a screen in the world where screens are what they are. We claim, and aim to show, that our analysis provides many insights about the screen-ness of screens that we can not gain through any other method of investigation. We also show that although our method is not empirical its results have many important implications for the empirical world.

Keywords: Phenomenology, theoretical investigation, information technology, information systems, personal computer, television, screen.

1. Introduction

Whether at home relaxing with the family, or at the workplace, or traveling, or engaged in entertainment, a growing majority of people find themselves increasingly in front of screens—television (TV) screens or personal computer (PC) screens in particular.

The last decades have witnessed a massive penetration of TV screens into people's day to day lives. It is a long way from November 1937, when the BBC made its first outside broadcast—the coronation of King George VI from Hyde Park Corner—which was seen by several thousand viewers, to the landing on the Moon in 1969, carried by satellite to an estimated audience of more than 100 million viewers (*Encyclopaedia Britannica* 1999), and to the funeral of Princess Diana in August 1997, followed by an estimated TV audience of 2.5 billion (ABCnews 1999), which represents more than 40 per cent of the world's population.

The PC screen seems to be experiencing an even more accelerated spreading. By 1985, there were 90.1 and 36.4 computers per 1,000 people, respectively in the USA and in the UK. Today those figures are around 580 and 441. From 1985 to 2000, that same figure for the whole of Europe went from 14.3 to 248.9, and for the world as a whole it went from 7.8 to 90.3 (*Computer Industry Almanac* 1999).

Although this phenomenon of the spreading of screens—TV and PC screens in this case—seems to have acquired a pervasive presence in the Western world, this pattern of spreading and implicitly colonizing the everyday world is no less significant in other cultures and in other regions of the world. In this paper, we want to inquire into the significance of our increasing engagement with screens for our understanding of ourselves and the world we live in. We contend that a suitable response to this question could in part rely on that which a screen is, qua screen. In this paper, we hope to demonstrate that the phenomenological method of investigation may be an appropriate way to proceed with such an inquiry.

2. Part I: An Introduction to Phenomenology

Phenomenology in the form of the phenomenological method of investigation is currently used in a wide range of fields, such as anthropology, sociology, psychiatry, biology, and so forth. Yet, we have seen very little of it in the field of information systems research (Boland 1978, 1983, 1985, 1989, 1991; Ciborra 1997; Introna 1997, 1993; Winograd and Flores 1987; Zuboff 1988). Even these studies to a lesser or greater degree use it only marginally or together with other approaches.

In this paper, we will attempt to demonstrate the possibilities of phenomenology in its original form as developed by Edmund Husserl (1859-1938). We will, however, extend the Husserlian formulation in the last phase of the method by using the arguments of Heidegger (1889-1976) (1962) on the opening up of concealed meanings, as proposed by Spiegelberg (1975, 1994).

Phenomenology as a movement began to take shape a century ago with the impact of the first works of Edmund Husserl.[1] This is in spite of the fact that both the word "phenomenology" and the intellectual activity it addresses were much older. Phenomenology "has long been on the way, and its adherents have discovered it in every quarter, certainly in Hegel and Kierkegaard, but equally in Marx, Nietzsche and Freud" (Merleau-Ponty 1962, p. viii).

The word phenomenology, which taken literally means the study or description of phenomena, has its origins in ancient Greek. Heidegger in *Being and Time* (1962, pp. 50-63) traced back the meaning of the two components of the word phenomenology—phenomenon and logos—suggesting the following preliminary conception of phenomenology: "To let that which shows itself be seen from itself in the very way in which it shows itself from itself" (Heidegger 1962, p. 58). For Heidegger, this formulation does not say more than the well-known maxim of phenomenology "to the things themselves!"[2] Heidegger concludes phenomenology does not designate its own subject-matter: "the word merely informs us on the '*how*' with which *what* is to be treated in this science, gets exhibited and handled" (Heidegger 1962, p. 59). This means that phenomenology is first and foremostly a method of investigation.

Its object is *the way* in which phenomena are treated, "*such a way* that everything about them which is up for discussion must be treated by exhibiting it directly and demonstrating it directly" (Heidegger 1962, p. 59). This directness is reached by the phenomenological method, which addresses the phenomenon as it is in itself for itself—in terms of its "thinghood" (Heidegger 1962, p. 59). This thinghood is the "beingness" or "is-ness" of a being, the "humanness" of humans, the "threeness" of three, and the "screenness" of the screen. This is-ness is the phenomenon phenomenology seeks to address—in its own terms.

2.1 Some Key Concepts of Phenomenology

For Husserl, the rigor of phenomenology was not that of exact sciences. "Scientific rigor was primarily the rigor of the deductive sciences familiar to the mathematician rather than that of the inductive natural sciences" (Spiegelberg 1994, p. 72).[3] This rigor rather means that the foundations of knowledge must be absolutely primary and self-evident

---

[1]*The Concept of Number* (1887), *Logical Investigations*, Volume 1 (1901), and *The Idea of Phenomenology* (1906).

[2]"To the things themselves!" became phenomenology's watchword, stressed by all major phenomenologists, namely Edmund Husserl, Martin Heidegger, Maurice Merleau-Ponty (1907-1961), and Jean-Paul Sarte (1905-1980).

[3]Husserl, inaugural lecture at Freiburg in 1917: "We often speak in a general, and intelligible, way of pure mathematics, pure arithmetic, pure geometry, pure kinematics, etc. These we contrast, as a priori sciences, to sciences, such as the natural sciences, based on experience and induction. Sciences that are pure in this sense, a priori sciences, are pure of any assertion about empirical actuality. Intrinsically, they purport to be concerned with the ideally possible and the pure laws thereof rather than with actualities. In contrast to them, empirical sciences are sciences of the de facto actual, which is given as such through experience. Now, just as pure analysis does not treat of actual things and their de facto magnitudes but investigates instead the essential laws pertaining to the essence of any possible quantity, or just as pure geometry is bound to shapes observed in actual experience but instead inquires into possible shapes and their possible transformations, constructing ad libitum in pure geometric fantasy, and establishes their essential laws, in precisely the same way pure phenomenology proposes to investigate the realm of pure consciousness and its phenomena not as de facto exists but as pure possibilities with their pure laws" (*in* McCormick 1981).

(*apodictic*). So, where are these "roots" of knowledge, these "pure laws," to be found? Husserl's first answer was in the things themselves, in the phenomena "in which all our concepts are ultimately grounded" (Spiegelberg 1994, p. 77). However, this was soon supplemented by another turn—a turn towards the subject. Husserl noted that a thing is always a thing for someone, thus, consciousness is the realm where things appear as that which they are in themselves, the *real* to be investigated.

To be conscious means to be conscious *of* something, i.e., to be directed toward something. Experience "always refer to something beyond itself, and therefore cannot be characterized independently of this...no straightforward sense can be given to an outer, external, world of objects which are not the objects of such experiences" (Hammond, Howarth, and Keat 1991, pp. 2-3). Intentionality means nothing more than this property of consciousness of being always already conscious. Because to be conscious is always to be conscious *of* something, intentionality is the consciousness as directed at something. Intentionality, addressed in this way, as structural to consciousness, is the ground from where phenomenological analysis proceeds.

That which appears in consciousness is that which is to be addressed without taking an *a priori* stand on its empirical existence or non-existence. For example when one identifies a particular object as a screen, *the screen-ness of screen must already be present in subject's consciousness*, otherwise this identification would not be possible. Phenomenology deals with this essence implied in the act of intentionality as such, i.e., it addresses this implied *what-is-ness*, in contrast with empirical existence of that addressed in such identification—which is rather a matter of *this-ness* or *there-ness*. For irrespective of its source—be it mental or sensory—"behind" every judgement about the particular there always already exits an essence that made such judgement possible in the first place.

Phenomenology has its object prior to the grounds of the empirical existence of a particular phenomenon, aiming at a pure description of it before any judgement about its existence can be made. Thus, the objective of phenomenology is to describe phenomena as they are in human consciousness—"the real has to be described, not constructed or formed" (Merleau-Ponty 1962, p. x). Phenomena are to be described faithfully and without presupposition, that is, as they primarily appear in consciousness, stripped out of historical contexts, scientific explanations, and philosophical interpretations—without presuppositions, free from prejudices, that is, not pre-judged.

Because phenomena in consciousness is the theme of phenomenology, and not their existence or non existence, the insistence on the presupposition of the world as empirically existent—as a condition for knowing—must itself be suspended. The existence of the world must be put between brackets, not because the philosopher should doubt it, but merely because its existence is not the theme of phenomenology as such (Biemel 1980, p. 627). The concept of "reduction," or "epoché," means this suspension of belief in the empirical existence of the world as the ultimate condition for its knowing. That which is to be apprehended is the pure phenomenon in consciousness, dropping all reference to the individual and to its particularities, as it manifests itself in consciousness, without any kind of evaluation, such as real, unreal, existent, non-existent, imaginary, etc. Reduction suspends the assumptions of the natural attitude, that is, suspends the taken-for-granted, everyday existence of the world, and returns to the things, as they are experienced in consciousness.

When we face an object, a situation, or an idea and "bracket" out its empirical existence and then describe it, we obtain an example. This description of the example is not yet the essence of the phenomenon, but it is a first reduction[4] toward the core concept of essence. But what is an essence for phenomenology?

Phenomenology studies essences (Heidegger 1962, 1977, 1978; Husserl 1960, 1964, 1970b; Merleau-Ponty 1962). The traditional signification of the word essence means *what something really is (in its own terms)*. When addressing that which makes a thing "is" what it is, we do not take into account those particulars that makes a thing a specific thing in time and space—a concrete empirical object: Rather, we focus our attention on those elements necessary for something to be part of a class of things we already take it as belonging to. When distinguishing something particular, identifying some concrete object, or characterizing some specific event, we implicitly admit to know in advance the kind of thingness to which the thing we are talking about belongs.

This initial meaning of the word essence has the character of an *a priori* necessity, a non-empirical, universal, and unconditionally valid condition (Husserl 1970b). However, we should point out that the concept of essence underwent some change in the work of Heidegger. He did not understand it simply as "what something is," but also as "the way in which something pursues its course, the way in which it remains through time as what it is" (Lovitt in Heidegger 1977, footnote 3). Heidegger re-addressed the issue of essence versus existence under the more fundamental question of the meaning of Being. The "what-Being" (essence) and the "that-Being" (existence) are undercut by the opening up of *Dasein's* essence "in its existence" (Heidegger 1962, p. 67). Thus, Heidegger tried "to develop a non-traditional concept of essence as "essential unfolding" (*wesen* as a verb)" (Polt 1999, p. 64).[5] It is this temporalized notion of essence that we would also adopt for our investigation.

Let's consider as an example a computer screen. Why is it that we are able to refer to a particular screen as a screen? To recognize particular screens *as* particulars implies to recognize that those screens are particulars of something else. Screens as particular must be delimited, actualized, concretized, specified, that is, they go beyond something that is common to all of them. This something common to all of them is that which is not particular but universal, i.e., that in which the essence of a screen is to be found. Thus, whenever we identify a thing as a particular thing (object, experience, event, and so forth) we have in fact unknowingly already entered the ground of essence.

Contrary to what a common understanding of the word essence suggests, to arrive at the phenomenological concept of essence, we can *not rely on empirical generalization*, comparing many examples and identifying the common features they have. This is so, mainly, for two reasons. First, the actualization of an essence in a particular context means an understanding in real terms that may add various non-essential elements because it happens within the domain of empirical existence. Second, what is common

[4]Husserl defended the reduction as a first step of a phenomenological investigation: only after the existence of the world has been bracketed should we proceed describing the phenomenon (Husserl 1960, section 20). This clearly made sense in his investigation into consciousness, *the Ego*, as it is. However, when ones investigates phenomena in the world it seems more advantageous to perform the reduction only as a second step, that is, only after a first description of the phenomenon has been done.

[5]This remark from Polt suggests the use of the English word essence as a verb: as *to essentiate*.

to any given quantity of examples is not necessarily the essence of the examples. The essence, of course common to all the examples, is common not only to the examples analyzed *but also to every potential* example of that phenomenon. This is so because the essence is such that without it there is no phenomenon.

Furthermore, the process of generalization itself already presupposes the existence of essence since

> the abstraction of the general idea "red" is arrived at by leaving out of account all those respects in which several red objects differ in order to hold on to that respect in which they are similar. But the concept of similarity (or even respect) which is in question here itself presupposes the very comprehension (of the essence of "red") which it is supposed to account for [Macann 1993, p. 9]

Therefore, essences are not generalizations. They are a different kind of common features inasmuch as they are the decisive elements in every particular example, whose actualization implies these decisive elements of the phenomenon in question. Essences are not actualized as something here or something there, because they do not exist in the actual world, but are in the very structure of consciousness, as foundations of knowledge and experience—as *a priori* and necessary features for knowledge and experience, as is clear in the case of generalization central to the empirical science.

The idea of a screen, against and in which all actual screens are confronted, is the original object, and does not necessarily follow from existence in any real world outside consciousness; it only remains as the necessary substrate for an object to be that which we designate it to be. This ideal object, intuited as it were, is the essence of all the actual objects we distinguish as part of a class, which is precisely defined by that same essence.

With these key concepts in mind, let us now turn to a discussion of the phenomenological method of investigation.

2.2 The Phenomenological Method of Investigation

Phenomenology is primarily a method of investigation. It is not a philosophical answer in itself but rather a way of questioning. As Heidegger (1977) wrote, this questioning builds a way, and this way is a way of thinking. The foundational basis of this thinking, of the method, is *evidence and logic*. Evidence here must not be confused with evidence in the empirical sense. Here evidence is that which is *evident in itself*, that which is impossible to conceive otherwise. In other words, to deny it would be to deny the very source of any empirical judgement already presumed. To ensure this interpretation we will use the term "self-evidence" but with this particular meaning in mind.

Only because consciousness is already evident for itself can consciousness logically conclude its own evidence. And it concludes logically because logic is the understanding in which consciousness is as such. Thus, for Husserl, evidence and logic are the indisputable grounds of thinking. Evidence and logic are in themselves self-evident, absolute primary, only relying on themselves to appear as themselves in the ways they are in themselves, that is, as necessary truths.

Like any other method, the phenomenological method of investigation is realized through a methodological circle; however, phenomenology strives to accept and to proceed only within the primary and foundational circle of human understanding: consciousness and *it's a priori* rules and procedures.

When applying the phenomenological method to uncover the essential nature, the is-ness, of the screen, we will closely follow the phenomenological method as it was synthesized by Spiegelberg (1975, 1994).[6] Nevertheless we felt that minor changes were needed on the basis of our ontological and epistemological assumptions, laid open as the investigation proceeds, and on the basis of the nature of the phenomenon revealed by analysis. In so doing, we consider the traditional phenomenological investigation of the etymology of the words, which identify the phenomenon, not merely as a step of the first phase of the method but rather as a whole second phase in its own right.[7] Such an adaptation is clearly supported by Heidegger's phenomenological investigations (Heidegger 1962, 1977, 1978).

Thus, the phenomenological method we will apply in our investigation has the following phases:

(1) Describing a particular phenomenon
(2) Analyzing the etymology
(3) Performing the phenomenological reduction
(4) Investigating essences
(5) Apprehending essential relationships
(6) Watching modes of appearing
(7) Interpreting concealed meanings

These seven phases constitute our full phenomenological method of investigation as Husserl developed it and as it was to a large extent synthesized by Spiegelberg (1975, 1994). In specifying these seven sequential phases, as we use them in this research, we want to stress their implicit union and their essential connections. The phases are united in the basic purpose of "giving us a fuller and deeper grasp" (Spiegelberg 1975, p. 57) of the phenomenon, which can only be achieved once all the seven phases were fully applied. Let us very briefly characterize each of the seven phases of the method with reference to our analysis of the screen.[8]

---

[6]*Encyclopaedia Britannica* (Biemel 1980, p. 630) classifies Spiegelberg's *The Phenomenological Movement—A Historical Introduction* (1994) as the "movement's first encompassing historical presentation." That work is an essential text both on the history of the movement, and on the technical issues pertaining to its method as well.

[7]For the same reasons, we decided not to separate the analysis of "the constitution of the phenomenon in consciousness" as a single phase, but to include it within the investigations of the fifth and of the sixth phases of our method—*Apprehending Essential Relationships* and *Watching Modes of Appearing*, respectively—as it revealed to be more insightful.

[8]Unfortunately, due to space limitations, we can not provide a detailed discussion here. We hope that the analysis of the screen will make the method clear. Also refer to Spiegelberg's account for more detail.

- ***Describing a Particular Phenomenon.*** This phase aims at a returning to the world as primarily and directly experienced, setting up the horizon of the phenomenon screen as "free as possible from presuppositions" and as intuitive as possible.

- ***Analyzing the Etymology.*** The task here is to trace back the origins of the words identifying the phenomenon screen. This analysis of the words is not destined to bring back the meaning of words, but rather to bring forth the meaning of the things themselves in the ante-predicative life of consciousness.

- ***Performing the Phenomenological Reduction.*** To perform upon the description of the first phase the reduction, *bracketing out* the features concerning the actuality—the particular presence in time and space—of any given screen. The phenomenon is reduced to a phenomenon in consciousness.

- ***Investigating Essences.*** This phase aims at reaching the elements strictly necessary for the phenomenon to be what it is. First, through generalization, the common features of screens are identified. Second, freely varying in imagination these common features, we strip them out of those elements that, despite being common, are not necessary, thus, leaving us the essence of the phenomenon screen.

- ***Apprehending Essential Relationships.*** This phase is an attempt to refine the essence of screen through *a priori* insight based on logic operations, which are to be performed within the essence of screen, and in what concerns its relationships with closely related phenomena. This aspect will be clearer when trying to relate the phenomenon screen with the ideas of data and world.

- ***Watching Modes of Appearing.*** The phenomenon screen is now to be investigated precisely in what concerns its appearances. Having identified the essence of screen, now the task is to pay attention to the ways in which such phenomenon appear: the aspects, contexts, perspectives, and modes in which it shows itself.

- ***Interpreting Concealed Meanings.*** This last phase of the phenomenological method, introduced by Heidegger (1962), is provided to give access to phenomena whose essence, whose meaning, has in itself concealment. We will show it to be particularly relevant for the case of the phenomenon screen. This phase involves decisive ontological and epistemological claims because that which is given in the phenomenological analysis is taken into account in the analysis of that which is doing that same analysis, and so on in circular movements.[9]

In Part, II we will apply the phenomenological method to uncover and explore the essence of the screen. The analysis will proceed carefully by following the phases outlined above. Due to the nature of the method there will by necessity be some repetition of formulations and reconsideration of statements and positions previously taken.

---

[9]To fully appreciate the method, we will make explicit core ontological and epistemological assumptions and this, in its turn, will enable us to bring forth concealed but relevant meanings of the phenomenon screen.

3. Part II: The Screen

3.1 Phenomenology of the Screen

A major part of our lives is becoming places located in front of screens. Is this significant for our understanding of ourselves and the world? We believe our phenomenological analysis will provide some insights into this question.

When phenomenologically investigating the screen, what we intend to think is not the *content* of television as such, or the kind of data we work with while facing a PC screen, but rather the screen as itself, in its *screen-ness*. It is the screen as a content of a specific understanding of the world and as a part, an enabler, or an element, of a concrete way of relating ourselves to and in the world that is the focus of our investigation.

3.1.1 Describing the Phenomenon Screen

From the start it is rather surprising what we encounter when starting the phenomenological analysis of the screen. When trying to describe a screen, a computer screen or a television screen, we immediately note that we never seem to look at a screen as a screen. We rather tend to look at screens in watching what appears on the screen. What seems evident when looking at a screen is the data presented on that screen—the text, images, colors, graphics, and so on—not the screen itself.

To try and look at a screen, and see it as a screen, not taking into account the particular data it presents, is apparently not an easy task. We are not familiar with this type of encounter with a screen. Rather our familiarity with screens are things—maybe surfaces—that function in particular contexts and for particular purposes, that is to say, we use screens as we act and relate ourselves to and in the world. This familiarization does not mean we consciously know what a screen is. Rather, that we are accustomed to screens, that is, we are accustomed in our daily life to perform the kind of activities in which screens are a part, are elements, participate in, or are present as just naturally "there." With Nietzsche, we note that "the familiar is that to which we are accustomed; and that to which we are accustomed is hardest to 'know,' that is to see as a problem, that is to see as strange, as distant, as 'outside us'" (1974, No. 355, p. 301).

However, in our phenomenological investigation, we take note to recover this strangeness—that is, that we seem not to see screens qua screens. Nevertheless, this strangeness is not the strangeness of a turned off screen, its strangeness is rather revealed in that we note its presence as a mere object, a piece of the furniture as it were. It might be this strangeness that often moves us to turn on the television or the computer screens we face. It is only when we look at the screen phenomenologically, as screen, trying to focus on itself as screen—but not as a turned off screen—that we enter the grounds of the screen as an "intentional object" of consciousness. What do we note?

Screens present, show, exhibit what is supposed to be the relevant data in each context, be it a movie while watching TV, a spreadsheet while working at the office, or a schedule while walking in the airport. Screens exhibit what was previously captured, processed, organized, structured, and finally presented on the screen. But what do we mean by "presented on the screen"? What is this "presented"? What is the data in question? Who presents it? Whom, where, and why?

The screen, as a screen, always finds itself at the center of the activity: in showing it attracts our attention, often also our physical presence, as it locates our activity. It is often the focus of our concerns in that environment, being at the office, working, or at home, watching a movie or the news. A screen gathers the attention of the people that surround it. Actions of those people are usually directly shaped by the presence of the turned on screen, by the kind of data it presents and by the understanding people surrounding implicitly assume of that data, which generates particular comportment and attitudes.[10]

So far the description of a screen is leading us into the notions of showing relevant data for and about each particular situation, of getting attention, of suggesting relevance, of acting as a mediation between ourselves and the world, and of gathering that which is appropriate in each particular context. We now have a first description of the screen. It is worth noting at this point that, although we do have a number of other words for screens such as display, output device, cathode ray tube, liquid crystal display, flat panel display, display window, and so forth, nevertheless, when engaged in everyday practices around them we tend to refer to them as screens.

We will now expand our investigation by an etymological analysis of the word screen, trying to uncover in this manner its meaning, and juxtaposing it with the description already presented.

3.1.2 Analyzing the Etymology of the Word Screen

Screen looks like a rather simple word. It is both a noun and a verb and its contemporary plurality of meanings can be collected along three main themes: projecting/showing (TV screen), hiding/protecting (fireplace screen), and testing/selecting (screening the candidates) (*The Oxford Paperback Dictionary & Thesaurus*[11] 1997, pp. 681-682).

The origins of the word screen go back to the 14th century. According to the *WWWebster Dictionary*[12] (1999), the contemporary English word screen evolved from the Middle English word "screne," from the Middle French "escren," and from the Middle Dutch "scherm." It is a word akin to the Old High German (eighth century) words "skirm," which meant shield, and "skrank," which meant a "barrier" of some kind.

The word screen still suggests another interesting signification, further away from us in history. It is a word "probably akin" (WB) to the Sanskrit (1000 BC)[13] words "carman," which meant skin, and "krAnti," which signifies "he injures." These meanings, possibly, are the ones from which the Middle Age words evolved. The Sanskrit origins

---

[10]Screens only function as screens when turned on. If a screen is turned off, it is just an object in the background. The screen comes to the foreground only when we attend to it for turning it on. When we push the on button, the screen locates our attention, we sit down, quit other activities we may have been performing, and watch the screen, as it is the place, the location, where that which is relevant for us in that particular time is to happen.

[11]From now on quoted as OPDT.

[12]From now on quoted as WB.

[13]Sanskrit—the language in which "The Vedas," the oldest sacred texts, are written—was an early form of an Indo-Aryan language, dating from around 1000 BC. The Indo-Aryan languages derived from Proto-Indo-European (before 3000 BC), from which also evolved Slavic, Baltic, Classical Greek, Latin, Germanic, and other families of languages. Old High German, Middle English, and Middle Dutch, belong to the *West* branch of the Germanic family. Middle French belongs to the Italic (Latin) family (Crystal 1987).

suggest that the notions of protection, shield, barrier, separation, arose as metaphors of the concept of skin, possibly of human (or animal) skin.

We should say that for our analysis none these meanings of the word screen has a definitive superiority. What is decisive is that the tracing back of the evolution of the meaning of the word screen enables us to penetrate the realm in which the word had its origins and in which it has evolved:

> What counts, rather, is for us, in reliance on the early meaning of a word and its changes, to catch sight of the realm pertaining to the matter in question into which the word speaks. What counts is to ponder that essential realm as the one in which the matter named through the word moves. [Heidegger 1977, p. 159]

The etymological analysis performed so far indicates that the word screen moved from Sanskrit meaning of skin and injury, along protecting, sheltering and covering, to the modern day projecting, showing, revealing, as well as electing, detecting, and testing. Now we may ask the following: Is there any central intent, distinction, or feature, common to all these specific meanings of the word screen? We believe the answer is *yes*. To defend such an assertion we will take up a different but related route, that of sound analysis.

The word "screen" is pronounced "skri:n."[14] It is close in its sound to the word "scream," pronounced "skri:m." It is just a final sound that distinguishes both words. The core sound of both words is the same. Do they both point to something beyond themselves?

The correspondence between sounds and meanings remains to a great extent an enigma. In spite of several attempts having been made to find specific accords between sounds and meanings, there is only "limited evidence on a few broad sound/meaning correspondences in language" (Crystal 1987, p. 175). However this state of affairs does not mean that this kind of sound analysis is senseless. Quite the contrary, that in some cases "speakers feel [that certain forms in language] *do* have a close relationship to objects or states in the outside world [means that] individual sounds are thought to reflect, or symbolize, properties of the world, and thus to 'have meaning'" (Crystal 1987, p. 174).[15]

When we look carefully at both of these words in the English and the Portuguese languages we can discern some interesting insights. The Portuguese word for screen is "écrã" (pronounced "Ekrã"), and for scream is "grito" (pronounced "gri'tu"). Quite different words at a first glance. However these two Portuguese words, as it is the case for the two English words referred above, have a common core sound. "kr" and "gr" are the same sound but for a very minor variation—the sound "gr" is almost the same of the sound "kr," only with a not so stressed "k"[16] (refer to Figure 1).

[14]We will use the standard International Phonetic Alphabet (IPA) in the sound analyhsis.

[15]*Onomatopoeia* is the most striking example, but other cases have also been researched as well (Crystall 1987, pp. 174-175).

[16]In Portuguese, it is possible to pronounce "grito" as "kri:'tu" without being misunderstood or incurring a worth mentioning mistake.

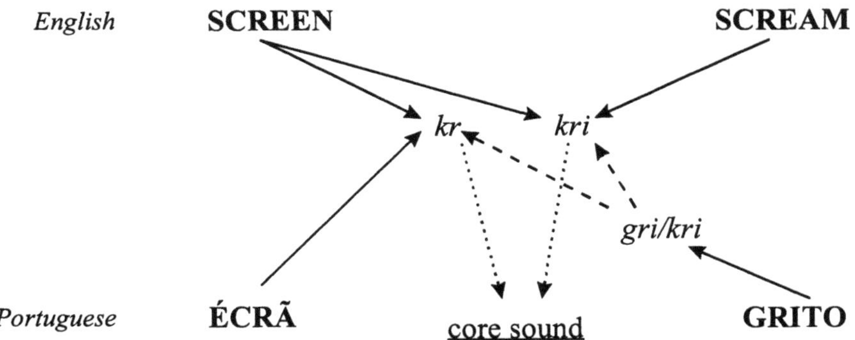

Figure 1. Sound Analysis of Screen and Scream

Thus, the question one needs to consider is if there is any meaning attached to the core sound "kr/kri"? By now, taking into account the analysis so far performed on screen, and the common meaning of scream, the answer appears intuitively: "to call for attention." "To call for attention" seems to be the meaning attached to the sound "kr/kri" for both words in both languages. A further scrutiny of the sounds in question strengthens such a conclusion.

The Portuguese word "écrã," a quite recent word, is clearly similar to the Middle French word "escren," referred to above. The corresponding current French word is "écran."[17] Is it the case that this initial "e," which the word screen does not have, has any specific meaning?

At a first glance the meaning of the "e" may seem a worthless question as it is a widely recognized principle of phonetic "that individual sounds do not have meaning: it does not make sense to ask what "p" or "a" [or "e"] mean" (Crystal 1987, p. 174). However, we must recall that we are not interested in words or in letters *de per si*, but rather on the things or concepts to which they refer.

Before continuing our analysis is important to note that although this kind of phenomenological analysis does share some concerns with linguistic analysis, it goes beyond it. This analysis is not destined to bring back the historical original meaning of those words, but rather to bring forth the meaning of the thing itself, around which the acts of naming and expression took shape" (Merleau-Ponty 1962, p. xv). What is at stake here is the bringing back of all the relationships implied in our experiencing of the phenomenon, the screen in this case. In the analysis of sounds, we intended to uncover or bring back some of the intentionalities implied in the way the words functioned. This is the reason for moving beyond the words, as words, to their sounds.

---

[17]Usually the Portuguese word for screen is "écrã," in *Dicionário Universal da Língua Portuguesa (1999)*. Some other dictionaries write it exactly as in French, "écran" (*Minidicionário Verbo-Oxford de Inglês* 1996). Still others do not include it at all, for example *Dicionário da Língua Portuguesa* (1989). The Portuguese word is an appropriation of the present French word "écran" that evolves from the Middle French word "escren." Both Portuguese and French evolved from the Italic (Latin) languages.

The "e" we are addressing has the sound "'e." So the appropriate question is: What does this "'e" sound mean? The answer is a surprising one. The letter "e," that represents the sound in question, is widely used in Portuguese as a prefix.[18] Yet, as we further inquire into the sound "'e" we note there is indeed a Portuguese word that only has that sound: the word "eh!" (*Dicionário da Língua Portuguesa*[19] 1989, p. 580). This word is a grammatical interjection,[20] a "designate of surprise, admiration, and calling"[21] (DLP, p. 580; translation ours).

This word "eh!" is commonly used in the Portuguese language in situations where someone wants to call the attention of someone else. Let us consider an example and the respective English translation: "Eh! Anda cá" means in English "Hey! Come here." We evidently note that in both languages the first word "eh"/"hey" is pronounced, that is, is destined, to call the attention of the one we want to "come here." The English word "hey," the most common translation of the Portuguese word "*eh" (Michaelis Illustrated Dictionary, Voume. I, English-Portuguese* 1958, p. 502), means exactly "calling attention" (OPDT, p. 349).

If this is so, one may ask what about the word screen? Maybe the word screen never had an initial "e," because the English language, as it evolved from the Middle Ages to the present day, seems to have had other solutions to emphasize the meanings at stake: the sounds "scr" and "i."

The Middle English word for screen was "screne." Contrary to what happens in the Portuguese language, in English the sound "scr" seems to have been widely preserved up to the present day use. The sound "skri:," the common core sound of the words screen and scream, adds to the sound "scr" the sound "i," which is precisely a key sound of the English word hey. This word, as with the French "hé" and the Portuguese "eh," means a *calling for attention*—but only the English word has the "i" sound. This sound "i" seems to be the one that in the English word screen that takes the place of the sound "'e" in the French and Portuguese words "écran" and "écrã," as they seem to have the same meaning.

Thus, we could conclude that from the Middle Ages to the present day, the evolution of the Middle French word "escren" and of the Middle English word "screne" followed the same path. Escren became écran, screne became screen, that is, in both evolutions, each according to its own context, the words moved toward stressing the "call for attention" intention implied in their usage.

The work done in this step leads to an idea that is in some important ways close to the one we had by the end of the previous step. This is the idea of screen as the bringing forth of (or calling forth for) attention and thereby implying relevance, since "calling attention to" always imply the supposition, correctly or incorrectly, of relevance. If the "attention" mentioned is our attention, to what does "relevance" refer? A first answer could be that relevance belongs to that presented on the screen (calling for our attention).

---

[18]It comes from the Latin and has the meaning of an outward movement (*Dicionário da Língua Portuguesa* 1989, p. 574), which is in line with the meanings of projecting and showing that the word screen gained in Portuguese, in French, and in English.

[19]From now on quoted as DLP.

[20]It comes from the Latin word "ehe-" (DLP, p. 580).

[21]Its French translation is "he!" pronounced just in the same way, "'e" as well.

If screen implies the calling for attention and such calling implies relevance, then we need to further analyze this implied relevance, as this seems to belong to the essence of screen. Is this relevance a matter of the content of that which is on the screen or does screen in its screen-ness (or screening) already presume it?

3.1.3 Performing the Phenomenological Reduction on Screen

At this point we must recall that, in order to reach the essence of screen, we should concentrate on the phenomenon screen as it appears in consciousness, not as thought, or as we assume it appears in an "outer empirical world." It is now important that we suspend belief in the existence of any particular empirical world, as a condition of our analysis. This is done by performing the phenomenological *epoché*. This means "reducing" the phenomenon screen to that which appears in consciousness, disregarding characteristics that value it as particular empirically "existent" things, while attempting to preserve its content as fully and as purely as possible. Nevertheless, this intentional object, the screen in consciousness, is not some pure isolated thing that has meaning in itself as such. It always claims to be, in its essence, a something in-the-world, not an isolated object in consciousness. Its being, as it appears in consciousness, is one of always already referring to its *functioning in a world* in which it makes sense, because it, and the other things and activities in the world,[22] mutually refer to each other—what Heidegger calls the "referential whole" (Heidegger 1962).

Having suspended the supposition of existence in the empirical world, thus in the existence of particular empirical screens as well, we discover that in consciousness screens qua screens still seem to exhibit attraction as that which it is. Without "calling for attention," screens would no longer be screens, merely surfaces or objects. Screens, in their screen-ness, are promises of bringing to present what is relevant and, simultaneously they hide their claimed physical being behind that same data, only calling our attention to the showing as such.

Screens, as screens, function in our flow of involvement in the world. Because this data is presented within our flow of involvement it is already relevant data—that is, data deserving our attention. The reduced phenomenon of screen appears in consciousness as something devised to attract—or rather already have—our attention and locate our action as acting beings in the, even now "suspended," supposed existent world.

This last argument can be made more clear by experiencing the kind of difficulty one has to go through in order to imagine a situation in which screens do not present relevant data at all. For example, a PC monitor at NYSE showing on a permanent basis the schedule of the trains of the suburbs of some African city; or the monitor of the cash registers of a supermarket showing air traffic control data. They may have an initially curiosity value but will quickly become ornaments in the background. These cases demonstrate the difficulty of imaging these displays *as screens* because to do it we would

---

[22]Throughout the discussion, we use the notion of world in the Heideggerian sence as a significant whole in which we dwell. Refer to section III of Part One of *Being and Time*, to Polt (1999, pp. 49-59), and to Dreyfuss (1991, pp. 88-107).

need to abandon the grounds of the essence of screen, and still force ourselves to use that same essence to understand an object that looks as having lost its *screen-ness.*

Screens display relevant data for us in each situation that engages us in the always supposed whole in which we relate ourselves. However, it is not the kind of data we immediately and intuitively grasp as human beings already in-the-world.[23] Rather this data is produced in such as way that it only appears for us (grabs our attention) in our "involvement whole" (Heidegger 1962) in which it refers to our activities and our activities refer to it, within a particular "form of life" (Wittgenstein 1967). For example, we can imagine what a man from the 15th century might think when confronted with a screen of an Automatic Teller Machine (ATM). That screen would be an object for him, but it would not be a screen because it would be impossible for him to conceive that particular screen in its *screen-ness*, i.e., in its essence. That screen would not be a screen for that man because that screen *was not* a screen for him—he would simply not recognize it in its *screen-ness*. Anything we do not recognize can be understood as many other things but for the thing that which it is.

Screens, in their screen-ness, claim a being in-the-world as focal interpretative surfaces, presenting relevant data for our involvement and action in the world. Screens promise to make evident our involvement in-the-world, because they present an already interpreted and reduced world to us, which is already consistent with our involvement in that world, our form-of-life.

This phenomenological reduced description of screen shows how closely intertwined are the ideas of *attention, relevance*, and *world* in the concept screen. However, this is not enough for a fully phenomenological characterization of the phenomenon screen. In order to reach the essence of a screen, we must now try to reach beyond the common elements to identify the strictly necessary elements for the phenomenon screen to be what it *is*.

3.1.4 Investigating the Essence of Screen

So far in our analysis we have relied on the *general-ness* of that which a screen is. However, this *common-ness* is not the essence of the phenomenon. We have already started to move beyond this commonness. As indicated before, essences are *not* generalizations. The essence of a screen is not that which is general to the screens so far addressed or to any screens that might be addressed. Essences are a different kind of common features inasmuch as they are the decisive elements in every particular example, whose actualization, in its turn, must imply existential elements of the phenomenon in question—thus, locating the example out of the domain of essences.

The essence, of course common to all the screens, is common not only to the examples analyzed but to all potential examples of that phenomenon, because the essence is such that without it there is no phenomenon. Thus, the essence can still rely on fewer elements than the common ones and suggest more than that which was revealed as common to the particular cases addressed. Imagination—"by discovering what one can and what one cannot imagine" (Hammond, Howarth, and Keat 1991, p. 76)—is the key

---

[23]Data on screens is not "natural information," according to Borgmann (1999, pp. 7-54).

of the following analysis, aiming at stripping out of screen those elements that, in despite of being common, are not necessary for a screen to be a screen.

We do not need empirical observation for discovering the answers we need because in every new variation in imagination we know the object we describe is an object of that same kind, a screen, if we recognize it as such, as a screen. Thus, the implicit criteria of recognition—*my ability to recognize the object as the object it is*—is the decisive way of this essential (*eidetic*) reduction.

First, we note that the same surface can be considered a screen and not considered a screen even if it displays the same data, as is clear from our example of the ATM above. If we have a mirror, with the size and shape of a screen, it displays data—the images it reflects—but we do not consider it to be a screen but a mirror. Yet we can have a screen displaying exactly the same image of that mirror and consider it a screen and not a mirror. So, what is the criteria that is implicit in this imagined experience?

Mirrors *reflect*, screens *(re)presents*.[24] This means the kind of data displayed by these different objects have diverse origins. In the case of a mirror, it is merely reflecting back what it receives. In presentation there operates a fundamental process of ordering. Presentation always assumes a theme—in the way that a jigsaw puzzle, to be a jigsaw puzzle, assumes a whole that will be its ordering criteria. Furthermore, the theme of the presentation assumes—or derives its meaning from—a form of life that renders it meaningful as a relevant presentation. As Wittgenstein argued, words do not refer to something because we agreed it; rather, they already have meaning because we share a form of life (Wittgenstein 1967, No. 241, p. 88). Thus, data presented on screens does not depend on the perceiving subject perspective but on the "themes" and "forms of life" in which it already functions as meaningful. Screens present selected data, that is, data that was previously selected to be displayed or data that is displayed because it is in accordance with a previously set theme for the presentation of data in that form of life. This last criteria means that the kind of data presented is the relevant data for the situation within the context where that screen makes sense being used *as a screen*. Screens always present relevant data—as themes in forms of life—and therefore gather and locate the attention of the people surrounding them.

Screens are not mirrors in that they do not reflect whatever they face. They present what is already relevant within the flow of purposeful action. The content/surface of screens is always already a presentation: the world made present according to a theme in a particular form of life. In watching, one could of course disagree with the relevance of the particular data being presented on the screen, but that evaluation itself is already relevant for the situation in which the viewer finds herself.[25]

We must also note that in selecting for presentation other possibilities are always implicitly excluded. Thus, the screen, as screen, "conceals" and "filters" in its revealing. This revealing/concealing attached to that which would be the phenomenon screen,

---

[24]In the discussion we will use the term presentation as it seems to us more effective in opening up the domains of meaning that we intuitively experience as relating ourselves to screens. However it will be technically more correct to use *representation* if we want to be consistent with Husserl's terminology. Refer to Husserl's *The Phenomenology of Internal Time Consciousness.*

[25]It may indeed be a crucial part of the context that precedes a relevant action of the subject; that is, a choice, a decision, a possibility for being, or even a stand on the grounds of meaning.

initially addressed when analyzing the etymology of the word screen, seems to keep on emerging as that which is essential to screen-ness. Let is further reflect on this.

Heidegger (1977) noted when investigating the concept of truth that for the ancient Greeks the word for truth was *aletheia*, meaning the simultaneous revealing and concealing of something. A revealing must in itself include a concealing of that which was not revealed. Likewise a concealing must include a revealing of that which was not hidden. The one always includes the other unless we admit the possibility of an absolute revealing of everything under all perspectives in every term and in all contexts, which human understanding evidently denies. Thus, to reveal implies to conceal. However, it also implies more.

If something was indeed revealed—and something else likewise concealed—in the presentation, then this possibility brought about discloses a notion of an *already there implicit agreement*. We must emphasize that our discussion only refers to screens, qua screens, which collect and attract attention. Obviously this agreement does not mean that one has to necessarily agree with the terms, conditions, analysis, or format of that which is revealed. The agreement is only that that which is revealed—on the screen qua screen—is present as that which attracts our attention as part of our ongoing activity in that form of life.

Where has this inquiry lead us to thus far? The essence of screen is being constituted in gathering attention by the presentation of relevance (and the concealment of irrelevance) in order to mediate our being in the world. This is to say that because we are acting in the world, always already engaged with our attention focused on something, the essence of screen is the presentation of relevance to mediate and (re)constitute the world. We now consider it appropriate to suggest the following essential description of a screen: *the essence of screen is to mediate our being in the world by presenting relevance in that world.* This concept should be kept open for further scrutiny because the notion of relevance, although implying the idea of hidden-ness, does not allow it to be brought forth. We also need to give more careful consideration to the idea of agreement that is emerging.

3.1.5 Apprehending Essential Relationships/Watching Modes of Appearing of Screen[26]

A screen always presents data of/in the world, that is, a screen presents the world in that the world on the screen is always accessed as the perceived world of my ongoing flow of activity—for example, the screen of the ATM is the world of my ongoing economic activity, the screen at the airport is the world of my ongoing travel activity. Thus, an understanding of the idea of screen implies the idea of world, as well as the ideas of attention and relevance. The kind of data that is displayed is neither reflections nor random data, but data generated by specific and usually complex and time consuming criteria or themes conceived by us. Thus, this relevance is dependent on a previous understanding of our activity in the world. It seems that a structural understanding of

[26]We present the results of these two phases together as their findings are clearly related.

screen implies at its essence the linkages between the ideas of presentation, relevance, and world—as noted before.

Can we now bring together the different stands of our analysis by rethinking our initial tentative concept of the essence of screen—a *mediation of our being in the world by presenting relevance.* Why does the screen in its essence not show invisibility, this hidden-ness, implicit in the presence of a screen? If the essence of a screen is to mediate our being in the world, that is, our activity in the world, by presenting relevance, why does the screen make itself invisible, that is, not relevant? Why in the essence of screen does this invisibility not show itself?

The answer to this process of questioning, consistent with the analysis so far performed, and enhancing our insight into the phenomenon of screen, is that this invisibility is itself essential to screen qua screen. The invisibility of screens is not only the invisibility of particular screens but the invisibility of the very essence of screen. This is to say, that screen, in its essence has concealment: they conceal that which is excluded *and*, more essentially, they conceal what they are for us, as screen. The invisibility we note in screens hides the very essence of the phenomenon screen, which would only be made accessible for us if we consider the meaning of this concealment, that is, the meaning of the invisibility of screens within our own understanding of being—our own being in the world.

3.1.6 Interpreting Concealed Meanings of Screen

In this final phase, we need to take into account that which is doing this analysis—we, as beings already in-the-world. The phenomenology presented must now become critical of itself as being presented by us as beings always already in the world. This phenomenological-hermeneutic circle involves a strong methodological claim:

> Since we must begin our analysis from within the practices we seek to interpret, our choice of phenomena to interpret is already guided by our traditional understanding of being. Since it deals with what is difficult to notice, this traditional understanding may well have passed over what is crucial, so we cannot take the traditional interpretation at face value. [Dreyfus 1991, p. 36]

By re-analyzing the phenomenological investigation under way, that is, by questioning the realm of human existence as it is humanly experienced, in the light of that which is given in that same phenomenological investigation, and so forth, in circular movements, this phase aims at the discovery of meanings that may not be immediately present to our intuiting, analyzing, and describing. This is what we now aim to do. This will not be easy, as we have to turn the phenomenological intuition "back" on ourselves, as it were. Let us recall the formulation that we initially proffered as the essence of that which is screen: a *mediation of our being in the world by presenting relevance.* But, what do we mean here by "mediation"? Why "mediate"? And why is "relevance" essential for this concept?

Our understanding of mediation is revealed within the ontological structure of our own being in the world (Heidegger 1962). Thus, this mediation has the significance of the ways in which we access and give meaning to our own being in the world, that is, to the world where we always and already find ourselves. This mediation appears against the background of understanding that *is* our being in the world. As a being-in-the-world for which itself is an issue, our being finds itself always thrown from the past and always projected into the future (Heidegger 1962). It is in this thrown-ness, structural to the being we are, that we act in the world, agreeing and disagreeing, choosing and creating meaning—that is, we mediate our own being in the world.

This means that this mediation is already the understanding of our being-in-the world, that which we are. So, having said that, we should ask if this mediation *truly* mediates. By asking this, we are saying that, whatever this mediation would be, *as* mediation, it must be agreed on as our manner of being-in-the-world—otherwise it would not mediate (here the concept of agreement emerges again).

Turning to the other elements of our initial formulation of the essence of screen, that is, toward "our being in the world" and "presenting relevance," mediation discloses what is already agreed upon: *precisely our being in the world.* How does a screen open an agreement of such a kind? Precisely by always and already presenting what is relevant in our flow of involvement, in our particular form-of-life. Thus, what is calling for our attention is not any potential agreement, or anything to be agreed upon as such, but an "already agreement" that is our being-in-the-world—where screens function qua screens.

Because this mediation is that which is *already agreed upon*, if it is to actually mediate, the concealed path of a screen is not mediation *de per si*, but rather *already agreement*. Already agreement is the essence in that it is how it essentially unfolds in the world. Already agreement brings the strands of our analysis together. Screens, in their screen-ness, attract those surrounding them because they are focal points of already agreement. They focus our understanding because they are already agreement on what is relevant, on the ways in which we access relevance, within our understanding in/of the world, in order to be what we are as consultants, train drivers, managers, and so forth.

To conclude, the hidden meaning of screen qua screen—when we consider the kind of investigation we perform as the beings we are in-the-world—reveals itself as already agreement. *The essence of screen is already agreement*. It is this already agreement that calls our attention, attracts us, makes us look at the screen in its screen-ness, and simultaneously condemns to forgetfulness that which was agreed upon, precisely because it is not an agreement but an already agreement—as such the thinking, the bargaining, the transacting, the negotiating, that typically precede an agreeing are pre-emptively excluded.

This essence as what it is in itself, understood within our being-in-the-world, is the way in which the screen pursues its course, "the way in which it remains through time as what it is" (Heidegger 1977, footnote 3), as already agreement. It is because this concealed meaning of already agreement is that which is most essential for screen, that it does not show itself, but hides itself, as it pursues its way in the world: concealing and spreading already agreement as its essence.

3.2. Some Conclusions and Implications

The essence of screen is already agreement. So what? Why and how is this investigation relevant? In this last section, we briefly address this issue, hoping that the potential of phenomenological investigations of this kind would be evident.

First, we should note that phenomenology is always and already a human activity in the world. We are in the world, and it is as we are in the world that we come to be, to know, or to act in that same world, becoming acquainted and involved with objects and other beings. Our familiarization with an object results "from experiencing it many times, which is a process that performs an unconscious induction all along" (Schmitt 1996, p. 141).

Because empirical objects and empirical events are recognized in accordance to the structure of consciousness—which they presume—they must be logically consistent and must be supported by evident foundational concepts—"matters of fact." As intentional objects, they rely on data from sensory experience and are a source for phenomenological investigation. But what is more, is that at its core the phenomenological way of proceeding is the capacity of consciousness to vary from examples to common-ness, from particulars to general, from existences to essences. This reasoning always occurs in both directions, from essences to actualities and vice-versa. It is precisely this variation that allows consciousness to identify what is shown in each domain.

Thus, the findings of a phenomenological analysis should be projected onto those matters of fact in which it has its sources, both as a coherent possibility and as a new horizon to understand human action in the world. Its results can be applied to specific situations, enhancing its understanding, and clarifying what is at stake. An important possibility of this method, in fact, is the possibility to relate to an "actual world" the investigation developed.

With this in mind, we will briefly present some of the consequences that seem more evident as implications that this investigation has for the empirical world. Unfortunately, we will be brief and can not fully develop and justify these here.

The power of *already agreement* can, for example, be seen with regard to our general view of television in everyday life. While there are many observations we can make here, we will refer to only one of these. What do we tend to think of people who live, on a permanent basis, without a television in their house? We tend to think of this as strange (possibly even somehow *dangerous*). Why is this so? Maybe we feel that these people do not share the already agreement—and the relevance implied—that the television is. We often refer to them as "living in another world" for we perceive the television as presenting that which is already agreed as relevant to those engaged in our world. Our analysis provides an explanation for such a view, as Fry (1993, p. 13) puts it, the television has arrived as the context—and those people seem to be out of context.

The power of television to reinforce what is presented just by the presentation itself has important consequences in our daily lives: "all that is important is revealed on television while all that is so revealed on television acquires some authority" (Adams 1993, p. 59). But this power does not belong to the essence of television but rather to the essence of screens, as the following example will show.

The kind of data about us that appears on a screen, at the bank, at the office, at the doctor of medicine, at a public department, is often taken as more valid and trustworthy than ourselves—as many of us have found out to our dismay. That the essence of screen is already agreement indeed helps to explain this: it is because the essence of screen is already agreement that that data is often taken as more valid and trustworthy than ourselves. Thus, this primacy of that which is on the screen over that which is not on the screen seems to be an issue that needs to be taken into account while addressing areas such as, among others, human resources management and marketing.

This argument gains a further impact when we consider the questions of change and of management of change. Screens, according to their essence, show already agreement. Thus, it seems reasonable to consider that they allow applications that rely on them to gain an edge of effectiveness—of power—when compared to applications with no screens. If this is so, then either implementation initiatives or contra-implementation initiatives can benefit by *acting* on screens.

But there is still another aspect important to change management with regard to the simultaneous revealing and concealing nature of screens. Screens, according to their essence, present agreement. However, in revealing agreement, they also conceal potential disagreement or resistance. This potential is always present on the edges of what is revealed/concealed, but always ready to turn screens into displays—to take from screens their self-evident place in the world. Screens in the world are highly fragile phenomena. To establish them may make it very difficult to lose them very easily.

Our analysis also seems relevant for thinking about innovation in a screen filled world. That which deserves to be questioned, it seems to us, is the likelihood that in a screen-intensive environment—that is, in increasingly screen-based "involvement wholes"—the possibilities for innovation would tend to be limited to paths and enigmas, accessible within the already agreement present on the *screen*. As such, they may hide that which the screen as a setting of context necessarily excludes, thus hiding opportunities for new articulations, new appropriations, and new understandings—which is exactly what innovation is about.

This section only served to highlight some of the possibilities of the empirical application of our phenomenology of the screen qua screen. These are taken as merely very tentative and explorative. They also point to some additional analysis that may be required to further consider the essence of the screen—such as the pages of a journal, newspaper, or magazine as screens. Unfortunately, due to the limits of a paper presentation, we can not pursue all of these possibilities here. Nevertheless, it is hoped that our presentation is sufficient to show the potential of phenomenology to enhance and develop our understanding of the complexity of our involvement in a world increasingly pervaded by screens.

Acknowledgments

We acknowledge the contributions of colleagues and reviewers for the substantial comments that lead to the improvement of the paper. Fernando M. Ilharco, in contributing to this paper, acknowledges the support of the Portuguese Catholic University, the Calouste Gulbenkian Foundation, and to the Luso-American Foundation for Development for the ongoing support of his Ph.D. research at the London School of Economics.

References

ABCnews. Http://archive.abcnews.go.com/sections/world/1997/97_diana.html, December 29, 1999.

Adams, P. "In TV: On 'Nearness', on Heidegger and on Television," in *RUA TV? Heidegger and the Televisual*, T. Fry (ed.). Sydney: Power Publications, 1993

Biemel, W. "Origin and Development of Husserl's Phenomenology," in *Encyclopaedia Britannica* (15th ed.). Chicago: Encyclopaedia Britannica, 1980, 1995 update.

Boland, R. J. "Information System Use as a Hermeneutic Process," in *Information Systems Research: Contemporary Approaches and Emergent Traditions*, H-E. Nissen, H. K. Klein, and R. A. Hirschheim (eds.). Amsterdam: NorthHolland, 1991, pp. 439-464.

Boland, R. J. "The In-Formation of Information Systems," in *Critical Issues in Information Systems Research*, R. J. Boland and R. Hirschheim (eds.). New York: John Wiley & Sons, 1983.

Boland, R. J. (1985) "Phenomenology: A Preferred Approach to Research on Information Systems" in *Research Methods in Information Systems,* E. Mumford. R. Hirschheim, G. Fitzgerald, and T. Wood-Harper (eds.). Amsterdam: North-Holland, 1985, pp. 193-201

Boland, R. J. "The Process and Product of System Design," *Management Science* (28:9), 1978, pp. 887-898.

Boland, R. J., and Day, W.F. "The Experience of System Design: A Hermeneutic of Organizational Action," *Scandinavian Journal of Management* (5:2), 1989, pp. 87-104.

Borgmann, A. *Holding On to Reality: The Nature of Information at the Turn of the Millenium.* Chicago: The University of Chicago Press, 1999.

Ciborra, C. U. "De Profundis? Deconstructing the Concept of Strategic Alignment," Department of Informatics University of Oslo, Oslo, Norway, 1997.

Computer Industry Almanac. Http://looksmart.infoplease.com/ipa/A0006115.html, December 23, 1999.

Crystal, D. *The Cambrige Encyclopedia of Language*, Cambridge, England: Cambridge University Press, 1987.

Detnews. Http://www.detnews.com/1997/diana/9709/05/0098.htm, December 29, 1999.

Dicionário Universal da Língua Portuguesa. Lisboa, Portugal: Texto Editora, 1999.

Dicionário da Língua Portuguesa. Porto, Portugal: Porto Editora, 1989.

Dreyfus, H. L. *Being-in-the-world: A Commentary on Heidegger's Being and Time, Division I*, Cambridge, MA: MIT Press, 1991.

Encyclopaedia Britannica. Http://www.britannica.com, December 23, 1999.

Fry, T. "Switchings" in *RUA TV? Heidegger and the Televisual*, T. Fry (ed.). Sydney: Power Publications, 1993.

Hammond, M., Howarth, J., and Keat, R. *Understanding Phenomenology.* Oxford: Blackwell, 1991.

Heidegger, M. *Being and Time.* Oxford: Blackwell, 1962.

Heidegger, M. "On the Essence of Truth," in *Basic Writings.* London: Routledge, 1978.

Heidegger, M. *The Question Concerning Technology and Other Essays.* New York: Harper Torchbooks, 1977.

Hegel, G. W. *The Phenomenology of Spirit.* Oxford: Clarendon Press, Oxford.

Husserl, E. *Cartesian Meditations: An Introduction to Phenomenology.* The Hague: Martinus Nijhoff, 1960.

Husserl, E. *The Crisis of European Sciences and Transcendental Phenomenology: An Introduction to Phenomenological Philosophy.* Evanston, IL: Northwestern University Press, 1970a.

Husserl, E. *The Idea of Phenomenology.* The Hague: Martinus Nijhoff, 1964a.

Husserl, E. *Logical Investigations.* New York: Humanities Press, 1970b.

Husserl, E. *Phenomenology of Internal Time Consciousness*, by M. Heidegger (ed.) and J. Churchill (trans). Bloomington, IN: Indiana University Press, 1964b.

Husserl, E. "Pure Phenomenology, Its Method and Its Field of Investigation," 1917, reprinted in *Husserl: Shorter Works*, P. McCormick and F. Elliston (eds.). Notre Dame, IN: University of Notre Dame Press, 1981.

Introna, L. "Information: A Hermeneutic Perspective," in *Proceedings First European Conference of Information Systems,* Henley on Thames, England, UK. 1993, pp.171-179

Introna, L. *Management, Information and Power*, London: Macmillan, 1997.

Macann, C. *Four Phenomenological Philosophers: Husserl, Heidegger, Sartre, Meleau-Ponty*. London: Routledge, 1993.

McCormack, P., and Elliston, F. (eds.). *Husserl: Shorter Works*. Notre Dame, IN: University of Notre Dame Press, 1981.

Merleau-Ponty, M. *Phenomenology of Perception*. London: Routledge, 1962.

Michaelis Illustrated Dictionary, Volume I, English-Portuguese. Comp. Melhoramentos de São Paulo, São Paulo, Brasil, 1958.

Minidicionário Verbo-Oxford de Inglês-Português-Inglês . Oxford: Oxford University, and Lisboa, Portugal: Press and Editorial Verbo, 1996.

Nietzsche, F. *The Gay Science*. New York: Vintage Books, 1974.

Oxford Paperback Dictionary & Thesaurus, Julia Elliot (ed.). Oxford: Oxford University Press, 1997.

Polt, R. *Heidegger: An Introduction*. London: UCL Press, 1999.

Schmitt, R. "Phenomenology" in *The Encyclopaedia of Philosophy*. New York: Macmillan, 1996.

Spiegelberg, H. *Doing Phenomenology*. The Hague: Martinus Nijhoff Publishers, 1975.

Spiegelberg, H. "Characteristics of Phenomenology," in *Encyclopaedia Britannica*, 15th ed. Chicago: Encyclopaedia Britannica, 1980, 1995 update.

Spiegelberg, H. *The Phenomenological Movement—A Historical Introduction*, 3rd ed. Dordrecht: Kluwer Academic Publishers, 1994.

Winograd, T., and Flores, F. *Understanding Computers and Cognition*. Reading, MA: Addison-Wesley, 1987.

Wittgenstein, L. *Philosophical Investigations*. Oxford: Blackwell, 1967.

WWWebster Dictionary. Http://www.m-w.com, December 17, 1999.

Zuboff, S. *In the Age of the Smart Machine*. New York: Basic Books, 1988.

About the Authors

Lucas D. Introna lectures in Information Systems at the London School of Economics and Political Science. He is also Visiting Professor of Information Systems at the University of Pretoria. His research interest is the social dimensions of information technology and its consequences for society. In particular he is concerned with the way information technology transforms and mediates social interaction. He is associate editor of *Information Technology & People* and co-editor of *Ethics and Information Technology*. He is an active member of IFIP WG 8.2, The Society for Philosophy in Contemporary World (SPCW), International Sociological Association WG01 on Sociocybernetics, and a number of other academic and professional societies. His most recent work includes a book, *Management, Information and Power* (Macmillan, 1997), and various academic papers in journals and conference proceedings on a variety of topics such as the phenomenology of cyberspace, theories of information, information technology and ethics, autopoiesis and social systems, and virtual organizations. He holds degrees in management, information systems and philosophy. Lucas can be reached by e-mail at l.introna@lse.ac.uk.

Fernando M. Ilharco is a Ph.D. student at London School of Economics and Political Science where he is developing a phenomenological investigation into information technology, strategy and the relationships between these two phenomena. From 1993 until 1997, when he started his doctoral studies, he lectured Information Systems at the Portuguese Catholic University, Lisbon, Portugal. In the academic year 1998/99, he taught at LSE, London. For some years now he has been responsible for a weekly column in the leading Portuguese daily newspaper *Público* where he comments on the political and social implications of IT. He holds a *Licenciatura* in Law, and an MBA. Fernando can be reached by e-mail at f.ilharco@lse.ac.uk.

19 DEVELOPING A VIRTUAL COMMUNITY-BASED INFORMATION SYSTEMS DIGITAL LIBRARY: A PROPOSAL AND RESEARCH PROGRAM

John R. Venable
Curtin University
Australia

Julie Travis
Curtin University
Australia

Abstract

This paper proposes that the worldwide, virtual community of information systems researchers develop a digital library system to support the needs of its members and those of the larger community that IS researchers serve. It describes an initial vision for an information systems digital library (ISDL) and the potential contribution of an ISDL to the worldwide virtual IS research and practice communities, in terms of improving the research function, contributing knowledge to the IS field, improving transfer of knowledge to IS practice, and improving the reputation of the IS field with respect to other academic disciplines. While the academic and professional domain of an ISDL would be information systems, the lessons learned could be applied to other domains. The paper argues that, to be successful, a virtual community-based approach is needed for the planning, analysis, organization, development, and operation of an ISDL. Taken together, the development of an ISDL and the institution of a virtual community-based approach provide rich opportunities for research into information systems that serve an entire community, development approaches that take community and consensus-building needs into account, and ways to restructure and improve upon the IS research community as an inquiring system. The paper briefly describes a framework for a research program and initial and planned research toward the practical realization of both an ISDL and community-based IS development.

Keywords: Digital library, group support system, soft systems methodology, information systems development, virtual community, virtual organization, scholarly discourse

1. Introduction

This paper proposes and discusses a research program into the creation of a digital library system by and for the IS community. A digital library is defined here as a computer-based system for the electronic storage of documents and publications in digital form and for their retrieval and presentation in human-comprehendible form. This definition is deliberately unrestrictive of the kind or format of the "document," e.g., including multimedia. An information systems digital library (ISDL) would be intended to serve the needs of IS researchers, publishers, teachers, students, and practitioners, as well as other interested parties.

Digital library systems (covering whatever subject matter, in this case, information systems) are themselves information systems. They are also much more than just technological artifacts. They have potentially profound implications for the people and organizations that work on, use, and rely on them and for society at large. They have the potential to profoundly change the way knowledge is created and disseminated, both within (and between) research communities as well as between researchers and other potential societal consumers—in the case of an ISDL, students and IS practitioners. Put into other words, they provide alternative (hopefully improved) ways to conduct scholarly discourse and technology transfer.

Digital libraries have recently become a hot topic for research. There are ongoing series of ACM, IEEE, and European conferences as well as special issues of *Communications of the ACM* and *IEEE Software* on the topic. The advent of digital library systems has been enabled by the simultaneous development of data (e.g., text) compression, storage, and search technologies, faster and higher capacity computers, and Internet technologies enabling organizational and societal access to such technologies. The development of digital libraries represents both opportunities and challenges to the IS field and research community. The opportunities lie in research in digital library technologies and more importantly in suitably fitting them to organizational and societal needs. Therein also lie the challenges.

The practice of and tools supporting scholarly discourse have evolved (or emerged) over time. Use of e-mail, discussion lists such as ISWorld, chat rooms and video conferencing, the WWW such as on ISWorldNet, various conference and publication websites, and electronic journals has altered the day-to-day practice of most researchers. The potential for these and other changes for IS research was described by Watson (1994). However, at this point in time, there has been little attention to digital library systems within the IS research community.

The field of IS provides much in the way of a theoretical basis for the establishment of suitable digital libraries within scholarly communities. As a human activity system (Checkland 1981), the system of scholarly discourse within the IS research community is an emergent one that is partly rationally designed and has partly evolved through accidents of historical timing, exercises of power, and serendipity. In *The Design of Inquiring Systems*, C. West Churchman describes different foundations and components for such systems (which include scholarly communities and discourse). Jürgen Habermas (1984) describes standards for the conduct of discourse by which the results of the discourse are legitimized for its participants. These ideas, together with many others that form the theoretical basis for the field of IS, could be (and, from the point of view of the authors, *should* be) brought to bear on the development of digital library systems and the roles that they might play within research communities.

This paper, therefore, proposes that the worldwide IS research community seize the opportunities for research and learning, as well as for improving scholarly discourse, by building a digital library system for itself and by studying the development processes and ongoing impacts.

The IS community can be characterized as a worldwide, virtual community with many and varied interests. As such, it is an example of the kind of communities that are evolving in the age of information. Such highly distributed, diverse communities present difficult challenges for the development of systems to support them. In order to adequately serve the needs of, and facilitate agreement among, the diverse stakeholders, which is essential to enable the success of an ISDL, we propose taking a virtual community-based approach to the planning, analysis, organization, development, and operation of an ISDL.

In the next section, we describe a vision for an ISDL, including possible features and the advantages that it would offer to the IS community. Section three describes a virtual community-based approach for developing an ISDL. Section four describes a framework for a multi-project program of research on both the ISDL and the community-based development approach. Section five discusses the potential benefits to the IS research community of building an ISDL. We then summarize and briefly give future directions.

2. A Vision for an ISDL

The authors' vision of an ISDL is one of a community-based service, which supports the IS community's goals and values. There are many possible (candidate) purposes, goals, objectives, functions, features, and services for an ISDL. In this section, we briefly introduce a vision of what we consider to be desirable for an ISDL system. We assert that we cannot and should not impose this vision on the community. Rather, we present it here only to motivate interest and begin discussion. In section 3, we propose and discuss a virtual community-based development approach that could be used to make specific decisions about an actual ISDL to be built.

2.1 The Central Idea

An ISDL could provide a single, unified source for flexible, full-text searching and retrieval of any kind of IS publication, at little or no cost, via the Internet. The central idea is to overcome the complex, chaotic situation illustrated in Figure 1, in which there are many means of distribution of IS research and knowledge, *none or even a few of which are sufficient by themselves*. Even existing digital library systems have problems of inadequate coverage, diverse user interfaces, and limited accessibility to subscribers. The current situation makes it difficult and costly to obtain publications (or the knowledge represented within them) and there is still a strong likelihood that some publications or knowledge will not be found, even with careful searches of multiple sources or distribution channels. Furthermore, easy access to web search engines is leading to reliance on lower quality material found on the web

We propose to augment, supplement, and possibly replace existing publication channels with an ISDL (see Figure 2), while providing additional services (described further below).

An essential feature of our vision is full-text searching. Full-text searching has become a practical way to overcome difficulties of keyword indexing systems (Witten et al. 1998). Full-text searching also offers the possibility to search publications' reference sections to locate those publications that cite a particular work, allowing one to conduct citation analysis and to identify related works.

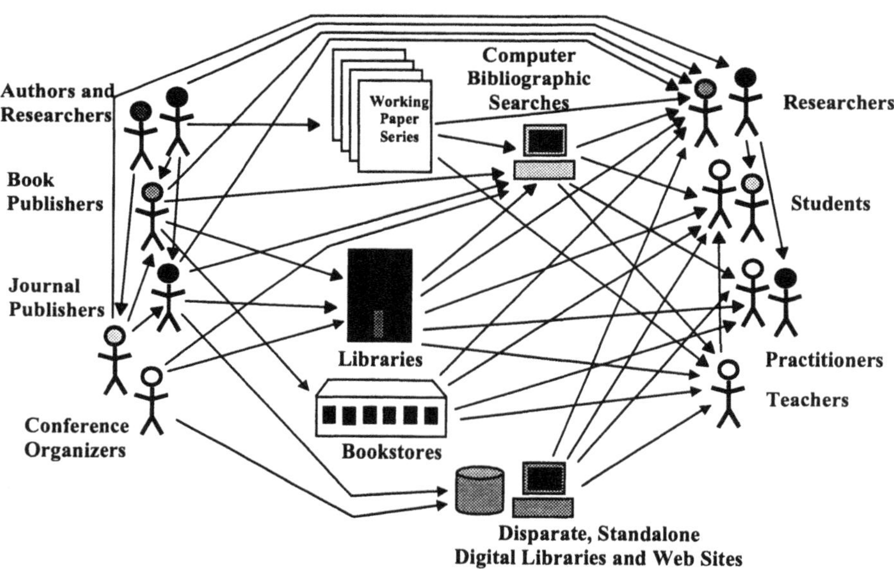

Figure 1. The Current, Very Complex Publication Distribution Situation (Simplified)

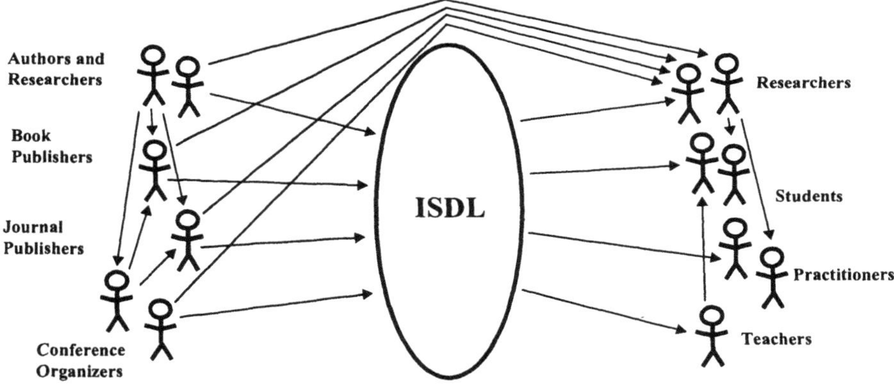

Figure 2. The Envisioned ISDL

Another essential feature of our vision is that the body of documents should be comprehensive, for example, including IS journals, conference proceedings, books, magazine articles, working papers, theses, web pages, even the content of discussion groups—whatever material is considered to be relevant and desirable to be searched and retrieved. In our vision, we distinguish between searchable publications and retrievable publications. While direct, free retrieval is preferable in an ISDL, it may not be practical in all cases. Where copyright or other interests prevent direct retrieval, support should be given for obtaining the publication indirectly, e.g., through physical library systems or at a cost from the publishers. Thus we might have comprehensive search support with less comprehensive retrieval support.

As a final element of our vision, we would like to see delivery of an ISDL service as a part of ISWorldNet, thereby making it freely available worldwide.

2.2 Candidate ISDL Features

Our main objectives for an ISDL would be to increase the probability of locating relevant publications and to reduce the costs of both searching for and retrieving relevant publications. However, an ISDL could potentially provide many other useful facilities. Table 1 proposes a number of possible features.

In our view, a particularly important potential area for an ISDL could be to greatly expand on the traditional role of libraries by providing a tighter connection with and support for scholarly discourse. For example, one could store/support reviews, ratings, or full discussions of particular publications within an ISDL (see White 1999) or have general discussions with references to publications available through the ISDL. The ability to reference publications from other publications and support tracing of their links (i.e., hyperlinks) within an ISDL would be a particularly useful way of supporting scholarly discourse. We further envision integrated, full-text searching of these ancillary annotations and discussions (in addition to the main publications). A digital library could also support the review process (Roberts 1999; Sumner and Shum 1996). These concepts follow directly from a view of the IS scholarly community as an inquiring system (Churchman 1971) or as a (virtual) inquiring organization (Courtney, Croadsell,and Paradice 1998). There is much interest in the library science community in changing the role of libraries in scholarly discourse (ARL 1999).

Table 1. Some Candidate Features for an ISDL

Topic Area	Candidate Feature
Improve search	• Publication coverage (book, journal, magazine, conference, working papers, web page, discussion lists, etc.) • Full-text searching • Meta-data searching (publication type, date, etc.) • Combined (full-text and meta-data) searching • Search interfaces (graphical, shopping cart, etc.) • Multi-session search/query saving/refinement • Automatic notification of long-term searches • Cooperative (multi-user) searching • Intelligent (machine) support • Librarian (human) support
Improve retrieval	• Direct retrieval of free publications • Specified level of free retrieval (e.g., title/author, abstract, first page only, full-text, etc.) • Take payments for non-free publications • Indirect retrieval of copyrighted publications from publisher/holder • Link to library systems for alternative retrieval
Publication collection	• Meta-data (author, title, etc.) form filling and automatic fetching into ISDL • Electronic/e-mail system interfaces • Automatic, periodic search and retrieval from known sites
Facilitate scholarly discourse	• Citation analysis via full-text searching • Hyperlinks between references and publications cited • Links to reviews and/or critiques of publications • Discussion of publications • Ratings of publications
Technology: Open systems	• Standard interfaces for electronic queries from other systems, • Query results returned, retrieval from ISDL, links to other retrieval systems (e.g., libraries' or publishers' systems)
Other	• Subject browsing • Virtual reality interfaces (e.g., for browsing) • Multi-media documents • Thesaurus/dictionary/terminology support • Teaching/collection building support • Multiple, user-selectable bibliography format support

2.3 Usage Scenarios

To better illustrate the potential usefulness of an ISDL, we will discuss how an ISDL might support some common research tasks or scenarios.

The primary purpose that we propose for an ISDL would be to support the scenario of researchers searching for and retrieving publications on a new (to the researcher) topic or concept. To do so, the researcher (or an agent for the researcher) would go the ISDL web site, enter a query using the full-text search capability, view and refine the results, and click on publications to be retrieved. Depending on the availability, the publication could be viewed and/or printed immediately, or could be received via other means, such as by fax or normal post. Possibly on-line payment might need to be made to a publisher. This scenario is much like for existing bibliographic search systems except that a more comprehensive body of IS literature would be supported, which would reduce the need to know how to use multiple systems with disparate user interfaces and would enable the location and retrieval of publications unsupported by any other systems.

This scenario could be extended over time to include storage and retrieval of previous search queries and results, spreading the search out over multiple sessions and refining queries over time, as the researcher's understanding of the topic and how to form a more successful query changes.

Another extension to this scenario is where the researcher has an on-going, long-term interest in a particular (re)search topic. In this case, the researcher might leave a query in an active mode. The ISDL would then periodically check all active queries against newly arrived publications and notify the person who owns the query about any relevant, newly arrived publications.

A potential difficulty with query search systems is information overload in the query results, i.e., where there are an overly large number of hits (publications that match the query constraints). The ability to refine the query and to spread out refining the query and viewing its results over multiple sessions (as noted above) addresses this issue. However, the researcher might also wish to constrain the query in ways other than on full-text search criteria. An ISDL could support this by allowing refinement of the search by constraining it both on the text and on the publications' meta-data, such as date of publication (e.g., no publications over four years old) or type of publication (e.g., refereed publications only).

Another way to reduce information overload might be to prioritize the publications automatically in some way. One way might be to prioritize the search terms in the query, whether on the text or on the meta-data. Another way (assuming that this feature is supported) might be to sort based on ratings given by other readers/raters. A third way (again assuming that this feature is supported) might be to sort query hits based on how many other publications cite a particular publication that matches the query.

Researchers are often interested in locating publications that reference a particular publication, particularly their own publications, not just for reasons of ego, but because they identify papers and authors with similar research interests. The full-text search capability of an ISDL, which also covers the reference sections of publications, would make this relatively easy. The researcher would just enter the title of the publication in the full text search.

Researchers often work cooperatively when conducting searches. An ISDL could provide facilities to share work spaces, either synchronously or asynchronously, to divide up search results among different people to determine relevance, or to forward queries and/or results to others.

A common, simple scenario is where one has become aware of a particular publication and wishes to retrieve that single, known publication. If the researcher found the reference while viewing a referencing publication on-line via an ISDL (and this feature was supported), one might simply click on a hyperlinked reference within the referencing publication. Alternatively, the researcher (or agent) could do a quick search by typing in the author and/or title and retrieve the appropriate query hit.

Another research scenario/task that could be supported by an ISDL is learning how other researchers view a paper or discussing a paper with other researchers, including the authors. Currently such discussions happen formally through discussants' comments, which may or may not be published, semi-formally at conference presentations, and informally through face-to-face or e-mail discussions. An ISDL could store discussants' written responses, conference presentation slides, or even audio and video. An ISDL could further provide support for discussions of particular publications, whether synchronous or asynchronous, and store those discussions for future reference. Such discussions could greatly increase the richness of the captured scholarly discourse and make it more widely available to others.

A common research scenario/task is to construct a list of known references. Support for individual researcher's reference lists could be provided within an ISDL. This could be very useful in constraining searches by including a query term to *not* return a hit for any publications on a researchers known reference list. Researchers could then easily add references to such lists. A researcher might even keep multiple lists, e.g., a list of "to be read" publications as well as a list of ones already read.

Another simple research task is obtaining a correct citation for a publication and incorporating it into the reference section of a paper. After conducting a search and locating a paper, an ISDL could provide its citation in whatever referencing system format the researcher desires, which could then be cut and pasted into the new paper's reference section. If one had a personal reference list, one could also copy it from there.

We have said little here about supporting the roles of teachers or students. Obviously many of the above scenarios apply to both of these roles. However, another task/scenario might be for a teacher to construct a list of references and to publish it for students to use in the context of a course. This might also be supported by an ISDL.

Finally, the creation of this sort of facility would create the need for various usability and housekeeping tasks that would also need to be supported, such as updating passwords, contact information, default settings, and so forth. Intelligent and tailorable defaults for search queries and the ability to change them easily would have a large impact on the usability of an ISDL.

In this section, we have discussed some task scenarios and how an ISDL might support them. The particular support would be dependent on the features provided by the ISDL, which would need to be designed and built. Some of them would be easy to provide and others more difficult. The above discussion also represents only a partial listing; many other scenarios could be supported.

3. A Community-based Approach to Developing an ISDL

There are many issues to be decided in the establishment of an ISDL. It is our view that the actual choices for the requirements, design, implementation, operation, and maintenance of an ISDL and for the resolution of important issues should be made in open discussion by the IS community as a whole. In this section, we discuss some issues and propose a method for addressing them.

3.1 Issues to Be Addressed

The most critical issue for the success of an ISDL system is to provide a suitable approach and features that will be accepted by enough of both the publication consumers and publication providers and publishers of Figure 2 to ensure a critical mass of users (Venable, Travis, and Sanson 1996a). Without enough consumers, there will not be adequate motivation for authors and publishers to provide publications. Without enough publications, there will not be adequate motivation for consumers to bother accessing an ISDL.

It is our belief that current technologies make an ISDL with the basic features described above easily technologically feasible. While technical issues are important, the primary obstacles are economic and socio-political issues. The various issues that would need to be decided in a community-based approach can be grouped into three main categories, as shown in Table 2.

Table 2. Categories of Issues for an ISDL

Economic and Socio-Political Issues:	• Ownership of the ISDL (IS community?) • Clients/customers to be served by the ISDL • Governance/organization • Funding for development, operation, and maintenance • Economic policy (e.g., fee charges, agreements with publishers) • Protection of the interests of copyright holders • Operational policies (access, etc.) • Relevant publications to be included
Functionality Issues:	• As in Table 1 • Prioritization and timing of delivery
Technical Issues:	• Centralized vs. Decentralized • Communication protocols • Security

While we believe that these issues need to be decided by the IS community, the authors' personal opinions are that ownership, clients, policies, and publication relevance should be as widely public and open as possible. We favor a model like that used for

ISWorldNet, in which a variety of people can be involved in as detailed a way as they choose. As for economic issues, we believe that the core functionality can be provided within the confines of normal discretionary research funding and that further development of more advanced functionality can be accomplished through grant funding and various forms of institutional research and teaching support. We believe that there will be no need to charge for services provided over the Internet. Nonetheless, these are only are opinions and we recognize the need for interested stakeholders within the IS community as a whole to choose solutions to these issues themselves.

3.2 Development Approach

There are many different methods that could be used in a community-based approach to address the above issues and determine an appropriate way to go forward. One could use SWOT analysis, critical success factors, cognitive mapping, the Delphi method, or even focus groups. However, it must be recognized that the highly distributed nature of the IS community presents difficulties in applying any approach. It seems that for the actual development, a prototyping approach would be the most useful to ensure user satisfaction with the interface and functionality by the different stakeholder groups.

We believe that the choices of development approach and the decision-making process to be used in addressing the above issues are also matters for the IS community to decide. However, we again have our own ideas as members of that community as to how to best proceed, which we describe here. The IS community needs to develop a better understanding of the above issues in order to move toward action by in building an ISDL. We need to be able to investigate the goals and desired functions of an ISDL in order to achieve a sufficient consensus to obtain a critical mass of adopters and to enable the system to be a success. In particular, we propose to use soft systems methodology (SSM), supported by a web-based group support system (GSS).

SSM (Checkland 1981; Checkland and Scholes 1990) is a well-known means for addressing "wicked" problem situations (such as requirements for an ISDL). The soft systems approach and its associated techniques (e.g., rich pictures, CATWOE, and root definitions) have proven useful for overcoming conflicting goals and complex problem situations to arrive at effective solutions to those problems. Furthermore, SSM is a systemic approach, which would invite wider consideration of the role of an ISDL within the IS research community and other academic practices. The use of the SSM conceptual modeling technique would provide better understanding of the functions and activities of the research community to better see where an ISDL could contribute and what other activities would be needed to enable an ISDL. SSM is also a flexible approach, into which one can incorporate other techniques not traditionally associated with SSM itself, such as the development methods discussed in the first paragraph of this section.

As an illustration of the relevance of SSM, in Appendix A we give examples of an initial exploration of the problem domain using some of the SSM techniques, which we developed from our own point of view. Participants in an SSM study would develop other versions from their own points of view.

SSM is generally applied by a person who is skilled both in the techniques and in facilitating groups to apply the SSM approach and techniques to take on the problem and to arrive at and implement their own solutions. Typically, this is accomplished through face-to-face contact with the stakeholders individually and/or in groups. Group support systems (GSS) have been proposed as a way to facilitate the group processes in SSM (Galliers et al. 1991, Venable, Travis, and Sanson 1996b). Research in the use of GSS to support SSM has concentrated on their use in supporting same time, same place (face-to-face) group meetings.

However, in the case of an ISDL, face-to-face contact is difficult, if not infeasible, because the communities of stakeholders are highly distributed, i.e., they are located worldwide. The obstacles to scheduling and traveling to meetings with all the stakeholders individually or to face-to-face group meetings may be too difficult to overcome. Regardless of the choice of development method from among those identified above, the worldwide nature of the IS and publishing communities complicates the problem of applying the method. Venable, Travis, and Sanson (1996b) proposed the use of a web-based GSS as a means of overcoming situations where the stakeholders are too widely distributed for the normal application of SSM. A web-based GSS would be accessible to any stakeholder who has access to a web browser.

4. A Research Program Framework

The research proposed in this paper fits within a larger planned research program. The proposition, development and use of technologies to address specific or generic problem situations constitutes systems development in IS research. Nunamaker, Chen, and Purdin (1991) argue that this is legitimate research and propose an overall framework for programs of research in that area, similar to Figure 3 (adapted from Figure 2, Nunamaker, Chin, and Purdin 1991). An ISDL would be an example of such a computer-based information system technology. Furthermore, in our opinion, this framework can apply to IS development methods, tools, and techniques. The development and evaluation of GSS support for SSM is an example of such research combining methods, tools, and IS support. Thus, the framework can apply to research in both the ISDL as an example of digital libraries more generally and in the GSS/SSM method that we propose for its development (or in whatever virtual community-based method the IS community chooses to apply). This section describes the framework and our research approach within that framework.

A research program nominally begins with questions and theory building about a particular domain or problem area. General, higher-level theories must be tied down to detailed theories and decisions about individual, lower-level components of some system to be developed. Detailed theories lead naturally into the applied design decisions of systems development, i.e., prototype and/or product development.

In our case, we began our research with the propositions that an ISDL could be built that would be helpful to the IS community and that a GSS might be useful for supporting the distributed application of SSM, based on theories of the applicability of GSS and knowledge of their general capabilities. However, theories of how GSS are applied in distributed situations and how they could support SSM techniques are not extant, so theories needed to be developed in this area. In our case, this meant determining what sort of GSS we should use and how it should be adapted to support the stages and individual techniques of SSM. For our initial research, we chose a web-based GSS, Discussion*Web*[tm] (McQueen 1999), made specific decisions about how we would apply Discussion*Web*[tm] to support SSM, and implemented them in a prototype.

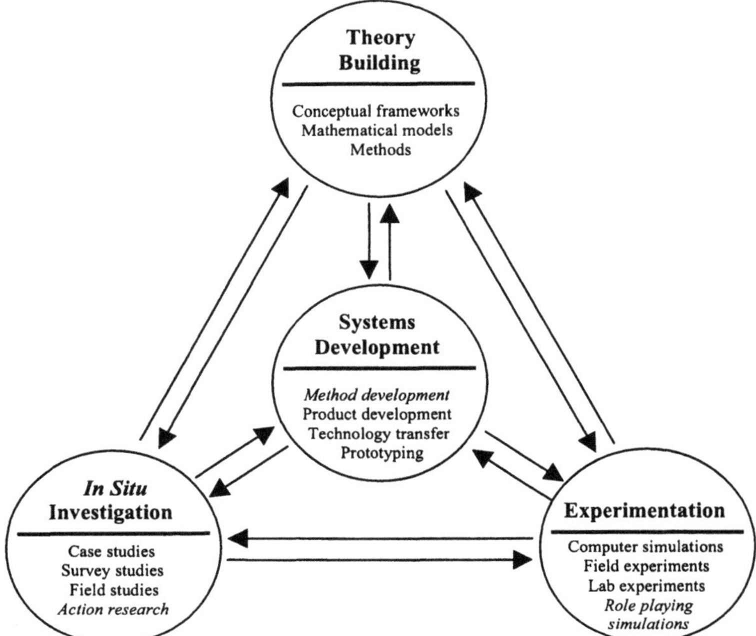

Figure 3. System Development in IS Research (Adapted from Nunamaker, Chen, and Purdin 1991)

Once the detailed design decisions are made and a prototype and/or product is developed, then it must be evaluated to see whether the theoretical advantages of the prototype and/or product actually apply. As shown in Figure 3, this can be accomplished in either or both of two streams or learning cycles: experimentally (in an artificially constructed situation) or *in situ* (in a real situation). Typically, a research program would first use experimental research to explore the problem space and determine any issues to feed back to theory building and/or systems development, before using the system/technology in live situations, which would contain an element of risk that the system/technology might be inappropriate or cause practical problems. Experimental applications often are carefully designed in order to address specific areas of theory or system development. In other cases, they more generally explore feasibility and general characteristics.

Our initial study of using a GSS to support the distributed application of SSM was an exploratory prototype pilot evaluation using a role-playing simulation (Galliers 1985). We have added this (in italics) into Figure 3. The design of our pilot simulation and some of the results of the study were reported in Venable and Travis (1999).

Experimental application of systems/technologies is not adequate for evaluating systems or technologies that interact in complex and/or unpredictable ways with real organizational situations. What Nunamaker, Chen, and Purdin called observation, we have termed "*in situ* investigation" in Figure 3, thereby expanding it to reflect the interventionist rather than passive nature of action research and have augmented Figure 3 to include action research (shown in italics) as a research method for *in situ* investigation.

We have developed an initial prototype ISDL (Venable 1999) to provide a discussion point and to make potential stakeholders aware of the possibilities that an ISDL might offer. The prototype ISDL addresses in particular the ability to combine meta-data and full-text searches, as well as entry of publications by authors and/or publishers. Figure 4 shows a top-level data flow diagram of the ISDL prototype. However, further work is needed before the prototype can be released for experimental use by the IS community. When released, the prototype will include an extensive sample of IS publications. Once development of the ISDL commences in earnest, one way to proceed is to continue to enhance this prototype significantly in accordance with requirements determined by the IS community at large.

In our opinion, it is absolutely essential to study technologies such as an ISDL or GSS support for distributed application of SSM further in real situations. Once we are satisfied with our prototype ISDL and GSS for SSM and with our experimental results, we plan to utilize an action research approach. Action research is particularly appropriate

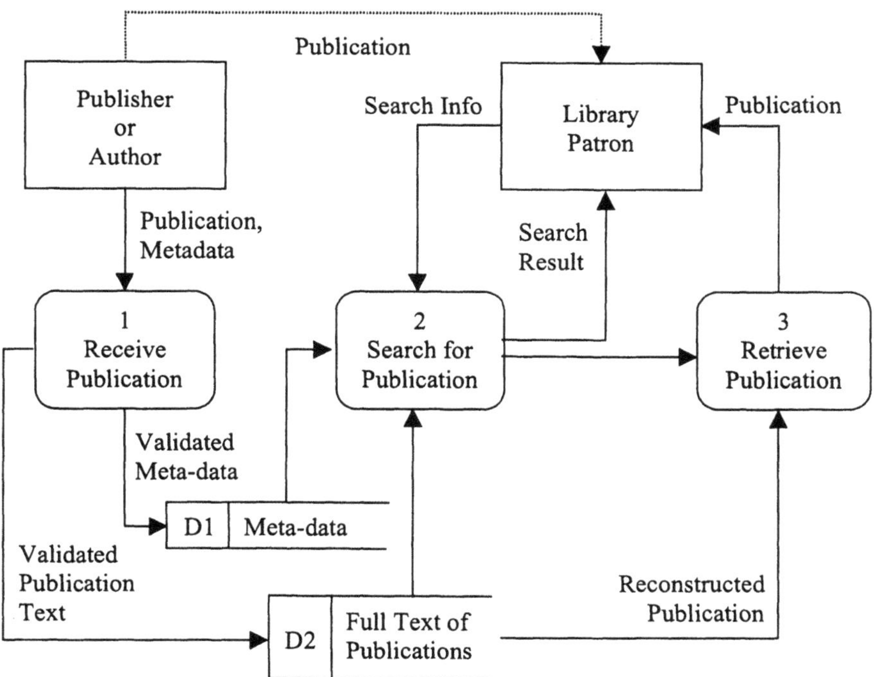

Figure 4. Top-level Data Flow Diagram of ISDL Prototype

in this situation due to our roles as interventionists and facilitators, especially if the IS community elects to utilize SSM. Ultimately, we believe that a research program combining both quantitative and qualitative approaches is appropriate in this situation (Jick 1979; Venable, Travis, and McQueen 1997). To that end, our pilot study of our GSS for SSM prototype was constructed to provide a basis for further quantitative study and measurement of improvement in subsequent versions of the GSS for SSM. As for the ISDL prototype, the ultimate measure of its success will be whether it is used or not. We also plan to study how it is used and to seek feedback on its strengths and weaknesses and on the usefulness of particular features as well as of the system as a whole. What we endeavor to create is an on-going relationship between developers and system users so that the ISDL will continue to be improved so that it better suits its users and the IS community.

5. Potential Benefits to the IS Community from Building an ISDL

We believe that the IS community would have much to gain by building and adopting an ISDL. One group of benefits are those that could accrue from the research and practice during the development process of an ISDL. First, we believe that the dialog engaged in during the process would result in the development of an improved vision and understanding of the community and field of IS. Secondly, conducting the process has the potential to reshape for the better the nature of the cooperation between the IS research community and publishers, as well as the cooperation between the IS research community and practitioners. Third, the process presents a good opportunity for ancillary research on organizational (and virtual organizational) issues relating to change, technology adoption, and development methods. Fourth, there would also be interesting opportunities for research on GSS success/failure, features, or social impacts.

Another group of benefits would come from the resulting availability and adoption of the ISDL itself. Having an ISDL would primarily improve accessibility of IS materials. It could also potentially improve IS teaching and technology transfer, as well as the process, content, and ultimately the results of scholarly discourse.

Finally, in our view, developing an ISDL could substantially increase the reputation of the IS community vis-à-vis other academic disciplines. If the IS academic community is able to successfully put the technologies and development methods that it espouses into practice, it would provide strong evidence of the utility of the knowledge that it creates and promotes both to business and to academia at large. A successful ISDL could serve as an example for improving scholarly discourse and technology transfer to industry, which could be followed by other academic disciplines.

6. Summary and Future Directions

In this paper, we have proposed that the IS community join together in a project to develop a digital library system for its own use. We have proposed some ideas for what an ISDL might look like and what benefits that one might provide to the IS community. We have emphasized that the most important thing is that community agreement will be necessary for the successful creation and adoption of an ISDL. We have also proposed a virtual community-based method by which the necessary agreement and further development might occur. Once the preparatory work for conducting the development is completed, the commitment and participation of an adequate cohort from the IS community is needed to realize the benefits that the creation and adoption of an ISDL by the IS community could offer.

Acknowledgments

The authors gratefully acknowledge the extensive helpful comments and suggestions from the anonymous reviewers and the track chair. Discussion*Web*™ is a trademark of Dr. Robert J. McQueen.

References

ARL. *Conference on New Challenges for Scholarly Communication in the Digital Era: Changing Roles and Expectations in the Academic Community*, Association of Research Libraries, 1990 (http://www.arl.org/scomm/ncsc/conf.html#P3; accessed June 1999).

Checkland, P. *Systems Thinking, Systems Practice*. London: Wiley, 1981.

Checkland, P., and Scholes, J. *Soft Systems Methodology in Action*. London: Wiley, 1990.

Churchman, C. W. *The Design of Inquiring Systems: Basic Concepts of Systems and Organizations*. New York: Basic Books, Inc., 1971.

Courtney, J., Croasdell, D., Paradice, D. *Inquiring Organizations*, 1998 (http://iops.tamu.edu/faculty/j-courtney/inqorg/inqorg.htm; accessed May 1999). Also published in *Australian Journal of Information Systems*.

Galliers, R. D. "In Search of a Paradigm for Information Systems Research," in *Research Methods in Information Systems*, E. Mumford, R. Hirschheim, G. Fitzgerald, and T. Wood-Harper (eds.). Amsterdam: North-Holland, 1985.

Galliers, R. D., Klass, D. J., Levy, M., and Pattison, E. M. "Effective Strategy Formulation Using Decision Conferencing and Soft Systems Methodology," in *Collaborative Work, Social Communications, and Information Systems (COSCIS)*, R. Stamper, P. Kerola, R. Lee, and K. Lyytinen (eds.). Amsterdam: North-Holland, 1991, pp. 157-179.

Habermas, J. *The Theory of Communicative Action* (Volumes 1 & 2), translated by Thomas McCarthy. London: Heinemann, 1984.

Jick, T. D. "Mixing Qualitative and Quantitative Methods: Triangulation in Action," *Administrative Sciences Quarterly* (24:12), December 1979, pp. 602-611.

McQueen, R. J. Discussion*Web*, 1999 (http://dweb.waikato.ac.nz/dw/; accessed October 1999).

Nunamaker, J. F. Jr., Chen, M., and Purdin, T.D. M. "Systems Development in Information Systems Research," *Journal of Management Information Systems* (7:3), Winter 1991, pp. 89-106.

Roberts, P. "Scholarly Publishing, Peer Review, and the Internet," *First Monday: Peer Reviewed Journal on the Internet* (4:4), April 5 1999 (http://131.193.153.231/issues/issue4_4/proberts/index.html; accessed May 1999).

Sumner, T., and Shum, S. B. "Open Peer Review and Argumentation: Loosening the Paper Chains on Journals," *Ariadne (The Web Version)* (5), September 1996 (http://www.ariadne.ac.uk/issue5/jime/; accessed May 1999).

Venable, J. "A Proposal and Prototype for an Information Systems Digital Library," *Proceedings of the Tenth Australasian Conference in Information Systems (ACIS'99)*, B. Hope and P. Yoong (eds.), 1999, pp. 1118-1128.

Venable, J. R., and Travis, J. "Using a Group Support System for the Distributed Application of Soft Systems Methodology," *Proceedings of the Tenth Australasian Conference in Information Systems (ACIS'99)*, B. Hope and P. Yoong (eds.), 1999, pp. 1105-1117.

Venable, J. R., Travis, J., and Sanson, M. D. "Requirements Determination for an Information Systems Digital Library," *Proceedings of the Seventh Conference of the International Information Management Association (IIMA '96)*, Estes Park, Colorado, December 1996a, pp. 35-46.

Venable, J. R., Travis, J., and Sanson, M. D. "Supporting the Distributed Use of Soft Systems Methodology with a GSS," *Proceedings of the 1996 Workshop on Information Technology and Systems (WITS '96)*, Arun Sen (ed.), Cleveland, Ohio, December 1996b, pp. 294-298.

Venable, J. R., Travis, J., and McQueen, R. J. "Researching the Use of a GSS to Support a Very Large Group," *Proceedings of the 1997 IFIP WG 8.2 Workshop on Organizations and Society in Information Systems (OASIS '97)*, N. L. Russo and S. Johnson (eds.), Atlanta, Georgia, December 1997, pp. 22-26.

Watson, R. "Creating and Sustaining a Global Community of Scholars," *MIS Quarterly* (18:3), September, 1994 (http://www.misq.org/archivist/vol/no18/issue3/ vol18n3art1watson.html; accessed May 1999).

White, J. *ACM Digital Library Enhancements*, 1999 (http://www.acm.org/dl/ dl_enhance.html; accessed May 1999).

Witten, I. H., Nevill-Manning, C., McNab, R., and Cunningham, S. J. "A Public Library Based on Full-text Retrieval," *Communications of the ACM* (41:4), April 1998, pp. 71-75.

About the Authors

John Venable is a tenured senior lecturer at Curtin University of Technology, Perth, Western Australia, where he is also coordinator of the Systems Research Group. He has lectured and researched in Information Systems since 1983. He obtained his Ph.D. in Advanced Technology (Computer Science and Information Systems) in 1994 from Binghamton University in Binghamton, New York. His prior degrees include a B.S. from the United States Air Force Academy and an M.S. in Management Science and an M.S. in Advanced Technology, both at Binghamton University. He has previously lectured in the business management schools at Binghamton University and Central Connecticut State University in the USA, in computer science departments at Aalborg University, Denmark, and the University of Waikato, New Zealand, and in the School of Information Technology at Murdoch University in Perth, Western Australia. John's main research interests are computer-supported cooperative work (CSCW), digital library systems, and information systems development, particularly in its practice, methods, and appropriate tool-based support. John can be reached by e-mail at VenableJ@cbs.curtin.edu.au.

Julie Travis is currently a tenured lecturer with the School of Information Systems, within the Curtin University of Technology, Western Australia. She has held positions within IS as systems analyst, business analyst, IS consultant and information and systems manager for 11 years before entering academia. She also has 12 years experience teaching and researching within universities, mainly in Australia and in the United States. She has also taught in programs in the UK, Hong Kong, Singapore, and Indonesia. Her research background and interests are mostly in the areas of IS research issues and methods, adaptive systems and business methodologies, facilitation and group support systems (GSS), organizational IS and learning, global IS and electronic commerce, (top level or early) information requirements determination (IRD), systems thinking, knowledge management, strategic planning and computer security management including risk analysis. Julie can be reached by e-mail at TravisJ@cbs.curtin.edu.au.

Appendix A
Preliminary Soft Systems Methodology Analyses

In this appendix, we present some initial ideas in the format of SSM techniques for rich pictures, CATWOE, and root definitions. These ideas are from our own viewpoint and in the actual application of SSM, we would need to facilitate others to present their own viewpoints.

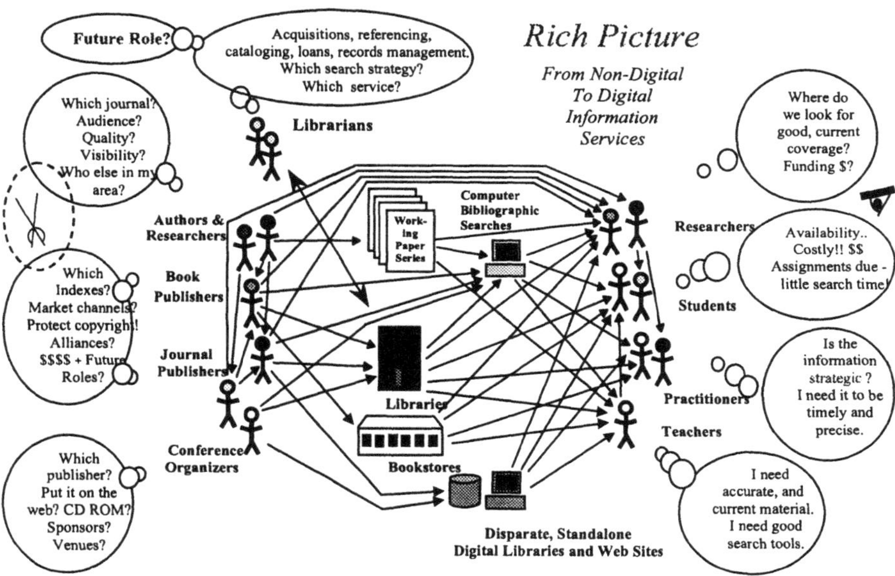

Figure A1. Rich Picture (Partial) of ISDL Context and Stakeholder Issues

CATWOE for an ISDL

Customers: Publishers, Authors, Conference Organizers, Students, Teachers, Researchers, Practitioners

Actors: Publishers, Authors, Conference Organizers, Research Librarians

Transformation: Too many, inadequate/limited publication search sources/distribution channels → Single, comprehensive publication search source/distribution channel

Unknown, relevant publications → Identified and retrieved publications

Poorly distributed publications → Well/more widely distributed publications

Students, researchers, teachers, practitioners lacking information → Better informed students, researchers, teachers, practitioners

An apparently chaotic, unstructured, difficult-to-follow scholarly discourse → a linked, better-structured, easier-to-follow scholarly discourse

World view: There is too much effort required to locate and obtain sufficient publications. Full-text searching can provide better identification of relevant publications. Web-based access is sufficient for enough of the community. Existing technology can be used effectively. Copyright needs to be protected.

Owner: The IS research and publishing community

Environment: Substantial web access at most locations, copyright regulations, competition among publishers, tight funding

Root Definition for an ISDL

An information systems digital library is a system owned by the IS community that would provide a single, comprehensive web-based distribution channel for publishers, authors, editors, and conference organizers to place IS publications for very low cost searching and retrieving by researchers, teachers, students, and practitioners, while protecting the copyright and commercial interests of the publishers and authors.

Part 7:

Reforming Automation

20 REPRESENTING HUMAN AND NON-HUMAN STAKEHOLDERS: ON SPEAKING WITH AUTHORITY

Athanasia Pouloudi
Brunel University
United Kingdom

Edgar A. Whitley
London School of Economics and Political Science
United Kingdom

Abstract

Information systems research is concerned with complex imbroglios of human and non-human components. As researchers, we need ways to represent the intricacies of the different stakeholders in such situations. Traditionally, it is assumed that representing the views of human stakeholders is relatively unproblematic, but that doing this for non-humans is far more complex. This paper addresses this assumption, drawing on the philosophy of science of Isabelle Stengers. It considers the case of the UK NHSnet project and focuses on two stakeholders in the project, one human (the patients) and one non-human (the encryption algorithm used to encode confidential patient data). As the case study shows, representing either stakeholder is equally problematic and the paper reflects on the implications of this for information systems research.

1. Introduction

In the past five years, a new theoretical approach has been included in the range of perspectives used to study the phenomena of information systems. This approach is typically known as actor-network theory (although there many problems with this name [Latour 1997, 1998] and the very act of naming [Law 1998]). Its origins can be found in the social studies of science (Bijker, Hughes, and Pinch 1987) and technology (MacKenzie and Wajcman 1999b), and in constructivist perspectives on the world (Callon 1998).

It is not easy to try to identify the core of what is meant by actor-network theory, or the sociology of translation, "[F]or the act of naming suggests that its center has been fixed, pinned down, rendered definite. That it has been turned into a specific strategy with an obligatory point of passage, a definite intellectual place within an equally definite intellectual space" (Law 1998 p. 2). Trying to fix on a single identity of the theory poses a danger to "productive thinking." It is possible, however, to identify certain characteristics found in most applications of these ideas. Of most importance to the information systems field is the emphasis that the approach puts on non-human actors, whether they be "natural" objects or man-made artifacts, although the distinction between humans and non-humans is much older. In the context of information systems, these are things we label as computers, networks, and organizations (Bloomfield and Vurdubakis 1994).

Actor-network theories suggest that non-humans are an essential component of any human activity system (if they are not, then society must be conceived as if it were constructed by human beings using their voices and naked bodies alone [MacKenzie and Wajcman 1999a, p. 24]). At one level, this is (with hindsight) an obvious contribution and one that can be accepted without much contention. Disagreements arise, however, when considering how this basic notion ("remember to include the non-humans") is operationalized.

One common technique is for the users of the theories to undertake a "semiotic turn" whereby all non-humans become semiotic devices that create texts alongside the texts created by the humans and these texts are analyzed with no consideration given to the nature of their producer. In so doing, however, they are open to criticism that they are ignoring the moral consequences of action (Walsham 1997, p. 473). Against this argument, Grint and Woolgar argue that actor-network theories are to be preferred to perspectives that implicitly or explicitly incorporate a technicist essence to technology and are thus guilty of technological determinism (Grint and Woolgar 1997).

A different kind of argument is presented by Collins and Yearley (1992a, 1992b) who point out that this approach takes Bloor's (Barnes, Bloor, and Henry 1996) principle of symmetry, whereby truth and falsity should be seen the same way, to many more dimensions. In so doing, it removes the human from the pivotal role of having special access to an independent realm (Collins and Yearley 1992a, p. 310). Given the overwhelmingly technicist view of much systems development discourse, it is perhaps unsurprising that many information systems researchers share the concern about displacing human agency from the center of information systems research.

Even if researchers are prepared to accept the moral questions associated with actor-network theories, a further concern is often raised by those considering using this approach (Latour 1999, p. 303): How are non-humans to be represented? How are they to be articulated? How do non-human actors speak? How can I be assured that I reliably report what they are saying? Implicit in such questions highlighting the problems of representing non-humans is the belief that applying these issues to humans is non-problematic. Thus, it is presumed to be unproblematic to have human actors speak and to report what they are saying reliably.

This paper will argue that this distinction is not as clear cut as one would expect. It will do this by first considering, from a methodological point of view, what is involved in representing or speaking on behalf of actors (both non-human and human). It will then present an empirical example that has been reviewed with the added sensitivity to questions of speaking authoritatively on behalf of humans and non-humans. As the example will show, the question of how different kinds of actors are presented is not straightforward. Finally, the paper will draw implications for information systems researchers.

2. Speaking with Authority

When researchers enter a research situation, one of their objectives is to be able to say something about the situation under investigation. The researchers hope to speak authoritatively about this situation, to represent what is going on there. They may do this by describing in detail the features of the one situation, they may also wish to draw on the results from this one instance to say things about a general class of situations, of which the research site is but one typical (or atypical) location. This section will explore the theoretical conditions that underlie the claim that the researchers are speaking authoritatively about the situation and, following from the previous section, it will consider this question in relation to speaking authoritatively about the human and non-human elements of the situation.

2.1 An Issue Based Focus

The first point to be explicitly stated relates to the fact that, even if we are dealing only with human actors, the subjects of the study can have many things to say about a whole host of issues completely unrelated to the research question under investigation. In practice, researchers are always filtering information in this way, excluding issues that are unrelated and focusing on those that address the question. Obviously, this filtering carries the risk that some of the excluded information will, in fact, be vital for a proper understanding of the situation (Whitley 1999), but it happens nevertheless. Moreover, the stakeholders involved will change over time (Pouloudi and Whitley 1996).

In operationalizing this filtering process, the researchers may choose to conduct semi-structured interviews based on a topic guide (Pouloudi and Whitley 1996) with the goal of obtaining a better understanding of a particular situation or process. When studying non-humans, a similar filtering needs to take place. The new computer system can have implications on the air-conditioning in the building, on the security procedures enacted by the organization, and on the relative efficiency of the business. Researchers, however, may only be interested in how the particular computer system transforms the culture of different subgroups, for example, by providing fora for allowing individuals to discuss issues of common interest (Hayes and Walsham 1999).

2.2 "Let the Facts Speak for Themselves"

The second issue to be addressed relates to the fact that non-humans (especially technologically based non-humans) do not normally have the power of independent speech. Humans can vocalize what they want to say about a particular issue. How can a non-human do the same? Without the power of speech, how can the "interests" of non-humans be represented? Ironically, the answer to this can be found in the many opponents of constructivist accounts of science and technology. Instead of focusing on the social processes that lead to the development and stabilization of new scientific facts (Collins and Pinch 1993), the critics argue that nature provides the answer to these questions: "let the facts speak for themselves" is a common refrain.

Thus, in one sense, we already have experts who can allow the non-humans we are studying to speak, we simply need to ensure that they ask them the right questions about the topic under investigation.

In the case of humans, the process of articulation appears to be much more straightforward. We simply ask them about the particular topic and record their responses. Of course, depending on our theoretical preferences, we can make this process more sophisticated, for example, by undertaking a detailed analysis of the terms used, by exploring the hidden agendas lying behind the spoken utterances. In situations where we are potentially dealing with many humans, we may have to rely on questioning a subset of the group we are interested in, but assuming our statistical sampling is rigorous, this should not be a problem.

2.3 Authors and Authority

One further clarification needs to be made at this point and from this we will be able to generalize what we mean by speaking authoritatively. To help with this, the paper draws on the work of Isabelle Stengers, who argues that there is ambiguity about what we mean by being the author of a scientific fact. Thus the researcher can either be "an author, as an individual animated by intentions, projects, and ambitions" or the researcher can be "the author acting as authority" (Stengers 1997, p. 160). As Stengers points out,

> every scientist knows that he and his colleagues are "authors" in the first sense of the term and that this does not matter. What does matter is that his colleagues be constrained to recognize that they cannot turn this title of author into an argument against him, that they cannot localize the flaw that would allow them to affirm that the one who claims "to have made nature speak" has in fact spoken in its place. [Stengers 1997, p. 160]

What the scientist needs in order to speak authoritatively is the second form of authorship, the second form of authority. Given that the world can generally be interpreted in a number of different ways, what is required is "the active invention of ways of constituting the world that is under interrogation, as a reliable witness, as a guarantor for the one who speaks in its name" (Stengers 1997, p. 161).

Moreover, the statement being made must also be interesting. Where "to interest someone in something does not necessarily mean to gratify someone's desire for power, money, or fame" (Stengers 1997, p. 83). Being interesting also does not mean simply entering it into a network of preexisting interests. Rather, for Stengers,

> To interest someone in something means, first and above all, to act in such a way that this thing—apparatus, argument, or hypothesis in the case of scientists—can concern the person, intervene in his or her life, and eventually transform it. An interested person will ask the question: can I incorporate this "thing" into my research? Can I refer to the results of this type of measurement? Do I have to take account of them? Can I accept this argument and its possible consequences for my object? In other words, can I be situated by this proposition, can it place itself between my work and that of the one who proposes it? This is a serious question. The acceptance of a proposition is a risk that can, if the case arises, ruin years of work. [1997, pp. 83-84]

Thus, in this view, researchers must be seeking to have the phenomena they are studying speaking actively about interesting questions. Here Stengers has been referring to non-human actors. The situation becomes increasingly complicated when the phenomena under study are humans (or rats and baboons) who "are capable of interesting themselves in the questions that are asked of them" (Stengers 1997, p. 172) because they are able to interpret the sense of the apparatus that is interrogating them into their responses. In these cases, the notion of a witness becomes very problematic.

> The scientist is dealing with beings who are capable of obeying him, or attempting to satisfy him, or agreeing, in the name of science, to reply to questions that are without interest as if they were relevant, indeed, even allowing themselves to be persuaded that they are interesting, since the scientist "knows best." [1997, p. 172]

One obvious example of these problems can be found in Stanley Milgram's infamous study that created the conditions under which normal individuals became torturers. For Stengers, this study did not produce reliable witnesses because it

> reproduced, in an experimental setting, the perplexity that human history constrains us to. Milgram's torturer-subjects knew they were at the service of science, and this knowledge had as a consequence that the experiment, which was supposed to restrict itself to bringing a behavior to light, without doubt contributed, in an uncontrollable way, to producing this behavior. If a living being is capable of learning, which is also to say of defining itself in relation to a situation, the protocol that aims to constitute this living being as a reliable witness in the experimental mode and thereby constrain it to reply in a univocal way to a question decided by the experimenter creates an artifact. [1997, pp. 172-173]

This principle can be restated in a different way. Questions of authorship, or the authority of a text, cannot be resolved by relying on situations in which the researchers "master all the inputs and outputs and leave the objects no other freedom than the ability to say 'yea' or 'nay'!" (Latour 1997, p. xvi).

This section has introduced a series of issues associated with representing and speaking on behalf of humans and non-humans. The complexity of these issues will be considered in the context of a national network for health service employees, introduced in the next section.

3. The NHSnet

The NHS Executive, the body responsible for the execution of health care policy in Britain (NHS Executive 1994b), launched the NHS-wide networking project in 1993, as "an integrated approach to interorganizational communications within the NHS" (NHS Executive 1994a, p. 6). The objective of this network has been to enhance communication and information exchange between various health care providers and administrators. Thus, the NHSnet is expected to support data communications that cover a variety of information flows across different levels. Its infrastructure is expected to cover a variety of business areas, including patient related service delivery, patient related administration, commissioning and contracting, information services, management related flows and supplies of NHS organizations (NHS Executive 1995).

The NHSnet has been available since 1996. Yet, despite the technological success of the project, and in particular its completion within schedule, its implementation has suffered from a lack of acceptance by the medical profession. Doctors remain skeptical mainly of the security that the network has to offer. These concerns have been overtly voiced, mainly by the British Medical Association (BMA), the national professional body of physicians in the United Kingdom, but also by computer security consultants. These parties fear that patient data may be misused by both NHS members (referred to as "insiders") and external parties (Willcox 1995). As a result of their concern, doctors, again through the BMA, threatened not to participate in the electronic exchange of data unless they could be convinced that patient privacy is safeguarded.

In response to the criticisms, the NHS Executive has stated that the proposed system will be better than the previous situation: data confidentiality was quoted as one of the shortcomings of the previous situation and one that the NHS-wide networking infrastructure would safeguard (NHS Executive 1994a). Recently, the network has been described as "the best medium for the transfer of clinical information" (NHS Executive 1998b). However, the concerns on confidentiality and patient-identifiable information and the debates about alternative solutions have been ongoing since the network became available (e.g., Barber 1998b; Turner 1998) and remain unresolved. Indicative is the slow uptake of the network by GPs: fewer than 10% of GPs were fully linked to the NHSnet in April 1999 (Clark 1999). Still, the Health Secretary stated that all computerized GP surgeries are expected to connect to the NHSnet by the end of the 1999 financial year.

The next sections consider two important stakeholders in the problematic history of the NHSnet. We look at patients, a human stakeholder, and at encryption algorithms, a non-human stakeholder. Both are key stakeholders in the debate about the security of the network and safeguarding the confidentiality of patient data. Importantly, they both raise some interesting and important issues about authorship and "speaking for" that are exemplified in the following discussion.

4. Patients

When it comes to healthcare provision, the most obvious stakeholder is the beneficiary, the patient. This section considers this process in the NHSnet case and provides an insight in the different approaches used by stakeholders to claim the right to speak for patients. Interestingly, the patient is also the party least involved in any discussions about how healthcare will be delivered and what processes will support this delivery. The interests of the patients, therefore, need to be represented by other stakeholders who claim to "speak with authority" on their behalf. Moreover, it is the argument that these stakeholders speak in the interests of patients that is used to give legitimacy to their views. It is interesting to note how this is reflected not only in the arguments of NHS members but also at a political level:

> But I know that one of the main reasons people elected a new Government on May 1st was their concern that the NHS was failing them and their families. [Foreword by the Prime Minister in *The New NHS,* Department of Health, 1997]

At the NHS Executive level, the representation of the interests of the patients is in recommending the use of the network: "Effective communications are vital to good patient care" (NHS Executive 1998a). The doctors' response, as we have discussed previously, was to react to the use of the network by arguing that it does not provide adequate safeguards for confidential patient data. Interestingly, not all GPs were aware of (and, therefore, concerned about) these shortcomings of the network. It has only been since the BMA raised the issue of confidentiality that doctors realized the risks for patients' privacy. The medical profession has since consistently argued that personal medical data can only be exchanged across a network that is safe from interception and received only by the professionals who need this information to provide care. The Data Protection Registrar also represents the interests of the patients, particularly since the enhanced provisions of the Data Protection Act of 1998 will shortly come into effect. In the case of the NHSnet, the Registrar has supported the BMA's concerns without becoming explicitly involved in the conflict over the network's use.

It should be noted that such smooth "nested" representation (the BMA representing the doctors or the Data Protection Registrar who represent the patients' interests) may be difficult to sustain as other interests of stakeholders are manifested. In the case of the NHSnet, the representation was effective because of the uniform reaction of doctors to the confidentiality issue:

> Each local medical committee decides whether it supports the BMA's position and so far each committee has universally supported the

> BMA's position on [the NHSnet] to the point that there was no dissent and that's because confidentiality is so closely linked to the general practitioners' hearts really. [Secretary to a group of local medical committees]

Some stakeholders question, however, the sincerity of the doctors' concerns about the patients. For example, employees of the NHS Executive have argued that the doctors, by adhering to the principles of data confidentiality, are in fact trying to keep close control over their patient information, which they have personally collected and which, therefore, defines—to an extent—their professional role. Other stakeholders, including some patients groups, argue that doctors resent sharing of patient information because the patient disclosed the information to a particular trusted GP. Making this information more widely available would damage the confidence of the patient and the consequently the doctor-patient relationship. The doctors fear that patients may then refuse to disclose sensitive information to the doctor, with unpredictable effects for diagnosis and hence the provision of appropriate care.

For this reason, the consent of the patient for the use of their private medical data has gained primary importance. However, since specific guidelines on consent arrangements are not available, doctors often rely on a notion of *implied consent* by the patient allowing doctors to share such data with other health professionals when appropriate. In other words, it is assumed that the patients rely on the professional *judgement* of their doctor in a given *context*. However, if this data is exchanged through an electronic network, and particularly if patient information is held centrally where healthcare professionals can access it, the notion of patient consent becomes extremely problematic. More importantly, the healthcare professionals (and support staff) accessing the information may be unable to view this information in relation to the context in which the patient disclosed the information. Thus, the relevance, or not, of some information may be difficult to judge (Introna and Pouloudi 1999). This is an indication that a non-human stakeholder, the network, can alter the representation context and its implications. This is more evident in this case study in the use of encryption, as illustrated in the next section.

5. Encryption Algorithm

Following the complex NHSnet debate on the confidentiality of personal medical data and the appropriate representation of patients, it is not surprising that the NHS Executive attempted to promote some formal mechanisms to safeguard access to such data. Thus, in response to the doctors' anxiety about confidentiality, and in order to avoid the cost of another spectacular system failure in the NHS (cf., Beynon-Davies 1995), the NHS Executive (and the government) have responded with a reconsideration of the security issue of the network. The Information Management Group of the NHS Executive commissioned Zergo Limited to "undertake a study looking at the ramifications of using encryption and related services across the NHS-Wide Network" (NHS Executive 1996). The Zergo report proposed the use of encryption to safeguard the privacy of medical records and was considered as "the solution" to the confidentiality problem by the NHS Executive:

> The measures we have put in place are to stop anybody who is unauthorized getting at data from and via, the [NHS-wide networking] system and one of the key parts of that system is a strong authentication challenge. [Statement by Ray Rogers, then Executive Director, NHS Information Management Group in Healthcare Computing 1996]

More specifically, the Zergo report put forward the use of the Red Pike encryption algorithm. This was devised by the *Communications-Electronics Security Group (CESG),* the information security arm of the *Government Communications Headquarters (GCHQ)* and the government's national technical authority on information technology and communications security. Thus, it would enable GCHQ to have access to the data transmitted over the NHSnet.

> The use of encryption for large and diverse applications such as the NHSnet is a new field and until recently none of the available encryption systems was sufficiently robust and comprehensive to suit NHS purposes. Implementation of Red Pike poses significant challenges but looks practicable. [NHS Executive 1996]

The suggestion raised new concerns within the BMA. Computer security consultants became actively involved. Ross Anderson, who has been consulting with the BMA on security matters, has been the one voicing most of the Association's concerns on their behalf and participating in the conflict with the NHS and the Department of Health:

> The Red Pike encryption algorithm is politically unacceptable, technically way out of date and won't command public confidence. [Ross Anderson, quoted in the *British Journal of Healthcare Computing and Information Management* (13:4), 1996, p. 6]

The NHS Executive has used encryption to speak for the issue of security and ultimately of confidentiality of patient data. Their suppliers have supported this view: "Firewall-to-firewall encryption could potentially act as an enhancement to NHSnet security and go some way to placating the BMA" (McCafferty 1996). In order to face the challenge, the BMA have formed alliances with privacy activists (e.g., Privacy International) and academics, on one hand, in order to raise the profile of the debate. On the other hand, they have created an alliance with security consultants, in order to challenge the technical features of the network as well. Thus, the BMA debated which encryption algorithm would satisfy the NHS needs best. Security and privacy specialists have also become involved in the debate to create awareness in the patient population about the dangers of the electronic exchange of healthcare data (Anderson 1996; Bywater and Wilkins 1996; Davies 1996). In this process, the encryption algorithm became a central non-human stakeholder, to whom the different human stakeholders attributed a number of (diverse, even conflicting) attributes, implying the inscription of diverse human stakeholders' interests, as the following quotations illustrate:

In making information secure from unwanted eavesdropping, interception, and theft, strong encryption has an ancillary effect; it becomes more difficult for law enforcement to conduct certain kinds of surreptitious electronic surveillance (particularly wiretapping) against suspected criminals without the knowledge and assistance of the target. This difficulty is at the core of the debate over key recovery. [Abelson et al. 1998]

Key recovery systems are particularly vulnerable to compromise by authorized individuals who abuse or misuse their positions. [Abelson et al. 1998].

In the case of Trusted Third Party (TTP), you do have to trust the central authority and if some of the data warehousing organizations with links to insurance companies etc. manage to get the key they will have access to all patient records. It's not clear from the Zergo report why TTP was chosen over RSA....If we do go TTP, what will be the legal liability, if there is a security lapse? [Representative of Scientists for Labour]

The level of confidentiality changes according to the social situation of the patient. [Kaihara 1998, p. 6]

[T]he security of the stored data is not just a matter of technology but also of administrative procedures. [Kaihara 1998, p. 7]

Ironically, total data encryption systems such as Pretty Good Privacy (PGP) are not compromised. But the NHS Executive has rejected systems such as PGP for the overall network. According to Ross Anderson, this is due to pressure from GCHQ which wants to control tightly the use of data encryption. The NHS Executive claims that systems such as PGP are unnecessary and cites cost and performance issues. [*British Journal of Healthcare Computing and Information Management* (13:3), 1996, p. 8]

From the point of view of security standards within the NHS the key issue is that, whereas confidentiality has always been seen as a key issue in the handling of patient information, it is sometimes not addressed as seriously as it should be. [Barber and Skerman 1996, p. 35. Note that both authors are security consultants to the NHS Executive.]

Although the concept of the confidentiality of personal medical data is well accepted by the general public and by health professionals, the detailed practice is under potentially serious attack by governments that want access in order to combat fraud or serious crime or to

> improve efficiency of services, by big business that wishes to improve its competitive edge or reduce its costs by utilizing detailed personal data in order to focus the promotion of its product and services and by health care organizations that do not keep their security measures up to the "state of the art" required by the information processing facilities available and the attacks on its personal medical data. All security measures need to be under constant review. [Barber 1998a, p. 25]

Despite the debatable role of encryption, the Zergo report was interpreted as a willingness of the government to take the doctors' concerns over security and confidentiality seriously. Thus, it signaled some progress in the resolution of the NHSnet debate at the time. Following the debate, the NHS Executive has now made explicit its view of the NHSnet as a "secure national network" (NHS Executive 1998a), effectively redefining the network. However, it did not manage to persuade doctors that the proposed system was secure, even though it was considered to be better than its predecessors, *ad hoc* manual and electronic exchange systems. The use of the NHSnet is still hindered by technical, organizational and cultural issues (*Computer Weekly News* 1999a, 199b, 199c).

The dialogues and conflicts between the NHSnet stakeholders indicate that some assume that encryption implementation warrants security. However, this assumption is problematic because it is a technical solution to a problem that is at the same time technical, organizational, political, and social. Furthermore, this assumption has repercussions for access to confidential patient data. In particular, if formal rules for such access are established at a national level, it is likely that professional discretion about disclosing personal medical data to appropriate recipients on a "need-to-know" basis will no longer be required. This impact has already caused the professional reaction of the medical profession as the role of the human stakeholder is diluted. More importantly, there is a danger that personal medical data become widely accessible to doctors at a national level, so that they can be accessed when necessary. While this is expected to have important benefits for the delivery of health care, it will be less evident to determine in which cases such data could be accessed and by whom. Any attempt to set specific rules would "remove the context" of professional judgement (and that could have severe implications, cf. Introna and Pouloudi 1999). Conversely, absence of rules would leave the system open to interpretation and possibly abuse of access rights, particularly if those accessing the information are not subject to rigorous professional obligations (Barber 1998b).

It is noteworthy that in arguing for or against the use of this non-human stakeholder, the issue of representation and speaking for the human stakeholder discussed previously, the patient, becomes relevant once more. Some security consultants describe the conflict on network security and encryption as a conflict, in essence, between the interests of the government and the end users (Abelson et al. 1998)—in this case the patients. The NHSnet security debate was manifested as a discord between the NHS Executive and the British Medical Association, representing the government and the patients respectively.

In information system research, it has been argued that non-human stakeholders, such as information and communication technologies, are not neutral, not least because they inscribe human values (Walsham 1997). In the case of the NHSnet, we can also observe how such inscribed values are interpreted in different ways by the stakeholders. In this section, we demonstrated how different stakeholders attributed different values and interests to the non-human stakeholder, that is the security mechanism, in particular the encryption algorithm, to be used on the network.

6. Summary and Discussion

This paper has illustrated the different ways in which humans and non-humans are represented through the examples of two stakeholders in the UK NHSnet project. One stakeholder was human (namely the patients), the other was non-human (the encryption algorithm used to secure patient data being transferred over the network). Table 1 lists the various candidates who claimed to represent the two stakeholders discussed in the paper, indicating how they articulated the concerns of the stakeholder and including a sample of what they articulated on the behalf of the stakeholder.

As the table makes clear, there are multiple authors seeking to represent the stakeholders. Each uses different forms to speak on behalf of the stakeholders and each says very different things about the stakeholder being represented. For example, in the case of the patients, the doctors, their professional body, and the Data Protection Registrar were all seeking to represent them authoritatively. The Data Protection Registrar has a legal requirement to speak on behalf of the patients, while the doctors, through their professional body, sought to take up the moral requirement to speak for them. Moreover, both the BMA and the Data Protection Registrar are viewed with some skepticism in some quarters. The BMA, in representing the interests of patients, could also be seen to be maintaining the privileged position of doctors in the health care network. Similarly, there is general concern about the role of any government agency in issues of personal privacy and this affects perceptions of the role of the Data Protection Registrar.

The situation for the non-human actor (the encryption algorithm) was also complicated. The different security consultants sought to legitimize their view of the technology by undermining those of the alternative representatives.

It is also interesting to note the situation in which this non-human was to be represented. The different consultants operated within a situation (which arose from the Zergo report) whereby the issue was one of *which* encryption algorithm to use that would balance the needs of the NHS and the rights of the patients, closing down the debate about whether encryption *per se* was an issue. Similarly, the formation of the Caldicott Committee obliged the BMA and its allied stakeholders to become less polemic to governmental proposals:

> The Caldicott Committee failed to lay down hard and fast rules for patient confidentiality but because it produced a list of "good intentions" it certainly made it harder for BMA and other concerned organizations like DIN to continue to breathe fire and brimstone about matters. In this the commission probably served its purpose well. [Chairman of the Doctor's Independent Network]

Table 1. Summary of the Different Perspective of the Possible Representatives of the Two Stakeholders Considered in the Paper

Stakeholder	Possible representatives	Means of articulation	Sample articulation
Patients			
	"Patients"	Not directly involved in debate	—
	Politicians	Public statements	Concerns that the NHS was failing them
	NHS Executive	Policy documents	Effective communication is vital for good care
	Doctors (BMA)	Lobbying. Refusal to use system	Inadequate safe-guards for confidential patient data
	Data Protection Registrar	Rulings on data protection issues	Limited public statements
Encryption algorithm			
	"Algorithm"	Performance of system	—
	Information Management Group	Policy	Strong authentication challenge
	Zergo Limited	Zergo Report	Propose Red pike encryption to protect patient data
	Ross Anderson (for the BMA)	Academic publications	Red pike is politically unacceptable, technically out of date
	Suppliers	Academic publications	Firewall-to-firewall security will placate the BMA
	Privacy activists	Public statements	Algorithm is subject to abuse by authorized individuals who abuse their position

It is commonly assumed that it is straightforward to represent humans but that representing non-humans is problematic. As this paper, and the examples in it, have shown, this is clearly not the case; representing either is equally problematic. This raises important questions about how we choose to represent human and non-human actors in information systems research, how we allow them to articulate themselves, and how the results of those articulations are used. In raising this question, the purpose is methodological (although there are also clearly "moral" issues associated with how much agency we grant to non-humans). The paper, therefore, presents the first stage toward answering the question of what is the best way for information systems research to be conducted so that the various stakeholders in the situation, whether human or non-human, can be represented. Having demonstrated the problematic nature of representing both humans and non-humans further research is needed to apply and evaluate different ways of allowing them to be represented authoritatively.

References

Abelson, H., Anderson, R., Bellovin, S. M., Benaloh, J., Blaze, M., Diffie, W., Gilmore, J., Neumann, P. G., Rivest, R. L., Schiller, J. I., and Schneier, B. *The Risks of Key Recovery, Key Escrow, and Trusted Third Party Encryption*, 1998 (http://www.cdt.org/crypto/risks98; accessed October 17, 1999).

Anderson, R. J. "Patient Confidentiality at Risk from NHS-wide Networking," in *Current Perspectives in Healthcare Computing Conference*, B Richards (ed.). Harrogate: BJHC Limited, 1996, pp. 687-692.

Barber, B. "Patient Data and Security: An Overview," *International Journal of Medical Informatics* (40), 1998a, pp. 19-30.

Barber, B. "Towards a Measure of Privacy," *British Journal of Healthcare Computing and Information Management* (15:1), 1998b, pp. 23-26.

Barber, B., and Skerman, P. "What Are Your Security Standards?" *British Journal of Healthcare Computing and Information Management* (13:6), 1996, pp. 34-35.

Barnes, B., Bloor, D., and Henry, J. *Scientific Knowledge: A Sociological Analysis*. London: Athlone, 1996.

Beynon-Davies, P. "Information Systems 'failure": The Case of the London Ambulance Service's Computer Aided Despatch Project," *European Journal of Information Systems* (4:3), 1995, pp. 171-184.

Bijker, W. E., Hughes, T. P., and Pinch, T. (eds.). *The Social Construction of Technological Systems: New Directions in the Sociology and History of Technology*. Cambridge, MA: The MIT Press, 1987.

Bloomfield, B. P., and Vurdubakis, T. "Boundary Disputes: Negotiating the Boundary between the Technical and the Social in the Development of IT Systems," *Information Technology & People* (7:1), 1994, pp. 9-24.

Bywater, M., and Wilkins, C. "Mystic Megabytes," *British Journal of Healthcare Computing and Information Management* (13:2), 1996, pp. 10-11.

Callon, M. (ed.). *The Laws of the Markets*. Oxford: Blackwell, 1998.

Clark, L. "NAO Report: A Surgical Strike on NHS IT Projects," *Computer Weekly News*, 6 May 1999.

Collins, H., and Yearley, S. "Epistemological Chicken," in *Science as Practice and Culture*, Andrew Pickering (ed.). Chicago: University of Chicago Press, 1992a, pp. 301-326.

Collins, H., and Yearley, S. "Journeys into Space," in *Science as Practice and Culture*, Andrew Pickering (ed.). Chicago: University of Chicago Press, 1992b, 369-389.

Collins, H. M., and Pinch, T. *The Golem: What Everyone Should Know About Science*. Cambridge: Cambridge University Press, 1993.

Computer Weekly News. "Executive Must Woo Doctors," 18 March 1999a.

Computer Weekly News. "Hospital Staff Boycott NHSnet Over Poor Performance," 15 April 1999b.

Computer Weekly News. "Opinion: NHSnet Scheme Suffered from Fatal Flaws," 18 March 1999c.

Davies, S. "Dystopia on the Health Superhighway," *The Information Society* (12), 1996, pp. 89-93.

Grint, K., and Woolgar, S. *The Machine at Work: Technology, Work and Organization*. Cambridge, England: Polity Press, 1997.

Hayes, N., and Walsham, G. "Safe Enclaves, Political Enclaves and Knowledge Working," paper delivered at the conference on Critical Management Studies, Manchester, 1999.

Introna, L., and Pouloudi, A. "Privacy in the Information Age: Stakeholders, Interests and Values," *Journal of Business Ethics* (22:1), 1999, pp. 27-38.

Kaihara, S. "Realization of the Computerized Medical Record: Relevance and Unsolved Problems," *International Journal of Medical Informatics* (49), 1998, pp. 1-8.

Latour, B. "On Actor-network Theory: A Few Clarifications," Centre for Social Theory and Technology, Keele University, United Kingdom, 1997.

Latour, B. "On Recalling ANT," in *Actor Network and After*, J. Law and J. Hassard (eds.). Oxford: Blackwell, 1998, pp. 15-25.

Latour, B. *Pandora's Hope: Essays on the Reality of Science Studies*. Cambridge, MA: Harvard University Press, 1999.

Law, J. "After ANT: Complexity, Naming and Topology," in *Actor Network and After*, J. Law and J. (eds.). Oxford: Blackwell, 1998, 1-14.

MacKenzie, D., and Wajcman, J. "Introductory Essay and General Issues," in *The Social Shaping of Technology*, D. Mackenzie and J. Wajcman (eds.). Buckingham, England: Open University Press, 1999a, 3-27.

MacKenzie, D., and Wajcman, J. (eds.). *The Social Shaping of Technology*. Buckingham, England: Open University Press, 1999b.

McCafferty, C. "Securing the NHSnet," *British Journal of Healthcare Computing and Information Management* (13:8), 1996, pp. 24-26.

NHS Executive. *A Strategy for NHS-wide Networking*. No E5155, Information Management Group, 1994a.

NHS Executive. *This is the IMG: A Guide to the Information Management Group of the NHS Executive*. No. B2126, Information Management Group, 1994b.

NHS Executive. *NHS-wide Networking: Application Requirements Specification*. No. H8003, Information Management Group, 1995.

NHS Executive. *The Use of Encryption and Related Services with the the NHSnet: A Report for the NHS Executive by Zergo Limited*. No. E5254, Information Management Group, 1996.

NHS Executive. *IMG: Programmes and Project Summaries*. No. B2232, Information Management Group, 1998a.

NHS Executive. *Information for Health-Executive Summary*. No. A1104, Information Management Group, 1998b.

Pouloudi, A., and Whitley, E. A. "Stakeholder Analysis as a Longitudinal Approach to Interorganizational Systems Analysis," paper delivered at the Fourth European Conference on Information Systems, Lisbon, Portugal, 1996.

Stengers, I. *Power and Invention: Situating Science*. Minneapolis: University of Minnesota Press, 1997.

Turner, R. "The Caldicott Committee Reports," *British Journal of Healthcare Computing and Information Management* (15:1), 1998, pp. 23-26.

Walsham, G. "Actor-Network Theory and IS Research: Current Status and Future Prosjects," in *Information Systems and Qualitative Research*, A. S. Lee, J. Liebenau, and J. I. DeGross (eds.). London: Chapman & Hall, 1997, pp. 466-480.

Whitley, E. A. "Understanding Participation in Entrepreneurial Organizations: Some Hermeneutic Readings," *Journal of Information Technology* (14:2), 1999, pp. 193-202.

Willcox, D. "Health Scare," *Computing*, October 19, 1995, pp. 28-29.

About the Authors

Athanasia (Nancy) Pouloudi is a lecturer in the Department of Information Systems and Computing at Brunel University. She has a Ph.D. in "Stakeholder Analysis for Interorganizational Systems in Healthcare" from the London School of Economics and Political Science, an MSc in "Analysis, Design and Management of Information Systems" from the same university, and a First Degree in Informatics from the Athens University of Economics and Business. Her current research interests encompass organizational and social issues in information systems implementation, stakeholder analysis, electronic commerce and knowledge management. She has more than 20 papers in academic journals and international conferences in these areas. She is a member of the ACM, the Association for Information Systems (AIS), the UK AIS, and the UK OR Society. Nancy can be reached by e-mail at Nancy.Pouloudi@Brunel.ac.uk.

Edgar Whitley is a senior lecturer in Information Systems at the London School of Economics and Political Science. He has a BSc (Econ) Computing and a Ph.D. in Information Systems, both from the LSE. He has taught undergraduates, postgraduate students, and managers in the UK and abroad. Edgar was one of the organizers of the First European Conference on Information Systems and is actively involved in the coordination of future ECIS conferences. He has published widely on various information systems issues and is currently completing a book on the socio-philosophical foundations of information systems. Edgar can be reached by e-mail at E.A.Whitley@lse.ac.uk.

21 IMPLEMENTING OPEN NETWORK TECHNOLOGIES IN COMPLEX WORK PRACTICES: A CASE FROM TELEMEDICINE

Margun Aanestad
University of Oslo and
The Interventional Centre
The National Hospital, Oslo
Norway

Ole Hanseth
University of Oslo
Norway

Abstract

New non-desktop technologies may turn out to be of a more open and generic nature than traditional information technologies. These technologies consequently pose novel challenges to systems development practice, as the design, implementation, and use of these technologies will be different. This paper presents empirical material from a project where multimedia technology was introduced into a complex medical work practice (surgery). The implementation process is analyzed at the micro-level and the process is found to be highly complex, emergent, and continuous. Using actor-network theory, we argue that conceptualizing the process as cultivating the hybrid collectif *of humans and non-humans, technologies and non-technologies (Callon and Law 1995) is a suitable and useful approach. This concept may capture the open-ended and emergent nature of the process and indicate the suitability of an evolutionary approach.*

Keywords: Telemedicine, multimedia, network, actor-network theory, cultivation, *hybrid collectif*.

1. Introduction

This paper reports from the development, use, and adaptation of a broadband multimedia network between two major Norwegian hospitals. The study is positioned at a stage where a potentially useful technology "out there" has already been identified and recruited by the hospitals. A rather simple (although not cheap) technology was installed, providing two parallel two-way audio and video streams. The focus at the time of the study was on trying to use it, adjusting and adapting it as necessary, before it eventually would become fully integrated and seamlessly functioning. However, the seemingly simple tasks of adapting and utilizing this technology turned out to be a challenging and non-trivial implementation process. The process through which the technology and the clinical practice became at least temporarily integrated was a continuous, iterative, and interwoven process where the activities of design, implementation, and use arenot clearly separable.

As the case stories show, complete integration and routinization of use is still far away. In fact, it is our belief that the present stage of the process may turn out to be the normal state of affairs for a long time still, due to the open and generic nature of the technology and the complex hospital organization into which it is introduced. This means that the word "implementation" may not be the best choice for an analytical approach. The traditional view on implementation as a rationally controlled, linear process consisting of separate design, implementation, and use phases has long been challenged, but in practice this view is still prevalent. In real life, implementation turns out to be a notoriously complex and unmanageable process, and failure stories abound. The introduction of notions such as *drift* and *improvisation* (Ciborra 1996b; Orlikowski 1996) points to the unplanned and unmanageable aspects of the processes. Other alternative metaphors that have been suggested are *care, bricolage*, and *tinkering* (Ciborra 1996a), *cultivation* (Dahlbom and Mathiassen 1993), and *hospitality* (Ciborra 1999). These metaphors capture other aspects, like the lack of total control, the presence of risk and ambiguity, and the importance of dedicated commitment and attention. Matthew Jones (1998), introducing the notion of *the double mangle,* emphasizes the temporally emergent nature of the interaction between human and material agency, and the inherently open and evolving character of the process. We think it is possible to gain valuable insight into implementation challenges from analyzing particular instances of "the dialectic of accommodation and resistance" or "the ongoing double dance of agency" (human and material agency) (Jones 1998). This empirical study aspires to analyze this mutual and multifaceted process at a micro-level.

We will have to employ concepts that encompass and cover the aspects that these modifications of the implementation concept suggest. The concept of a translation process from actor-network theory may be a path to explore. It is offered as an alternative to *diffusion* by Bruno Latour (1987), and has been used to describe IT diffusion (McMaster, Vidgen, and Wastel 1997) and implementation (Linderoth 1999).

Latour (1987) argues that when a technology diffuses, it is not an artifact with fixed meaning that is adopted in the same way by each user. In each single case, the adopter has to figure out how to use the technology in her work or life. This implies a reinterpretation—translation—of the meaning assigned to the technology by its designers and previous adopters and at the same time a reinterpretation—translation—of her understanding of herself and her work tasks. In this way, Latour sees adoption and diffusion as mere co-design of both users and technologies.

Broadband multimedia technology is not adopted inside hospitals by individual users in isolation. Rather the opposite: the technology is adopted as a shared infrastructure for the "whole" medical community. But in the beginning, the technology must be adopted by a small user community that subsequently grows. In this process, the technology and the users are both continuously changed and aligned to each other. Together they constitute a collective of humans and non-humans (Latour 1999), a *hybrid collectif* (Callon and Law 1995). The implementation of broadband multimedia technology throughout the hospitals can be seen as a such a collective that is growing in terms of members and where each member as well as the collective as a whole learn and improve their performance. Managing the implementation of this kind of technology could then be seen as the cultivation of such a collective.

The case stories illustrate the details of such a cultivation process by using the concepts *translations, enrolment,* and *alignment.* Our focus will be on the detailed and mundane activities that occur in a given practical situation, focusing on the relation between the new technology and the activities of the human actors.

Integral to the actor-network perspective is the view that there is not one single, universally "right" solution or resulting condition of an implementation process. One may envision several equally probable end results, depending on the character of the network(s) involved, the actors' strategies, the success of enrolment attempts, the appeal of translations, and the strength of the inscriptions. The situated and emergent nature of the translation process is evident from the empirical data, and we think this open-ended view of any implementation process is an essential and crucial feature of reality. Consequently, we argue that this view should be integrated into systems development practice.

2. Multimedia Networks Entering a Hospital

2.1 Project Background

The intention of the telemedicine project was to explore the potential of networked multimedia technology in minimal invasive surgery (also called "peephole surgery"). Minimal invasive surgery is a collective term for a range of medical procedures where the "invasiveness" of the procedure is minimized. Minimal invasive surgery of the abdomen (laparoscopic surgery) avoids an ordinary open surgery by making three to five small holes (with a diameter of 5mm to 10 mm) and using instruments that are entered through these "ports" to perform the operation. The surgeon's view of the workspace is mediated through video images on a monitor and, consequently, is easily available for capture and transmission across communication networks. However, the central role of the images (as a substitute for direct vision in ordinary, open surgery) motivates the very strong requirements to the image quality. Important information may be related to subtle texture and color variations, as well as delicate structures (e.g., vessels), and these details may be lost if the video is compressed too heavily. These demands to image quality dictated the choice of technology in the project. Broadband technology, in this case a 34 Mbit/s ATM network (Asynchronous Transfer Mode) and MPEG2 real-time coding of video and audio, facilitates use of high quality video transmission. Additional information on the project may be found in Hanseth et al. (1999).

Most of the descriptions in this article relate to the process within one of the two hospitals. The focus is thus on the local adaptations and learning "within" one of the project partners, more than on the project process as a whole. The site is a research and development center, and attracts large numbers of visitors. In order to facilitate the information and education work, a permanent set-up of audio and video transmission was installed (independently of the telemedicine project). The presence of this local infrastructure turned out to be an important asset in the later telemedical activities. Several image signals from each of the two operating rooms are connected to a control room with mixing facilities. In the control room up to 16 images from the two operating rooms are displayed on small monitors and microphones and loudspeakers provide two-way audio communication with the operating rooms. This control room is also connected to two local lecture halls as well as to the external ATM and also to the ordinary ISDN network.

2.2 Research Method

The study is an empirical study situated within the interpretative strand of information systems research (Klein and Myers 1999; Walsham 1995) and has thus mainly employed qualitative methods.

The main method for data collection has been participatory observation. The authors have been members of the project organization and the first author spent approximately 50% of her time for one and a half years (from the summer of 1998 to the end of 1999) as a participant observer at one of the sites. The other site has also been visited several times for parallel observations. The first author participated in the technical support work for most of the transmissions in this particular project. This work includes planning, actual preparation, and set-up, as well as assistance during transmissions. Keeping a formal log over the activities and experiences and writing up an evaluation report have been important tasks in gaining and revising an understanding of the process. Personal diary notes have provided additional detail on the detailed case stories below.

Semi-structured and free interviews were conducted with a wide range of personnel groups, including surgeons, operating theater nurses, anaesthesia nurses, technicians, students, and lecturers. The interviews were mainly focused on bringing out experiences, opinions, and feelings after specific transmissions and to a lesser degree around the project and the multimedia technology in general. Findings and conclusions have been discussed informally (after most of the transmissions) and formally (in discussion meetings) with the staff involved. Other qualitative methods were used as well, including observation of medical work, video recording of work, and analysis of textual documents. General participation in the daily life and interaction with the rest of the personnel has proven useful to gain an understanding of the values and the resources of the medical practice, as well as the challenges and problems that are encountered.

2.3 New Technology Meets Surgical Work

The project intended to perform experimental development of technological solutions for minimal invasive surgery. The focus was on determining sufficient and cost-effective image quality. However, "using" the new technology soon turned out to be more than selecting the optimal bandwidth and encoding algorithm. The project changed from the careful execution of well-planned experiments into a process best characterized as improvisation, tinkering or bricolage (Ciborra 1996a, 1996b; Orlikowski 1996). The actual use became different from the planned use, which was education of gastro-intestinal surgeons through transmissions of live operations and recorded material. Several other medical specialities became involved and transmissions of meetings, lectures, and demonstrations became an important part of the use.

One of the main reasons for this *drifting* of the project is the fact that the work practice into which the technology was introduced (surgery) is a complex practice with many requirements. Patient safety is of utmost importance, which implies low tolerance of disturbances and problems introduced by the technology. There are also certain demands on where the equipment can be placed, how it can be used, and requirements to design with regard to cleaning and sterilization. Thus the technology had to "learn" or "discover" how to comply with these new criteria. At the same time, the technology was not "allowed in" before it could perform satisfactorily. This situation creates a "deadlock": real use was the only way to learn, but real use was not allowed until learning had taken place. Simulations are not feasible, as they either become too simple or too expensive and time-consuming. The way out of the deadlock in this project was to build iteratively on earlier experience from simple settings and to generate increasingly demanding opportunities for further learning. Use evolved from simple and small-scale transmissions to more demanding, more "marketed" (and, therefore, more high risk) situations. Thus transmissions of lectures (non-critical) have direct relevance for transmissions of surgery (critical), as it generated valuable experience with the technology.

In addition to the learning-driven change of project focus, the fact that different medical disciplines cooperate in practice and need to share a common infrastructure is important. Focusing on separate discipline-specific systems is thus not desirable. What needs to be developed is a common and shared resource. This was realized in the project and steps were taken to expand the scope of the experimental use. Such an expansion is not without challenges, as the involvement of other departments and institutions implies an increased complexity of the translations necessary to enrol all actors.

2.4 Actants in Networks

A central feature of actor-network theory is that humans and non-humans (i.e., technology, organizations, institutions, etc.) are treated symmetrically. Technological and social elements are considered tied together into networks, based on the assumption that technologies are always defined to work in an environment including non-technological elements, without which the technology would be meaningless and it would not work. In the same way, as humans, we use non-human objects (technologies and other artifacts) in all our dealings in our world; our existence in the world is based upon the existence of these objects. Accordingly, neither humans nor technological artifacts should be considered as pure, isolated elements, but rather as heterogeneous networks. When any actor acts, this very actor is always such a network, not a single element. An actor is always a *hybrid collectif* (Callon and Law 1995). In the same way, elements in a network are not defined only by their "internal" aspects, but rather by their relationships to other elements, i.e., as a network. This further implies that elements in such a network are not initially defined as human, social or technological; they are referred to by a common term: actant. These assumptions do not deny any differences—or borders—between what is human or social and what is technological. However, these borders are seen as negotiated, not as given.

According to actor network theory, stability, technological and social order, is continually negotiated as a social process of aligning interests. As actors from the outset have a diverse set of interests, stability rests crucially on the ability to translate (reinterpret, represent, or appropriate) others' interests to one's own. Through translations, one and the same interest or anticipation may be presented in different ways, thereby mobilizing broader support. A translation presupposes a medium or a "material into which it is inscribed." Translations are "embodied in texts, machines, bodily skills [which] become their support, their more or less faithful executive" (Callon 1991, p. 143). Design is then a process where various interests are translated into technological solutions as well as organizational arrangements and procedures to be followed to make the technology work properly. In this process, existing technology will be reinterpreted and translated into new ways of using it. To make the technology work, all these elements must be aligned, i.e., cooperating toward a common goal. This is achieved through a translation process, which, if successful, may lead to *alignment*. An aligned network is in a kind of equilibrium or stable state (at least temporarily). The alignment attempts occur through *enrolling* the different and heterogeneous *actants* in this *network* by translating their interests. As large actor-networks are aligned, they may become irreversible and hard to change (Callon 1991; Hanseth and Monteiro 1997).

3. The Anatomy of Translation

The natural starting point for an analysis of the introduction and adaptation of the technology and the organization is to focus on the *work that is required* in order to facilitate the meeting between the technology and the work practices. In actor-network terms, a "successful" implementation is *a stabilized network*, where the actants are *aligned*. The following three stories from particular instances serve as illustrations to the ongoing, mutual process of alignment or stabilization efforts. The medical work practice has its demands to an acceptable or desirable state of affairs, but so has the technology (illustrated by the first story). These different demands may peacefully coexist in a stable network, or they may initially be in conflict and thus require "translation work" in order to enrol and align the different actants into the same stable and working network. The second story in particular focuses on this translation work. The situatedness and emergent character of the mutual learning process is evident. Most of it is unplanned and it occurs in real life and in practical use situations, i.e., transmissions. This emergent character of the process, together with its openness and dependence on adjacent networks, also implies lack of control, and the third story illustrates the fragility and vulnerability of the achieved alignment.

3.1 The Demands from Medical Work Practice

For transmissions of live operations, the previously installed permanent set-up was in place in the operating room. Video from wall- or roof-mounted cameras would be transmitted during the preparations and early phases of the operation, and images directly from other imaging equipment when the actual surgical work started. The technician could watch and choose between all of these image sources in the control room. Verbal communication between the surgeon(s) and the viewers were facilitated through use of permanent microphones and loudspeakers in the roof above the operating table. The transmission must be prepared by connecting the imaging equipment to the network (this involved connecting a dedicated piece of cable from the monitors to a wall-mounted panel) and turning on the camera system, which has to be done inside the operating room. This feature together with a red "on air" light, was introduced in an early phase (before this project) to give the operating room team control over the situation. Before this particular transmission, an additional wish from the team had been accommodated: The image from the receivers, which earlier had been shown only in the control room, was displayed on a monitor inside the operating room. Now the surgeon did not need to ask explicitly who was watching, but the whole team could see whether there were any receivers, and who they were.

Here we already see several translations of different actants' interests. The placement of the cameras reflects a focus on the surgeon's work (as opposed to nursing work, which takes place in larger areas of the room). The operating room team's wish for control over the communication situation was translated into physical pieces of technology: the red light, the camera control facility, and the additional monitor showing an image of the receivers. Or viewed the other way round, the "technology" has to be able to provide the team with the required control, in order for it to be accepted and allowed into the operating room (this is an example of "learning" on behalf of the technology). The wish to transmit images and sound to the receivers leads to the enrolment of several physical devices (cables, microphones, and loudspeakers). Also, the telemedicine project leader needs to enrol the technicians and nurses to actually perform the required work of connecting and turning on equipment in order for the transmission to be possible or the network to be aligned.

A procedure known as endoscopic retrograde choleangio pancreaticography (aptly abbreviated to ERCP) would be transmitted to a group of medical doctors specializing in endoscopy. This procedure consists of accessing the duodenum and the pancreatic channel for diagnosis and treatment of gallstone-related problems, through a gastroscope entered via the mouth. An ERCP involves the simultaneous use of several image sources, at least the image from the gastroscope and a x-ray cine (a sequence of separate x-ray images taken during a brief period, in fact an "x-ray video"). In this particular procedure, a so-called "mother- and child-scope" was to be used. This is a gastroscope where a second imaging scope (the "child") is extended into the pancreatic channel, generating a third image.

The surgeon and the rest of the team were informed by one of the support technicians that the procedure would not only be transmitted to the other (partner) hospital, but to three additional hospital as well (expanding the project network). For some reason, the patient did not show up at time and the operation was delayed for one hour. Then the lack of routines for informing the receivers was discovered. First, the telephone was used to inform one of the receiver sites, then a PowerPoint slide announcing the expected starting time was made on a computer and transmitted to the whole network. By then, several of the potential receivers had left due to other duties at their local hospitals. When the patient arrived, the surgeon took on the responsibility of informing and asking for consent.

The gap that occurs between the receivers waiting for the transmission to start and the missing patient is a threat to an aligned network and needs to be remedied. Several parallel activities occur: a search for the patient is conducted and the possibility of summoning the next patient in the queue is checked out. The technical team use telephone and a PowerPoint slide to inform the receivers. However, the attempts are not entirely successful in ensuring that the receivers stay, as the demands from other networks (local hospital's work schedule) turn out to be stronger. A crucial part of the alignment needed for a successful transmission (i.e., an obligatory passage point) is the patient consent: if the patient had rejected the idea, the transmission would have to be cancelled. Also, the operating room team needs to agree to become enrolled into the network and to take part in the transmission. This is achieved through the telemedicine activities being defined as a formal activity of the department and accepted as such.

When students are present in the operation room, they may detect the surgeon's changing focus on the different image sources (gastroscope or x-ray image) without this being explicitly mentioned. The receivers across a network do not see more than one image at a time, and the technical personnel (who have access to all of the images) do not know the procedures in detail. Consequently, at several instances, the receivers did not get all the relevant information. In order for the image switching to be appropriate, either the surgeon had to mention explicitly that he was looking at the x-ray cine or the gastroscopic image, or the technicians had to detect where the activity seemed to happen (e.g., switch when the x-ray cine sequence was started). In some instances, the receivers would ask to be shown a particular image. During this session, a students' facilitator was present in the operating room, assisting the surgeon and taking care of some of the verbal explanations. When this problem was discovered, he started to comment on the surgeon's use of images more explicitly, which aided the technical personnel's switching task.

The discrepancies between what the surgeon sees and what the network transmits to the receivers is a problem that must be solved. The means for doing this may be several. As indicated, either the surgeon must perform an additional task (explicit verbal comments), additional training of technicians can be done, or the organization can provide extra personnel resources. The intended result of any of these translations is an aligned network where the receivers see what they want and need to see. In any of the scenarios, the human actants must relate to the requirements and overcome the limitations of the technology. One may also envision other solutions where the technology takes on more responsibility for solving this problem. One solution could be to transmit all image sources in parallel all the time, and leave to the receivers to watch whatever image they want. This would require high bandwidth and several monitors. Another solution could be a tracking system for the surgeon's direction of gaze, and an automated image switching facility. Both of these possible translations were too expensive for the project according to its rather short-term goals.

At the point in the procedure where the child-scope was extended from the mother-scope and activated, an additional image was generated and displayed on the surgeon's monitor in a Picture-in-Picture (PIP) manner. This additional image was not included in the image that was transmitted to the control room and further to the receivers, and one technician had to enter the operating room to detect why the child-scope image was not included. The solution consisted of finding another place to tap the signal, something that introduced another problem. The monitors allow video signals to be tapped either in a so-called Y/C format or as composite video. To allow for the transmission of the higher quality Y/C signal to the control room, some special adapter cables had been made previously. In this particular instance, only composite video was available if the PIP function was required and the lower quality signal had to be chosen. This meant that, in addition, the technician had to find other cables and substitute them for the special cables.

Here we see the previous translation of the wish for optimal image quality into the physical material of adapter cables. Or conversely, we could say that the technology has to "learn" how to fulfil the demands for high image quality. However, another wish from the medical community, namely to transmit the PIP image, gets higher priority. The technology introduces some obstacles to this wish and a technician and other cables have to be enrolled to create the new stable network where the receivers can watch the PIP image.

There was some noise from the suction machine (which was an old one and not ordinarily used), and some activities occurred in order to minimize this disturbance. Disturbances arise also from the fact that discussions between technicians on the audio lines between the sites are sent to the operating room. This included messages such as sub-optimal use of microphones on the other side needed to be corrected, camera position not optimal, discussions of whether to try specific solutions to problems, or whether to terminate the connection after the attendees had left. In many cases, the technicians chose to make separate telephone calls to avoid disturbing the operating room team and the receivers. In general, the operating room personnel encounter a more "messy situation" during transmissions (several additional cables on the floor, a technician may enter and do some work).

As we see, not all actants are completely enrolled into the network and noise and disturbances are the result. One way to avoid these specific problems is through use of the telephone (enrolment of the telephone into the network). Another potential solution would be to provide a separate audio channel for control messages (again, technology would have to "learn" to comply with demands from the organization). New silent suction machines may be enrolled or "intelligent" microphones may be provided (which renders voice well and filters out noise).

3.2 The Work of Alignment

The need for learning and gaining experience with the technology induced the search for different use areas. One opportunity for a different kind of use came with a regional meeting of ENT (ear, nose, and throat) specialists. Both hospitals are regional hospitals for one region each (which together comprise approximately half of the population of Norway). Much of the health care administration occurs *within* a region, which implies less contact *between* regions. A regional seminar was planned at the same time in both regions and the network between the two hospitals provided a unique opportunity for the two groups to perform a joint meeting. This required some activity from the project members on both sites in order to enrol the local ENT departments. The network had previously been used in one meeting, and then the support was purchased from an audio-visual consultant company. Thus this meeting was the first transmission of a high profile meeting managed by the local support team.

The meeting would consist of lectures and discussions and the support personnel were provided with a program for the meeting (time slots, name of speaker, title of lecture). This, however, was not sufficiently detailed information. A separate meeting between the technical support personnel and the organizers was needed to clarify additional issues, such as: At which of the two sites was each lecturer going to be located? What kind of presentation was going to be used (electronic, such as PowerPoint, or videos, where the signal could be tapped directly, or analog, such as ordinary slides or overheads that had to be filmed by an additional camera)? What level of interactivity was expected (would one cordless microphone to be passed around suffice)?

After this clarifying round, the technical planning work could start. The equipment needs were thought through and the number of video cameras, video selectors and distributors, monitors, microphones, audio mixers, and cables was established. Some of this had to be purchased or borrowed, cables had to be made, and the operation of the equipment had to be planned. Each of the four members of the technical support team was allocated to tasks (one in the control room, three in the lecture hall) and the responsibilities were spelled out. This involved getting access (and the keys) to the lecture hall before the meeting for set-up and tests. One person operated the projector in the lecture hall and switched between incoming video and two sources for local presentations (video or PC-based). He also started the videotapes when appropriate. Another person operated the video selectors and audio mixers, choosing to send to the other side either camera on lecturer (or on screen), camera on audience, local video signal, or projector signal (for the PC-based presentations). He also mixed and controlled the audio (in and out), muting microphones when needed and adjusting the sound level. The third person in the lecture hall operated the camera on the lecturer and the screen and also adjusted the room light in order to achieve a good image quality. The person in the control room (some distance away from the lecture hall) was responsible for the connection to the other side and for appropriate switching when an ordinary ISDN videoconferencing session from a third party was to be connected and transmitted to both hospitals. A lot of "ordinary" articulation work (Strauss et al. 1985) comes in addition: for instance, arranging with the support personnel at the other site for testing (especially sound testing is highly interactive), the physical work of carrying and connecting equipment, telephone conversations to clarify details, tidying up after meeting, and locking rooms, etc.

We see the myriad of minute details that are necessary for the transmission to succeed, i.e., for the network to be stabilized (aligned) when it needs to be. The work of the support personnel can be seen as translating the interests of the different users and user groups with the interests (or requirements and needs) of the technology. This process of translation involves work. The amount of work is extensive; for this particular meeting, which lasted for three hours, 47 hours of technical support work was needed before, during, and after the transmission. If lecturers could be induced to take the detour to digitize all the material they wished to present in a lecture or a meeting, the network would look different. The additional technology and support work necessary catering to the analog parts would not have to be enrolled, and a simpler network would result, which might have been easier to align. This simplification will, of course, depend on the success of the previous translation (digitized material), but the important point is that there is a choice between several routes toward alignment (different translations of interests). Due to the availability of technicians (low-paid community workers), this solution was chosen instead of educating and enforcing lecturers to go digital, where the risk for failing to enrol them might have been greater.

3.3 The Fragility of Alignment

The third case is an example of the problems that may arise when the alignment of the network breaks apart. This is another lecture transmission with yet another kind of medical content, with different requirements to the technology. This was going to be a "low-risk" session, i.e., a short and simple transmission for a small group. The weekly lunch lecture for radiologists on one of the sites was to be transmitted to the other site. Several transmissions had been carried out for this group before and the amount of work in planning was not as high as it had been on the first and novel cases.

Due to external factors (the lecture halls at both sides were occupied by others), the time for set-up and testing before transmission became very short (less than 30 minutes). Therefore, all the needed equipment (cameras, monitors, microphones, cables, etc.) had been prepared and loaded on a cart and one of the technicians went to install it in the lecture hall. Another stayed in the control room to work on establishing the network connection (which on "bad" days might take hours). The third (and most experienced) technician was ill that day.

The technician in the lecture hall reported that he could not get the local video projector to work and another projector was requested, searched for, and installed before he discovered that he had used a wrong remote control and the projector worked after all. Then it was discovered that the image received from the other side was displayed in black and white in the lecture hall. The time did not allow for extensive fault searching and the initial remedies did not help. Finally, one of the microphones generated a lot of noise, the source of which was also not found. Such unfortunate and accidental pile-up of problems could have been managed with either more time or more personnel, but in this situation the breakdowns were not corrected before the transmission was about to start. The lecturer (on the other side) had to stop after a couple of minutes and start his lecture anew because the sound in the lecture hall was turned down and no technician who knew how to operate the equipment was present. The technician in the control room had to be summoned by telephone and had to run to the lecture hall, which was several floors away, in order to correct the audio problems. The image was displayed in black and white and the lecture hall was too dark for the other side to see (through the audience camera) who was present, as the room had to be dark in order for the projected images to be visible.

Another point of disappointment was the image quality of the x-ray images that were the focus of the lecture. X-ray images that are projected as ordinary (analog) slides are of a sufficient quality. When these images are filmed by a digital video camera and transmitted as a video signal, the quality is not satisfactory (at least according to some radiologists). This had been done earlier, but the technicians had not been told that this was inadequate and they had consequently not tried to find other solutions for transmissions of analog x-ray images. In the earlier lectures that had been transmitted for this group, the radiological images were not the main part of the lectures.

Here we see so-called "external factors" affecting the transmissions. These indicate the presence of other, potentially stronger networks. These networks determine the availability of infrastructural resources (Kling 1987), which here were lacking: access to facilities, systems for resource reservation, adequate time, systems for information exchange or de-briefing, large enough support organization, presence of skills and competence. There is a need also to align with these other networks (the "infrastructure"), as they may influence this particular network. The interdependence with the rest of the world is also evident if we take a look at the more practical and technical details.

We see that the stability of an aligned network may be fragile and vulnerable: the failure of aligning even small parts may turn out to be disastrous if these details are obligatory passage points. One slider on the audio mixer must be enrolled into the network and aligned with the other actants' interests if the meeting is to be possible at all. The transmission of sound is an obligatory passage point. In order for this tiny slider to be aligned (put in the right position), a telephone (and a telephone network), somebody who makes the call, a person with audio-technical competencies (actually a musician), and his physical presence in the lecture hall are all necessary. The presence of an image with sufficient quality is necessary (i.e., another obligatory passage point) if the rationale for the existence of the network (i.e., the joint lunch meeting) is to look at images. "Behind" the presence of a sufficiently high-quality image lies a host of other actants that need to be aligned. For example, a few of these actants are the cables where the whole image signals is let through, not only the black/white part, a working video projector, a sufficiently high-quality capture mode (previous digitization of radiological images would have been appropriate), technicians' knowledge of the importance of images for radiologists, and so on *ad infinitum*. The network never ends.

4. Open Network Technologies Meet Complex Work Practices

The project from which the three snapshots presented above are selected was the involved hospitals' first encounter with broadband multimedia technology. Accordingly, some experimental activity was required in order to figure out how to use it properly. One might possibly believe that the use of the technology becomes routine and the degree of experimentation will decrease as experience is gained, and accordingly that the kind of problems and challenges described above will not play any major role after this initial experiment-oriented phase. We do not think that is the case. Rather the opposite: we believe that the use of this kind of technology in organizations like large hospitals will regularly be of this experimental character. This is so because of

- the open and networking character of the technology,
- the complexity of the work practices into which the technology is introduced, and
- the potential "revolutionary" benefits of the technology.

Multimedia network technology is open, or generic, in the sense that it might be applied in virtually any kind of work. But this open and generic character implies that it is not "given" how the technology should be used in the specific settings where it is adopted. The technology is not designed specifically for surgery; accordingly, its designers cannot tell the surgeons how they should use it. They have to figure this out themselves—through a learning process involving trial and error.

Medical work is very complex. There are many different tasks carried out by different kinds of medical personnel. (There are, for instance, 44 different medical specialities in the Norwegian health care sector.) Each task (like laparoscopic surgery) is complex in itself. In addition, different activities are highly interdependent. They are linked directly in the sense that one department is delivering services to another (support services such as labs, radiology, and anaesthesia), patients are treated by many different departments almost concurrently, etc.

Broadband multimedia technology is supposed to be very useful in almost any kind of service as it opens up possibilities for collaboration among medical personnel independent of time and space. It is envisioned that medical expertise might help in treating a patient independent of the physical location of both patient and expertise. This is supposed to improve most medical services and open up more efficient ways of organizing the health care sector. Reaching this point is a long walk. And every single step requires experimentation, as illustrated above. Within each area, like surgical laparoscopy, the technology may be used in many different ways and for many different purposes. Further, because the work practices of different departments and personnel groups are interdependent, the practices have to be changed in a coordinated fashion. And the technology and services developed are serving as a shared infrastructure for the health care sector, as was illustrated for surgery, ENT, and radiology in the examples above. This means that existing technological solutions and working practices must be changed to fit new requirements as one moves to each new use area. For each step, technology and working practices must be translated and aligned as illustrated in the first snapshot above. This alignment requires considerable work. Because the technology is constantly changing, it will permanently have an immature character and be highly fragile. We believe this continuous process can be seen as cultivating a collectivity of humans and non-humans.

5 Cultivating the *Hybrid Collectif*

Conceptualizing the organizational implementation process as cultivation of a hybrid collectif may lead to a more natural focus on the translation work so crucial for success. In the case stories, both the organization and the technology can be looked upon as "children" or new members to be socialized and educated: they both have a lot to learn. From the point of view of the medical community, some rules are absolute and the technology has to learn and obey them. The new technology is not allowed to adversely affect the patient outcome and it has to perform satisfactorily (sound and image transmission quality). Its way of working must be adjusted to fit with the organization (physical placement of cameras and microphones, timing of transmissions, expanding its "skills" to allow flexible mixing of several image sources used). Also, if we take the opposite point of view (from the technology's position), we see absolute rules where the organization needs to comply. A speaker who wants to communicate with the other side must use a microphone, the image signals to be obtained and displayed need to have the right format, physical entities need to be introduced into the operating room. The descriptions of actual usage show that the translations are not only performed by the humans. The process is not only one-way (users inscribing their interests into the technology). Mutual translations and inscriptions occur. Routines within the operating room are altered to accommodate the technology's interests (e.g., the nurses get additional tasks in order to care for the demands of the microphones and cameras). Also, modifications and adaptations of the technology are carried out to facilitate the users' demands on the services of the network. There is a concurrent design and redesign of technology, individuals, and work practices; all parts in the network are involved.

We think the temporally emergent nature of the interaction between the human and non-human actants is a salient feature of such a cultivation process, as well as the inherently open character of the process. Conversation analysis shows the irreducibly interactional structuring of talk (Suchman 1994). It is tempting to look for a parallel to this phenomenon in design and implementation. The parallel could be that the dialogue between an organization and the technology is irreducibly interactional and consequently not possible to plan or specify in advance. Thus the concept of cultivating the *hybrid collectif* to be an adequate approach toward analyzing this process in detail.

References

Callon, M. "Techno-economic Networks and Irreversibility," in *A Sociology of Monsters: Essays on Power, Technology and Domination,* J. Law (ed.). London: Routledge, 1991.

Callon, M., and Law, J. "Agency and the Hybrid Collectif," *The South Atlantic Quarterly* (94:2), 1995.

Ciborra, C. "Hospitality and IT," in *Proceedings from the Twenty-second Information Systems Research Seminar in Scandinavia (IRIS'22),* T. Käkölä (ed.), Keuruu, Finland, August 1999.

Ciborra, C. "Improvisation and Information Technology in Organizations," in *Proceedings of the Seventeenth International Conference on Information Systems,* J. I. DeGross, S. Jarvenpaa, and A. Srinivasan (eds.), Cleveland, OH, December 1996a.

Ciborra, C. "Introduction: What does Groupware Mean for the Organizations Hosting It?" in *Groupware and Teamwork: Invisible Aid or Technical Hindrance?*, C. Ciborra (ed.). Chichester, UK: Wiley and Sons, 1996b.

Dahlbom, B., and Mathiassen, L. *Computers in Context: The Philosophy and Practice of Systems Design.* Cambridge, MA: Blackwell, 1993.

Hanseth, O., Buanes, T., Eide, K., Hustad, R., Johansen, M., Røtnes, J. S., Wörkvist, B., and Yvling, B. *DIMedS—Development of Interactive Medical Services, Distant Learning in Surgery and Radiology Using Nroadband Networks. Final Report, Pilot Regular Teaching.* Telia, Norway, 1999.

Hanseth, O., and Monteiro, E. "Inscribing Behavior in Information Infrastructure Standards," *Accounting, Management and Information Technologies* (7:4), 1997, pp. 183-21.

Jones, M. "Information Systems and the Double Mangle: Steering the Course between the Scylla of Embedded Structure and the Charybdis of Strong Symmetry," in *Information Systems: Current Issues and Future Changes,* T. J. Larsen, L. Levine, and J. I. DeGross (eds.). Laxenburg, Austria: IFIP, 1998, pp. 287-302.

Klein, H. K., and Myers, M. D. "A Set of Principles for Conducting and Evaluating Interpretive Field Studies in Information Systems," *MIS Quarterly* (23), 1999.

Kling, R. "Defining the Boundaries of Computing Across Complex Organizations," in *Critical Issues in Information Systems Research,* R. Boland and R. Hirschheim (eds.). Chichester, UK: Wiley, 1987.

Latour, B. *Pandora's Hope: Essays on the Reality of Science Studies.* Cambridge, MA: Harvard University Press, 1999.

Latour, B. *Science in Action.* Boston: Harvard University Press, 1987.

Linderoth, H. C. J. "Don't Close the Big Black Box—Close the Little Ones. A Study of Telemedicine," paper presented at the Fifteenth Nordic Conference on Business Studies, Helsinki, August 19-21, 1999.

McMaster, T., Vidgen, R. T., and Wastel, D. G. "Towards an Understanding of Technology in Transition: Two Conflicting Theories," in *Proceedings of IRIS 20, Social Informatics,* K. Braa and E. Monteiro (eds.), Oslo, 1997, pp. 171-180.

Orlikowski, W. "Improvising Organizational Transformation Over Time: A Situated Change Perspective," *Information Systems Research* (7:1), 1996, pp. 63–92.

Strauss, A., Fagerhaugh, S. B., and Wiener, C. *The Social Organization of Medical Work.* Chicago: University of Chicago Press, 1985.

Suchman, L. "Do Categories have Politics? The Language/Action Perspective Reconsidered," *Computer Supported Cooperative Work* (3:3), 1994, pp. 398-427.

Walsham, G. "The Emergence of Interpretivism in IS Research," *Information Systems Research* (6:4), 1995, pp. 376-394.

About the Authors

Margunn Aanestad holds a B.Sc in biomedical engineering and a M.Sc in applied natural sciences. She is currently pursuing a Ph.D in systems development. Her research field is within telemedicine, or use of network technologies within health care. Her case studies focus on utilization of broadband network technologies in minimal invasive and image guided therapies. Margunn can be reached at margunn@ifi.uio.no.

Ole Hanseth has worked most of his career in industry and applied research before moving to his current position as an associate professor at the Department of Informatics at the University of Oslo. He also held a part time position at Göteborg University. The main focus of his research has been comprehensive, integrated, and geographically dispersed information systems—information infrastructures—in both the private and public sectors. Recent journal publications on these issues have appeared in *Computer Supported Cooperative Work; Information Technology and People; Accounting, Management, and Information Technology; Science, Technology, and Human Values; Systèmes d'Information et Management; and Scandinavian Journal of Information Systems.* Ole can be reached at ole.hanseth@ifi.uio.no.

22 MACHINE AGENCY AS PERCEIVED AUTONOMY: AN ACTION PERSPECTIVE

Jeremy Rose
Manchester Metropolitan University
United Kingdom

Duane Truex III
Georgia State University
U.S.A.

Abstract

Recent theoretical debates in the literature have taken up the themes of social and technological determinism in the context of actor network theory and structuration theory. This paper explores (computerized) machine agency from an action-based perspective. How is it that information technologies affect our actions, how can we marshal this property, and what can we do about the results if we don't like them? In order to gain some purchase on these questions, we distinguish between two styles of analysis and between two social systems or networks. Cross-sectional analysis is distinguished from longitudinal analysis. The use system, *which enmeshes social practice and IT in our everyday activities is distinguished from the* development system, *which is responsible for putting the IT in place, maintaining, and updating them. In the majority of workday situations, cross-sectional analysis of the use system leads to the appearance of material agency. However, longitudinal analysis of the development system tends to locate agency in the design decisions of the developers. These analytical distinctions lead to a new conceptualization of machine agency as perceived autonomy from the development system. Unlike previous accounts, this view is consistent with both structuration theory and actor network theory. This allows continued access to these powerful analytical vehicles and enables the strong analysis that is the precursor to effective action.*

1. Introduction

Some recent contributions to the IS literature involving actor network theory and structuration theory focus on the relationship between the social and the technical. The debate has sometimes been characterized as a bipolar one, with technological determinism set against social determinism (Jones 1999; Markus and Robey 1988). More recent contributions based on structuration theory emphasize the recursive interaction of the social and the technical; thus Orlikowski (1992) gives us a structurational model of technology in which the technical is both constituted by, and constitutive of, the social. Actor network theory posits associations of human and non-human components without artificial distinctions between the social and the technical. A difficulty that both these well-established bodies of theory share, in different ways, is the notion of material (non-human) agency.

The concept of agency has a number of different facets. It may encompass actions and the freedom to choose those actions; intentionality, will, and power; causality, consequences, and outcomes (which may be intended or unintended); and decision making. Actor network theory and structuration theory offer rather different accounts of agency (compared in Table 1). In structuration theory, agency is the ability of humans to "make a difference" or cause an effect (Giddens 1984). Giddens distances agency from intentionality (since unintended consequences may engender effects as well as intended ones) and links it instead with power as "transformative capacity." In the structurational cycle of structure and agency, agency is a particularly human phenomenon that non-humans cannot possess. How could material objects act in the context of their understanding of structure (memory traces) and in doing so re-enact that structure? Clearly this is not possible. In structuration theory, material objects are resources to be employed by human agents. However, the ability to constitute (cause an effect upon) the social, as theorized by Orlikowski, implies technical (non-human) agency: the effect caused by the technical is the constitution of the social. Structuration theory is unable to resolve the problem of how technical artifacts, such as computer systems, seem to engender profound consequences.

By contrast, a central tenet of actor network theory affords non-humans the same status as humans in a principle of generalized symmetry. In an actor network, technical artifacts may be actors in the same way that humans are. Latour (1991) writes forcefully of the "testimony of non-humans." Elsewhere he indicates that sets of actants[1] may behave like machines which have themselves characteristics of agency, volition, and autonomy.

[1] In *Science in Action* (1992), Latour defines an "actant" as "both people able to talk and things unable to talk have spokesmen. I propose to call whoever is and whatever is represented as actant....I introduced the word 'actant' earlier to describe what the spokesman represents. Behind the [texts, instrument, laboratory]...what we have is an array allowing new extreme constraints to be imposed on 'something.' This 'something' is progressively shaped by its re-actions to these conditions" (pp. 83-84; 89).

Table 1. Agency in Structuration Theory and Actor Network Theory

	Agency	
Oxford English Dictionary:	"The means of action through which something is done." Agent: 1. Person who does something or who instigates some action. 2. One who acts on behalf of another. 3. Something that produces an effect or change.	
Associated concepts	**Structuration theory**	**Actor network theory**
Focus of theory	Reconciliation of structure and agency	Interplay of actants; the building and maintaining of the networks
Agents	By definition, human actors; machine agency not allowed	Actors, actants—either human or other; machine agency inferred and co-equal
Action	Explicit and assumed of humans who have freedom to choose actions	Explicit and required of actors, inanimate objects and machines still have (limited) freedom to act
Power	Enabler of action—machines are resources that extend the power of humans	Exhibited in networks—non-humans also have power
Intentionality or volition	Associated with agency but not required; human agency can exist with or without intention	Explicit and required to construct and maintain networks, inferred of non-human actors
Causality, effects, consequences	Intended and unintended consequences of action, transformative power, re-enaction of structure	*Interessement*, enrolment, translation leading to formation of associations, black-boxing, irreversibility
Decision-making capacity	Assumed of humans	Assumed of humans; ascribed to non-human actors
Structure	Network of rules and context, wholly human constructions; "memory traces" explicitly not technology (material resources used by people).	Associations, networks of humans, non-humans, black boxes
Relationships	Interactions between people	Associations in network
Independence, self-government	Human, societal	Human and non-human, network

> The simplest means of transforming the juxtaposed set of allies into a whole that acts as one is to tie the assembled forces to one another, that is to build a machine. A machine...is first of all, a machination, a stratagem, a kind of cunning, where borrowed forces keep one another in check so that none can fly apart from that group....it is important to note that the skills required to go from a [tool] to a [machine] are symmetrical....Complicated negotiations have to go on so that provisional alliances do not fly apart. [Latour 1992, p. 129]

However, many commentators are unable to accept the proposition that material things should have the same powers of agency as humans (Walsham 1997). Moreover, Latour would probably regard social and technical as unhelpful modern conceptual distinctions: "We are never confronted with science, technology and society," he asserts, "but with a gamut of weaker and stronger *associations*" (Latour 1992). Some aspects of agency in structuration theory and actor network theory are compared in Table 1.

The comparison table demonstrates that structuration theory and actor network theory, although often lumped together, have different and quite incompatible accounts of agency and the relationship between the social and the technical. Nor are these accounts able adequately to resolve difficulties central to researching information systems in their social contexts.

In trying to solve this problem, Jones (1999) proposes a middle course. Inanimate objects should be allowed agency, but without the particularly human component of intentionality. He offers a "double mangle" model of social and technological interaction. The "mangle of practice" describes human agency's efforts to adapt intransigent material agency to its own ends in an emergent process. In double mangling, according to Jones, human agents "channel material agency to shape the actions of other human agents," or "marshal material agency to direct the actions of other human agents" in a "double dance of agency." In IS, we must relate this to human actors building and using computers systems. In this paper, we seek to build upon this account of material agency and the relationship between the social and the technical. IS is an applied discipline and we are primarily interested in action; our interest in theory is that it helps to promote better actions. In exploring these issues, we seek to move the debate in the direction of explicit IS concerns (which require explanations not of material agency in general, but of computerized machine agency in particular) and toward action. Hence, the purpose of the paper is to build on previous accounts of machine agency (in the context of understanding the relationship between the social and the technical) and offer new conceptualizations that facilitate rich analysis as a precursor to action.

The paper is structured as follows. The next section of the paper develops an understanding of analysis as the key link between theory and action. This helps to identify weaknesses in the double mangle model as a vehicle for action. In order to develop better understandings of machine agency, we then make two sets of distinctions about analysis. The first distinction concerns the type, or style, of analysis; the second concerns the its object or focus. We first distinguish between cross-sectional analysis and longitudinal analysis styles, then between analysis of the social system (network) that develops a computer system and the one that exploits or uses it. The distinctions are brought together to provide a framework for the deconstruction of two examples. The first is the example used by Jones in his double mangle paper. The second takes the form of a thought experiment. Using this analysis, a new conceptualization of machine agency is developed, compatible with both structuration theory and actor network theory, leaving the would-be analyst/actor in a stronger position. Finally, consequences for practice are elicited.

2. Theory and Action

A simple framework for relating theory and action is given first (Figure 1). It is derived from Checkland and Scholes (1990). Neither theory nor practice (action) is grounded (i.e., demonstrable without relation to the other). Theory must, in the end, be based on practice, whereas practice cannot be carried out in isolation from theories of the world. Analysis links the two. Analysis (using theory) of a set of phenomena permits action: I take the world to be like that, therefore I will act like this. Often this movement of thought is conflated into common sense or instinct and its parts are not really discernible. Theory is derived from practice via analysis. Theoretical discussions can take part in isolation from practice, but theory ultimately is not sustainable if it cannot be shown to relate to practice. Thus theory is the emergent result of analyzing practice; practice the emergent result of analysis governed by theory. In our applied discipline, for the development and use of information systems, it is necessary not only to have well-informed and defensible theory, but theory that enables rich analysis and thus purposeful action.

Theories that allow us, then, to proceed to sensible actions do so by improving our powers of analysis. If we can make more sense of situated local phenomena, because of the power of the theoretical constructs at our disposal, we have a better chance of choosing a sensible course of action to deal with them.

In assessing the usefulness of the double mangle model for practice, we need to evaluate its analytical power. Both structuration theory and actor network theory can help us to make better sense of empirical situations. Structuration theory is a well-elaborated set of related concepts with well-demonstrated ability to help analyze phenomena in the IS arena (Barley 1986; Brooks 1997; Jones and Nandhakumar 1993; Walsham 1993). Actor network theory also offers valuable insights into the IS domain (Latour 1996; Walsham 1997). However, the double mangle model, with its conditional acceptance of

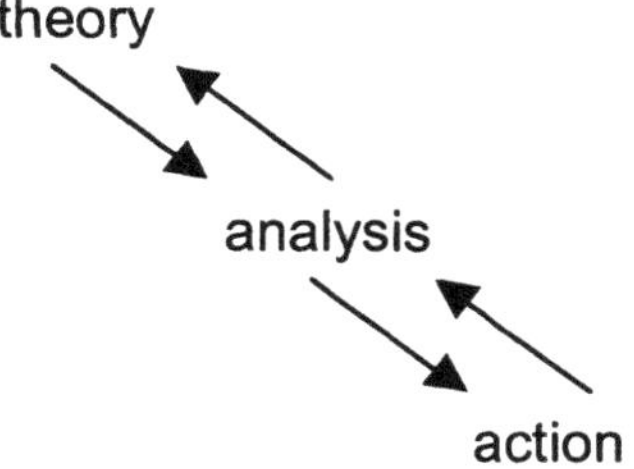

Figure 1. Theory and Action

material agency, is not compatible with structuration theory. According to Giddens, the reflexive evolution of structure and agency is located in the minds of knowledgeable human actors, not embodied in artefacts. Artefacts are material resources which agents may employ (Jones 1997). Unfortunately, the double mangle model is not really compatible with actor network theory either. If only privileged human actors possess intentionality, then we must reassess whether the non-human actants can enrol, translate, *interesse*, delegate, inscribe—all actions seemingly requiring of intentionality. In adopting a compromise, a "blended middle" (Latour 1991), Jones (1999) constructs a tenable theoretical explanation. Unfortunately, the would-be analyst using Jones' model cannot employ the concepts of structuration theory or those of actor network theory with safety. These well developed bodies of theory become inaccessible. The double mangle model offers no similarly well articulated concepts to replace those which are undermined. Clearly the analytical power of the double mangle model is less well developed than that of structuration theory or actor network theory and the position of the analyst/practitioner is, therefore, weakened rather than strengthened. The implication is that we should search for understandings of machine agency that leverage the analytical power of the existing theory bases rather than diminish it.

3. Distinction 1: Analysis Style—Cross-sectional Versus Longitudinal

Analysis is central to the ability to take meaningful action. The theory behind meaningful action both enables rich analysis and is derived from it. However, there are many forms of analysis. In order to illustrate the effect that the choice of analysis may have on a given theoretical concept (in this case machine agency) or the interpretation of an empirical situation, a simple distinction between cross-sectional analysis and longitudinal (historical) analysis is made (Figure 2). The distinction is derived from Franz and Robey (1987) and is not intended to characterize exhaustively modes of analysis, but simply to enable the illustration of the effect of different kinds of analysis on theory and action. Cross-sectional (single period, point-in-time) analysis concentrates on the event or phenomenon at the expense of its history. A cross-section of a situation investigated at a single point in time reveals webs of related phenomena, or variables in particular states at the given moment. A systems analyst employing a traditional systems development methodology will largely concentrate on forming a picture of what is happening at the time of investigation. This may provide an in-depth analysis of a given situation, but it does so at the expense of forming a historical picture of why those phenomena are they way they are and not otherwise. In contrast, longitudinal analysis concentrates on understanding the evolution of a phenomenon over time. It may do so by taking smaller slices of the phenomenon's history, but over many points in time, or continuously. The amount and type of data captured and examined is an open issue. Some analysts may choose to concentrate on the web of exchanges, events, or relationships leading from development to development. Thus Latour's (1992, p. 104) study of the evolution of the Diesel engine traces its progress from source idea, through development, diffusion, and Diesel's suicide, to acceptance in the market place.

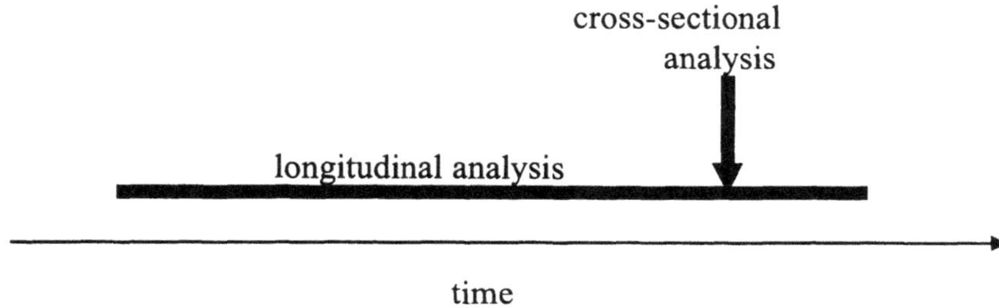

Figure 2. Longitudinal versus Cross-sectional Analysis

4. Distinction 2: Analysis Focus—IT Use versus IT Development

As the style of analysis may vary, so may the object, or focus, of enquiry. A common way of thinking separates the development of IT from its use (Figure 3). Machines may be analyzed in use, as human actors incorporate them into the task structures of their lives. Barley's (1986) study of computer tomography focuses on the use of the machine in its hospital environment. Machines may also be analyzed as they are developed (as in the Diesel engine example quoted above). Then "We study science in action and not ready made science or technology; to do so, we...arrive before the...machines are black boxed" (Latour 1992. pp. 258).

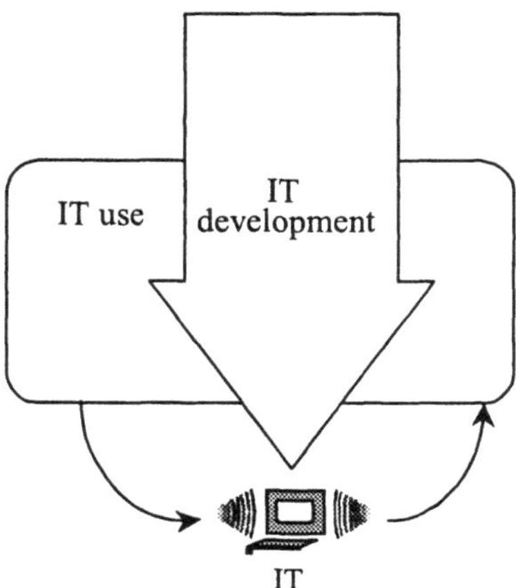

Figure 3. IT Use and Development

Use and development cross over, overlap, and become enmeshed as developers investigate use for design purposes and as users provide feedback to developers or seek to influence development. Thus the distinction, while well established, is artificial (Jones and Nandhakumar 1993; Orlikowski and Robey 1991). In terms of structuration theory, we may think of a two social systems concerning use and development, which may be represented as sets of social practices, involving the recursive interaction of structure and action, routinized over time and space. Giddens does not theorize IT, but it must be embedded in the discourse that relates structure to action. Alternatively, in the context of actor network theory, we may think of use and development as networks. In the development network, the machine is the end product, whereas in the use network, it is a black-boxed component. The distinction between use and development is made here for analytical convenience; in practice, development and use are inextricably enmeshed.

5. Analysis of Machine Agency

Mapping the distinction between analysis styles on to the distinction between analysis foci gives a framework for the deconstruction of empirical examples (Table 2).

Table 2. Framework for Deconstruction of Machine Agency Examples

		Focus of Analysis	
		Use	Development
Style of Analysis	Longitudinal		(shaded)
	Cross-sectional	(shaded)	

Of particular importance to the deconstruction of these examples will be the difference in perspective on machine agency obtained by longitudinal analysis of the development system as compared with that obtained by cross-sectional analysis of the use system (represented by the shaded areas in the table). These perspectives provide a story line that guides our initial deconstructive reading of the examples to follow. Deconstruction, a technique for disassembling socially constructed meanings presented in texts, or phenomena represented as texts, is also an analysis technique used to decenter authorship of texts and to surface multiple, often inconsistent, readings. In the present instance, we choose a type of deconstruction that turns and contrasts a text while holding its meaning in a deferred or not quite complete state. This is similar to Derrida's (1992) notion of *différance* Hopper's (1987) linguistic "emergence" in that concepts kept "in play," repeatedly turned and reconsidered, may yield fresh insights and the surfacing and unfreezing of implicit assumptions. Our use of the deconstruction technique is similar to that of Beath and Orlikowski (1994). It is also consistent with the deconstruction techniques suggested by Boje and Dennehy (1994) and Truex, Baskerville, and Travis (forthcoming). We examine meta-narratives associated with both actor network theory and structuration theory and local narratives arising from the two cases chosen for illustration. Each meta-narrative and each case constitutes a text that is examined in the context of the set of concepts identified earlier. Deconstruction turns a step further as conclusions and constructs are held in abeyance while core analytical concepts (agency vs. autonomy) are themselves considered more deeply. Ideas arising from the suspension of closure eventually suggest a path through the knotty issues raised by Jones' double mangle model.

5.1 Deconstruction 1: Jones' Lotus Notes Application

Jones (1999) gives the example of a Lotus Notes application that was judged, by some of its users, to be unacceptably slow. He asks,

> Are there inherent characteristics of technology, such as "speed" which inevitably lead to certain conclusions? Or are these apparent characteristics simply the playing out of broader social forces, reflecting, for example, decisions made in particular configurations of organizational power relations?

If we take a cross-sectional analysis of the use system for the Lotus Notes application at the point in time that it is implemented (which is more or less Jones' analysis), the speed of the application may appear a property of the technology. This property will have an outcome: it makes the application easy or difficult to use. Thus the variables "speed" and "user satisfaction" seem linked together. Users, as Jones hints, may socially construct their criteria for satisfaction. The application is a black box, apparently finished, outside the user's considerations. It is a fait accompli, sitting on the users' desks, and operates in a way that causes more, or less, satisfaction. In this analysis, the Lotus Notes application has a kind of agency: it makes a difference or causes an effect. If we focus on the use system while performing cross-sectional analysis, what tends to emerge is strong machine agency. At a given point in time, and with criteria set by users' expectations, the Lotus Notes application exhibits a property, speed, that is satisfactory or unsatisfactory.

Analyzing the development system longitudinally, however, produces a different impression. Socially constructed design decisions affecting the speed of the application may include the functionality of the application, the way it is programmed, the amount of data involved, the characteristics of the host software, the choice of hardware, the communications network topography and configuration, the physical communications media, the network router configurations, the number of users and quantity of network traffic, and many other things. In this second analysis, the application speed is the result of a complex series of resource and design decisions—human agency rather than machine agency.

Turning this a bit further, if we focus on the development system and perform longitudinal analysis, then the characteristics of the machine look like emergent properties of the developers' decisions. Since none of the machine's characteristics are its own, but the result of human design decisions that are all, in principle, rescindable, a decision to award the machine agency appears an arbitrary one. It is based on the assumption that the machine's development has finished. The property, speed, of the Lotus Notes application, is a result of previous decisions about communications network configuration and hardware purchase, as well as the local decisions of the application programmers. User satisfaction as a property of system response time now appears an unwarranted type of technological determinism.

But let us examine these assumptions a bit further against the concepts identified in Tables 1 and 2. Who are the agents in this case? From a structurational context, it is a set of humans who create, install, and care for the application: the net administrators who install and keep it running, the database designers and administration staff, and the more-skilled and less-skilled user community. From the structurational cross-sectional/use perspective, we are principally concerned with the relative level of the skill-set and experience in the use of the application as possessed by any given user. If experienced and well-trained, they may get the system to do more and respond more quickly. Because of that relatively more advanced skill level, they may feel more in control and accepting of the response time behavior of the machine. The less well-trained or less experienced user is more at the mercy of the system and might be less inclined to accept the speed or response and other operational foibles.

Once we admit the whole of the longitudinal and development perspective, the human aspect of the system becomes more apparent to either the more or the less-skilled user. The system appears less arbitrary, distant, and inscrutable and hence other possible courses of action are open. One may contact the network administrator, the developers, or DBA for assistance. Or, given the particular operational setting, one may choose to accept the slowness because the given development and administrative setting is judged to be unfavorable to affecting a change. Decisions made by human actors are constrained by the structure of previous design and business decisions and are, therefore, reflected in the ways the software behaves as they help define the structural limits of the whole system.

From the actor network theory perspective, we now must look beyond the human actors and add other actants (agents) to the list. Those may include the application itself, the network, other users of the network, and those decisions and decision makers that made the commitment to using Notes (rather than other possible collaborative software tools). In an actor network theory cross-sectional view we must examine the whole of the decisions and decision contexts that brought about a network of operational events. Here the user may now consider the range of response options when they perceive response time as diminishing. Since "the machine" has a perceived agency and the interaction of the machine and other mutually dependent components (agents) is understood, the user may, at the time of use, phone up the network administrator and request a higher priority or extra privileges on the queue (override of the rule set) or a check to see it the network is down at that particular time. This may translate to a greater set of options in the hands of an actor network aware user and might, therefore, be considered when we are designing such systems.

5.2 Deconstruction 2: The Ubiquitous Cash Dispenser

The second example is a thought experiment after the fashion of Introna and Whitley (1997). The content of the experiment concerns an automatic teller machine (the familiar cash dispenser).

> A cash machine dispenses me money. On day one it works perfectly and with the money dispensed I am able to buy lunch for my friends. I return the next day but it is out of order. On the third day it is working, it seems, but refuses to give me money.

Cross-sectional analysis of the cash dispenser in use may yield an impression of machine agency. It certainly has an effect on my subsequent behavior. Either I can buy lunch for my friends, or they pay for themselves. It dispenses, it is unable to dispense, or it refuses to dispense—implying some form of intentionality. It makes simple decisions prescribed by its rule base. It appears largely independent and self-governing. In the structurational story, it is a resource I employ to facilitate my lunch date interactions with my friends. Of course, this theory base forbids us to ascribe it agency. In the actor network story, it is a black box actant in the lunch network that does not need to be further investigated unless it becomes a problem. It "speaks" for the bank (yes, you may have some money). However, longitudinal analysis of the development system paints a different picture. Connotations of intentionality in the machine recede. The "refusal" to dispense now seems a kind of linguistic shorthand. The autoteller has been designed to interface with the computer that records the state of my account. The bank's rule system forbids withdrawals when an account is overdrawn and the developer has programmed this rule into the software. Intentionality here resides with the programmer. However, the programmer is only articulating the bank's rule system; the programmer's only contribution is to buy into the rule system without questioning or reinterpreting it, so perhaps intentionality resides more collectively with the bank. In the development system, agency is distributed around many humans, perhaps around many development systems (the machine needs silicon chips, a VDU, the communications network to which it is connected, a secure power source, security devices, precise machine engineering, someone to restock it with money and consumables). One might ask whether, since the locus of agency does not rest with any *one* human (or even one development system), does it rest in the machine? This is essentially an emergent theory of machine agency: agency is the emergent property of previous design decisions. However, accounts of collective agency in social theory deal perfectly adequately with these phenomena without invoking emergent properties. The programmer builds software (action) according to his understanding of the bank's rule system (structure), thus re-enacting social practice, which replicates over time and space. Many such interactions constitute the fabric of social interaction (structuration theory). Alternatively, many actants (programmer, programming language, rule system, cathode-ray tube, power supply engineer, etc.) are enrolled with interests in common in the network of technical and social components that produces the machine (actor network theory). If the machine now denies me money because it is not working, one of those components is to blame. A mechanical component has failed, there is a software glitch, the communications lines are down, or the machine's armor failed to resist a vandal. Cross-sectional analysis of the use system shows me inconvenienced by the machine's apparent failure to do its job. The effect again is to make my friends pay for lunch. Analyzing the development system, however, shows something is unintentionally awry. Someone must, over the longer term, respecify an engineering process, debug the program, organize rerouting of the communications channels, or install better security.

Let us amend our thought experiment. What happens if the automated bank teller is replaced with a human teller? The human, as a knowledgeable actor, possesses agency; however, *the outcomes are the same*. If I am overdrawn, I still don't get any money (but perhaps I may renegotiate my overdraft limit). The human teller may make a mistake (software error), be sick (broken), or be assaulted by a bank robber (vandalized). Similar analysis is possible. There is an institutionalized rule system that is formalized by the programmer of the machine agent and taught to the human agent. In each case, the outcome is the result of the rule system rather than the agent.

Now what happens if we develop the cash dispenser with more functionality so that it resembles the human bank teller? Let us now give it a voice recognition interface, the ability to hold a rudimentary conversation and recognisz my signature, a large rule base encapsulating the bank's procedures, and a powerful inferencing engine. Let us give it the ability to assess my credit-worthiness and call up its manager for decisions it can't handle—in fact, the ability to do what the human teller can. This is a version of the Turing test: if the machine can do what the human can do, it must be attributed agency.

6. Discussion

Since we cannot perform theory-free analysis, the theoretical base must influence the analysis. Structuration theory emphasizes the agency of the knowledgeable human actor, whereas actor network encourages the co-equality of non-human agency. Therefore, a structurational analysis of an empirical situation will tend to produce a strong view of human agency, whereas an actor network theory analysis of the same situation will encourage the possibility of strong machine agency. Action on the basis of structurational analysis is likely to be focused on human actors, whereas action on the basis of actor network analysis may target humans and machines. However, the style and focus of analysis also play a large part in determining what theoretical stance will be taken (deriving theory from action) and how an empirical situation might be viewed (proceeding to action via theory). The first feature to be noted from the preceding examples is the relativity of the concept of machine agency. Machine agency can appear quite strong as long as the machines are taken as black-boxes and observed in use over a short period, but strong agency tends to disappear when the development system is considered historically. This is illustrated in Table 3.

Giddens' discussion of agency involves three factors: the capacity to make a difference, power, and intentionality. All three can be observed in the preceding examples. The cash dispenser clearly affects peoples' future actions—making a difference—and so exerts power. Power is related to the scale of the effect. My pen, a simple machine, may stop working but the effects on me are negligible; the cash dispenser may cause me considerable inconvenience, but failure of the brakes on my car may end

Table 2. Framework for Deconstruction of Machine Agency Examples

		Focus of Analysis	
		Use	**Development**
Style of Analysis	**Longitudinal**		Weak machine agency
	Cross-sectional	Strong Machine Agency	

my life. In the case of the cash dispenser, the refusal to dispense might be considered a form of intentionality, at least in one analysis. Intentionality attributed to machines has quite a strong hold on our everyday thoughts ("I braked but the car wouldn't stop," "my laptop won't connect to the network," "the central heating's playing up today"). However, some agency-related concepts are more essentially human than others. In each case, we can analyze these concepts in a different way by looking at the design system. The consequences and effects that the machine engenders will probably still seem like attributes of the machine; however, intentionality may be more likely to be seen as a property of the human designers.

The last question to be considered is the distinguishing feature of agency. What does the human teller have that the cash dispenser does not which allows it more agency? Most of us can concede that a machine can act, can demonstrate its power, and that its actions have effects and consequences. What do we have to add to the machine to allow it to be viewed as an agent in the same way as the human? Perhaps the clue is in the human's ability to respond to an overdrawn client. The cash dispenser has one response, which is programmed: no cash. The human has variety in the way (s)he handles the task and, in the last resort, the ability to override the rule systems, or operate in a wider context. To make the machine resemble the human in terms of agency, the designer must greatly enhance the variety of its responses to different circumstances and allow it to interpret the underlying rule systems with the flexibility that the human displays. It must be given independence, the ability to govern itself, decision-making powers, intentionality, and volition. These are the more human characteristics of agency, which we find it harder to ascribe to machines. We term this (rather human) side of agency *autonomy*.

7. Machine Agency as Perceived Autonomy

Now we consider underlying models of machine agency. The separation of development and use of IT immediately offers one version of machine agency: sequential "discontinuous separation of design and use" (the phrase is from Orlikowski 1992). In this version of technological agency, the machine is first designed, then has agency. In development, the system is in the hands of the developers, who exert agency over its components. Agency is the emergent property of the development process and becomes embedded in the completed machine. Now the machine can have an effect on the people who use it. However, this view of development and use is a particularly impoverished one. It derives from the ubiquitous project-based view of information systems development. A computerized information system is developed in a discrete block of time (project); when it is finished, it is implemented and is then in use. Analyzing formal project activities takes little account of their wider social context and history. In life cycle models, on-going development work is simply re-labeled "maintenance." No account is taken of "interpretive flexibility" (Orlikowski 1992)—the capacity of users to shape the systems with which they work. Better understandings of the relationships between development and use recognize that they are recursively and reflexively dependant upon each other. Development cannot take place without consideration of the use system; use is a form of feedback on development; machines evolve in an emerging process which is dependent on both overlapping systems.[2] Jones' double mangle model incorporates this understanding. Human agents seek to "channel material agency to shape the actions of other human agents," or to "marshal material agency to direct the actions of other human agents," in a "double dance of agency." Machine agency is at the interface of the two social systems: use and development. Its distinguishing quality is autonomy. Two interlinking uses of the word autonomy may be distinguished. The first is its natural language meaning of "independent, self governing." Now we refer to sets of concepts that are related to agency, but comprise its more human side. The second, in this theoretical context, is the sense of autonomy from the machine's development system. However, this kind of autonomy is partly a function of the analyst's approach. If the analyst leaves the machine as a black box by concentrating on cross-sectional analysis of the use system, the machine is likely to exhibit more of the characteristics of autonomy: strong agency. If, however, the analyst chooses to deconstruct the black box, via longitudinal analysis of the development system, certain features of its agency are more likely to be attributed to its human designers. Therefore, we find that autonomy in the machine is not integral to the machine itself, but strongly dependent on the way it is perceived.

This relative concept of machine agency as perceived autonomy is consistent both with structuration theory and actor network theory. In structuration theory, Giddens theorizes social systems as integrated routinized practices, in the recursive cycle of structure and agency, connected by time-space edges. All of these concepts can be applied to the analysis of the social systems of use and design. When an analyst perceives autonomy in a machine, (s)he perceives the temporary, embedded, emergent outcome of the development system. Development and use can also be analyzed as actor networks. In the use network, the machine is a black box; the deconstruction of the black box is its development network.

---

[2]When I write to my bank complaining about the charging system for cash dispensers (for which I am normally part of the use system), I deliberately attempt to locate myself in the machine's development system.

8. Conclusions

Understanding the relationship between the social and the technical, the human and the machine, is at the heart of the study of information systems—an applied discipline in which action is related to theory via analysis. The recent theoretical debate revolving around actor network theory and structuration theory highlights the fundamental incompatibility of these two theoretical positions regarding the question of non-human agency. In actor network theory, non-humans and humans alike are afforded symmetrical agency, whereas in structuration theory only humans can be agents. Jones' double mangle model tries to resolve this difficulty by offering non-humans limited agency. Jones adopts a compromise approach in which he seeks to amend the theory base to solve the problem. He sees the inherent contradiction and incompatibility of the two theoretical accounts of machine agency, and tries to create a theoretical solution that will help explain IS phenomena. Whatever its theoretical merits, this model presents difficulties for researchers who want to do practical things in IS. It is wholly compatible neither with structuration theory or actor network theory, nor does it offers similarly well-elucidated concepts. As such, it weakens the analytical power of the researcher, and hence the capacity of the actor to take action. It leaves the would-be analyst and practitioner in a position that is essentially inactionable, while abandoning the relative security offered by the structuration and actor network frameworks.

Our deconstructions allow us to understand that the starting theory base, and the style and focus of analysis, inevitably influences, and to some extent governs, how we perceive the agency of a machine in any empirical situation. Our perception is relative, depending on whether we choose to take the machine as given (black box) or deconstruct its development. Agency is a complicated association of concepts, not a crystal-clear dictionary definition, and we find it harder to ascribe some of these concepts to machines than others.[3] For instance, people can generally cope with the idea that machines act, and their actions have consequences, but find it harder to attribute to them our more human characteristics such as intention or volition. We develop the concept of autonomy to refer to these human elements: self-government, independence, intentionality, volition, and decision making. When we perceive the more human aspects of agency in the machine, we perceive the machine as more autonomous. That perception deserves to be taken seriously by developers and managers of IT infrastructures, despite the fact that it may change if we study the historical development of the machine in more depth. Restated more simply, the closer one is to the moment, and to the situation of use (a user faces a machine that does not respond to their will), the more likely one is to attribute autonomy to the machine. A more formal description (inspired by Latour) is: every black box (IT system) is a network waiting to be deconstructed. Nevertheless, there may be numerous instances when we either choose not to open, or where we believe we are constrained from opening, the black box.

Our analytical devices and deconstructions lead us toward a rethinking of machine agency: machine agency as perceived autonomy. This conceptualization is compatible with both structuration theory and actor network theory and, therefore, leaves the would-

[3]The personification of inanimate objects has been the study of cultural anthropologists and art historians (c.f. Langer 1953; Skillman 1981; Turkle 1984).

be analyst/actor in a stronger, not a weaker, position. In undertaking analysis that leads to useful action we need to have our cake *and* eat it. Focusing on the development story exposes the roles of the human agents who make information systems and encourages a socially determinist view. Focusing on the information systems in use ignores the story of the machines' construction and encourages a technologically determinist interpretation: now we mainly look at the finished machine as it constrains and enables social interactions. The trick is to have both stories in mind at the same time. What this buys the analyst is a way to imagine and anticipate user response to systems under design. What it buys the users is a way to understand the underlying conceptualization of the system.

This approach differs substantially from our initial point of reference, Jones' double mangle model. Jones tries to resolve incompatibilities between actor network theory and structuration theory at the cost of inconsistency with both theories. We offer instead a position that is consistent with both actor network theory and structuration theory. Rather than accepting or denying machine agency, we allow that machines may have a *perceived autonomy*. Perceived autonomy is a property derived partly from what people think the machine has and does and partly from what is designed into the machine. When we distinguish between analysis styles—both cross sectional and longitudinal—and the use system versus the development system, we acknowledge contexts that can never really be wholly separate. We offer a way to keep the questions of design and use in constant play; that is, to forestall the freezing of the conception of a given system while remembering that specific properties of a machine are never fixed but are always emerging (Truex, Baskerville, and Klein 1999). What we give designers and users, therefore, is a way to keep both arguments in focus at the same time while drawing upon two well understood bodies of theory and of practice. What should the developer, manager, or user of an information system do with these insights? The trick becomes living with the anomalies and remaining in the difficult dialectical center, rather than trying to resolve one problem at the expense of impoverishing the wider analytical repertoire. We should use the analytical power of the competing frames of reference to keep the different perspectives, of machine agency in focus *at the same time.* If notions of agency depend upon theoretical and analytical perspectives we do not suggest ironing them out to arrive at the "correct" theoretical stance; rather, we harness the power of their dialectical tensions in order to arrive at richer understandings. Both actor network theory and structuration theory are different routes to those understandings. The recursive emergent nature of development and of use should be both recognized and exploited.

References

Barley, S. R. "Technology as an Occasion for Structuring: Evidence from Observation of CT Scanners," *Administrative Science Quarterly* (31), 1986, pp 78-108.

Beath, C. M., and Orlikowski, W. J. "The Contradictory Structure of Systems Development Methodologies: Deconstructing the IS-User Relationship in Information Engineering," *Information Systems Research* (5:4), 1994, pp 350-377.

Boje, D. M., and Dennehy, R. F. *Managing in the Postmodern World: America's Revolution Against Exploitation.* Dubuque, IA: Kendall /Hunt Publishing Co., 1994.

Brooks, L. "Structuration Theory and New Technology: Analyzing Organizationally Situated Computer-aided Design," *Information Systems Journal* (7), 1997, pp 133-151.

Checkland, P. B., and Scholes, J. *SSM in Action*. New York: Plenum, 1990.

Derrida, J. (ed.). *Différance. Margins of Philosophy*. Chicago: University of Chicago Press, 1982.

Franz, C. R., and Robey, D. "Strategies for Research on IS in Organizations: A Critical Analysis of Research Purpose and Time Frame," in *Critical Issues in Information Systems Research*, R. J. Boland and R. Hirschheim (eds.). Chichester, UK: Wiley, 1987.

Giddens, A. *The Constitution of Society*. Cambridge, UK: Polity Press, 1984.

Hopper, P. "Emergent Grammar," *Berkeley Linguistics Society* (13), 1987, pp. 139-157.

Introna, L. D., and Whitley, E. A. "Imagine: Thought Experiments in Information Systems Research," in *Information Systems and Qualitative Research*, A. S. Lee, J. Liebenau, and J. I. DeGross (eds.). London: Chapman and Hall, 1997.

Jones, M. "Structuration and IS," in *Rethinking Management Information Systems*, W. L. Currie and R. D. Galliers (eds.). Oxford: Oxford University Press, 1997.

Jones, M. "Information Systems and the Double Mangle," in *Information Systems: Current Issues and Future Changes*, T. J. Larsen, L. Levine, and J. I. DeGross (eds.). Laxenburg, Austria: IFIP, 1999.

Jones, M., and Nandhakumar, J. "Structured Development? A Structurational Analysis of the Development of an Executive Information System," in *Human, Organizational and Social Dimensions of Information System Development,* D. E. Avison, J. E. Kendall, and J. I. DeGross (eds.). Amsterdam: North-Holland, 1993.

Langer, S. K. *An Introduction to Symbolic Logic*. New York: Dover Publications Inc., 1953.

Latour, B. *Nous n'avons jamais été modernes*. Paris: Editions La Découverte, 1991.

Latour, B. *Science in Action*. Cambridge, MA: Harvard University Press, 1992.

Latour, B. "Social Theory and the Study of Computerized Work Sites," *Information Systems Research,* Special Issue, "IT and the Structuring of Organizations," W. J. Orlikowski and D. Robey (eds.), (2:2), 1996, pp 143-169

Markus, M. L., and Robey, D. "Information Technology and Organizational Change: Causal Structure in Theory and Research," *Management Science* (24:5), 1988, pp 583-598.

Orlikowski, W. J. "The Duality of Technology: Rethinking the Concept of Technology in Organizations," *Organization Science* (3:3), 1992, 3, pp 398-429.

Orlikowski, W. J., and Robey, D. (eds.). "IT and the Structuring of Organizations," Special Issue, *Information Systems Research* (2:2), 1991, pp 143-169.

Skillman, Y. "Ancient Inspirations and Contempory Intepretations," *Roberson Museum Exhibition Series*, D. Truex (ed.). Binghamton, NY: Roberson Memorial Inc., 1981, pp. 5; 152.

Truex, D. P., Baskerville, R., and Travis D. "Amethodical Systems Development: The Deferred Meaning of Systems Development Methods," *Accounting Management and Information Technology* (9), forthcoming, pp 1-27

Truex, D., Baskerville, R., and Klein, H. "Growing Systems in Emergent Organizations," *Communications of the ACM* (42:8), 1999, pp 117-123.

Turkle, S. *The Second Self*. New York: Macmillan, 1984.

Walsham, G. *Interpreting Information Systems*. Chichester, UK: Wiley, 1993.

Walsham, G. "Actor-Network Theory and IS Research: Current Status and Future Prospects," in *Information Systems and Qualitative Research*, A. S. Lee, J. Liebenau, and J. I. DeGross (eds.). London: Chapman & Hall, 1997.

About the Authors

Jeremy Rose won an exhibition to read English at Cambridge and subsequently trained to be a musician at the Royal College of Music in London. After working for some years for the Rambert Dance Company and Music Projects London, he retrained at Lancaster, gaining his MSc in Information Management with distinction. As a senior lecturer in Business Information Technology in the Faculty of Management and Business at the Manchester Metropolitan University, he collaborated with Peter Checkland on research projects. More recently he has been working with colleagues at the University of Aaalborg. While completing his Ph.D. at Lancaster, he has published in management, systems and IS forums. His research interests include IS development and evaluation, systems methodology, structuration theory, actor network theory, BPR and knowledge management, the health service, and inter/intranet development. Jeremy can be reached by e-mail at J.Rose@mmu.ac.uk.

Duane Truex is an assistant professor of Computer Information Systems at Georgia State University and a Leverhulme Fellow at the University of Salford, England. His first degrees were in the arts and he spent his early professional life in symphony orchestra and museum management. As a critical social theorist, his work in information systems research has been generally concerned with emancipatory issues of information technologies. He also writes about information systems development and IS research methods. He has published in *Communications of the ACM*, *Accounting Management and Information Technologies*, *Information Systems Journal*, *IEEE Transactions on Engineering Management*, the IFIP transactions series, and in various other proceedings. Duane can be reached at dtruex@gus.edu.

Part 8:

Transforming Automation

23 SOME CHALLENGES FACING VIRTUALLY COLOCATED TEAMS

Gloria Mark
University of California, Irvine
U.S.A.

Abstract

Numerous challenges face geographically distributed teams who are expected to perform as physically colocated teams: to provide deliverables, meet project schedules, and to generate feasible and even innovative problem solutions—all from a distance. Limitations due to technology and distance make it difficult for geographically distributed teams to develop necessary social processes in order to function as a "well-formed" team. This paper discusses three challenges for these teams, based on results of an empirical study of virtually colocated teams, along with a reexamination of past empirical studies. These challenges are: (1) achieving a high standard of participation, which affects impressions, interaction patterns, and trust; (2) developing an appropriate culture, for motivation and cooperation; and (3) integrating the remote team suitably into members' local working spheres. In the conclusion, future research directions and technology requirements to help meet these challenges are discussed.

1. Introduction

This paper is concerned with the relatively new phenomenon emerging in organizations where geographically distributed people meet together and collaborate as a team. The term "virtually colocated team" is becoming increasingly more visible as a research topic due to recent trends of corporate mergers, global markets, and interdisciplinary teamwork. Different technologies are being investigated as to their potential to support real-time communication and interaction, as well as asynchronous collaboration in such teams. Some technologies investigated in business settings, which have highlighted the problems and benefits for teams, include: desktop conferencing (Mark, Grudin, and Poltrock 1999), e-mail, telephone, and document exchange (e.g., Sproull and Kiesler 1991; Zack 1993); chat (Bradner, Kellogg, and Erickson 1999; Fitzpatrick et al. 1999); and an audio-only media space (Ackerman et al. 1997).

The goal of this paper is to describe challenges facing geographically distributed teams with long-term agendas who use these technologies. The challenges are derived based on an empirical study, together with a reexamination of past empirical studies. Numerous challenges face these teams, who are expected to perform as physically colocated teams: to provide deliverables, meet project schedules, and generate feasible and even innovative problem solutions. And yet all this must be done at a distance. Team members stem from different departments, organizations, countries, and sometimes even different companies. Sometimes teams meet face-to-face on a regular basis, sometimes rarely, often not at all. How can team members be expected to be motivated to attend meetings, to develop trust, or even to adopt the technology when social pressures from a distance are weak? Even management and technical support for these teams at the local level may be weak: managers may consider such teams to be a part-time activity; local sites may lack gurus, champions, or even compatible hardware.

The topic of virtually colocated teams is distinguished from a general discussion of the problems involved in distributed work. The key word in this paper is *teams,* and the focus is on how the development and sustainability of social processes in the group are affected by the use of technology. The challenges that these teams face are directly tied to the limitations that the technology imposes on the communication of relevant social information, believed to be essential to the effectiveness of teams.

It is widely believed that a well-functioning team needs to forge common goals, working procedures, and rules of interaction (e.g., Weick 1969). Researchers are accumulating evidence that suggests that the ability of a distributed team to function depends on such factors, and not only on the capability of the technology to enable communication and data-exchange. The next section describes the role of interaction in developing perceptions of membership of a social/working group. Section 3 describes the methodology of the empirical study. Section 4 explains the challenges, illustrating them with examples found in virtual meeting interaction. The conclusion outlines further research directions.

2. Interaction and Group Formation

This paper discusses how limitations due to technology and distance make it difficult for geographically distributed teams to develop certain social processes in order to function as a "well-formed" team. However, a well-formed team, especially one that is geographically distributed, is hard to define. Originally Lewin (1948) proposed that a group becomes a social system when members' goals and their means to attain these goals become interdependent. Rabbie and Horwitz (1988) elaborate on this idea by adding that not only is the perceived interdependence of individual members essential to forming a group, but they must also experience a common fate. The notion of managing mutual interdependencies is a central idea that has been proposed to define cooperation in a group (Schmidt and Bannon 1992). However, an alternative view by Tajfel and Turner (1986) explains that members can still perceive themselves as a group even though their goals may not be interdependent. Rather, the process of emphasizing and defining social category differences defines a group for individuals.

Interaction would seem to facilitate perceptions of interdependence or a distinct social categorization for the group. Most investigations of group formation have been performed with groups who are physically colocated, i.e., whose primary contact is face-to-face. The value of interaction has been characterized as enabling individuals to slowly merge their attitudes and develop a new set of "group" attitudes, identifications, and behaviors, which becomes a collective, or congruent structure (e.g., Newcomb 1968; Weick 1969). Hackman (1976) describes that interaction and common stimuli play a major function for group members in influencing their affective and informational states. Interaction is especially critical in the initial stages of group formation. Its function has been explained to evaluate others' discrepant and conforming behaviors (Moreland and Levine 1989) and also to exchange information in order to seek commonalities (Tuckman and Jensen 1977).

With teams that are virtually colocated, it is not clear how interaction should be assessed. With the use of computer media, communication in groups changes along a number of dimensions, e.g., the overall amount increases (Hiltz et al. 1986), back-channel responses decrease (O'Conaill, Whittaker, and Wilbur 1993), and meetings last longer (Gallupe and McKeen 1990). Such changes certainly must alter the nature of interaction, but how they do is still largely unknown. Even though interactivity has been proposed as a mechanism responsible for promoting engagement, and possibly stable membership in on-line groups (Rafaeli and Sudweeks 1997), the nature of this interactivity still needs to be clarified. One aspect of interaction identified in a virtually colocated group is that the effectiveness of information exchange has been shown to correspond to the relational links of the members (Warkentin, Sayeed, and Hightower 1997). It does appear, though, that computer-mediated communication partners strive for clear interaction, as shown by their adoption of compensatory strategies to reduce confusion, at least in text-only systems (Herring 1999).

That interaction plays a role in shaping a group's culture, norms, cooperative attitude, and identity is hardly disputable. But now we enter an era where individuals are meeting from a distance and their interaction is both enabled and constrained by technological means. This paper attempts to define challenges for teams whose primary means of communication and operation is through technology. The claim put forward here is that group attributes are affected as a result of this communication, and the focus is specifically on participation, culture, and integration into other work.

3. The Study

Most of the arguments presented in this paper are based on an empirical study, performed in 1998, of virtually colocated teams done at The Boeing Company. Part of the results from this empirical study are reported in Mark, Grudin, and Poltrock. In 1996, two years before the study began, mergers with two other large corporations nearly doubled the number of employees in the organization and expanded their geographic distribution to the entire country. This redistribution led to the development of geographically distributed teams. The motivation of the study was to understand better the problems and benefits for such teams meeting regularly using desktop conferencing. The arguments in this paper are further developed with empirical results of other studies which examined synchronous communication among distributed actors and in distributed teams.

In the study, four geographically distributed teams of eight to 15 members, which had existed for six months or longer in the company, were observed during their weekly meetings for a three-month period. All teams used desktop conferencing technology combined with telephone conference calls to support their meetings. The desktop conferencing enabled application sharing. Most team members participated in the meeting from their desktop. In two of the groups, some participants gathered in a conference room, where a shared PC or a laptop connected to a projector enabled everyone to see. An ethnomethodological approach was used: for two groups, the observer regularly connected to the desktop conferencing meeting from her own desk; for the other two groups, the observer sat in the conference room where part of the team met face-to-face. The observer took notes (recording was not permitted) and supplemented the observations with on-line questionnaires, meeting agendas, minutes, chat, attendance records, and in-depth interviews conducted with key members of the teams.

The four teams included a Scientific Problem solving team, a voluntary team that solved real problems for the company; a Technical Working group, whose goal was devising standards for the company Web; a distributed Staff, whose manager was located in another city and whose team goal was information exchange; and a Best Practice team, with the goal of designing vehicles better, faster, and cheaper. Most meetings lasted between one and a half to three hours, and the meeting formats varied in the groups. In the first two teams, presentations (Power Point) were generally followed by discussion. The Staff mostly discussed agenda items, with an occasional presentation. The Best Practice team reviewed the status of action items assigned to team members and decided upon new ones.

4. The Challenges

Three different challenges were selected for virtually colocated teams. These three challenges are important for practitioners to consider in trying to design and implement appropriate technical support. All of the challenges that face such teams cannot be listed in this paper. Rather, key examples of team behaviors that are profoundly affected by computer-mediated interaction have been selected. The argument is that not only does technology use produce immediate behavioral effects in the group, but more importantly, these observable effects will have long-term consequences on the team development and effectiveness and, therefore, must be considered seriously. The challenges that will be discussed include: (1) achieving a high standard of participation, which affects impressions, interaction patterns, and trust; (2) developing an appropriate culture for high motivation and cooperation; and (3) integrating the remote team suitably into one s current working sphere. These challenges are not independent, but intricately related. For example, limited participation and discussion shapes the way that culture will be developed.

The challenges exist at different levels of granularity. This is to be expected, as team work is composed of a combination of individual and group effort, amidst a background of organizational constraints. The participation challenge exists for individuals in the team. Social cues transmitted or restricted by the media, and clarity of the channel, affect individual perceptions. Developing a suitable culture to promote engagement, and motivation in the group, is a challenge at the team level. On the other hand, integrating work from a geographically remote team that is organizationally distant into one's current work is a challenge that may only be solvable through organizational change.

4.1 The Participation Challenge

Virtually colocated groups communicating in real time via audio and video conferencing face a definite challenge in making participation easy for members. The awkwardness of participation not only impedes the current discussion, but also results in long-term cumulative effects for the group. Several interaction phenomena observed with audio and video communication will be reviewed first.

Ruhleder and Jordan (1999) found that transmission delays in video conferencing negatively impacted group interaction by resulting in collisions, unnecessary rephrasing, and misapplied feedback. Even though video cues were available, these problems still occurred. The authors point out that when different actors hear different conversational segments due to delays, the actors' different expectations and interpretations inhibit the construction of a shared meaning.

In a study of an audio-only media space, Ackerman et al. found that interaction suffered from people not knowing who was present and that turn-taking was awkward. This finding was confirmed by Mark, Grudin, and Poltrock, who also found that actors had difficulties with identifying remote speakers. In the large distributed organization, actors often did not know to which organizations the speakers in their teams belonged. Difficulties in turn-taking were also observed by Bowers, Pycock, and O'Brien (1996) in audio communication in a collaborative virtual environment.

An example of the kind of awkwardness in turn-taking, and the problem of not knowing identities, or who is present, is shown here, from distributed teams in the current study. These kinds of exchanges are heard often during the team meetings:

Remote: I'd like to make a suggestion.
Leader: Is this Anita?
Remote: Yes.

M: Everyone clear so far? [One "yeah" is heard].

Dan: I have a quick question. Paul, are you still here?

Al: Who is this speaking?

Carol: I'm Carol, I work at ..., my area is data exchange.
[long pause]

Carol: Is everyone still there?
[a few say yes]
Carol: Because I didn't hear the background noises and didn't know if everyone is still there.

In face-to-face conversation, communication mechanisms exist that manage conversational flow through a rich and varied set of behavioral cues (Duncan 1972). These include intonation, gesture, gaze, and back-channel responses, which serve to turn conversation into a smooth dance among partners, without bumping into each other, so to speak.

It is questionable whether video images provide a real advantage for coordinating interaction. Isaacs and Tang (1993) found video advantageous for communicating nonverbal expression and reasons for pauses, yet found it disadvantageous for turn-taking and floor control. Similarly, Sellen (1995) found that turn-taking and other conversational aspects were not facilitated by adding video to audio. She concluded that there are no observable effects on turn-taking and synchronization when adding video to audio.

In the short term, these effects of the technology impact the group by limiting the richness of discussion. Isaacs et al. (1995) found that, in attending distributed video presentations, audience members were less likely to ask follow-up questions or clarifications, compared to face-to-face interactions. In the current study, it was discovered both from observations, and confirmed from interviews, that team members at remote sites find it hard to interject. Nonverbal cues serve for them as signals that a conversation turn has ended; without such cues, remote members prefer to err in favor of not speaking, and thus not interrupting.

However, few researchers have considered the long term impacts for a group in experiencing communication delays, awkwardness in turn-taking, not knowing who is present, who the speakers are, the meaning of pauses, and whether others are listening. In the long term, they create expectations for a certain type of interaction that can easily become a pattern for the group. It is not even clear whether distributed groups can develop norms to regulate conversation flow and establish positive interaction patterns. In longitudinal studies, Ackerman et al. found that group members developed norms to regulate their conversational exchange, but Mark, Grudin, and Poltrock and Ruhleder and Jordan did not, even among people who were experts in using the technology. One difference is that in the Ackerman et al. study, cooperation was continuous, among a well-defined set of team members. In the latter two studies, communication in the media space was discontinuous, with larger numbers of people. This may be more representative of the type of virtual meeting interaction that can be expected in large organizations.

Another type of consequence as a result of the awkwardness of participation is that it inhibits spontaneous behavior. Granted, in many face-to-face formal meetings, especially those run by strict meeting protocols or facilitation, spontaneity is not encouraged. In the teams studied at Boeing, however, during most parts of the meetings, free discussion was encouraged—for example, in posing questions to a presenter, in offering relevant information to the group, or in problem-solving discussions. Yet spontaneity was often not observed among the members and its absence was described in the interviews.

Many of the team members in the study expressed in interviews that it was difficult to interpret others' on-line conversations. They remarked that facial expression helps to establish a context for them, to understand in which light they should frame a comment. Understanding the context in turn helps them to formulate a response (Clark and Schaeffer 1987). As the users described:

> I get extra feedback of the body language of a person. Having met that person, I have that in the back of my mind [during a desktop conferencing session]. Without it, something is missing.
>
> Reflective looks means they are thinking. Silence on the line doesn't. People may say things sarcastically, but the expression on-line is confused. Many signals that you have in face-to-face are lost.

The "sterile" type of interaction that forms in the group subsequently affects the potential for further group development. It creates expectations of the kinds of interaction possible in the group; for example, that after a ten-minute monologue, no one in the group may be listening. Even worse, the limitations on the interaction and richness of discussion can lead to the formation of stereotypes of other group members, as opposed to a more multi-faceted view of them. Research has shown that frequency and length of interaction is negatively associated with the reduction of bias toward others (Wilder and Thompson 1980) and interaction encourages partners to disconfirm stereotypes of each other (Cook 1985). This idea was recognized by a team member, who stated that face-to-face interaction helps him form a better model of the other:

> It has value. When a person speaks, I attach a face and personality to him. When I hear a voice, I'm dying to associate the face with a voice. The only way to get a goodness of interaction is best to represent the person. You have to interact with this person. Your stereotype of the person doesn't work on the program, the person works on the program.

Additionally, "impolite" behaviors such as non-response, silence on the line, and interruptions are often attributed to the individual, and the role of the technology in contributing to this is downplayed. This phenomenon is known as the fundamental attribution error, the tendency for people to attribute the cause of behavior to an actor, rather than to situational factors (Ross 1977). This misattribution further impedes the development of the group as it can convey rudeness, or lack of involvement, as shown in this example:

Leader: Susan, anything else?
[long pause]
Leader: Susan left us awhile back.
Susan: No, I haven't. [Susan then explains that she put two URLs in the chat window. The leader replies that he did not have the desktop conferencing shared application running]

The team leader did not see what Susan was doing, i.e., that she was actively participating in the meeting by putting URLs into the chat window. He had instead assumed that she had left the meeting.

Another example occurred when one member of the Scientific Team began drawing on the shared whiteboard of the desktop conferencing system. It was very slow and took time for the remote members to receive the image. The remote members commented that the Scientific Team members, who were located in the conference room, were excluding them from the discussion, when in fact they were watching the member draw on the whiteboard.

Another long-term consequence of the interaction difficulties is on the formation of trust in the group. Trust involves forming an expectation of reciprocation. Trust is explained by Gambetta (in Hinde and Groebel 1991, p. 5) as

> a particular level of the subjective probability with which an agent assesses that another agent or group of agents will perform a particular action, both before he can monitor such action (or independently of his capacity ever to monitor it) and in a context in which it affects his own action.

Being able to assess or predict the actions of others involves learning about their behavior, preferably in as many varied situations as possible in order for the predictions to be robust. The lack of nonverbal behavior in computer-mediated communication, combined with inhibited participation, hinders people in developing robust predictions of others' behaviors. It is easy to understand how trust can be broken down in an audio-mediated conversation. If a delay occurs in the transmission and the speaker does not hear a response, she may form an assumption about the others' reaction to her remark. If the speaker misspeaks, the facial expression is not visible in order for the speaker to repair the mistake (Duncan 1972).

Many of the team members expressed their intuitions about this relationship between their computer-mediated communication and building of trust:

> You build relationships in face-to-face meeting. I notice that in the council time, the work in the group, and the social time, this all builds relationships. When you build trust with...top directors, then they are more likely to deal directly with you. It's easier to get through to them, and making contact with them is easier.
>
> Trust is not a [desktop conferencing] issue. You build trust by working with the individual.
>
> [Desktop conferencing] is possible only after trust is established. Problems are amplified by not meeting face-to-face. The real content of communication is not on the screen. If someone says something and then pauses, you can imagine anything.

> Everyone has a public and hidden agenda. You can't get hidden agendas from [desktop conferencing]. Here [face-to-face meeting] I can see everyone, look at the body language.
>
> Until roles and activities are well-shared, people will be suspicious. As experience with team members grows, the predictability grows. For example, are people disclosing agendas that are not apparent?

On the other hand, a few members do not expect that the distributed team should develop a high level of trust. One member differentiates between a high level of trust which is necessary for his physically colocated team because he depends on them, and the virtually colocated team, which is "several organizational boundaries away." Another describes:

> The level of trust that we have is appropriate for the group. It is not a close-knit team, as...a basketball team. Therefore, it is not so necessary.

Several members described that trust for them meant authenticating the knowledge of the others on the team, i.e., that they had competence in the context of their field. Keeping in mind that many of the members had not met each other and come from different organizations within the company, it is hard for them to judge the level of expertise and competence of others. Restricted verbal conversation is not sufficient for learning about others' knowledge.

4.2 The Culture Challenge

Groups, as Schein (1990) points out, can have their own cultures. Although different definitions of culture have been offered for groups and organizations (e.g., behavioral regularities—Goffman 1967) and the general atmosphere in which people interact (Tagiuri and Liwin 1968), Schein's definition of the shared assumptions and beliefs that are held by members of a group, that is a learned product of group experience is used here. Now when discussing distributed groups, we are faced with the problem of establishing what a definable group with a history means. There may be a roster with a list of members, but who actually attends scheduled meetings may be variable and even unknowable, especially when presence is not announced. Schein discusses the importance of having sufficient common experiences to have developed a shared view. This creates a challenge for the distributed group whose members themselves may not be able to define who is a member of their social unit. This is especially the case if attendance is not regular and if the group does not have a long history of meeting face-to-face.

On the other hand, assumptions of mission and operation can be formed implicitly, and patterns of behavior can set in quickly. It can be misleading to presuppose that teams, because they are geographically distributed and victim to some of the interaction difficulties described earlier, lack a culture. There is a danger of fooling ourselves if we believe the solution to effective teaming is simply that of presenting a set of recommendations that the group can follow. By following these recommendations, the group can use the technology more efficiently and conduct better meetings. Wrong. We first have to discover whether a set of assumptions exists for the members in order to change them. And change, once culture is set in place, does not come easily.

Culture in a group is affected by a number of factors: the group's history, its experience, its structure, the larger organization, the composition of its members, and the environment. However, additionally in a virtually colocated group, the nature and usage of the technology also plays a major role in shaping the group's culture. Deep-seated assumptions in a group can be manifest through its handling of artifacts, its products or creations, or overt behavior of the group members, as Schein describes. In a virtually colocated group, when, due to the technology use, patterns set in of interruptions or lack of engagement in the meeting, although these are not favorable cultural manifestations, they are observable traits by which the group operates. They often reflect deeper assumptions by which the group functions. To illustrate this, the case of the Best Practice team is presented. The members, who had been meeting for about nine months, developed the behavior of multi-tasking during the weekly desktop conferencing meetings. The meeting format was to review action items. A spreadsheet scrolled down on the shared display and the team member responsible for the action item would report on its status: done or due date extended. The meeting had very little interaction among the members, influenced by the action item reporting practice. The members soon began to do multi-tasking until their action item was reached or until they were called upon to comment about the action item. The following meeting excerpt was quite typical for their interaction (see Mark, Grudin, and Poltrock 1999):

Mark:	Dan, that's your action item.
Dan:	Sorry, I didn't catch that.
Mark:	Jack, what's your comment? [long pause] Jack, are you there?
Jack:	I had my mute button on.
Mark	Next is the rotocraft area, but he is not here.
Rob:	We need communication. 80 hours is not the problem. I waste 80 hours talking on the telephone. It requires a tremendous amount of extra effort to clearly communicate. When someone says we're having a telecon, then we need to be on the telecon....People need to be on the telecon. We need a schedule.
Joe:	We all sit here and what are we doing? Is everyone trying to be on these telecons?....We have to decide on issues that we feel are important. Things can be done but we need to talk to each other and use these telecons.
Mark:	Thank you, Joe. [Mark moved on to the next action item]

The multi-tasking resulted in a lack of engagement in the discussion and, thus, low commitment to the meeting. As is shown here, the members sometimes did express frustration, yet the group leader did not try to counteract the problem. One possible interpretation of the group's practice is that they lacked a culture. Instead it might be argued that the practice of multi-tasking during the meeting, and the corresponding lack of engagement, *was* the group culture. Contrary to the broader organizational culture, which promoted participatory decision making and where group meetings were expected to have high engagement and contributions of the members, the assumption of this group was that only minimal involvement was necessary during a virtual meeting.

Although multi-tasking occurred to some extent in other groups that were investigated, in no other group did the lack of engagement approach the level found in this group. This suggests that lack of engagement is not a behavior that is found in all virtual meetings. Multi-tasking was also reported by Isaacs et al. in a video broadcast setting; however, unlike a team setting, audiences would not necessarily be expected to be highly interactive with the speaker.

In fact, the members gained benefits by multi-tasking. Since their main task was reporting on the status of an action item, they needed only to be peripherally (or not at all) aware of others' reports in the meeting and could conduct other work in their offices in parallel. On some occasions, they even met with others in their offices during the virtual meeting. Yet overall, this group, who was expected to design a virtual colocation program within their division, did suffer by the lack of involvement. The meetings took more time because people had to refresh members about previous discussions. Information was unnecessarily repeated. Sometimes hostility was expressed by different members concerning lack of engagement and miscommunication. Since the action items were interrelated, directed toward achieving a common goal, learning about others' contributions could often have helped members with their own action items. Frequently the team leader asked people if they knew certain information; it was unclear whether the silence on the line was due to not knowing the answer or simply due to lack of involvement. For example:

Leader: Has everyone the information? [long pause] I assume silence means yes.

The team displayed very different types of overt behaviors when they met face-to-face, in contrast to their virtual meetings. The different overt behaviors reflect different expectations that the members had for their face-to-face and for their virtual meetings. For example, in one face-to-face meeting, where members flew in from all over the U.S., the group used a facilitator. The purpose of the meeting was different than the typical virtual meetings of reviewing items; its goal was to prepare a charter. In contrast to the virtual meetings, the discussion was quite free, with many interjections that the facilitator had to cut off. There were jokes, clarifications, debate, nonverbal gesturing and facial expressions in response to comments in short, a high degree of interaction and engagement by the members.

One way that the technology may have influenced the interaction style is through the provision of a shared reference in the desktop conferencing; i.e., all members had the same view of the information. In the face-to-face meetings, members often referred back to points discussed earlier. Here we can see an influence of the technology: the shared application defines the focus, or view, for the group. In a face-to-face group, people are focused not only on the information, but on each other. (Also, the members in the virtual meeting may not have paid attention to earlier items, which could also explain why they did not refer back to points.)

It might be argued that the multi-tasking, and corresponding lack of engagement, made the most sense for the group members, given the action item reporting format of the meeting. However, other models of interaction were possible in the group and the claim is that the group would have benefitted more had the members paid more attention at meetings. The argument is that the interaction style that the group adopted was shaped by different factors: the technology and the meeting format *interacted* with other factors, such as the history of the group and composition of the members.

If we examine the group's history, we find that, after the company merger, the team was formed officially by a vice president and assigned the goal of designing vehicles better, faster, and cheaper. The team was intentionally composed of representatives from all geographical areas of this company division, one person per site, spread across the entire United States. It is important to consider that the members stem from the different heritage companies and also different organizations within each of the former companies. The different members thus come from organizations that had different company cultures (e.g., participatory and tree-structured decision-making models) and different language codes (e.g., different acronyms). Since the group began, they had met about four times face-to-face. Given the members' differences, it is perhaps not surprising that this type of behavioral pattern set in. One team member even expressed that the meeting, and the technology use, was not being used to its full potential. A final product so far had not been created by the group.

Thus, developing a shared set of assumptions to promote positive group behaviors is a challenge for a distributed team. Culture forms from a number of influences and it is important to consider that the technology, and how it is used, comprise a key factor in shaping a distributed group's culture.

4.3 The Integration Challenge

The third challenge of virtually colocated teams is to discover ways that they can integrate remote team colleagues into their current working spheres. Work is continuous and complex. People move fluidly from one task to the next. The problem with virtually colocated teams is that the end of meetings often signal "out of sight, out of mind." Remote people and tasks from remote teams are thus not part of the continuity of each other's working spheres. Rather, as the team meetings in Boeing were, they fulfilled specific functions at specific points in time. In the words of the manager of the Staff, "it compromises the ability to get involved in things at the right time."

What does integration mean? It refers to more continual communication than just weekly or bimonthly formal meetings, with occasional telephone or e-mail exchanges in between. It refers to learning more about the other members on the team: their organizational homes, their expertise, their backgrounds, which currently is not the case in the teams studied. It refers to the exchange of results with selected team members to get feedback when the results are ready, instead of waiting for the formal meeting time. It refers to building a shared repository of common team materials, e.g., a shared workspace, accessible to all from remote locations. Again, currently none of the teams have such a shared space for team materials. Web sites exist, but it is not easy for team members to upload information. Integration refers to engaging in informal as well as formal communication, which can even lead to building relationships outside of the formal meeting structure. In all the interviews conducted, only one member from the Scientific Team reported contact with another remote team member outside of the formal meeting time, based on finding common interests. The Technical Working group used chat during the meetings, where informal communication took place. A member reported that it helped the team bond. It may in fact have served more a purpose of "breaking the ice" during the meeting than of bonding.

The lack of integration of the remote team work into each others' current working spheres became manifest in different ways in the organization. First, one consequence for the lack of integration is found with adoption of the technology. Remote team members had problems in adopting the desktop conferencing technology at Boeing (Mark, Grudin, and Poltrock 1999). These problems included downloading the software, securing the appropriate computers to run the software, and learning how to use the technology. One reason cited was that the team was viewed as a part-time activity and, therefore, assigned less importance than other tasks with people with whom one is colocated. This was also true of the members' local site managers. They viewed the team as part-time and were reluctant to fund appropriate equipment needed for the software.

Second, another consequence we might expect of the lack of integration is that members may not give high priority to the assigned tasks for the virtually colocated group. Signs of this were from the large numbers of action items in the Best Practice team that were delayed. However, a comparison of task completion in the members' face-to-face group would need to be done to determine this.

5. Discussion and Conclusions

This paper described three challenges for teams with long-term agendas who are virtually colocated. Much previous research into the effects of computer-mediated communication in groups has focused on identifying the immediate communication difficulties that people experience. It is hoped that the challenges outlined in this paper can spark more longitudinal research into the effects of technology and distance on team building. Dissecting the process of team development identifies a number of social processes in the group, such as the development of participation patterns, culture, norms, refinement of roles and procedures, cohesion, identity, motivation to perform, trust, and leadership.

In considering team participation, not only does interaction via computer-mediated communication present immediate interaction difficulties, but it also has consequences on the long-term development of group processes. Not knowing who is present, not knowing identities or organizations, misattributions of behavior, inhibited participation from remote members, and the lack of nonverbal behavior to help interpret on-line speech, all affect the group development. I have argued that they contribute to expectations and patterns of certain participation behaviors, as well as impeding the building of trust.

Virtually colocated teams can have their own unique cultures, shaped by the use of the technology. The assumptions shared by the group members may be different for a group's virtual meetings with technology than a group's face-to-face meetings. In the case of the Best Practice team, the lack of engagement of the members was much higher than in the other teams observed. It seems reasonable to argue that the structured meeting format and shared reference of the desktop conferencing, combined with the group members' origins in different organizations, all contributed to the members' multi-tasking behaviors. Although it brought the members benefits, the price in the long-term was a set pattern where low commitment and lack of engagement in the meeting was accepted. Certainly multi-tasking brings advantages: it enables people to attend and monitor more meetings while working in parallel. Yet practitioners need to reflect on whether multi-tasking should be promoted for peripheral meeting participation rather than for people who are core group members.

All of the groups in the current study met at specific points in time for formally structured meetings. All of the groups had long-term agendas. Two groups had major tasks to accomplish for the company. Again we see a challenge that is really a paradox. Virtually colocated team members are expected to produce high quality deliverables on schedule. Yet group members limit their main involvement to the formal meetings. Many teams would in fact benefit from more continuous communication and involvement with each other and with the task. This is not to suggest more formal involvement, but rather more regular informal communication, which has been shown to provide value for teams (Whittaker et al. 1997).

The integration challenge applies at two levels. First, it addresses how people can seamlessly integrate their remote team activities into their current activities. Secondly, more broadly, it is a challenge to make the remote sites of the newly merged company more of a presence in the company. Ultimately, the goal would work toward creating one company culture.

The diversity of team members in virtually colocated teams increases the difficulty in integration. Team members stem from different organizational units. The organizational rewards are given at the individual level: members of cross-unit teams do not get salary reviews; members of local teams get salary reviews. Further, team members give distributed groups less priority than groups at their local sites. This was reflected in the interviews and also in people's reflections on the level of trust they expect among virtually colocated partners. These views are held not only by the team members, but also by their managers, many of whom were reluctant to provide funding for the equipment necessary to run the desktop conferencing software.

How can the improvement of technology make a difference?

There are a number of ways that technology and coaching efforts can support virtually colocated meetings and research is already underway in many areas. One method is to display nondisruptive signals on-line showing people's status in the meeting. Other methods involve setting new roles. The problems of not knowing who is present, who is speaking, turn-taking, and interpreting silence on the line can be alleviated through appropriate forms of facilitation, as Mark, Grudin, and Poltrock discovered. It is worth considering a role of "producer" for the team who can insure that members receive appropriate visual and audio information about others during a meeting. Success has already been reported with the use of producers in interactive television (Benford et al. 1999).

The challenges of developing a positive team culture, and integration, are somewhat related. Poltrock and Engelbeck (1997) articulate the requirement of supporting opportunistic interactions through a combination of e-mail, telephone, and awareness mechanisms. Synchronous and asynchronous awareness mechanisms can help inform team members of each others' current and past activities, can spark discussions, and provide them with a context for their own work (e.g., Fitzpatrick et al. 1999). Chat has been shown to have value in supporting informal communication in distributed teams (Bradner, Kellogg, and Erickson 1999; Fitzpatrick et al. 1999). Shared repositories of information for the group would be beneficial, such as a shared workspace, where all can enter and download material. In this way, members need not wait for the scheduled meeting to view the results. They can have the results at-hand when they are ready, and could even discuss them prior to the meeting. In fact, meetings consisting of action item reports may not even be necessary, if the action item status could be made available to all in a shared workspace. In this case, the meeting time could be used for other purposes, such as creating a final product, as one member proposed.

Research is called for to understand new roles necessary in such teams: for example, appropriate forms of leadership and facilitation. Although it is costly to create new roles, if communication and technology use becomes more efficient, then it becomes an overall gain. Research is also called for to understand the role of face-to-face meeting in distributed teams. Studies of group development suggest that the early phases of group formation is the critical time for interaction: at this time, information exchange, the accommodation of the group to the individual (and vice versa), and settling of roles takes place. However, in some cases it is impractical for geographically distributed teams to meet face-to-face. Research is needed to better define face-to-face schedules, such as when "booster shots" may be needed. The meeting itself may be restructured so that certain tasks can be better accomplished face-to-face, while other tasks, e.g., those having fewer dependencies among the members, can sufficiently be accomplished from a distance.

A number of other challenges for virtually colocated teams, such as developing leadership and facilitation, collective goals, coordination, and technology adoption and use, were not addressed in this paper. There is a long road ahead for such research. Technology is shrinking the world, but the understanding of how teams can adapt to the new technologies is not yet keeping pace.

References

Ackerman, M. S., Hindus, D., Mainwaring, S. D., and Starr, B. "Hanging on the 'Wire: A Field Study of an Audio-only Media Space," *ACM Transactions on Computer-Human Interaction* (4:1), 1997, pp. 39-66.

Benford, S., Greenhalgh, C., Craven, M., Walker, G., Regan, T., Morphett, J., Wyver, J., and Bowers, J. "Broadcasting On-line Social Interaction as Inhabited Television," in *Proceedings of the Sixth European Conference on Computer-Supported Cooperative Work*, S. Bødker, M. Kyng, and K. Schmidt (eds), Copenhagen, September 12-16, 1999, pp. 179-198.

Bowers, J., Pycock, J., and O'Brien, J. "Talk and Embodiment in Collaborative Virtual Environments." in *Proceedings of CHI'6*. New York: ACM Press, 1996.

Bradner, E., Kellogg, W. A., and Erickson, T. "The Adoption and Use of 'BABBLE': A Field Study of Chat in the Workplace," in *Proceedings of the Sixth European Conference on Computer-Supported Cooperative Work*, S. Bødker, M. Kyng, and K. Schmidt (eds), Copenhagen, September 12-16, 1999, pp. 139-158.

Clark, H. H., and Schaefer, E. F. "Collaborating on Contributions to Conversations," *Language and Cognitive Processes* (2:1), 1987, pp. 19-41.

Cook, W. W. "Experimenting on Social Issues: The Case of School Desegregation," *American Psychologist* (40), 1985, pp. 542-60.

Duncan, S. "Some Signals and Rules for Taking Speaking Turns in Conversation," *Journal of Personality and Social Psychology* (23:2), 1972, 283-292.

Fitzpatrick, G., Mansfield, T., Kaplan, S., Arnold, D., Phelps, T., and Segall, B. "Augmenting the Workaday World with Elvin," in *Proceedings of the Sixth European Conference on Computer-Supported Cooperative Work*, S. Bødker, M. Kyng, and K. Schmidt (eds), Copenhagen September 12-16, 1999, pp. 431-450.

Gallupe, R. B., and McKeen, J. D. "Enhancing Computer-mediated Communication: An Experimental Investigation into the Use of a Group Decision Support System for Face-to-face versus Remote Meetings," *Information and Management* (18), 1990, pp. pp. 1-13.

Goffman, E. *Interaction Ritual*. Hawthorne, NY: Aldine, 1967.

Hackman, J. R. "Group Influences on Individuals," in *Handbook of Industrial and Organizational Psychology*, M. D. Dunnette (ed.). Chicago: Rand McNally., 1976.

Herring, S. "Interactional Coherence in CMC," *Journal of Computer-Mediated Communication* (4:4), June, 1999 (http://www.ascusc.org/jcmc/vol4/issue4/).

Hiltz, S. R., Johnson, K., and Turoff, M. "Experiments in Group Decision-making: Communication Process and Outcome in Face-to-face versus Computerized Conferences," *Human Communication Research* (13:2), 1986, pp. 225-252.

Hinde, R. A., and Groebel, J. *Cooperation and Prosocial Behavior*. Cambridge, UK: Cambridge University Press, 1991.

Isaacs, E. A., Morris, T., Rodriguez, T. K,. and Tang, J. C. "A Comparison of Face-to-face and Distributed Presentations," in *Proceedings CHI'95 (Denver)*. New York: ACM Press, 1995, pp. 354-361.

Isaacs, E. A., and Tang, J. C. "What Video Can and Can't Do for Collaboration: A Case Study," *Proceedings of the Multimedia, 1993*. New York: ACM Press, 1993, pp. 199-205.

Lewin, K. *Resolving Social Conflicts: Selected Papers on Group Dynamics*. New York: Harper, 1948.

Mark, G., Grudin, J., and Poltrock, S. "Meeting at the Desktop: An Empirical Study of Virtually Colocated Teams," in *Proceedings of the Sixth European Conference on Computer-Supported Cooperative Work*, S. Bødker, M. Kyng, and K. Schmidt (eds), Copenhagen, September 12-16, 1999, pp. 159-178.

Moreland, R. L., and Levine, J. M. "Newcomers and Oldtimers in Small Groups," in *Psychology of Group Influence,* P. Paulus (ed.). Hillsdale, NJ: Lawrence Erlbaum, 1989, pp. 143-186.

Newcomb, T.M. "Interpersonal Balance," in *Theory of Cognitive Consistency: A Sourcebook,* R. P. Abelson (ed.). Chicago: McNally, 1968, pp. 10-51.

O'Conaill, B., Whittaker, S., and Wilbur, S. "Conversations Over Video Conferences: An Evaluation of the Spoken Aspects of Video-mediated Communication," *Human-Computer Interaction* (8), 1993, pp. 389-428.

Poltrock, S. E., and Engelbeck, G. "Requirements for a Virtual Colocation Environment," in *Proceedings of ACM Group '97* (Phoenix, AZ, November 16-19). New York: ACM Press, 1997.

Rabbie, J. M., and Horwitz, M. "Categories versus Groups as Explanatory Concepts in Intergroup Relations," *European Journal of Social Psychology* (18), 1988, pp. 117-23.

Rafaeli, S., and Sudweeks, F. "Networked Interactivity," *Journal of Computer-Mediated Communication* (2:4), March 1997 (http://www.ascusc.org/jcmc/ vol2/issue4/index.html).

Ross, L. "The Intuitive Psychologist and His Shortcomings: Distortions in the Attribution Process," in *Advances in Experimental Social Psychology*, Volume 10, L. Berkowietz (ed.). New York: Academic Press, 1977, pp. 173-219.

Ruhleder, K., and Jordan, B. "Meaning-making Across Remote Sites: How Delays in Transmission Affect Interaction," in *Proceedings of the Sixth European Conference on Computer-Supported Cooperative Work*, S. Bødker, M. Kyng, and K. Schmidt (eds), Copenhagen, September 12-16, 1999, pp. 411-429.

Schein, E. H. *Organizational Culture and Leadership*. San Francisco: Jossey-Bass, 1990.

Schmidt, K., and Bannon, L. "Taking CSCW Seriously: Supporting Articulation Work," *Computer Supported Cooperative Work (CSCW), An International Journal* (1:1-2), 1992, pp. 7-40.

Sellen, A. J. "Remote Conversations: The Effects of Mediating Talk with Technology," *Human-Computer Interaction* (10), 1995, pp. 401-444.

Sproull, L., and Kiesler, S. *Connections: New Ways of Working in the Networked Organization.* Cambridge, MA: MIT Press, 1991.

Tagiuri, R., and Litwin, G. H. (eds.). *Organizational Climate: Exploration of a Concept.* Boston: Division of Research, Harvard Graduate School of Business, 1968.

Tajfel, J., and Turner, J. C. "The Social Identity Theory of Intergroup Behavior," in *Psychology of Intergroup Relations,* S. Worchel and W. Austin (eds.). Chicago: Nelson-Hall, 1986, pp. 7-24.

Tuckman, B. W., and Jensen, M. A. C. "Stages of Small Ggroup Development Revisited," *Group and Organizational Studies* (2), 1977, pp. 419-427.

Warkentin, M. E., Sayeed, L., and Hightower, R. "Virtual Teams versus Face-to-face Teams: An Exploratory Study of a Web-based Conference System," *Decision Sciences* (28:4), Fall 1997, pp. 975-996.

Weick, K. E. *The Social Psychology of Organizing.* Reading, MA: Addison-Wesley, 1969.

Whittaker, S., Swanson, J., Kucan, J., and Sidner, C. "TeleNotes: Managing Lightweight Interactions in the Desktop," *ACM Transactions on Computer-Human Interaction* (4:2), 1997, pp. 137-168.

Wilder, D. A., and Thompson, J. E. "Intergroup Contact with Independent Manipulations of In-group and Out-group Interaction," *Journal of Personality and Social Psychology* (38), 1980, pp. 589-603.

Zack, M. H. "Interactivity and Communication Mode Choice in Ongoing Management Groups," *Information Systems Research* (4:3), 1993, pp. 207-239.

About the Author

Gloria Mark is an assistant professor at the University of California, Irvine, in the Computing, Organizations, Policy, and Society (CORPS) group, in Information and Computer Science. She was previously at the German National Research Center for Information Technology (GMD) in Bonn, Germany. She has been active in the field of CSCW, having investigated a variety of different types of technologies in use: electronic meeting rooms, collaborative hypermedia, shared work spaces, and collaborative virtual environments. Recently she was a visiting researcher at The Boeing Company, where she conducted a study of virtually colocated teams. Her research interests include virtual colocation technologies, and the cognitive and social aspects of technology use. Gloria can be reached by e-mail at gmark@ics.uci.edu.

24 MOA-S: A SCENARIO MODEL FOR INTEGRATING WORK ORGANIZATION ASPECTS INTO THE DESIGN PROCESS OF CSCW SYSTEMS

Kerstin Grundén
University of Borås
Sweden

Abstract

Scenarios using the MOA-S model are proposed as a method for discussion and design of the future work organization of CSCW systems. CSCW technology has the potential to enable design of the organization with more flexible communication patterns than previously available. However, we still lack methods for the design process of CSCW systems. A case study is presented of the development and test of a prototype in Lotus Notes for dealing with social insurance matters within the Swedish Social Insurance Board. The results of the case study are structured in accordance with the MOA-S model. The MOA-S model could then be used as a design tool in order to initiate and support a dialogue about the remaining design of the work organization in the CSCW system.

Keywords: CSCW, Lotus Notes, social insurance board, work organisation, work situation, efficiency, quality, systems development work, scenarios

1. Introduction

The research field of computer supported cooperative work (CSCW) is a rather mature and broad research field with a history of more than 10 years. The focus of this research field on cooperative work among the people is taken as the departure point for an analysis of the design of CSCW technology support. The field of CSCW includes technology such as video communication, mainly supporting informal communication, as well as complex administrative systems for more formal communication, allowing communication regardless of time or geographical location. There is a lack of detailed empirical studies of consequences for the work processes when CSCW technology is used (Bannon and Schmidt 1995, p. 13). There is also a lack of design methods for CSCW systems. The traditional life cycle model for system development work is not as relevant for the development of CSCW systems. Traditional system development work has a more rigid character; most of the detailed specifications of the system are done in the early stages of the process. Supporting cooperation and coordination with information technology shows characteristics that differ from traditional system information systems (Grudin 1994; Kyng 1991).

The design of CSCW systems could lead to more flexible communication patterns within the organization. New organizational ideals, often characterized as a paradigmatic change of organizational thinking, have emerged. These new organizational forms are characterized in the literature as boundless (Ashkenas 1997), flexible (NUTEK 1996), process-oriented, customer-oriented, and learning-orientated (Argyris and Schön 1996). Most of these taxonomies are, however, to some extent overlapping and no clear definitions exist (Andersen and Chatfield 1996). Many public sector organizations by tradition have a rather large, complex and bureaucratic structure. However, even in the public sector there are many efforts to restructure the organization and technology in order to achieve quality and efficiency gains. The central level of the Swedish social insurance board (SIB) has formulated a very large project, "Försäkringskassan 2005" (SIB 2005), including a vision of the future work. The development and test of the ELAKT prototype was part of this project. SIB 2005 also included many sub-projects within the following fields: analysis of the work activities, competence, technology, planning and control (Försäkringskassan 1998). The vision that was formulated for the project included a customer orientation, e.g., more flexible ways of interacting with the customer.[1] The customer could execute some current work tasks, such as information through phone systems or Internet communication, with new functions for customer service. The use of the new technology also was expected to lead to a reduction of the time needed for dealing with social insurance matters. The technology was seen as the main work instrument. According to the vision, the organization should be less authoritarian than today, more process-orientated, and more of a learning organization (Jälmestål, Sköld, and Wahlqvist 1998).

The use of CSCW technology has the potential to facilitate the design of such an organization as it allows for a flexible structure of communications in the organization. The change of an organization from bureaucratic and hierarchical communication patterns toward increased flexibility means that the organizational culture needs to be changed as well. There is a need for design methods to facilitate such a process by facilitating

[1] Notice the fact that the concept "customer," rather than "client," is used in the vision of SIB 2005.

discussions and dialogues of different organization alternatives within the CSCW system. In this article, scenarios using the MOA-S[2] model see (Figure 1) are proposed as such a design method that especially focus on ways the work situation and efficiency and quality aspects could be affected by different design alternatives for the organization and the technology. MOA-S is used as a frame of analysis in a case study where the ELAKT prototype for electronically dealing with social pension matters was tested at a local office within the Social Insurance Board of Sweden. The model is also introduced to systems developers within the organization.

In the next section, the case study of the ELAKT prototype is briefly presented. The results of the study are structured according to the MOA-S model. This scenario is expected to be a useful tool in further development work to initiate and structure a dialogue among the personnel, especially focusing the design of the future work organization in the CSCW system.

2. A Case Study of the Test of the ELAKT Prototype

The methodological approach of the empirical work was a qualitative case study (Merriam 1994). Several personal visits to the organization at different stages of the test work were made. Qualitative interviews were conducted with the main actors (system developer, project leader, the staff at the local office that was testing the prototype). The interviews with the test personnel were conducted as group interviews with two or three informants. The interview situation then became more socially dynamic when compared to traditional interviews with one informant. According to Kvale (1997), the interview then also becomes a learning situation for the participants , allowing different aspects of the issues to be analyzed and discussed. The interviews took about an hour, on average. Most of the interviews were tape-recorded and transcribed verbatim. The researcher also attended one meeting where the test personnel were educated in the ELAKT prototype by the system development personnel. Some written documents have been studied, such as manuals and descriptions of the prototype.

One aim of testing a prototype is to receive the best knowledge possible about the consequences that will arise when the system is introduced for ordinary work. The informants, therefore, were asked what they expected as consequences of using ELAKT in terms of their work situation and quality and efficiency aspects.

The fact that the prototype was tested of the staff of just *one* local office means that the potential of the technology to change the work organization among different offices was not realized during the test. The test work was done in parallel with the ordinary work, which means that the potential of changing the work organization within the office was not realized. Rather, the informants were asked about expected changes to the work organization, their work situations, and efficiency and quality aspects.

---

[2]MOA is an acronym derived from the title of the author's dissertation: *Människa, Organisation, ADB-system* (Human Being, Organization, EDP-system) (Grundén 1992).

3. The Development of the ELAKT Prototype

The ELAKT prototype system consists of a relation database in Sybase with a graphical user interface in Visual Basic. The relation database is used for keeping track of current matters. Statistics can also be produced from this database. The relation database is connected to Lotus Notes, which is used as a document database. The documents are stored in Lotus Notes. Information from the relation database can be mirrored and presented in Lotus Notes. A root document is created for every person with a current or completed pension matter. All information needed for handling pension matters can be stored electronically in the ELAKT prototype.

The development of the ELAKT prototype was part of a Lotus Notes project initiated by the National Social Insurance Board (in Swedish *Riksförsäkringsverket*, RFV) in the spring of 1996. Five different regional social insurance offices took part in the project. The ELAKT system was the only system developed for dealing with pension matters. The most common applications of Lotus Notes were conference systems and discussion databases. Another common application was a reference database that stored formalized information such as manuals or law sections. Lotus Notes was also used as an intranet system in order to distribute internal information. One office also tested using Lotus Notes to facilitate distance work. In August 1997, the final report was published with some evaluation of the different applications. Some of the experiences with Lotus Notes were very positive. Lotus Notes was seen as *one* solution to attain simplicity, security, accessibility, and a comprehensive view within the field of social insurance work. The report also references other positive experiences with Lotus Notes in Swedish public administration, such as the administration office of the Swedish Parliament and the national tax department. Some general negative experiences were, however, distinguished. The project was just a test and those involved felt the future use of the system to be very insecure, a fact that reduced their enthusiasm. The project also had low priority from the central level and from the central EDP department at Sundsvall. According to the final report from RFT, the following consequences were epxected from the ELAKT system:

- A complete and always available electronic record for each social pension matter in which the documents could be viewed in chronological order.
- Work methods would be more flexible, and not limited by geographical or organizational boundaries.
- Other personnel would be able to attain information about the status of the matter and the responsible person for each social insurance matter more easily than was currently possible.
- Security was expected to increase, as different people could receive different authorities.

The evaluation was, however, not based on real tests of the system.

A regional system development department of Bohuslän (RSDD) developed the ELAKT prototype. RSDD had at that time about six systems developers to give support and education regarding IT issues to about 15 local offices. RSDD also worked with systems development of information systems. RSDD employees worked very closely with the users. They characterized their style of systems development work as an organic style.

The systems development work was seen as very important for the employees of RSDD. They wanted to improve their special competence in the field while developing efficient systems for the organization. The development of the ELAKT prototype was an important project for them. The system development work was part of the central project SIB 2005, but the future of the ELAKT prototype was very insecure and there were simultaneous efforts from the central level to develop other similar systems. In fact, the test work of the ELAKT prototype was interrupted in the middle of 1999. Decisions were made by the central level of management to develop another similar system using different software. This group had shown very little interest in the development work of the ELAKT prototype. But the manager of the RSDD group had been very positive about the development and struggled to have the prototype accepted. Even if the ELAKT prototype would not be used in ordinary work, the development and test work was seen as valuable. The development and test of the prototype increased knowledge about the technology and about new ways of organizing work when dealing electronically with social insurance matters. Resumed testing of the ELAKT prototype, especially focusing on work organization aspects, is planned for the beginning of 2000 with another local office being added to the project. A real test of different distributions of work tasks among two offices would then be possible.

4. The Planned Future Work Process Compared with the Work Process Today

Today, work processes for dealing with pension matters are initiated by an application for social pension from the client. Most of the local offices have separated further dealings into two different departments. The investigation department does the necessary investigative work. Contacts can be made with consultants or medical doctors. When the investigative work is finished and a final decision is proposed, the matter is transmitted to the person that will report on the case to the social insurance committee. A date for the meeting is booked. The personnel at the pension department are informed about the matter and start the accounting work. This work is completed when the final decision in the social insurance committee is made. If the proposal from the investigation department is allowed, information about the conclusions in the matter is submitted to the central information system in Sundsvall and the pension is paid to the client. If the claim is denied, the client cam make a new proposal, after a specified time; a new matter is initiated and a similar work process takes place.

Today all documents needed for handling pensions are paper documents. The documents are physically stored in an archive. When external actors such as consultants or medical doctors need information about the document, a copy must be made and sent through the post. The person currently responsible for the matter stores the documents. When the work is transmitted from the investigation department to the pension department, the documents are physically transferred to the new person responsible.

The overall functions of the current work process are similar to the work functions of electronically handling pension matters. When the local office receives an application for social pension, certain information about the client is registered in the relation database and a root document, as well as a mirror document, is created. It is also possible to create specific documents that belong to the pension matter. Documents can be scanned into the electronic document, transmitted as a Word, Excel, or Jetform document, or directly written into the database. All information that had been stored in a paper journal can now be stored electronically.[3]

All further dealings are then registered in the ELAKT system. The investigation department initially deals with the matter. When they have finished, the matter is transmitted to the person that will report on the case to the social insurance committee. That person transmits the authority to the pension department for further dealings with the matter in ELAKT after the date for the meeting is noted. The investigation and the pension departments can then deal with the matter in parallel until the final decision is made. The person responsible for the matter can delegate authorization to read the whole electronic journal or part of it to consultants/advisors, deputies, or the insurance doctor. Even the switchboard operator could be authorised to read some elementary information of the journal.

The main changes have more to do with work organization and methods than functions. The information about the matter is registered, stored, retrieved, and transmitted electronically instead of by paper documents. This implies a potential for future organizational changes that could lead to consequences for the individual regarding work content and quality and efficiency aspects.

4.1 Initiation of the Test Work

The ELAKT prototype was demonstrated for several participants from different local offices in the region of Bohuslän during the spring of 1997. Several offices expressed interest in participating in the test. The main advantage of the ELAKT system was interpreted by the participants as being a way to keep the pension documents in better order than was currently possible. The system also seemed to reduce the need of sending documents by post to the different actors handling the pension claim. It would also be easier for actors at different geographical locations to look at the same document in ELAKT.

The local office in Kungälv was chosen as a test office by the project management for several reasons. The office was situated very near the system development office. The cooperation among the staff was positive and they, as well as their manager, were interested in technical issues. One of the staff members was a system administrator of the pension system. The test work of ELAKT started in the spring of 1998. About four people at the local office in Kungälv took part in the test work in the beginning. They were from the department of investigation work and the department of pension handling. They attended two days of education in the ELAKT system.

4.2 Experiences of the Test Work

The test of the prototype was done in parallel with the ordinary work. Some of the pension matters were administered manually as well as being registered in the ELAKT

[3]According to current law, decisions still need to be stored as a paper document.

system. Some of the documents were scanned into the ELAKT system. The test personnel tried to do most of the test work during times when the ordinary workload was lighter. According to the interviews, the test personnel were very enthusiastic and positive about the ELAKT system. The test work was causing them more work than just handling the pension matters in the ordinary way, but they thought it was very interesting to be able to take an active part of the development of the system. They were aware of the fact that another system might be chosen for the future work, but they did not see this as a big problem. Instead they thought they learned a lot from the test work and it allowed them to try a quite different way of doing their work, a way that they thought would be a model for their future work.

There have been some technical problems, such as too a long response time. There also have been some problems with the scanner, but most of the problems seem to have been solved by the system developers.

The work organization of the local office has not been changed due to the test work, but the individual work situation for the test personnel has changed. They have to do more work than before. They have increased their knowledge about how to handle pension matters electronically.

The test personnel receive system support primarily from one of the system developers, who seems to be very enthusiastic and engaged in the work. He is continually in contact with the test personnel by phone. They can ask questions of more general technical character using a discussion database in Lotus Notes.

In order to follow the test work and discuss any changes of the system, meetings are held with the system developers, the project leader and the test personnel. These meetings have not been as frequent as is usual in development work primarily because of the frequently used discussion database and the ELAKT database. Using the ELAKT database, the system developer and the test person can look at the same document at the same time. According to one of the system developers, the uncertain future of the ELAKT system due to the decision being made in the central part of the organization has affected the speed of the development of new functions and changes to the system.

5. Expected Consequences of Using ELAKT in Ordinary Work

5.1 Consequences for the Organization of Work

All of the informants think that the distribution of pension matters could be more flexible among the different organizational units as well as among the social insurance staff. You don't have to move the physical documents; you just change the authority. The workload for one office with a shortage of personnel capacity at a given time (e.g., due to vacancies or illness) could then be reduced and some pension matters be moved to another office. The system could produce current statistics of the workload of each office (and each of the pension handlers) which could be used for decisions about changing the distribution of the matters.

The use of consultants, advisors, and medical doctors could also be more flexible, according to the informants, as they could be given the authority to read special insurance matters in the ELAKT system and discuss the matter by phone with the person responsible. One of the system developers expects the work to be more specialized than today, both on the office level and on the individual level. The competence of the staff will then be used in a more effective way.

5.2 Consequences for Individual Work Content

The test personnel are somewhat worried about the fact that some personal contacts could be reduced or replaced with electronic communication. The social interaction among specialists and the handlers of the insurance matter could be reduced, as most of the information about the client is stored in the electronic system. Meetings could be reduced and important information of a more informal character could be lost, which could affect the quality of the decisions that are made.

One of the system developers expects future work to be more specialized for personnel dealing with complicated social insurance matters. If the work becomes more specialized, the formal work group could be spread around several offices, which could affect the personal contacts among the group members. The personnel dealing with less complicated insurance matters that can be completed during a client visit could be generalists.

One of the system developers expects the work with electronically handling social matters to be more formalized and controlled. There could be controls in the programs of the contents of the documents indicating that specific documents are needed. It will also be easier for managers to receive information about the number of matters being handled by each employee.

All of the informants think that more technical knowledge will be needed in the future work.

5.3 Expected Consequences for Quality and Efficiency Aspects

Changes to the organization of work and individual work content also can affect quality and efficiency aspects. Will there be any change to the quality aspects, such as the contents of the decisions that are made or integrity and security aspects, due to the use of this CSCW technology? If consultants and experts could be contacted more easily, the quality of the decisions that are made in the handling of social insurance matters could be increased. But if important informal information is lost due to less personal meetings, then the quality of the decisions could deteriorate.

Most of the informants expect that electronically dealing with pension matters will be safer than today. One of the test personnel compares the security aspects of electronically dealing with pension matters to electronically dealing with patient journals at hospitals. The authorization could be much more restrictive with pension handling, as fewer personnel need to have information about the contents of the matter, as compared with the journal of a patient.

Will the efficiency aspects, such as the time needed for each pension matter or the division of work among the workers, change when the pension matters are handled electronically? The test personnel expect the time for handling a pension matter to be reduced. Currently, much time is spent searching for documents that have "disappeared." There will be much less paper in the archives. They also think the system will reduce the need to make copies of documents and send them to external actors such as experts when handling a pension matter. Using the ELAKT system, the actors could instead look at the same document at the same time. When a client makes a demand for a copy, the information could be printed from the system rather than having to make a copy of each document. But the test personnel think that the paperless office is a myth. It is easier to read a paper copy than to read on the screen. There could also be a feeling of security in having the information on a paper document. It is easier to bring a paper copy if you are going to discuss with a colleague.

6. New Design Methods for Developing CSCW Systems Needed

The traditional life cycle model for system development work is not as relevant for the development of CSCW systems. Traditional system development work has a more rigid character; most of the detailed specifications of the system are done in the early stages of the process. The focus is mainly on the technical aspects of the system. Supporting cooperation and coordination with information technology shows characteristics that differ from traditional information systems (Grudin 1994; Kyng 1991). There is thus a need for a revision of traditional system development strategies; a new tradition of methods for CSCW is emerging. Still, this tradition is very immature. Very often this tradition uses disciplines such as sociology and ethnography as a means for understanding the social interaction in the work.

A pilot implementation could be a successful method for video communication, as shown by Grundén (1997). For more complex CSCW systems with more programming work, prototyping as a system development method could be a relevant approach for the development process. The specification and test of a prototype is a kind of experimental situation. It is easy to make changes to a prototype. The test of a prototype in real settings could generate experiences with the potential for future alternative ways of using the technology, for example, different ways of organizing work using CSCW technology. The test of a prototype could be seen as a learning process to generate knowledge about future design of the work organization. It is important to integrate such experience from the test work into the system development process. Such work could be systemized in different ways. The experience from the test work about future design alternatives for the work organization could be communicated during special meetings between the system development personnel, the test personnel, and other important actors.

Scenario is a commonly used method for future planning situations and has its background in military war games. Scenarios could be described as future histories, descriptive narratives of possible alternatives that need to be plausible, credible, and relevant (Fahey and Randall 1998). Within a modern, flexible organization, there is a need to identify, discuss, and evaluate many different strategies and their consequences. Scenarios could be used as a design method in order to identify, discuss, and evaluate future possible organizational and technological alternatives (both formal and informal in character) and their consequences for the individual work situation and efficiency and quality aspects. Scenarios contrast with traditional functional analysis and specifications. The use of scenarios could raise more questions in the discussions than they settle, but they could also be used as a method for supporting new thinking about design. Systems design could be seen as designing scenarios of interaction (Carroll 1995). The use of scenarios could facilitate strategic conversations that could contribute to continuous organizational learning (Schwartz 1998). Special qualities of the method are the contribution to a higher level of group understanding and qualitative causal thinking (van der Heijden 1996). Bardram (1996) uses scenarios as a technique within the strategy of organizational prototyping to enable discussions of different ways of organizing the organization and the technology. Another similar method is organizational games (Ehn and Sjögren 1991), but this method only focuses on organizational issues.

7. The MOA-S Scenario Model

The design of the technology and work organization can affect the work contents and efficiency and quality aspects. Grundén (1992) introduces a human-oriented perspective of an organization that emphasizes important relationships among ideas about coordination and control, the design of the EDP system, and the organization and possibilities for individual development in the work situation. In Figure 1, quality and efficiency aspects are added to the perspective.

The relationships between the boxes of the figure are very complex as changes of work often generate political processes with power struggles among the actors. The model could be used as an *evaluation model* of technology and/or organizational changes or a *scenario model* for describing possible outcomes of coming changes.

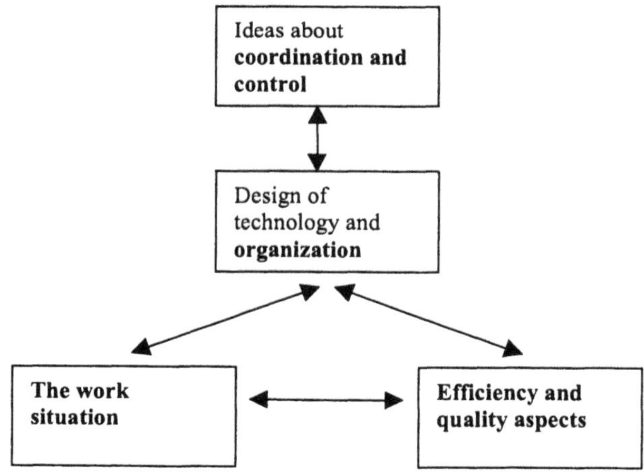

Figure 1. The MOA-S Model

7.1 Design of Technology and Organization

Both the designs of the organization and the CSCW technology can be influenced by ideas of the management regarding control and cooperation aspects. As Leifer (1988) describes, a bureaucratic and hierarchically structured organization, has more complex, centralized computer systems than a more decentralized organization, which as has a more decentralized computer system. In the terminology of Mintzberg (1984), the administration of the Swedish Social Insurance Board could be described as a combination of a machine bureaucracy and a professional bureaucracy with mainly centralized computer systems. As the CSCW technology has the potential of supporting very flexible and also informal communications for an organization, it is an interesting future research issue to investigate to what extent this potential will be realized in the social insurance board.

7.2 The Work Situation

Some important aspects of individual work content can be influenced by changing the organization or the information system (Grundén 1992). The implementation or change of information systems very often changes the patterns of social interaction and co-operation of the organization. The contacts between personnel can be reduced or increased. Change of media for communication can affect the quality of the communication. Face-to-face communication includes more informal information than does mail communication. Different communication media also require different co-ordination mechanisms. Dealing with pension matters using CSCW technology can change the nature of communication and coordination mechanisms among the actors.

The autonomy of the work is one of the most important aspects (Gardell 1986) that also could be affected. According to traditional Taylorism, the human being is seen more or less as a machine and an object for control. Other organization theories emphasize to greater extent the importance of human values. In a professional bureaucracy or in an "adhocracy" (Mintzberg 1984), the workers have more influence on their work situation, based on their professional knowledge, compared with a machine bureaucracy. The CSCW technology can be used in order to increase or decrease the control of the work, depending on the design of the system.

The workers' knowledge of the work could also be changed when new technology is used. When a big information system was implemented in the Swedish social insurance organization during the 1980s, the knowledge of the personnel became more formalized, abstract, and computer oriented. Some of the old knowledge of the workers of the insurance system has been lost (Josefsson 1985).

When an information system is introduced, the time for dealing with a matter very often is reduced. More matters can be dealt with, within a shorter time period than before. The workload for the individual is then increased, which could result in a work situation that is more stressful and less varied than before. If the work situation instead is organized so that other work tasks can be substituted, the variation will increase.

7.3 Efficiency and Quality Aspects

There is a strong pressure on the public sector to make the work as efficient as possible and produce services with high quality. Efficiency can be defined as the amount of utility or goal fulfilment that is reached compared with other resources. One aspect of efficiency is productivity: the volume of the production compared with the used time or salary costs. A study by Johansson and Ulfvensjö (1990) shows that efficiency was the motive for most of the organizational changes within the Swedish local government sector. This is similar to the efforts within the Swedish Social Insurance Board.

The use of CSCW could contribute to increased efficiency if the time for dealing with pension matters could be reduced. If more contacts with actors at different geographical locations could be done using the electronic database, then time for distribution of paper documents could be reduced. If the time for dealing with a pension matter is reduced, the client has received a better quality of service. The quality of the decision making regarding the pension matter could also be affected if the information is stored in an electronic database. The availability of the information could be better when the information is stored electronically and the programs could execute more controls on the information. Other quality aspects, such as integrity and security, could also be affected when CSCW technology is used.

7.4 The Notion of Work Organization

The concept of work organization has changed over time and among different researchers (Macheridis 1997). According to Mintzberg, organizational structure can be defined as the way of coordination and distribution of work tasks in an organization. Organizational structure could be more or less consciously planned. In organizational theory, a general distinction is between the formal and the informal organization. According to a classical definition by Litterer (1963, p. 10), the formal aspects of an organization are aspects that are consciously planned. One example of the formal organization is an organizational scheme. The informal aspects are conceived of as the aspects of organization that are not formally planned but more of less spontaneously evolve from the needs of people.

One example of the informal aspects is informal social networks in the organization. An institution, as defined by Berger and Luckman (1989), can be characterized as a combination of formal and informal aspects. An institution of the organization is manifested when patterns of actions become a habit among several actors.[4] Such patterns of actions could serve as norms that control the human behavior of the organization. According to Grundén (1992), the EDP system could be conceived as an institution of the organization. As an institution evolves as a combination of formal and informal aspects, there are more organizational alternatives and organizational choices compared with the use of more traditional technology. Each organizational choice has a different implication for the work situation of the individual and for the efficiency and quality aspects. However, it is more difficult to specify such new patterns of communication in advance

---

[4]Conventions, as described by Mark, Fuchs, and Sohlenkamp (1997), could be treated as one aspect of institutions.

compared with the use of traditional information technology. The CSCW technology is often used in more informal settings and facilitates more informal communication compared with the old technology (Kraut et al. 1993). When CSCW technology is introduced in an organization, drifting of the use is a common phenomenon (Ciborra 1996).

8. The MOA-S Scenario Model for Integrating Work Organization Aspects into the Design Process of CSCW Systems

The results from the case study show that the test work has generated experiences and knowledge about possible ways of using the system in future work. These experiences need to be systemized and integrated into the development process in order to enable the articulation and design of the future organization. Several discussions and decisions about the future design (both formal and informal aspects) of the work organization when the CSCW technology is used need to be made before electronically dealing with pension matters can be fully integrated in the ordinary work. Scenario is a technique that could be used for such a discussion. The MOA-S could be used as a model specifying different scenarios, such as aspects of the design for distribution of work tasks, control and coordination, social interaction and cooperation, and the paperless office.

Below, different possible scenarios regarding different design alternatives of technology and organization are discussed and some of the consequences of different alternatives, are analyzed using the MOA-S model as a frame of reference. Such a discussion could be integrated into the design process in order to facilitate the future design of the technology and the organization. Such work then needs to be evaluated in order to analyze the relevance of the scenario model for the design work of CSCW systems.

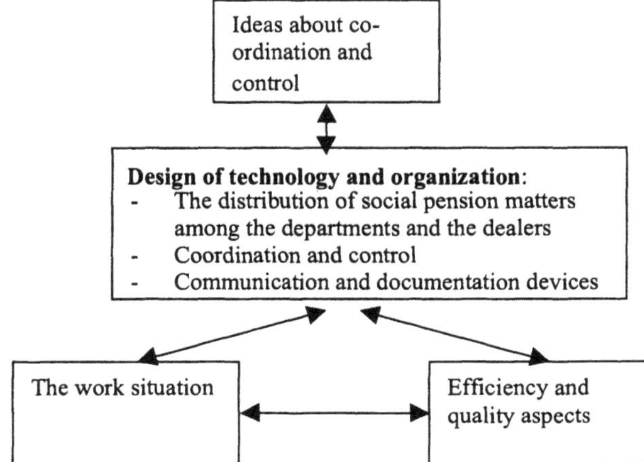

Figure 2. The MOA-S Model as a Scenario Model for Electronic Pension Handling

8.1 Distribution of Social Pension Matters Among Departments and Handlers

The use of electronically dealing with pension matters will require decisions about the actual distribution of pension matters among different offices, departments, and individuals. The potential of distributing the pension matters in a more flexible way among different parts of the social insurance organization will not be realized until several offices use the system.

This will require a change in the work organization. One way is to gather the specialists within a field in one office and distribute all matters in a field to that office. Alternatively, the specialists could be distributed to many offices. The CSCW technology could also be used for communication and cooperation with consultants and doctors at different geographical locations.

The consequence for the work situation is that the stress of a too heavy workload could be reduced, which could improve the work situation of the individuals. This could then be a way of using the staff more efficiently, as the workload could be held more constant. This way of distributing the matters in a more flexible way will probably also require a change in the organizational culture toward a more cooperative and wholeness-oriented attitude. Otherwise there could be struggles among organizational units and among the staff. Important issues to solve are deciding the personnel that will participate in decision making about the distribution of pension matters and the criteria to be used in making the decisions (e.g., statistics about current workload).

The individual work content could be more specialized for personnel dealing with more complicated matters. If the specialists in a field are gathered in one office, the personal contacts with the clients would probably be reduced, which could affect the quality of service. Alternatively, the specialists could be distributed to many offices. Contacts with the members of the work group could be made using the CSCW technology, but the personal meetings will probably be reduced.

8.2 Coordination and Control

By monitoring the number of matters and the way they are executed. CSCW technology makes it possible to control work more thoroughly than is possible today. One control aspect is the authorization of personnel to read, change, and write into the electronic documents. The authorization to access electronic documents could be made more restrictive than today. Coordination mechanisms could also be changed when electronically dealing with pension matters. The control of the coordination of authorization could increase if the transfer of authorization from one handler to another is given by the electronic system according to programmed rules instead of the more information communication currently in use.

Another aspect of control is the degree of formalization and standardization of the work (Grundén 1992). The work could be more formalized and standardized when using the electronic database when compared with working with paper documents. There could be controls in the program specifying what kinds of documents should be included as well as the contents of the documents. Statistics could also be produced regarding the number of matters for each office, department, and individual.

The consequence for the work situation is that the authorization will be more restrictive, which could lead to improvement of such quality aspects as integrity and security. But this is not only a technical matter: it is also a social and cultural matter. A social culture that prevents misuse of authorizations will be needed (Huff and Finholt 1994).

The degree of formalization of the work can affect quality aspects of the service and the individual work situation. There could be advantages to a high degree of formalization in dealing with the matters because of the fact that all clients should have as equal and comparable treatment as possible. On the other hand, a high degree of formalization could lead to a more rigid treatment of matters that do not fit the rules in the programs and individual adjustments are more difficult.

If formalization in dealing with matters will be affected, then the work situation and the knowledge of the individuals also would be changed. The more of knowledge of insurance rules programmed into the software, the less the person handling matter needs to keep in mind, which in turn can affect the learning possibilities and autonomy of the work situation. Technical knowledge would seem to be more important in future work, a fact that could reinforce a tendency of more formalization and standardization of the work. That was one consequence of the early stages of computerization within the social insurance organization studied by Josefsson.

8.3 Communication and Documentation Devices

When CSCW technology is used in the work, the communication devices between the personnel could be changed. Different communication devices could affect the quality of communications and the efficiency of professional performance. Different ways of communicating are face-to-face interaction or human-computer interaction.

The future use of electronic systems could reduce the amount of paper documents produced and distributed around the system. A reduction in the amount of paper documents also would reduce the manual work of storing paper documents and keeping them in order, which could increase the efficiency of the work. Documents stored in the electronic system also could be more available as they probably are easier to retrieve than paper documents stored in an archive.[5] The vision of the paperless office was formulated decades ago, but many offices still seem to have a lot of paper documents (Hedberg et al. 1986).

The consequence for the work situation is that face-to-face interaction involves more informal information than, for example, e-mail interaction. The choice of suitable methods of social interaction and cooperation must be made so that important information is not lost because of it. Even if there are technological facilities that enable communication regardless of time and space, there probably also will be a need for personal meetings because of the quality aspects of the communication. The result from the interviews indicates that informal communication is important in the current work process.

---

[5] There could, of course, be a risk in the electronic system that documents are registered and stored in the wrong document due to human failures, but there probably will be programmed controls to prevent this situation.

To what extent will the theoretical potential of the paperless office be realized in practice when pension matters are dealt with electronically? This aspect of the design has very much to do with informal ways for people to interact with the electronic system. Different people have different individual habits requiring the production of paper documents. Some people prefer to read information from a paper document instead of from a computer screen, especially when the information is discussed in personal meetings. The actual number of hours per day working with a computer terminal is also important. Health problems such as muscle pains or eye stress could be a consequence of working too many continuous hours in front of the computer screen. The actual security and availability of the future electronic system could also affect confidence in the system and the number of paper document produced.

9. Introduction of the Social Insurance Board to the MOA-S Model

The MOA-S model as a scenario model for integrating work organization aspects in to the design process was introduced and discussed at a three hours seminar at the Social Insurance Board in September 1999. The model was presented as a frame of reference for discussing different design alternative of the technology and organization of the electronic pension handling as described in the previous section.

Participants at the seminar were systems developers from the whole region of Västra Götaland. In 1999, a significant organizational change of regions in Sweden caused the earlier region of Bohuslän to become part of the new region Västra Götaland. All systems developers with the Social Insurance Board of Västra Götaland are now situated in an office at Gothenburg. One of the people interviewed at the local office of Kungälv also participated.

The presentation seemed to be relevant for the participants and generated a lively discussion. The manager of the systems analysts mentioned that the presentation had raised many new and important issues, especially regarding the design of the future organization of the work and that they were interested in further cooperation in the development work process. So far, the systems analysts had mainly focused on the technical aspects of the system. A central issue of the design of CSCW systems is which parts of the design of the organization systems developers should be involved with and which parts should be left to the users of the systems. Traditionally, most of the organizational aspects are left to the users. However, it is important that there be a dialogue between the users and the systems analysts about organizational aspects of the systems at different stages of the development process. The MOA-S model could be used as a development tool for initiating and promoting such discussion.

10. Future Work Planned for the MOA-S Model

In the beginning of 2000, a test of the ELAKT prototype is planned with the Kingälv office and an additional local office. It will then be possible to test and discuss different distributions of the work tasks among the two offices. The use of the MOA-S model will be tested in order to support that work and to evaluate and further develop the model and the integration with the systems development process. The model will be presented and discussed both with the systems developers and the users of the systems in order to support a dialogue about the design of the work organization and the consequences for efficiency and quality aspects as well as the work situation.

11. Conclusion

The study shows that there are important discussions and choices regarding the work organization that need to be made during the development process. Due to the flexible character of the CSCW technology, more organizational alternatives of the design of the work organization can be taken into account than with the use of more traditional technology, especially regarding the distribution of work tasks, control and coordination aspects, social interaction and cooperation, and communication and documentation devices. There is a need for coordination between the design of the technology and the organization and to integrate such decision-making into the development and implementation process. Different designs of the technology and organization could lead to different consequences for the work situation of the individual and efficiency and quality aspects. There is a need to integrate a discussion and analysis of these aspects into the design process. The scenario model MOA-S is expected to be used as a method for articulating, structuring, and integrating such discussions into the design process. In order to contribute to the development of relevant models for the design of CSCW systems and organizations, further research work is planned in order to test and evaluate the use of such scenarios in real contexts.

References

Andersen, N. B., and Chatfield, A. "Using IT for Creating the 21st Century Organization.," keynote presentation, *Giga's European Business Process and Workflow Conference*, Amsterdam, October 28-30, 1996.

Argyris, C., and Schön, D. *Organizational Learning II.* Reading, MA: Addison Wesley, 1996.

Ashkenas, R. *Den gränslösa organisationen* (The Boundaryless Organization). Lund, Sweden: Studentlitteratur, 1997.

Bannon, L., and Schmidt, K. "CSCW: An Initial Exploration," *Scandinavian Journal of Information Systems* (5), 1995, pp. 3-24.

Bardram, J. E. "Organizational Prototyping: Adopting CSCW Applications in Organizations," *Scandinavian Journal of Information Systems* (8:1), 1996, pp. 69-88.

Berger, P., and Luckman, T. *The Social Construction of Reality.* New York: Anchor Books, 1989.

Ciborra. C. (ed.). *Groupware and Teamwork.* Chichester: John Wiley & Sons, 1996.

Carroll, J. M. *Scenario-based Design: Envisioning Work and Technology in Systems Development.* Chichester: John Wiley & Sons, Inc., 1995.

Ehn, P., and Sjögren, D. "From System Description to Scripts for Documentation," in *Design at Work: Cooperative Design of Computer Systems*, J. Greenbaum and M. Kyng (eds.). . Hillsdale, NJ: Lawrence Erlbaum Associates, Inc., 1991.

Fahey, L., and Randall, R. M. *Learning from the Future: Competitive Foresight Scenarios*. New York: John Wiley & Sons, Inc., 1998.

Försäkringskassan. *Förteckning över planerade projekt i Utvecklingsprogrammet* (A List of Planned Projects in the Development Program), internal report, 1998..

Gardell, B. *Arbetets organization och människans natur* (The Organization of Work and the Nature of Human Beings). Stockholm: Arbetsmiljöfonden, 1986.

Grudin, J. "Groupware and Social Dynamics: Eight Challenges for Developers," *Commications of the ACM* (37), 1994, pp. 93-105.

Grundén, K. *Människa, organisation, ADB-system. Ett människoorienterat perspektiv på systemutveckling* (Human Beings, Organizations, EDP Systems: A Human-Oriented Perspective on Systems Development), doctoral dissertation, Department of Sociology, University of Gothenburg (monograph no. 47). Lund, Sweden: Studentlitteratur, 1992.

Grundén, K. "Local Initiatives with Pilot Implementation of Desktop Video Communication in the Swedish Social Insurance Board," in *NOKOBIT-97: Norsk konferense om organisasjoners bruk av IT* (A Norwegian Conference About Organizations Use of IT), N. M. Nielsen (ed.), Department of Economy, (Informatics), University of Bodö, June 12-13 1997.

Hedberg, B., Elling, M., Jönsson, S., Köhler, H., Mehlmann, M., Parmsund, M., and Werngren, C. (eds.). *Kejsarens nya kontor. Fallstudier om datorisering på kontor* (The Emperor's New Office: Case Studies of Computerization). Stockholm: Liber, 1986.

Huff, C., and Finholt, T. *Social Issues in Computing: Putting Computing in Its Place*. New York: McGraw-Hill Series in Computer Science, 1994.

Jälmestål, C., Sköld, L., and Wahlqvist. "Projektrapport strategier för kompetensförsörjning," delprojekt 1 och 2, Försäkringskassan 980929. Internal report;not published.

Johansson, A., and Ulfvensjö, L. *Kommunalt organisationstänkande: en förstudie av kommunernas förändringsarbete under 1980-talet*. Örebro, Sweden: Högskolan i Örebro, 1990.

Josefsson, I. "Kampen om innebörden i begreppen" (The Struggle About the Meaning of the Concepts), *Språk och erfarenhet*. (Language and Experience), I. Josefsson (ed.). Stockholm: Carlsson & Jönsson, 1985.

Kraut, R. E., Fish, R. S., Root, R. W., and Chalfonte, B. L. "Communication in Organizations: Form, Function and Technology," *Readings in Groupware and Computer-Supported Cooperative Work: Assisting Human-human Collaboration*," R. M. Baecker (ed.). San Mateo, CA: Morgan Kaufmann Publishers Inc, 1993.

Kvale, S. *Den kvalitativa forskningsintervjun* (The Qualitative Research Interview). Lund, Sweden: Studentlitteratur, 1997.

Kyng, M. "Design for Cooperation: Cooperating in Design," *Communications of the ACM* (34:12), 1991, pp. 65-73

Leifer, R. "Matching Computer-based Information Systems with Organization Structure," *MIS Quarterly*, March 1988.

Litterer, J. A. *Organizations: Structure and Behavior* (Volume 1). New York: John Wiley, 1963.

Macheridis, N. *Arbetsorganisation. En historisk granskning av konventioner* (Work Organization: A Historical Examination of Conventions), Working Paper, Institute of Economic Research, School of Economics and Management, Lund University, 1997.

Mark, G., Fuchs, L., and Sohlenkamp, M. "Supporting Groupware Conventions through Contextual Awareness," in *Proceedings of the Fifth European Conference on Computer Supported Cooperative Work*, J. A. Hughes, W. Printz, T. Rodden, and K. Schmidt (eds.). Dordrecht, Netherlands: Kluwer Academic Publishers, 1997.

Merriam, S. H. *Fallstudien som forskningsmetod*. (The Case Study as Research Method). Lund, Sweden: Studentlitteratur, 1994.

Mintzberg, H. *Structures in Fives: Designing Effective Organizations*. Englewood Cliffs, NJ: Prentice-Hall, 1984.

NUTEK. *Towards Flexible Organizations*. Stockholm: Swedish National Board for Industrial and Technical Development (B 1996:6), 1996.

Schwartz, P. *The Art of the Long View Planning for the Future in an Uncertain World.* New York: John Wiley and Sons, 1998.

van der Heijden, K. *Scenarios, Strategies and the Strategy Process.* Nijenrode Research Paper Series No. 1997-01, Centre for Organisational Learning and Change, University of Nijenrode, 1997. (Available at http://www.library.nijenrode.nl/library/publications/nijrrep1997-01/1997-01.html)

About the Author

Kerstin Grundén has a background as a system analyst and programmer. She completed her doctorate in 1992 at the Sociological Department of the University of Gothenburg in Sweden. From 1992 until 1995, she worked as a researcher at a center for research about the public sector at the University of Gothenburg. In 1995, she became an assistant professor in informatics at the University of Borås. Since 1997, she has been working as an assistant professor in informatics at the School of Health Sciences at the same university. Kerstin can be reached by e-mail at kerstin.grunden@hb.se.

25 CONSTRUCTING INTERDEPENDENCIES WITH COLLABORATIVE INFORMATION TECHNOLOGY

Helena Karsten
University of Jyväskylä
Finland

Abstract

Interdependence construction is the gradual formation of mutual relationships between people. In this study, the area is narrowed to interdependencies at work, in long term projects or groups. Viewing interdependence relationships dynamically, as social practices, it is possible to appreciate the complex and situated nature of this formation. The main goal of the study is to develop a theoretical account of the dynamics of the intertwined processes of interdependence construction and collaborative technology appropriation and use. The main dimensions of this account are: (1) how interdependence is constructed and established as a social process, (2) how information and communication are involved in these processes, and (3) in what ways collaborative information technology can contribute to or hamper these processes. Three earlier case studies are revisited using it. The theoretical approach opens up an extensive research program of interdependence construction in relation to collaborative information technology appropriation and use.

Keywords: Interdependence, collaboration, collaborative information technology, structuration theory, information, communication

1. Introduction

Interdependence construction is the gradual formation of mutual relationships between people. In this study, the area is narrowed to interdependencies at work, in long term projects or groups. Thus, the interdependencies to be studied here concern doing something together over a period of time. Interdependence construction and its reverse, interdependence dismantling, take place continuously in organizations, for example, when task forces or project organizations are formed, or when work tasks are reorganized.

Traditionally, interdependence relationships have been looked at from two perspectives, either as interdependence between people or as interdependence between work tasks (Mintzberg 1979: workflow, process, scale, and social interdependencies). In separating work tasks and people, these perspectives provide only relatively narrow and clear cut views on what could be assumed to be a wide variety of forms and appearances. They also may remain insensitive to the complex and situated nature of interdependencies (Weick 1979). Most importantly, they may lead to a static view of what may be a very dynamic relationship.

In this study, therefore, the focus is on social practices: the relationships are not considered separately, but the attention is on people engaging in action and interaction. Interdependencies are then seen as constantly constructed and reconstructed social practices (Giddens 1984), that is, repetitive, patterned, and reciprocal action and interaction (Weick 1979, p. 46). Interdependence construction is then creating or reconstructing patterns of action and interaction where two or more people are mutually dependent on each other.

Interdependence construction may be expected to take different forms when the relationships are mediated. The mediator of interest in this study is collaborative information technology (CIT). CIT is a label used to denote the kinds of asynchronous groupware where the designers' intent is to provide support for coordination and collaboration through group access to technological capabilities such as shared repositories, discussion forums, and communication facilities (Orlikowski 1995). The purported "collaboration-inducing facilities" of CIT have been identified as related to their capabilities to support high levels of interaction, many-to-many communication and information sharing, in a group of known users, across hierarchical, divisional, or time-geographic boundaries (Coleman 1996; Dyson 1990).

When interdependence relationships are seen as social practices, the focus of interest shifts to the process through which these relationships are formed and reconfigured. The role of technology, in this case "installations of CIT," needs to be considered in parallel, since, as is widely argued (Button and Sharrock 1997; Haraway 1991; Joerges 1988; Latour 1993; Sproull and Kiesler 1991), the social process and the technical system cannot be considered separately. The view in this study is that CIT can be used to support interdependence construction via its capabilities. These capabilities do not necessarily lead to interdependence construction, but rather they may become heavily involved in a variety of ways as they become woven into the social practices of the users. It is contended that a richer understanding of interdependence construction in relation to CIT appropriation and use can be achieved by carefully dissecting their intertwined dynamics. The goal of this study is to form a theoretical account of these dynamics.

2. Earlier Studies

Several studies acknowledge that CIT appropriation and use is influenced by the "fundamental and sometimes subtle social processes in work" (Kling 1991, p. 84), which, when ignored, can contribute to the failures. These social processes are said to include user innovativeness (Swanson and Ramiller 1997), learning through use (Attewell 1992), improvising (Orlikowski 1996) or bricolage (Buscher and Mogensen 1997), among others. If these social processes are seen as adjustments and modifications both in technology and in people, they include mutual adjustments in user organization and the CIT system (distributed processes of co-evolution, Rogers 1994) and establishing congruence (Prinz, Mark, and Pankoke-Babatz 1998) in the user group, and between users and designers.

Prinz, Mark, and Pankoke-Babatz also claim that congruence cannot be achieved unless designers and users achieve a common understanding of users' tasks, work processes, and system design. In their three-year case study, they found that after two years of CIT use, when the work patterns had become seemingly firmly established, the group members began to report problems that concerned coordinating their work patterns. Their interpretation was that at this point the group members were beginning to recognize the consequences of their mediated interdependencies. "This suggests that only with continued system use, the users gradually become aware of how others' actions were affecting their own system use, and they adjusted and accommodated their behaviors accordingly" (Prinz, Mark, and Pankoke-Babatz 1998, p. 377). Thus these studies also point to the relevance of studying the intertwined processes in parallel even though they do not explicitly tie interdependence construction to capabilities of CIT.

Very few empirical studies have so far discussed the microsocial dynamics (Barley 1990) of how interdependencies are constructed or reconfigured and how these processes are tied to CIT appropriation or use. Three detailed, longitudinal studies, by Orlikowski and others (Orlikowski 1996; Gallivan et al. 1993), Ngwenyama (1996, 1998), and myself (Jones and Karsten 1997; Karsten 1995, 1999; Karsten and Jones 1998) are taken here as examples. The studies by Orlikowski (Zeta) and Ngwenyama (Eiger) do not focus explicitly on it either, but, in both cases, intensified interdependence construction is visible through the increased density and complexity of organizational relationships and the emerging novel kinds of interdependency relationships. The data in my own case, CCC, also covers CIT appropriation and interdependence construction. Due to space constraints, each of these is presented only briefly below. For more information, the reader is referred to the articles mentioned.

In Eiger, a Lotus Notes application was introduced to automate processes of managing software development that had been distributed across three continents, to take advantage of time differences. Ngwenyama reports a number of expected and unexpected changes. The groupware application, ADM, was based on the application development methodology that was well established in the company. Initially, the methodology was implemented in a quite straightforward way in Notes. This had unexpected consequences as the work practices could deviate considerably from the inscribed methodology, and the application had to be modified. Also, as the team members learned of the capabilities of the program, they started to make proposals for embedding layers of intelligence into the application. The ADM application served both as a medium of work and as a medium of social interaction. Lateral communication increased between designers and programmers across the three continents. Many of the events in CCC during the period studied were colored by a deep recession and slow recovery from it.

In Zeta, a software company, a Lotus Notes application, ITSS, was introduced to manage helpdesk calls. In addition to planned changes, a number of ongoing local improvisations took place, in response to deliberate and emergent variations in practice. The specific changes that Orlikowski reported include changes in the nature and texture of work (from tacit, private, and unstructured to articulated, public and more structured); patterns of interaction (from face-to-face and reactive to electronic and proactive); distribution of work (from call-based to expertise-based); evaluation of performance (from output-focused to a focus on process and output as documented); forms of accountability (from manual and imprecise to electronic and detailed); nature of knowledge (from tacit, experiential, and local to formulated, procedural, and distributed); and mechanisms of coordination (from manual, functional, local, and sporadic to electronic, cross-functional, global, and continuous). The specialists started to take shared responsibility for the whole team performance by contributing where they could, and by offering their expertise for use by others outside their team. They began to intervene in each others' work as prompted by the organization but also at their own initiative. By this, their interdependence changed its nature to become proactive and perhaps even coercive.

The third case, CCC, is about a small Finnish computer consulting company. Much of the events in CCC were colored by the deep recession in Finland during 1991-1993, by several changes in company management, including three different managing directors, by radical overhaul of expertise by several consultants, and by gradually increasing participation by the consultants. Notes use became gradually focal to new organizational practices, including mutual help in winning and coordinating projects, and a consensual management style, with all the consequent changes. In terms of coordinating and winning projects, the main results were increased horizontal and vertical coordination. Mutual visibility led to horizontal coordination and eventually to a higher degree of horizontal integration through an increase in joint projects. The applications also assisted vertical coordination, by providing a project history which could be consulted if a project had to be transferred to another consultant.

In the cases above, the relationship between CIT appropriation and interdependence construction reveals its complexity. The relationships appeared to have to do, among other things, with certain kinds of use practices, which included disclosure of information beyond the immediate users, (construction of) mutual responsibility for the work in the user community, and immediate access to the information, regardless of the physical locale. The users appropriated the technology, because it gave the kind of information and communication tools that were useful in their work and established them as members of the community. Nevertheless, these cases indicate a need to study the interdependence relationship more closely, as the earlier accounts do not discuss what exactly were the capabilities and use practices in these installations, how they were connected, and how the interdependence relationships emerged in relation to the new capabilities and the changed use practices.

To achieve such an understanding of the social processes of interdependence construction, structuration theory (Giddens 1984) is used as the basis for a theoretical account of interdependence construction, which is then connected via information and communication to CIT capabilities. As this is a first presentation of the emerging theoretical account, the emphasis is placed on laying out its rudiments. Relating this work to the vast bodies of relevant literature (such as computer mediated communication, knowledge management, information access, surveillance, etc.) has been left to a large extent for later refinements of the model.

3. Interdependence Construction as Social Process

3.1 Overview of the Emerging Theory

As the earlier studies show, the relationship between interdependence construction and CIT appropriation and incorporation is complex. This work introduces rudiments of a theory that seeks to encompass that complexity and outline the social processes involved (see Table 1 for a first characterization of the emerging theory). Drawing on Giddens' (1979, 1984) structuration theory, four interrelated aspects of interdependence construction—social integration, time-space distanciation, institutionalization, and system integration—are brought forward.

Table 1. The Four Aspects of Interdependence Construction and Their Relationship to Information and Communication (Following Giddens 1979, 1984)

Social Integration
Origins of interdependence construction lie in social integration, in the systemness of interaction or interdependence in action. This is possible not only in face-to-face situations, but also via situated mediated communication.

Time-space Distanciation
For a socially integrated system, stored resources provide a means for time-space distanciation, for extending beyond the present time and place. Stored resources are managed with related information and communication, which are in this way anchored into a context and made significant.

Institutionalization
Institutionalized relationships are routine and regular. Institutionalization builds on active and chronic reproduction and on past interactions. If the history of interaction is available in an accessible form, it can increase the transparency of the practices and provide a growing archive of information that can be referenced to further ground the relationships.

System Integration
Practices of reciprocity in information access and maintenance between interacting communities may increase their mutual closeness via ownership and responsibility for information. Surveillance and disclosure allow for control and visibility from afar. Together, these change the nature of the systemness of interaction, extending it from social to system integration.

According to Giddens, these four aspects can rely on information and communication in special ways, as outlined below. The role of collaborative information technology could then be to act as a mediator for communication and as a provider of information storage, plus possibly as a provider of other CIT-specific capabilities. These three dimensions—the four aspects of interdependence construction; information and communication that may play a role in it; and CIT as providing mediation and storage—do not form an exhaustive framework to explain evolving groupware use. My goal in bringing them up is to draw attention to the gradual and complex nature of interdependence construction, i.e., to the subtleties of interaction involved, the specifics of information and communication in this, and to how the capabilities and uses of CIT may be involved in the process.

Each of the three dimensions of the emerging theory will be treated differently in the following discussion. The four aspects will be directly drawn from Giddens and related social theory, and they form the basis for further work. The role of information and communication is discussed based on Giddens, but the discussion is extended in the light of IS research. The third dimension of the theory—what all this has to do with collaborative information technology—will be first visited when information and communication are discussed. However, the three cases will provide much more richness into this dimension and, therefore, their role is emphasized, as the goal is to achieve empirically grounded insights into the relationships between the social processes, information and communication, and CIT introduction and use.

3.2 The Four Aspects of Interdependence Construction

The starting point in studying change processes is the basic idea of how something remains the same and something else may change. In Giddens' (1979, 1984; Jones 1999) structuration theory, stability and change are approached from the idea of the duality of structure: each action draws upon the structures that enable and constrain it and by this each action also contributes to reproducing or changing the structures (Giddens 1984, pp. 25-29; 297-304). Structures refer here to the intersubjective structures of social existence, held by the individuals participating in these social practices. Structures indicate to an individual how she or he should act as a member of a particular community. Recently, Giddens has called these structures "conventions" (Giddens and Pierson 1998, p. 87), which gives a useful, even though simplified, common sense idea of them.

Giddens extends his theory of social systems then to how the individuals become members of the community. He uses the sociological terms of social and system integration to examine mechanisms of interdependence construction. Social and system integration both refer to how social systems—such as work groups, organizational departments, or professional associations—gain their systemness in relation to the people within the system and in relation to other social systems. That is, social and system integration are the counterpoint of the duality of structure. Together, they seek to explain how it is possible to have at the same time discernible stable patterns of action in a community and a possibility for change. Time-space distanciation is used to explain how these patterns survive beyond the present time and place. Institutionalization then is the process by which the patterns of action become regularized and routinized. Together, these four aspects explain how social practices such as interdependence relationships are created and established. However, these aspects are not phases, there is no sequence between them. Human actions that contribute to these aspects of interdependence are parallel and interspersed. Therefore, interdependence construction is a process where all these aspects are present, more or less. Before going on, each of the aspects will be discussed in more detail.

Social integration is what makes a group of people a school class or a work team. Giddens' starting point in discussing social integration (Giddens 1979, pp. 76-81; 1984, pp. 28; 89; 191) is face-to-face communication, which is generally considered the primary, immediate mode to which other, mediated modes are compared (Berger and Luckmann 1967, pp. 43-48; Giddens 1984, pp. 64-72). In meeting face-to-face, in situations of co-presence (Goffman 1972), the other person is accessible as a person with a physical body, with bodily expressions, giving the interaction a density in reciprocity that is difficult to achieve in other kinds of interaction. Social integration is then the systemness of this densely reciprocal interaction, the interdependence of action, between the co-present actors. That is, social integration concerns patterning of interaction, knowledge about how the others will act, and the potential predictability of interaction.

Time-space distanciation (Giddens 1984, 256-262) refers to the ability of social systems to exist beyond the immediate here and now; that is, how interdependence relationships carry beyond the immediate interaction, how they persist. He defines it as the "stretching of social systems across time-space, on the basis of mechanisms of social and system integration" (p 377). That is, the systemness of interaction between people and between social systems, resulting from its dense reciprocity, is a necessary mechanism for social systems to be able to exist beyond the immediate here and now. The greater the time-space distanciation of social systems, the more their institutions "bite into time and space" (Giddens 1984, p. 171).

The systemness of interaction gradually becomes routine, and its discernible, even distinct, patterns become regular. Berger and Luckmann call this *institutionalization*, which they define as "reciprocal typification of habitualized actions by types of actors" (Berger and Luckmann 1967, p. 72). The institutionalized practices of interaction exhibit the structural properties of the particular relationships, which constrain and enable their reproduction. At the same time, they exist as such only as a result of active and chronic reproduction. To quote Giddens (1984, p. xxi): "The structural properties of social systems exist only in so far as forms of social conduct are reproduced chronically across time and space." Continuity is the key here.

The fourth dimension of how interdependence relationships become established is *system integration*, which has been characterized as systemness of interaction outside the conditions of co-presence (Giddens 1984, p. 377). An earlier characterization (Giddens 1979, pp. 76-81), however, defined system integration as reciprocity between groups or collectivities, without regard for physical presence or absence. My understanding is greatly influenced by the earlier version, even though the later version seems to be adopted in IS research (e.g., Lyytinen and Ngwenyama 1992; Ngwenyama 1998). The later definition, however, has the danger that it may encourage a misunderstanding of (computer) systems integrating collectivities across space and time. For the current discussion, the emphasis on reciprocity between collectivities as system integration is focal. Therefore, in this work, system integration is used to refer to systemness of reciprocity between groups or collectivities.

3.3 Information and Communication in Interdependence Construction

3.3.1 Social Integration and Situated Mediated Communication

The face-to-face interaction that was deemed necessary for social integration is not without problems. It is intensive, synchronous, and time-consuming as compared to mediated interaction, which can be asynchronous, can span a longer period of time and space, and where periods of communication can be interspersed with other activities when suitable. The question then is to what extent mediated communication can replace or supplement face-to-face communication when a higher level of social integration is desired.

Giddens (1984, p. 68) admits that it may be possible to simulate some of the "intimacies of co-presence," some of the closeness of face-to-face encounters, in mediated communication to facilitate social integration. Letters carry some of the presence of the letter writer, and in telephone conversations, the other party can be heard. This observation gives a starting point for also studying computer mediated (and hence also CIT mediated) communication in terms of social integration.

In the IS literature, the primacy of face-to-face interaction, and the problems with mediated communication in terms of density in reciprocity have commonly been addressed under the label of media richness. Daft and Lengel (1984, 1986) initiated discussions of the bandwidth that would be sufficient for various kinds of interaction. Prinz, Mark, and Pankoke-Babatz (1998) gave a common explanation of this dilemma by stating that the difficulty that electronic groups face during system use is that they lack the *social information* that groups generally gain through formal and informal face-to-face interaction.

However, empirical data (e.g., Dennis and Kinney 1998; El-Shinnawy and Markus 1998; Kock 1998) show that social integration also is possible with computer mediated communication, even though it might be more difficult than in face-to-face communication. These results suggest that sufficient social information can be conveyed in electronically mediated interaction. Moreover, the results indicate the importance of *situatedness of interaction*. Situatedness here means that interaction takes place in the context of particular, concrete circumstances (Suchman 1987, p. viii). In situated interaction, the group is able to share both task related information and communication. When they can also exchange social information, they are able to adjust their interaction, which then, over time, can achieve the density and systemness in reciprocity, necessary for interdependence.

3.3.2 Time-space Distanciation and Stored Information

Giddens (1984, pp. 256-262) connects time-space distanciation to his theory of power. Power is defined as the capacity to achieve outcomes and it is generated in and through the reproduction of structures of domination. Giddens emphasises storage of resources as a medium for domination. Stored resources, both material and symbolic, bind time-space involving "the knowledgeable management of a projected future and recall of an elapsed past" (p. 261), that is, with stored resources the social relations can be carried beyond the particular situations. In other words, one reason why social systems such as work teams and school classes persevere even when the people are not together in the same place at the same time is because they are tied together by stored resources such as the object they are working on (e.g., a joint memorandum, a piece of art) and their tools for that, "their room" with its facilities, or a person whom they consider their leader. This view of Giddens seems defensible especially when stored resources are seen to include also symbolic resources, such as reputation or mission.

Retention and control of information, among other things, contribute to storage of resources. In Giddens' words: "the storage of authoritative and allocative resources may be understood as involving the retention and control of information and knowledge whereby social relations are perpetuated across time-space" (1984, p. 261). He emphasizes that "information storage...is a fundamental phenomenon permitting time-space distanciation and a thread that ties together the various sorts of allocative and authoritative resources" (p. 262).

By discussing "stored information" instead of stored data, Giddens can be seen to draw attention to the *contextuality* and assigned *significance of the information*, gained by its connection to the stored resources, as opposed to detached pieces of data. Equally, we can discuss mediated, stored communication instead of stored messages, when communication is tied to the information that communication is about. By this, separate messages gain their significance as parts of communication about this piece of information, as *contextual communication*. Access to and control of these is significant: contextual information and communication also implies access to the stored resources.

Giddens points out that the stored information requires a means to carry, recall, and disseminate it (that is, for storing and communicating it), in addition to skills for interpreting it. The dissemination of information is influenced by the technology available for its production. Giddens uses the example of mechanized printing (p. 262), which conditions what forms of information are available and who can make use of it; that is, its accessibility. The one who has access to the information and who can control it, has access to the stored (material and symbolic) resources. With technical aids, such as CIT, this accessibility can be interpreted in a very concrete way to mean access to the data and messages in the databases, embodying the information and related communication.

It can be claimed that by use of CIT, the stored information and communication can become *highly accessible*, as compared to, for example, when it is stored in paper files and folders in an office. First, because reading and browsing the stored information and communication is not necessarily noticeable to others, thus learning by lurking, i.e., by legitimate peripheral participation, is possible (Lave and Wenger 1991). Second, because access to CIT can be implemented in such a way that users can use it at their own discretion, when and where it suits them (Connolly and Thorn 1990). Third, because the stored information and communication can be permanently and publicly accessible.

This third reason warrants some elaboration. Information and communication, the data and the messages, can be stored in various forms, but the major way of conveying information in organizational life is still by writing. To quote Goody (1987, p. 280), "The written language (reaches) back in time." Written artifacts can at any time be mobilized as a referential object for clarifying ambiguities and settling disputes: "while interpretations vary, the word itself remains as it always was." Schmidt (1997) also draws attention to the permanence and public character of written records: "They are, for all practical purposes, unceasingly publicly accessible." Information in files and folders in the office are also permanent and public records. Information in CIT differs from that in files and folders not only due to the accessibility, but also because other ways of presenting the information, such as video or audio, are tied together. For these reasons, CIT can be seen as potentially significantly relaxing the conditions of access.

3.3.3 Institutionalization and History of Interaction

Institutionalization of interdependence relationships, that is, how the interaction becomes routine and regular, depends, on one hand, on its active and chronic reproduction. On the other hand, it also depends on past interactions, on its own history. The patterns of interaction contain traces of these past interactions in the form of structures that people employ in conducting the interaction. The history of interaction can also be more transparently available, for example, a written account, to be used as a resource in carrying out the interaction. This history of interaction may provide an opportunity to ground the practices of interdependence further, as the interdependent participants can have more background and may be better informed about how to go on (mutual knowledge, Giddens 1984, p. 375).

CIT can contribute to institutionalization in several ways. Practices can become increasingly *transparent* if both present and past actions are visible in the stored information and communication (cf. with the idea of informating by Zuboff 1988). Also, the *archive* of information can gradually become substantial and in this way become a significant source of information.

CIT mediated interaction can be a special case in at least two ways. One way is that the messages can remain as entries in the databases, and in this way the flow of communication can become stored information. The messages can be either connected to each other by message header information such as time stamps, or they can appear as threaded messages, showing the first entry of a discussion and replies and comments connected to it. By tracing the messages, an account of the history of the discussion can be constructed.

3.3.4 System Integration and Reciprocity

Issues of reciprocity and of surveillance and disclosure relate closely to system integration. They can play a role in the move from social to system integration as the interacting parties can be communities and not only individuals. From the perspective of the interacting communities, they enable confidence to be maintained in the other party.

Reciprocity may include mutual access to, and disclosure of, relevant and significant information, and as a possible consequence, mutual maintenance of the information. Zuboff (1988, p. 356) has interpreted mutuality to imply equality of access and the presence of sufficient depth of intellectual skill so that those who have access to data also have access to their meaning. Others (such as Giddens 1984, p. 127; Poster 1990, 1995) have emphasized not equality but the negotiated nature of the forms of reciprocity in each case.

When information is maintained together, the nature of information use and thereby also practices of reciprocity may change, and mutually maintained information can become even more closely tied to action and interaction. By mutual maintenance, the quality and credibility of the information can gradually improve, in terms of its assigned purpose (VanHouse, Butler, and Schiff 1998). Perhaps mutual maintenance is then the key in explaining how ownership and responsibility for information evolve.

Negotiating access to information involves at the same time also negotiating the extent and boundaries of *surveillance* (Clement and Wagner 1995). Similarly, the extent of *disclosure* illustrates compliance to being subject to surveillance. Giddens sees surveillance as unidirectional, creating a non-equal relationship between two collectivities, as one group can control the other group by it. Thus with surveillance, the nature of the relationship changes, Giddens argues, from solely a social relationship to one including the system dimension by the visibility and control aspects from afar. In terms of time-space distanciation, this means persistence of the social system also in relation to other social systems.

4. Review of the Cases

4.1 Social Integration and Situated Mediated Communication

These elements and issues of the emerging theory of the relationship of interdependence construction and CIT will now be used to revisit the three cases described earlier. This brief and far from thorough "analysis" is mainly aimed at illustrating ways in which the theory can be used to inform reading and understanding the cases.

Eiger provides an example of the way that situated mediated communication contributed to social integration. In addition to communicating about tasks at hand, users in the USA, Asia, and Europe became more informal in their mutual relationships, exchanging weekly chit chat and updates about their lives. Ngwenyama (1998, p. 141) refers to this as social integration. His usage of the concept differs from that of Giddens, for whom social integration lies in interdependence of action, not only in acquiring social information.

Social integration in the Giddensian sense can be discerned by going further into the systemness that gradually emerged, as the familiarity then spread to the task-related communication, which became gradually freer, and where opinions were exchanged across continents. For example, the Asian team members, who did the programming, became more visible as people with considerable skills, and their views were taken more into account. The Asians felt more a part of the team, as they were more involved in discussions, and not just receiving orders: "Now we know what they are doing and they know what we are doing." Thus their interaction had gained its systemness in a recursive fashion: familiarity spread from social to task-related interaction, which became more interactive leading into more familiarity and reciprocity.

Neither Zeta nor CCC provided clear examples of social integration with solely mediated communication. In both cases, there was also a physical locale where the people could and did meet. They were not fully dependent on any single medium to communicate. In Zeta, however, the people started to prefer electronic communication because it was experienced to be less intrusive. As the electronic interaction reduced face-to-face interaction during the course of the day, it was then compensated for by arranged get-togethers.

In CCC, patterning of interaction with Notes was slowed because it took 13 to 14 months before all people in the group were users. This was different from Eiger and Zeta, where the application needed to be used by the whole group, or not at all. The systemness of interaction in Notes discussions emerged gradually, as Notes started to contain whole discussions, and not just to supplement discussions in meetings. Notes was used to discuss joint decisions, such as hardware and software investments, to gather everybody's views, and to reach an agreement. This practice of taking everyone's view into account gradually became so well established that, for some issues, the third managing director mentioned to avoid Notes as it was "too democratic." A major reason for this democratic use, as stated by several consultants, was that, with Notes, one could participate in the discussion at one's leisure and the pace allowed both quick and slow, quiet and verbose, people to enter their views. The situatedness of the interaction became, in a way, stretched over time.

The suggestion that social integration was supported by the situatedness of the communication together with social information, rather than the particular media employed, found backing in all of the cases. As long as all members of the group had access

Table 2. The First Dimension: Social Integration

Dimension of interdependence construction: social integration	*What kind of information and communication is involved?* Situated communication with social information.
Specifics of CIT: Provides mediated channel for situated communication, which can include social information.	*Social integration in the cases:* Electronic communication became sufficiently dense and reciprocal in all three cases to support social integration. Situated interaction was present in all cases, but social information only in Eiger, where there was no opportunity for face-to-face meetings. *Additional insight provided by the cases:* **Eiger**: Social information was tied reciprocally to task-related communication. **Zeta**: Mediated communication preferred as less intrusive. Acknowledgment of need for different, complementing channels of communication. **CCC**: "Democratic participation" became possible when the media allowed all to participate. Situatedness of interaction became in a way stretched over time. Different media and changes of media gave possibilities for disintegration.

and used the same media for discussing joint issues, the particular media choice was not decisive for interdependence construction. Moving from one medium to another, however, took place in passages of transition during which the group could also show symptoms of disintegration. For example, in CCC, there was a period when those not yet using Notes feared that the discussions and information in Notes would somehow serve only the Notes users, whom they already considered to be in the "inner group" and holders of much information. The "outsiders" feared that the same information would then not be distributed to them. This division disappeared when, with access to sufficient technical equipment, everybody could read the entries regularly.

4.2 Time-space Distanciation and Stored Information and Communication

The significance of contextual communication is perhaps best illustrated in Zeta. Information about the phone calls from customers and the help desk specialists' communication around these was significant for creating and managing interdependencies between the specialists. With this, requesting and giving help about a specific problem were possible. Also in Eiger, the information and communication about the software project at hand were significant for coordinating the work of the teams in different countries. Likewise, only within the context provided by the application development methodology could the information gain its specific significance.

In Eiger and in Zeta, all information and communication regarding the software project and the customer calls had become highly accessible to all concerned, as they had started to use the CIT at the same time for the same purpose. The consequences of this accessibility give interesting insights into how the interdependency relations were reconfigured with CIT. In Zeta, the specialists began to assist each other proactively, without request, based only on what they saw in the Notes databases. The specialists also began to put together summaries of common problems they had solved and make them available in other help desk groups. These new practices reflected their awareness of shared responsibility for calls, and they could be interpreted as involving considerable interdependence creation. They can also be interpreted as reconfiguring the interdependence relations from reactive to proactive, and using CIT to facilitate the changed demands for information accessibility.

In Eiger, the initially very strict rules in the ADM application dictated what information and communication were significant. Gradually, the actual practices took over and the rules in the ADM application were modified to correspond to these. In parallel, the practices became adjusted to take advantage of the free communication and the interaction started to include social chat. That is, the ADM application with the related information and communication made the resources not only accessible, but also discussable, and these discussions started to include social aspects as well. In this way, communication gained new dimensions. The interaction became denser in reciprocity and thereby (by definition) more interdependent.

In CCC, the situation was different. The key resources were project leads and current projects. However, some did not see these as significant for themselves, as they were already fully employed. Only during the recession, when the information about leads and projects was reinterpreted as reflecting the success of the whole company, did it became significant for all, as the consultants were also shareholders in the company and, therefore, interested in its fate. Thus not only availability of the information in Notes databases, but also the possibility of reading the information as an *overview* (Robinson 1991) of the situation provided an incentive to read and maintain information. An overview is an aid to understanding constantly changing context. It is the best way of situating action—realizing the agenda set out in the plan—within this flux of context. CIT facilitated this by providing not only higher accessibility but also by offering different views into and summaries of the information.

As we saw, high (levels of) accessibility of stored information and communication had a number of consequences. The information and communication were likely to gain attention as they were significant. Access also brought division among those who had access—directly or indirectly—and to those who did not. When the information was useful, joining those who had access became more appealing than staying outside. Other studies (VanHouse, Butler, and Schiff 1998) have also shown that when information and communication are significant and highly accessible, they are also likely to be more controllable, and this control can contribute to improved quality and accuracy of the information. Also, practices to assess its credibility can be established.

Table 3. The Second Dimension: Time-space Distanciation

Dimension of interdependence construction: time-space distanciation by stored resources	*What kind of information and communication is involved?* Contextual, stored information and communication, significant due to the connection to stored resources.
Specifics of CIT: CIT relaxes the conditions of access significantly: high accessibility of information and related communication. Permanence and publicity of written information and communication. Overview aids comprehending the information as a whole.	*Time-space distanciation in the cases:* Significant, often large collections of information became used as a resource increasingly or in a novel way. This contributed to the persistence of the interdependent group. *Additional insight provided by the cases:* **Zeta**: Awareness of shared responsibility of information; reconfiguring interdependence relations from reactive to proactive. **Eiger**: Resources became discussable and interaction gained new dimensions, which resulted in denser reciprocity. **CCC**: CIT-supported possibility for overview. The perceived advantages of having high access to information and communication encouraged start of use.

4.3 Institutionalization and History of Interaction

In CCC, the history of interaction was significant for the emergence and establishment of collaborative decision making practice and compliance to joint decisions. Individual consultants could (and did) act against the decisions made in meetings, even against those they had participated in making. Several measures were taken to make the decision making and implementation more effective. Meeting agendas and minutes were entered in Notes. After about two years of use, a consultant stated that the meeting minutes had become a very valuable resource for managing the decision making process in the company: if well recorded, they would provide the backing that was sometimes needed for implementing decisions. However, what probably was more influential for the openness of managing the company was the gradual transfer of discussions from meetings into Notes databases. Thus the consultants gradually started complying with the decisions, as the processes became more transparent and more established because the accounts of the discussions and decisions were permanently and publicly available in Notes.

In Zeta, an incident history field was implemented in the call record application ITSS. The person updating the incident record was asked to enter what was done and what would be required next. ITSS then appended a time stamp and an identification of the person who had done the update. Nothing entered into the history field could be deleted. This led to some self-censorship on the part of the specialists, as the whole history of the call could be easily read by others. The work process became documented and an audit trail was generated by which the specialists would became accountable not only for output but also for work in progress. The supervisors monitored the work by reading what was entered in the call records. Also, informal norms for free text fields gradually evolved. These norms reflected a recognition of the database as a shared resource and an observation that its value lay in making the contents of incident records reusable. The practices of interdependence via call records became gradually so established that they became taken for granted and a basis for further changes in work practices. Examples of these are the previously mentioned case of proactive help and the new practice of making model cases of common types of problems.

In Eiger, as the interdependence relations became more pronounced, the interaction via CIT became more cryptic and shortened. In parallel, the number and length of messages decreased, and fewer iterations were required to settle an issue, as understanding among the participants improved. To quote one designer (in the USA): "The more you know the programmer (in Asia), the less you may need to write effectively with him." Ngwenyama (1998, p. 138) interpreted this development as building up a shared context of meaning, despite the geographic and cultural distance. It can also be read as creating significant stocks of mutual knowledge, as the participants in the discussion had past interactions available. A similar phenomenon also took place in CCC, where messages started to appear in the midst of a project record. This was possible because of reading through all new entries with the *scan unread* command. When read in this way, instead of opening each database separately and finding changed entries, the process of informing became one undifferentiated flow, an ongoing conversation around documents and issues, quite similar to what had taken place before in the meetings. A consequence of this practice was also that gradually the databases became "an incomprehensible mess" for those reading them in some other way.

Both in Eiger and in CCC, the institutionalized practices thus tended to further integrate the participants by the coded form of the interactions that acted as a barrier to outsiders. Even though the interaction was inscribed in the database entries, they were not entered for the purpose of later use as a consultable record, as in Zeta. There the clarity of entries and agreed conventions became focal, as the entries became resources to be used in future problem solving. In this way, they became *actively constructed history*. In terms of the nature of institutionalization, this difference between task at hand and consultable record is important. In Eiger and CCC, the practices became institutionalized as "our practices" and they provided the difference between "us" and "them." In Zeta, the institutionalization was more formal and regulated as the aim was to establish person-independent systems of interaction.

Table 4. The Third Dimension: Institutionalization of Practices

Dimension of interdependence construction: Institutionalization of practices	*What kind of information and communication is involved?* Permanently and publicly available history of interaction is focal in institutionalization of practices.
Specifics of CIT: Increased transparency of practices as actions are shown in the stored information and communication.	*Institutionalization of practices in the cases:* Significant stocks of knowledge were built as the history of interaction started accumulating. Interdependent work practices became established, using these as their key resource. The institutionalized practices tended to further integrate the participants by the coded form of interactions that acted as a barrier to outsiders. *Additional insight provided by the cases:* **CCC**: Institutionalization of participatory decision making due to increased transparency, given by history of interaction being permanently and publicly available. **Eiger**: The more pronounced the interdependence relationships, the more cryptic and concise the CIT-mediated interaction became, and the fewer iterations were needed. **Zeta**: Permanence of record resulted in accountability not only of completed work but also work in progress. Reusability of call records made it important to construct them robustly. Actively constructed history.

4.4 System Integration and Reciprocity

In Eiger, neither access to information nor surveillance by using the information were open for negotiation. Senior management wanted to ensure that access to information was restricted and a complex set of authorization structures was designed to limit different classes of users to specific areas of the application and the databases. All team members had read-access to design documents, but update-access was determined by project responsibilities. However, the designers in the USA and the programmers in Asia were free to communicate by e-mail. They negotiated the nature of their reciprocity by discussing it in a separate communication channel. While previously the finished software design had been simply sent to Asia to be implemented, with this separate, non-controlled communication, the design could be rendered open and negotiable. As the designers and programmers learned more about each others' competencies as they engaged in, for example, the reasoning behind particular designs or implementation, mutual appreciation for each others' skills and trust in them became clearly visible in the e-mail exchanges.

In Zeta, access to the ITSS database was free. A consequence was that the specialists became aware of this mutual surveillance or "big brother." The possibility of scrutiny focused specialists' attention on (and possibly modification of) what impression they conveyed of themselves in electronic text. Orlikowski refers to this self-regulation as a form of "participatory surveillance" (Poster 1990). A subtler point brought up by Orlikowski, as informed by Foucault (1979, pp. 202-203), is that by being knowingly electronically visible, the specialists participated in defining the constraints of power to which they were subjected.

As in Eiger, the Zeta specialists could not influence the access that others had to their "textualized work." What they could influence, however, was what they would disclose themselves. The ITSS call entries provided a "brag-record" for high performers, a showcase for their efforts, embellished or not. Orlikowski interpreted this as a subtle shift in the texture of work, into an interest in symbolic artifacts that describe execution of work, immediately and continually available through the technology. The negotiation of mutuality had proceeded relatively smoothly within the original department, but when access to the ITSS database was planned to be given to others, as well, concerns arose about access to and use of the information. The solution was to make only edited versions of key topics available. Both consciousness of self and consciousness of "us" thus lead to highly *managed disclosure*.

In CCC, the negotiation of mutuality was perhaps the clearest of the three cases because the consultants had full control over what they entered into the databases. At first, only a few of the consultants entered their project information regularly. A key Notes champion also entered information on behalf of non-Notes users until everybody had developed a regular usage pattern. Awareness of the information in the databases also led to awareness of how they were becoming increasingly visible to each other. At first, only information that was considered useful to the others was entered. As the information in the databases was kept confidential, those who had initially entered only a minimum of information gradually came to reveal more of their activities. Disclosure in Notes always lagged some distance behind the face-to-face interdependence practices, but the experiences during the recession had convinced even the more reluctant ones of the benefits of being "in the know."

Table 5. The Fourth Dimension: System Integration and Reciprocity

Dimension of interdependence construction: System integration and reciprocity	*What kind of information and communication is involved?* Mutually accessible and maintainable information, together with the negotiated forms of reciprocity, may result in ownership and responsibility for the information. Visibility and control between collectivities is managed by negotiating the extent of surveillance and disclosure of information.
Specifics of CIT: CIT can provide technical means for information access, disclosure, and maintenance. Means for managing reciprocity.	*System integration in the cases:* By *managed disclosure*, the users could influence how they were surveiled and hence how they gave others the opportunity for control. The possibility for *mutual control* turned into mutual trust in CCC and Eiger, but into a more formal system integration in Zeta. *Additional insight provided by the cases:* **Eiger**: Programmers and designers negotiated the nature of their reciprocity by discussing it outside the application. **Zeta**: Awareness of mutual surveillance led to self-regulation. Possibility to determine the extent of own disclosure drew attention to symbolic artifacts that describe execution of work. **CCC**: Need for mutual control was used to invite disclosure. Experiences of mutual maintenance with less risky information gradually led also to disclosure of confidential information.

5. Discussion and Conclusions

The four aspects of interdependence construction were taken from structuration theory, which again rests on a considerable body of social research. Therefore, it seemed relatively safe to assume that structuration theory might offer the kind of processual approach that the problem area invited. Also, structuration theory attempts to bridge the everyday social practices to what is established or institutionalized in a social system, and in this way it had promise for studying interdependence relationships. The main impetus for choosing structuration theory was the comments about mediated communication and role of stored information in time-space distanciation with which Giddens had peppered his text. Since he has not followed himself on these lines, I took these as an invitation to give it a try. Even though structuration theory is very rich and it does not yield to appropriation easily, it proved fruitful for the work in this study. It was possible to construct a comprehensible (even if not comprehensive) account of the main processes of interdependence construction with it and to relate these to information and communication.

Information and communication appeared to be implicated in each of the four aspects as follows: First, social integration can take place not only in circumstances of co-presence but also via situated mediated communication with sufficient social information. Second, stored resources with related information and communication make time-space distanciation possible, as they become persistent and highly accessible. Third, history of situated interaction contributes to the institutionalization of interdependence relationships, as it gives a basis for comparing present to past action. And fourth, practices of reciprocity in relation to information and those of surveillance and disclosure contribute to system integration by allowing visibility and control between groups.

All of these aspects of interdependence construction appeared also to be linked with each other. Increased social and system integration meant increased reciprocity of practices, which is needed for time-space distanciation and institutionalization of practices. Information and communication play several roles in these connections. For example, access to significant information is not only a way to the stored (material) resources the information is about but also to the actions of other people in the group. Available history of past interaction can contribute to building significant stocks of mutual knowledge which can then not only ground the relationships but also support coordinated action. When information and communication is significant and highly accessible, it is also likely to be more controllable. Accessible information is also discussable, and these discussions contribute to increased reciprocity. The cases gave ample evidence of how these connections worked in practice.

The collaborative information technology that was used in each of the cases was Lotus Notes. However, the installations differed significantly in terms of number of users, applications in use, way of appropriation (discretionary in CCC, mandatory in the two others), extent of use, etc. Regardless of these differences, several common features or phenomena were found in all cases. Among others, possibility for situated interaction with related social information, high accessibility, accessible account of the past interactions, and managed disclosure were identified. Revisiting the cases appeared fruitful, as they gave the work empirical grounding and also opened unanticipated vistas.

To sum up, the theory introduced here provides a way to understand the complex relationship between (emergent) collaboration and collaborative information technology. By it we can move away from the dilemmas of causality between collaboration and CIT. Also, the theory provides a more specific characterization of the relationship than approaches that focus merely on emergent processes or collaborative IT in general. The key insight of the theory is to connect these by information and communication, and thus it is able to provide a plausible account of their relationship. Further work on the theory will both relate it to current IS/CSCW research and use it to inform empirical studies.

Acknowledgments

Matthew Jones, Andrew Brown, Geoff Walsham, Kalle Lyytinen, Juhani Iivari, Matti Rossi, Uta Pankoke-Babatz, Markus Ritterbruch, and Helge Kahler gave very valuable feedback while I worked on the first versions of this article. I have given short presentations of (some of) these ideas in Cambridge, Copenhagen, and Jyväskylä. I thank all participants for lively and challenging discussions. The reviewers of the WG8.2 conference also gave very useful criticism: many thanks for their efforts.

References

Attewell, P. "Technology Diffusion and Organizational Learning: The Case of Business Computing," *Organization Science* (3:1), 1992, pp. 1-19.

Barley, S. R. "The Alignment of Technology and Structure through Roles and Networks," *Administrative Science Quarterly* (35), 1990, pp. 61-103.

Berger, P. L., and Luckmann, T. *The Social Construction of Reality*. Garden City, NY: Doubleday, 1967.

Buscher, M., and Mogensen, P. H. "Mediating Change: Translation and Mediation in the Context of Bricolage," in *Facilitating Technology Transfer through Partnership: Learning from Practice and Research,* T. McMaster, E. Mumford, E. B. Swanson, B. Warboys, and D. Wastell (eds.). London: Chapman & Hall, 1997, pp. 76-91.

Button, G., and Sharrock, W. "The Production of Order and the Order of Production: Possibilities for Distributed Organizations, Work and Technology in the Print Industry," paper delivered at the European Conference on Computer-supported Cooperative Work (ECSCW'97), Lancaster, UK, 1997.

Clement, A., and Wagner, I. "Fragmented Exchange: Disarticulation and the Need for Regionalized Communication Spaces," in *Proceedings of the Fourth European Conference on Computer-supported Cooperative Work (ECSCW'95),* H. Marmolin, Y. Sundblad, and K. Schmidt (eds.). Dordrecht, The Netherlands: Kluwer, 1995, pp. 33-49.

Coleman, D. *Electronic Collaboration on the Internet and Intranets*. San Fransisco: Collaborative Strategies, 1996 (http://www.collaborate.com/intranet.html).

Connolly, T., and Thorn, B. K. "Discretionary Databases: Theory, Data, and Implications," in *Organizations and Communication Technology*, J. Fulk and C. Steinfield (eds.). London: Sage, 1990, pp. 219-233.

Daft, R. L., and Lengel, R. H. "Information Richness: A New Approach to Managerial Behavior and Organization Design," in *Research in Organizational Behavior (Volume 6)*, B. M. Staw and L. L. Cummings (eds.). Greenwich, CT: JAI Press, 1984, pp. 191-233.

Daft, R. L., and Lengel, R. H. "Organizational Information Requirements, Media Richness and Structural Ddesign," *Management Science* (32), 1986, pp. 554-571.

Dennis, A. R., and Kinney, S. T. "Testing Media Richness Theory in the New Media: The Effects of Cues, Feedback, and Task Equivocality," *Information Systems Research* (9:3), 1998, pp. 256-274.

Dyson, E. "Why Groupware is Gaining Ground," *Datamation*, March 1, 1990, pp. 52-56.

El-Shinnawy, M., and Markus, L. "Acceptance of Communication Media in Organizations: Richness or Features?" *IEEE Transactions on Professional Communication* (41:4), 1998, pp. 242-253.

Foucault, M. *Discipline and Punish*. New York: Vintage Books, 1979.

Gallivan, M., Goh, G. H., Hitt, L. M., and Wyner, G. *Incident Tracking at Infocorp: Case Study of a Pilot NOTES Implementation*. Working Paper No. 149, Center for Coordination Science, Sloan School of Management, Massachusetts Institute of Technology, Cambridge, MA, 1993.

Giddens, A. *Central Problems in Social Theory: Action, Structure and Contradiction in Social Analysis*. London: Macmillan, 1979.

Giddens, A. *The Constitution of Society*. Cambridge, England: Polity Press, 1984.

Giddens, A., and Pierson, C. *Conversations with Anthony Giddens: Making Sense of Modernity*. Cambridge, England: Polity Press, 1998.

Goffman, E. *Interaction Ritual*. London: Allen Lee, 1972.

Goody, J. *The Interface between the Written and the Oral*. Cambridge, England: Cambridge University Press, 1987.

Haraway, D. *Simians, Cyborgs, and Women: The Reinvention of Culture*. London: Free Association Books, 1991.

Joerges, B. "Technology in Everyday Life: Conceptual Queries," *Journal for the Theory of Social Behavior* (18:2), 1988, pp. 219-237.

Jones, M. R. "Structuration Theory," in *Re-thinking Management Information Systems*, W. J. Currie and R. Galliers (eds.). Oxford: Oxford University Press, 1999, pp. 103-135.

Jones, M. R., and Karsten, H. "Collaborative Information Technology and New Organizational Forms: A Case of a Consulting Firm," paper delivered at the Pacific Asia Conference on Information Systems (PACIS'97), Brisbane, Australia, 1997.

Karsten, H. "Converging Paths to Notes: In Search for Computer-based Information Systems in a Networked Company," *Information Technology and People* (8:1), 1, 1995, pp. 7-34.

Karsten, H. "Relationship between Organizational Form and Organizational Memory: An Investigation in a Professional Service Organization," *Journal of Organizational Computing and Electronic Commerce, Special Issue on Organizational Memory Systems* (9:2), 1999, pp. 129-150.

Karsten, H., and Jones, M. "The Long and Winding Road: Collaborative IT and Organizational Change," paper delivered at the Conference on Computer Supported Cooperative Work (CSCW'98), Seattle, WA, November 16-18 1998.

Kling, R. "Cooperation, Coordination and Control in Computer-supported Work," *Communications of the ACM* (34:12), 1991, pp. 83-88.

Kock, N. "Can Communication Medium Limitations Foster Better Group Outcomes? An Action Research Study," *Information & Management* (34:5), 1998, pp. 295-305.

Latour, B. *On Technical Mediation: The Messenger Lectures on the Evolution of Civilization, Cornell University, April 1993*. Working Paper ISRN LUSADG/ IFEF/WPS–93/9–SE. Lund, Sweden: Lund University, 1993.

Lave, J., and Wenger, E. *Situated Learning: Legitimate Peripheral Participation*. New York: Cambridge University Press, 1991.

Lyytinen, K., and Ngwenyama, O. "What Does Computer Support for Cooperative Work Mean? A Structurational Analysis of Computer Supported Cooperative Work," *Accounting, Management, & Information Technology* (2:1), 1992, pp. 19-37.

Mintzberg, H. *The Structuring of Organizations*. Englewood Cliffs, NJ: Prentice Hall, 1979.

Ngwenyama, O. *Breakdowns and Innovations in Computer Mediated Work: Groupware and the Reproduction of Organizational Knowledge*, unpublished manuscript, University of Michigan Business School, Ann Arbor, MI, 1996.

Ngwenyama, O. "Groupware, Social Action and Emergent Organizations: On the Process Dynamics of Computer Mediated Distributed Work," *Accounting, Management, & Information Technology* (8:4), 1998, pp. 123-143.

Orlikowski, W. J. *Evolving with Notes: Organizational Change Around Groupware Technology*, Working Paper 186, Center for Coordination Science, Sloan School of Management, Massachusetts Institute of Technology, Cambridge, MA, June 1995.

Orlikowski, W. J. "Improvising Organizational Transformation Over Time: A Situated Change Perspective," *Information Systems Research* (7:1), 1996, pp. 63-92.

Poster, M. *The Mode of Information: Poststructuralism and Social Context.* Chicago: University of Chicago Press, 1990.

Poster, M. *The Second Media Age.* Cambridge, England: Polity Press, 1995.

Prinz, W., Mark, G., and Pankoke-Babatz, U. "Designing Groupware for Congruency in Use." paper delivered at the Conference on Computer Supported Cooperative Work (CSCW'98), Seattle, WA, November 16-18, 1998.

Robinson, M. "Computer Supported Cooperative Work: Cases and Concepts," in *Readings in Groupware and Computer Supported Cooperative Work*, R. Baecker (ed.). Palo Alto, CA: Morgan Kaufman, 1991.

Rogers, Y. "Exploring Obstacles: Integrating CSCW in Evolving Organizations," in *Proceedings of Conference on Computer Supported Cooperative Work (CSCW'94),* Chapel Hill, NC, 1994, pp. 67-77.

Schmidt, K. "Of Maps and Scripts: The Status of Formal Constructs in Cooperative Work," paper delivered at the GROUP'97 ACM Conference on Supporting Group Work, Phoenix, AZ, 1997.

Sproull, L., and Kiesler, S. *Connections: New Ways of Working in the Networked Organization.* Cambridge, MA: MIT Press, 1991.

Suchman, L. *Plans and Situated Action.* Cambridge, England: Cambridge University Press, 1987.

Swanson, E. B., and Ramiller, N. C. "The Organizing Vision in Information Systems Innovation," *Organization Science* (8:5), 1997, pp. 458-474.

VanHouse, N. A., Butler, M. H., and Schiff, L. R. "Cooperative Knowledge Work and Practices of Trust: Sharing Environmental Planning Data Sets," paper delivered at the Conference on Computer Supported Cooperative Work (CSCW'98), Seattle, WA, 1998.

Weick, K. E. *The Social Psychology of Organizing.* Reading, MA: Addison-Wesley, 1979.

Zuboff, S. *In the Age of the Smart Machine.* New York: Basic Books, 1988.

About the Author

Helena Karsten works in the Grouptechnologies Program at the Department of Computer Science and Information Systems at the University of Jyväskylä. Her long-term interests have been appropriation and use of collaborative information technologies, and structuration theory. A current project is Globe, together with Kalle Lyytinen and Ojelanki Ngwenyama. In Globe, the concern is in managing large projects at Valmet, a major paper machinery supplier. Helena can be reached by e-mail at eija@jytko.jyu.fi.

Part 9:

Transforming into New Shapes of Technology

26 THE ROLE OF GENDER IN USER RESISTANCE AND INFORMATION SYSTEMS FAILURE

Melanie Wilson
University of Manchester Institute of Science and Technology
United Kingdom

Debra Howcroft
University of Salford
United Kingdom

Abstract

Using the case study of a nursing information system, this paper demonstrates the applicability of a social shaping approach to software development for deconstructing the success/failure divide and providing a means to understand how failures occur within their social and organizational context. In contrast to many previous approaches to failure, we suspend disbelief concerning the inherent superiority of the dominant artifacts and challenge the "survival of the fittest" theory of production favored by technological determinists. Hence in this paper, we view success and failure as social constructions, the result of hindsight, and the victory of one version of the technology over contending accounts.

Two areas considered to be of central relevance to the case study are gender and nursing practice, which are posited as mutually defining. Presenting an argument for the substantial influence of gender on computer usage in organizations, and associating this to a gender perspective on nursing, the scene is set for the empirical research into how these complex and sometimes contradictory issues are played out in the local setting. In so doing, the dwindling struggle by sponsors of the case study to persuade the users to make the technology a success is witnessed. The role of gender in furnishing many nurses with a hostility to computers and the belief in the incompatibility of their roles as care givers and computer users are examined. Also attested is their ability to resist convincingly the implementation of undesirable technologies.

1. Introduction

In their influential account of the history of the development and implementation of computer-based systems, Friedman and Cornford (1989) formulate the drive to innovate as a reaction to commonly occurring failures of information systems development (ISD) in organizations. In this respect then, failure has played a pivotal role in shaping the dynamics of information and communication technologies (ICTs). It would seem that when we peer toward the horizon of information systems development, failures loom large (Flowers 1996; Lucas 1984). Consequently, the volume of work by academics and practitioners dealing with this phenomenon from the perspective of the relationship between organizational change and information technology has recently increased and includes, among others, Drummond (1996), Flowers (1996), Fortune and Peters (1995), Latour (1993), Sauer (1993), and Vaughan (1996). In addition, several writers have applied such a perspective to IT failures in the health service (Benyon-Davis 1995; Bloomfield 1995; Newman and Wastell 1996; Robinson 1994).

Although there is no "unified framework for understanding information systems failure" (Sauer 1993, p. 3), such "analyses that trade on the image of a predictable, controllable world" (Bloomfield and Vurdubakis 1995, p. 2) *by definition* neglect to problematize failure. In order to more fully understand information systems failure in organizations, it has been argued that we "need to appreciate and account for the way analyses of [failure]...operate within specific social contexts and professional milieus and are both an influence on, and shaped by, the cultural beliefs, norms and values that surround them" (Bloomfield and Vurdubakis 1995, p. 1). Accordingly, specifics of the organizational and social context in which failure takes place are of interest in this paper. The area of social life deemed significant in the case study described below is gender, given that women constitute the vast majority of end-users of nursing information systems (NIS).

The case is made for the value of a gender perspective for understanding organizations and the division of Latour, as well as technology and users. The literature review combines feminist writings on both technology and organization while challenging essentialist and determinist ideas about women's inherent incompatibility with technology. This is accomplished by identifying an archetypal female role in the workplace. Hence we look at the social, cultural, and gendered nature of nursing as it occurs in the hospital setting and examine how this relates to automated systems and user attitudes to them.

The paper is structured as follows. In the next section, we argue the appropriateness of a social shaping approach to IS failure. The third section examines gender approaches to IS in organizations and highlights gaps in the literature. The subsequent section looks more closely at gender and IT, specifically in relation to nursing practice. The case study begins in the fifth section with a description of the research approach adopted, while the case study proper is contained in the sixth section. The paper concludes with a discussion of the findings and their implications.

2. The Record of Hospital Information Systems Failure

The fate of any NIS must be considered in the light of accounts of previous hospital information systems failures. The prognosis is not a favorable one. Well publicized failures include the computer-aided despatching system at the London Ambulance Service (Benyon-Davis 1995; Dutton et al. 1995; Flowers 1994; LAS 1993; Newman and Wastell 1996; Robinson 1994; Watts 1992); the failed integration of computer systems at Wessex Health Authority (Kelsey and Brown 1993); and nurses' difficulties with data collection for NHS information systems (Brindle 1995).

While the NHS is not alone in its poor track record of implementing IT (Galliers 1994; Keen 1994), it does appear to be especially blighted by a lack of success and consistency (Grindley 1992):

> *The accident prone NHS Information Management Group which has overseen a succession of NHS computer disasters, is on the brink of another meltdown* (Observer 29.3.98).

In relation to NIS, seemingly surmountable problems still persist (BCS Nursing Specialist Group and IMG 1995; Redmond 1983). This negative perception is deepened by the lack of evidence that the resultant systems (costing annually £220 million) have benefitted patient care (Audit Commission 1992).

Given the highly political nature of IS development in the NHS, admissions of doubt about a project's chances of success are unlikely on the part of original sponsors. It has often been remarked that setting out on these projects is risky because, once a large amount of money has been committed, those who sponsored the system have a lot to lose by an admission of failure (Sauer 1993). The escalator theory explored by Drummond (1996) suggests that projects proceed even when disaster looms partly due to politics, organizational culture, and psychological issues (Dutton et al. 1995). One such issue is that desisting from a project entails a writing off of prior activities and investments - an admission of being wrong (Quintas 1996; Keen 1994).

Of particular importance to this paper is the observation that *hostility* to an IS may also contribute to its downfall. A potential for a conflict of interests and differing perspectives with relation to the new technology is revealed on examination of the stated objectives of the NHS IT strategy. The resource management initiative took place within "a long-term, systematic, though uneven and variegated imposition of "scientific management" within the NHS" (Flynn 1992, p. 36). Further, Keen emphasizes the potency of the desire for central control over data and the way in which the power of IT prevails in this regard. That the introduction of the internal market reforms was rooted in "more fundamental attempts to reshape and reposition the NHS in the minds of employees, patients and the public in general" (Bloomfield, Coombs, and Owen 1994,

p. 135) is clearly significant for nurses as it affects them in a variety of ways and is fundamental to the outcome of the case study described below.

3. A Social Shaping Approach to IS Failure

Before discussing the role of gender in user resistance, we outline our framework concerning the nature of IS failure in organizations. We contend that this is necessary because of the gap that exists both in the IS literature concerning the sorry record of information systems development and in areas of social shaping research that fail to problematize common sense notions of "successful" innovations.

Although managerialist writers and those who favor technical solutions would very much like failures to be rare (Robinson 1994), it appears that they are perhaps as frequent an occurrence as success (Lyytinen and Hirschheim 1987). Much of the research by practitioners and academics concerned with explicating this poor record of ISD (Friedman and Cornford 1989; Laudon and Laudon 1998) has entailed the identification of social and technical "factors" (see Flowers 1996; Fortune and Peters 1995; Sauer 1993; Vaughan 1996) with associated solutions for their eradication. Such approaches have been criticized for their prescriptive orientation (Hirschheim and Klein 1989) and their cookbook solutions (Dutton et al. 1995). In addition, their erroneous simplification of the complexity of organizational life (Knights and Murray 1997) into problem situations to be solved betrays the many rationalist and managerialist assumptions (Robinson 1994) underpinning this type of writing.

In contrast, the reconceptualization of IS failure made possible by recent social studies of technology is not only desirable to overcome the weaknesses of technologically determinist and procedural analyses (Dutton et al. 1995), but is also useful for a new vantage point from which to see more clearly the process of technological development itself. This is because the controversy that surrounds failure reveals processes that are otherwise obscured in the case of successful projects (Bijker and Law 1992). By the same token, however, failure cannot be researched in isolation from stories of technological achievement (Bloomfield and Vurdubakis 1995).

Further, there is the notion that distinct "relevant social groups" will define technological problems differently and there will be disagreement over what constitutes success and failure (Pinch and Bijker 1987). This suggests that the definition of failure is a social one and not shared by all groups involved in technology development or use (Lyytinen 1988; Robinson 1994). Hence the terms success/failure contain within them the value-judgements of which they are an outcome. It is more appropriate, therefore, to ask *for whom* does a failure present itself as such? In place of these terms, social shaping has long since used "stabilization" (Callon 1993; Law and Bijker 1992; Law and Callon 1992) to describe in agnostic terms (Latour 1993) the process by which artifacts come into being and are displaced in the world. Such an alternative approach recognizes that the designation of failure to an innovation is the result of hindsight.

Our aim, therefore, is to produce an account which consciously attempts to employ a framework that entails the following elements:

- A rejection of managerialist and rationalist accounts of organizational life, thereby paving the way to take into consideration the significance of relevant social groups. This then encourages a less rigid view of failure and success, perceiving them as two sides of the same coin and potentially coexistent.
- An acknowledgment of the richness and complexity of organizational life that avoids a "black boxed" approach to social and technical issues, preferring to discuss the two in one breath.
- A resistance of the static approach to systems development, opting instead to emphasize the dynamic nature of a process whose parameters are difficult to delineate. This implies that failure be perceived as negotiable and a system rather than a discrete event.

Before moving on to explore the account of failure provided in the IS literature, it is necessary to look at another aspect of organizational culture that extends beyond the walls of institutions and into society at large: gender. This will serve to deepen our grasp of the world of the (mainly female) users of the NIS.

4. A Gendered Approach to IS in Organizations

Our main concern in this paper is an examination of the role of gender in the outcome of IT adoption and stabilization. In the previous section, we presented a case for a two-pronged approach to understanding failure: by asking *how* and *for* whom does the technology present itself as having failed and by analyzing the social and organizational setting in which failure takes place. Our assumptions are that gender is a vital social factor shaping organizational life and that it is inconceivable that the interaction of nurses (largely a female workforce and occupation) with information systems is not in some way shaped by the gendered spheres we inhabit. A historically contingent fact of life is that gender relations do not just involve difference "but inequality and power - male domination and female subordination" (Webster 1996b, p. 2).

Historically, a belief in the gender neutrality of technology has dominated (Knights and Murray 1994). However, over the last decade or so, within the computing domain, discussion concerning the inequity between men and women both within industry and education has opened up (see, for example, Frenkel 1990; Klawe and Leveson 1995). Nevertheless, the issue of gender is largely *under-theorized*. Recently, researchers into gender and computing have argued for the necessity of an alliance with social science (Lander and Adam 1997; Star 1995) in order to overcome the narrowness in perspective entailed in the "add-more-women" goal Adam 1997; (Grundy 1996). There are a number of limitations with this approach since, first, it assumes success is constituted by the victory of computer systems projects and thus entails a managerialist slant and, second, it is a product of liberal feminism and technological determinism with computers being seen *per force* to be a good thing. Thus resistance or rejection is deemed undesirable. Admittedly, given the feminist origins of much of this work, there is not a demonizing of users (Oliver and Langford 1987). Rather, inconfidence and cultural bias, among others, are relevant and necessary explanations employed to combat notions of women's technological "ineptitude."

Consequently, social studies of technology focusing on issues of gender (Cockburn 1983; 1986; 1988; Cockburn and Omerod 1993; Mackenzie and Wajcman 1985; Wajcman 1991; Webster 1996a, 1996b) can offer a starting point for understanding the organizational and broader societal context of nursing information systems development and implementation. The theoretical and empirical focus will be on those activities concerned with the sexual and social division of labor, the organization of work by management, and the allocation of skill labels, skilled status, prestige, and rewards (Webster 1996b). Against notions of technological and biological determinism, we set out to show that both technology and gender are socially constructed and mutually defining. Further, we offer explanations as to the disadvantage suffered by women in their relation to technology. Our intention in so doing is to play a part in reducing the power of "common sense ideas" that typically involve underestimating women's technological ability and are partially a consequence of how technology is defined in society. The social construction approach (accompanied by a materialist grounding) is intended to persuade that women's relationship with technology is not a fixed entity but rather due to social convention. One consequence of this is that the relationship is open to change—a non-incidental consideration given that "gender research...is clearly a political project" (Alvesson and Billing 1997, p. 11).

Within the exploration of the phenomenon of failure described above, users are brought into the foreground. In this paper, a case is made for the significance of users' perceptions of and responses to technology in determining the fate of an IS. In keeping with the agnosticism advocated by practitioners of social shaping, acceptance or rejection of the technology is not adjudged as good or bad. Rather, we seek to understand the subjective rationalization of their actions by the users especially with regard to the role of gender in this process. A sub-issue of the gender research is constituted by an exploration of the gendered sphere of nursing, deemed necessary if we are to get closer to the inner world of nurses. This is discussed in the next section.

5. Gender, Nursing Practice, and IS

Having established the need to examine the social and organizational context in which failure takes place, we develop some issues relating to nurses as potential users of information systems. It is not our intention to universalize the observations and explanations proffered here, for we believe culture to be a key element in any explication of the relationship of women with technology. This means by definition that divergent accounts of nurses and information systems are both possible and welcome (cf. Bjerknes, and Bratteteig 1987; Bowker, Timmermans, and Star 1995).

In the UK-specific literature, two main influences on nursing practice are delineated: hands-on care versus professionalization. Since the former has its roots in traditional roles of women in society, it is tentatively posited as pertaining to the female sphere, while the latter, with its association with rationality, scientism, clinical intervention, and standardization, is seen as derived from a more masculine domain. This proposition is examined in relation to the empirical research and the manner in which this plays out in the local setting described.

Nursing is evidently a gendered job (Davies and Rosser 1986), not only because women make up the vast majority of workers (Corby 1997), but because of the centrality of care (the customary duty of women) to the work they carry out (Brechin et al. 1998).

In addition, the gendered occupation of nursing is associated with the notion of "a good woman" (Davies 1995). In contrast to the expertise of doctors, which is seen as scientific and the result of acquired knowledge, the role of the nurse blurs with that of the ideal woman:

> Some people believe "good nurses" have the right qualities through instinct, luck or an accident of birth, and need only a bare minimum of instruction, while others think nursing should be a graduate profession with every nurse taking a degree. [Salvage 1985, p. 51]

Thus we might say that health work is divided between the high status of curing, interventionist work (traditionally carried out by mainly male doctors) and the lower status supportive, caring work (carried out by nurses, who are mostly women) (Wagner 1993). This, we argue, makes sense only if conceived as a reflection of different and unequal roles found in society at large. Further, for nurses, proximity (both physical and emotional—epitomized by the phrase hands-on) is deemed essential if good care is to be provided (Bowker, Timmermans, and Star 1995). Given that technology is often associated with masculine culture, we suggest that the implications for nurses' relationships with computer technology are likely to be adverse: that is (1) using a computer distances the nurse from the patient, thereby preventing hands-on care being delivered, and (2) given the association of women with caring and not science (the prevail of doctors), it is unlikely that nurses will feel computers are within their realm of capabilities. The perceived negative effects of technology, combined with the lack of confidence on the part of nurses, are liable to influence the acceptance or rejection of nursing information systems in hospitals. Although popular perceptions do not anticipate that these "angels" will fight for improvements and resist unpopular changes, they nevertheless have a combative—albeit hidden—tradition (Bagguley 1992). This is portentous for the success/failure outcomes of information systems implementations.

6. Research Approach

The focus of the case study is directed to the care planning function of the Zenith Nurse Management System and its users at the Eldersite Hospital. A case study illustration is included below since it enables the researcher to ask penetrating questions and capture the richness of organizational behavior (Gable 1994). This was particularly relevant given the size and diversity of the organization in question. This approach is also recommended in instances where there is a desire to gain insight into emerging topics (the "how" and "why" questions), but there is no need to control behavioral events or variables (Benbasat, Goldstein, and Mead 1987; Yin 1989).

Multiple techniques of data collection were used. Since the research was primarily descriptive, interviews were the primary source of data collection. These have previously been identified as one of the most important sources of case study information (Yin 1989), enabling the respondents to propose their own insights as a basis for further enquiry. The interviews were semi-structured in nature. However, while not wishing to be bound by a rigid questionnaire that ensured the same questions were asked of all interviewees in the same way, an interview questionnaire was nevertheless used, both to act as an *aide memoire* and to give some structure and consistency to the process. A copy of the interview guide is available from the authors on request. The questions aimed at eliciting views and information on issues directly raised by Zenith's implementation and presence, as well as the concerns developed in the first part of the paper.

The interviews took place during a ten-month period across the various hospital sites with a cross-section of those members of staff who were deemed to be affected by the introduction of the Zenith system. These included 15 nurses (three males), eight IT project nurses (six males), three key members of the IT team, and one member of the Zenith design group. In relation to the nurses, individual and personal statistics (in terms of gender, grade, number of years in post, and union membership) were gathered to ensure that a cross-section of staff had been represented. The interviews were both taped and notes taken by hand, and in line with qualitative research approaches, witnesses were encouraged to comment freely when important issues were raised. The field research continued until nothing new was being learned from the interviews and a state of "theoretical saturation" had been achieved. In terms of data analysis, the transcripts were organized according to topics or issues raised in the discussion. These were classified into a number of main categories, which were then further broken down into more detailed seed categories.

In addition to interviews, the study also entailed observations of nurses entering details to the care plans and informal evaluations of the NIS, along with an analysis of the various texts and representational practices associated with IS training and use. Indeed, much of the story which unfolds below was pieced together through "benefits realization" and update reports and correspondence written by members of the nursing implementation team or the IT manager. In terms of the documentation, a number of reports spanning a four-year period and concerning issues ranging from initial operational requirements through to sign-off were consulted.

7. Case Study

Zenith Nurse Management System is a database system whose purpose is "to support the decisions of managers and clinicians by providing informed, sensitive and timely information in ways which are effective and understandable." The care planning function consists of a database of Care Libraries, which can be edited individually and free text added to produce a printed and standardized document. These are intended to replace the hand-written notes used by nurses in the recording of their intended care delivery for patients.

The installation of Zenith formed part of a broader implementation project management. Three years prior to the study, the product had been purchased from a small software company, following a year-long extensive evaluation and procurement process and reported to reflect "the best available at the time." It was initially piloted with the intention that it met the recommendations of the Audit Commission in the use of the Care Planning and Rostering modules and as part of the Resource Management Project.

The project was steered by a Nursing Implementation group, chaired by the Director of Nursing. In this respect then, the importance of a high level of support for the Zenith system was recognized from the outset. The implementation plan aimed to have full usage throughout the 100 wards of the Eldersite hospital within one year. This was seen as essential if hospital resources were to be used effectively, since nursing costs accounted for over 40% of revenue expenditure.

As is the case with many tales of IS in the NHS, the story of Zenith had begun with the desire and perceived need for standardized health care practice and methodological financial management (Keen 1994). Given the fact that many of the organizational changes taking place at that time were already quite unpopular with the nurses, the association in people's minds of new technology and these upheavals boded badly for the project nurses' efforts to enrol nurses to using the NIS.

7.1 The Job of Nursing and Record Keeping

Having established the importance placed upon user response to the technology in question, it is now necessary to understand the possible ways in which the Zenith system might change the way nurses do their job and the nurses' feelings toward the computers. We will begin by delineating their views of their role in the hospital and contextualizing their reaction to the new technology.

The scope of characteristics required to do the job well suggests a broad role for the nurse. Not surprisingly, the interviewees found it difficult to pin down what their job entails. The broadest response to the question, "What do nurses do?" was simply and assertively: "Care!" This was described by one nurse as "wanting the best for them." Some viewed themselves as the administrators of care prescribed by the doctors. In Mental Health the mainly male nurses were there to "help people come to terms with their illness." Finally, many expressed the sentiment that nurses tended to be "put upon." This was succinctly and forcefully communicated by an experienced Sister from Urology:

> *If no-one else will do it, it will be a nursing job.... We're easily pushed around. People play on our conscience and always have done.*

Interestingly, psychiatric nurses interviewed (who were all males) were much more positive about their own standing in relation to other professions and generally in themselves. They felt highly valued, while highlighting the fact that Psychiatric nurses tend to be far more appreciated and listened to than their colleagues in General Nursing. As one staff nurse from a Mental Health ward endeavored to explain:

> *General Nurses do as they're told. Psychiatric nurses tell people what to do. I don't know whether it's the type of person you get doing Psychiatric nursing or it comes through with the training and the experience you get.*

These observations are significant if related to nurses' attitudes to the introduction of IS on the wards, which were shaped by the way they feel they are treated in other respects. Nurses were viewed as information providers: it was recognized by Audit Commissions and Trust Management that nurses were with the patients 24 hours a day, seven days a week and during this time collected lots of information. Therefore, they could now be required to carry out this task using NIS for the purpose of resource management. Hence, their eventual rejection of the Zenith systems should not simply be seen or explained in isolation. Rather, it incorporates and is a result of a resentment of being put upon generally: i.e., "if no one else will do it, ask the nurses." After all, this is their experience of work being delegated down from the doctors. The significance for the role of gender enters since a strong case can be made that this treatment of nurses and their own sense of being undervalued is related to the job being gendered—and gendered female.

Furthermore, according to the region's Project Nurses, another reason why there is hostility to the systems is that it takes nurses away from care. For many nurses, the Zenith system, as part of the administrative tasks they had to complete, took them away from hands-on care. This was partly because the location of terminals (in the nursing station or a rest room) meant that patients had to be interviewed and assessed in their bed, and then the nurse would leave the bedside to enter the information in the computer. This was explicitly counterposed to spending time talking to the patient by many of the interviewees. Even where the nurses did not immediately draw up a care plan following patient assessment, they felt they were having to prioritize record keeping over direct care. That nurses saw this as a choice they were forced to make is underlined by the following comment:

> *Care for the patients—that's what we're here to do. We're not computer programmers.*

7.2 Kardex, Assessment Documents, and Continuation Sheets

A good deal of duplication in record keeping was due to the continuing use of Kardex—the traditional form of card-based record keeping for nurses—alongside the care plans. Indeed, all the wards still used Kardex, and for nurses this was *the* record to trust, the anchor to their work even though there were varying types of documents. The care plan never became the sole means of recording nursing work. This is a significant point for the way the Zenith system was resisted. As one staff nurse put it:

> *We're neither doing one thing nor the other, we're doing a bit of both.*

Information from the interviews suggests that nurses used up to four main documents to process the patient's stay in the hospital: the assessment document was used first to gather all the main details of the patient's state of health, as well as their vital statistics; this was then used as the basis for the care plan—the nurses responded to the diagnosis, detailing the care required; the Kardex was utilized to minute activities carried out—as per the plan; and finally this was backed up by the continuation sheets—which are bound. As one staff nurse recounts, the nurses were very loyal to this register:

> *We write it all up in the Kardex to cover ourselves....Kardex is the main stay of everything. If we were without Kardex we'd be lost.*

By contrast, the Zenith care plans were criticized for either having too little or too much detail. Some believed that the standard care plans made nurses think less about what they were doing, and thus de-skilled them to an extent. On the other hand, where a plan existed, the nurses would have to type in all the details themselves, which was no mean feat since few were trained typists. Hence, it could take longer to produce an automated care plan, especially since the nurses had to wait for the rather slow network in order to call up the Patient Administration System and get the patient's details. Even when nurses had time to spend on administration, there would often be a queue for the one terminal on the ward. In addition, the nurses still had to take written notes from the patient upon their admission to the ward and then type in the details to the care plan, thereby duplicating the amount of their administrative work—perceived by many as the most unattractive aspect of the job. Finally, the shortage of time faced by nurses meant that the Zenith system did not "make visible" how hard the nurses worked: if they were not busy, they had time to type up the plan (in retrospect, rarely an advance); if they were busy, they could not do a plan. However, management had made it clear that the automated care plan would be taken as proof of work carried out. No plan meant they had not done the work.

7.3 User Resistance

Expectations of the information technology were rather high before the installation of the system with nurses believing that it could solve their problems, relieving them of unpopular administrative tasks and freeing up their time to deliver high-quality patient care. In addition, promises were made on behalf of the system, in terms of its potential, which would be realized if users committed themselves. Efforts were made to enrol nurses to the system by participation in implementation committees, via training sessions, and benefits realization seminars. Yet, it is doubtful that these were adequate to secure support for the system. First, participation was limited because of the selection process for user representation. It had been assumed that high-ranking nurses on the ward would make the most suitable candidates, but these were often older nurses, less familiar with computers, and hostile to the information system. Second, the training strategy was cascade training—with sisters as the ward tutors who were to initially train others. This was a problem since the sister was likely to be very busy and not familiar with computers prior to Zenith training. Third, the benefits realization sessions were attended voluntarily and, therefore, unlikely to make any impression on those who were hostile to the whole IS project since they simply elected not to attend. This is where a gender angle is important: they did not regard this expertise as technical in nature.

Despite the committees, user representation, and participation, many nurses felt that they were not " really" consulted about what they wanted: they had neither been asked nor listened to. This led to the outright refusal by some to use the system. Of course, all of this was intermingled with their own fear and lack of confidence concerning information technology. That is not to say nurses have no technical expertise. While the vast array of sophisticated machinery with which the nurses mediated patient care on a daily basis is a testament to their technical capabilities, they *themselves* did not acknowledge this. In relation to the system, alternative means of persuasion such as outright coercion were evidenced, but not applied as systematically as some supporters of the system would have liked. Resistance, through non-usage, was both possible and effective: the outcome of political priorities, as well as gender-related assumptions about technology.

7.4 Management Response: Rhetoric of Retreat and the Achievement of Failure

Given the level of hostility described in this case study, it would not surprise the reader that the system was eventually withdrawn.

7.4.1 Dissent Legitimized

Nearly three years after the installation, the nurses' opinions were finally detailed in a report on the Zenith system. However, there is a good possibility that the looming decision of whether to carry on with the system (a significant cost of $26,000 annually to the NHS Trust) promoted a less survival-oriented report from the project team. Whereas previous benefits realization reports were intended to convince the reader of the need to continue with the project, this particular one raised the question of whether it was worthwhile to continue and mobilized the nurses' views to do so. Included in the report were a few suggestions of what could be done to improve the situation. This entailed the setting up of a focus group of Clinical Nurse Specialists to meet bi-monthly and to carry out ward audits of the care planning system, monitoring the functionality, identifying problems, and assessing requests in relation to reports. But this proposal now concentrated on identifying failures with the system.

7.4.2 Dissent Mobilized

Six months on, the IT manager had prepared a report for the Information Management and Strategy group at Eldersite. It is clear from this report that the impetus to make a decision about the continued implementation of Zenith arose from the suppliers demand for the aforementioned outstanding structured support fee. Eldersite were in fact in dispute with the suppliers and were receiving no external software support at this time. In the meantime, the hospital had joined a consortium "in order to ensure that the Trust were being protected." In response to Eldersite's refusal to pay the maintenance fee, the suppliers declined to release the new "improved" version of the Zenith system. The report is said to be based on a specially organized Zenith workshop where users' views had been represented by managers and end-users. Significantly perhaps, Mental Health (who liked the system and were said previously to produce plans for 99% of patients) were not represented "due to former commitments. However their views were sought separately." It is at this stage that problems with the *system*, not just the users, were enunciated.

7.4.3 System Withdrawn

Later that year, coinciding with the end of the three year roll-out period of the RMI project, in a "sign off" report from Eldersite, the decided failure of the NIS is described as non-achievement. However, something is gleaned from the ashes:

> *The principal achievement of the Resource Management Project within Eldersite has been the implementation of organizational change.*

8. Discussion

The case study raises a number of important issues related to gender, resistance, and failure.

8.1 Failure

Although the Zenith system was eventually regarded as a failure, there are several ways in which the project may be considered a success. First, Zenith was a success for its sponsors for a good proportion of its life before becoming a failure. Although the system had been a failure in the eyes of many nurses for some time, it was only when management and the implementation team decided to construct it as thus that the system was officially dubbed a failure. Hence, failure is here more appropriately viewed as a *process* rather than an event.

Second, Zenith was a success in other hospital trusts and was also a success in the Mental Health Unit. But this variation cannot simply be attributed to "mismatch." Indeed, according to the Mental Health staff themselves, care plans were *least* appropriate on their wards because of the specificity of individual illness. Yet it is here that they have the most success. This, the IT manager writes, is because of the commitment of the staff and their willingness to adapt their working practices to the exigencies of the system.

This leads to the third success of the Zenith project: it had the effect of accustoming nurses to the use of computers and to keeping records of their work. This final point is significant for the future of systems development at Eldersite and further afield. Innovation in systems design entails not only the computers and software but also the new routines and organizational behavior that are required. The implementation of organizational change—a criteria of success for the RMI—is vaunted since nurses have accepted the fact of life of keeping electronic records.

Finally, from different perspectives, the *withdrawal* of Zenith was a success. Evidently, through fear of revealing the perceived technical incompetence on the part of senior nurses, the withdrawal of the system was a victory. However, an unwillingness to conform to a system (suspected of furthering the auditing and costing practices associated with the RMI) also forms part of the explanation for user resistance through the eyes of opponents to this government policy. In sum, successful resistance achieved the nurses' objective: the withdrawal of the Zenith system.

A key point in the Zenith case study is the proven validity of arguments by social constructivists that for a technology to stabilize, the relevant social groups must be persuaded that they need to "pass this way" to solve their problem or accomplish their task. In the case of the Zenith system, an alternative route to recording nursing care was kept open. The automated care plans were *not* a substitute for all the other documentation that had preceded the installation of Zenith. Kardex remained the preferred record of delivered care. This preference was due in no small part to the persistence of an established culture and routine centered around the Kardex and reinforced by the care plan's negative qualities: its physical location, aesthetic aspec,t and limited access to it due to a dearth of PCs. And, given the necessity, due to shortage of time, to choose between giving hands on care or writing records, nurses elected care.

8.2 Gender, Care, and Resistance

The role played by gender in the demise of the Zenith system is both fundamental and indirect. Gender did not appear to have any explanatory power among the nurses since they took for granted that women should be the care givers in society. This is precisely our connection of gender with the nurse as computer user. The dominant culture at ward level prioritized physical proximity to the patient and a caring approach.

The dissatisfaction among nurses in relation to their jobs hinged on their inability to deliver emotional and physical care. It is far from clear how this frustration could ever have been answered by an IS. In the example of the Zenith system, the hands-on culture was deemed incompatible with the use of computers. Conditions in general appeared to have deteriorated and, for many, the Zenith system had made the job of nursing harder at Eldersite because it was more "involved." The changes inaugurated by Zenith's installation had increased the proportion of administration tasks in their daily lives, reducing the proportion of time spent on direct patient care. In return, no great improvement in the nursing care of the patient resulted.

Interestingly, the apparent dominance on the ward of the hands-on culture does not preclude resistance, although it seems to have had a powerful shaping influence over the way that resistance was carried out. The demands of the ward and the hierarchical nature of nursing imply certain constraints curtailing nurses' behavior. At the time of the empirical research, there was a high level of compliance with unpopular policies, according to the nurses at Eldersite. Yet disagreement festered beneath this surface. Thus, the senior nurses' seemed especially angered at the way their own reluctance to endanger patient care was used against them. Many of the nurses from different grades on the General wards resented the devaluing of their work by other health care professionals. Their treatment and lack of self assurance stands in stark contrast with those in the Mental Health Unit, as described below.

It has been argued that the predominance of a culture of hands-on care, being intimately linked with femaleness, would exclude or be unfavorable to IT if viewed as masculine culture. Indeed, many nurses—especially more senior nurses—expressed the view that the term "technology" held problems for them, preferring the term "machines." Further, the allegiance to technology was associated with climbing the ladder to management, thereby confirming the "away from care" view of technology on the part of nurses. Given that hands-on care was valued above all else by nurses, this was unlikely to make the technology appear attractive. In fact, this association would have condemned the system still further in the eyes of many nurses on the wards.

The lack of self worth, evident in some of the things the nurses' recounted, concerned not only IT but also their standing. It is significant that the male nurses interviewed did not experience technophobia or inconfidence, even when they were older and/or skeptical of Zenith's usefulness. They appeared more self-confident and had a feeling of worth that was rarely observable among those in General Nursing. This is perhaps due to the nature of the work.[1] It is impossible here to say whether this state of affairs has arisen because of the nature of the work or because it is where male nurses are predominantly found. No doubt there is a dialectical relationship between the two.

9. Conclusion

It seems impossible to even begin to understand the lives of nurses without recourse to the existence of gendered spheres. Yet, this is an invisible phenomenon, being taken for granted as a natural state of affairs. In this paper, it has not been our intention to reveal a transparent causal relationship between gender and failure/resistance, but rather to indicate the centrality of "occupational segregation by sex" (Cockburn 1988, p. 29) for characterizing and patterning user relationships with information systems. Because of the crucial role of care to nursing practice, the latter has come to be thought of as the prerogative of women in our society, establishing a triangulation of mutual constructions between females, care, and nursing. That is not the same as saying all actors are thus structured entirely within this framework, such that, for example, there are no male nurses, or that they nurse in a fundamentally different way. But it does tell us why men are an *exception* and the social price individuals pay for acting out of synch with these so-called "natural" laws of behavior. Indeed, it is the *invisibility* of the gendered nature of nursing that is so interesting and menacing at the same time. The tendency in our culture toward dichotomous classifications of the world ensures that the association of technology with masculine culture, although not stated explicitly, will make IT alien to nurses who, when carrying out their duties effectively, act out the archetypal female role. Hence the corresponding construction of technology and "objectivity" (or intervention or cure, depending on the context) as male. Given the archetypal gendered roles in health care, the division of latour that pervades most organizations is extreme in its consequences for nurses who most definitely inhabit the female sphere. This implies that the priority of physical and emotional proximity to patients will mean hands-on care over-rides all other concerns including the use of IT if it is seen to detract from this. Gender then, in this organization, plays a significant role in the fate of IS. We believe this may be generalized to other institutions, even in a more subtle and less clear-cut fashion.

---

[1]"Curing" mental health patients is achieved in the main by verbal diagnosis and discussion. Nurses on psychiatric wards were providing interventionist health care, as well as looking after people's emotional and physical welfare. In this respect, their activities were not so distinct from the roles of the doctors working in that area. Thus they felt valued.

The combination of feminist writings on technology and organizations, applied to the empirical research, resulted in observing close to how technological expertise is mutually constructed by the particular definition of technology. Again, nurses *themselves* tended to call the sophisticated artifacts that they handled with ease machines, even though these may have been run by sophisticated software. Perhaps the association of computers with new technology increases the likelihood of this distinction being drawn. Whatever the case, nurses are, to all intents and purposes, technical experts, although they may not consider themselves to be. This further suggests that women's relationship with technology is indeed due to social convention and evolves over time.

The example of nurses' counteraction illustrates how resistance is shaped by the environment in which it takes place. We saw in the empirical study that despite the stifling images of acquiescence, nurses can and do resist measures they do not like. Ironically, perhaps it is precisely that they do so *because* they perceive themselves as angels, the guardians of the health service. So although the image can be a stifling one, it can also have the opposite effect, igniting rebellion, when the service appears threatened by official policy. In finishing, if we shift from an agnostic view and take a partisan view of nurses' ability to increase their control of the workplace and achieve a certain level of autonomy, then rhetoric about professionalism used to enrol nurses by management is seen in a poor light. It serves to cut off qualified nurses from other health workers and it acts a barrier, rather than a bridge, to autonomy because it can be mobilized to exert political and moral control over nurses. Finally, its uncritical advocating of compliance with IS fails to assess the validity of such systems and their potential effects on the working lives of women.

References

Adam, A. "What Should We Do with Cyberfeminism?" in *Women in Computing*, R. Lander and A. Adam (eds.). Exeter, UK: Intellect Books, 1997.

Alvesson, M., and Billing, Y. D. *Understanding Gender and Organizations*. London: Sage Publications, 1997.

Audit Commission. *Caring Systems: Effective Implementation of Ward Nursing Management Systems*. London: HMSO Press-on-Printers Ltd., 1992.

Bagguley, P. "Angels in Red? Patterns of Membership Amongst UK Professional Nurses," in *Themes and Perspectives in Nursing*, K. Soothill, C. Henry, and K. Kendrick (eds.). London: Chapman & Hall, 1992.

BCS Nursing Specialist Group and IMG. *Benefits Realization*, monograph on Nursing Information Systems (NIS) Information Management Group of NHS Executive, London, 1995.

Benbasat, I., Goldstein, D. K., and Mead, M. "The Case Study Research Strategy in Studies of Information Systems," *MIS Quarterly*, September, 1987, pp. 369-386.

Benyon-Davies, P. "Information Systems 'Failure' and Risk Assessment: The Case of the London Ambulance Service Computer-aided Despatch System," in *Proceedings of the Third European Conference on Information Systems*, Athens, 1-3 June 1995, pp. 1153-1170

Bijker, W. E., and Law, J. (eds.). *Shaping Technology/Building Society, Studies in Sociotechnical Change*. Cambridge, MA: MIT Press, 1992.

Bjerknes, G., and Bratteteig, T. "Florence in Wonderland: System Development with Nurses," in *Computers and Democracy*, G. Bjerknes and others (eds.). Avebury, UK: Aldershot, 1987, pp. 279-295.

Bloomfield, B. P. "Power, Machines and Social Relations: Delegating to Information Technology," *National Health Service Organization* (2:3/4), August/ November, 1995, pp. 489-518

Bloomfield, B. P., Coombs, R., and Owen, J. "A Social Science Perspective on Information Systems in the NHS," in *Information Management in Heath Services*, J Keen (ed.). Buckingham, UK: Open University Press, 1994.

Bloomfield, B. P., and Vurdubakis, T. "Risk, Blame and Agency: Deliberating IT Failures in Organizations" unpublished article, CROMTEC (Centre for Research on Organizations Management and Technical Change), 1995.

Brechin, A., Walmsley, J., Katz, J., and Peace, S. *Care Matters: Concepts, Practice and Research in Health and Social Care*. London: Sage Publications, 1998.

Brindle, D. "Paperwork Hey to Nurses' Strike Action. Failure to Collect Data can Hit Employers but Not Patients, *The Guardian*, April 18, 1995.

Bowker, G. C., Timmermans, S., and Star, S. L. "Infrastructure and Organizational Transformation: Classifying Nurses' Work," in *Information Technology and Changes in Organizational Work*, W. Orlikowski, G. Walsham, M. R. Jones, and J. I. DeGross (eds.). London: Chapman & Hall, 1995.

Callon, M. "Variety and Irreversibility of Networks of Technique Conception and Adoption," in *Technology and the Wealth of Nations: The Dynamics of Constructed Advantage*, D. Foray and C. Freeman (eds.). London: Pinter, 1993.

Cockburn, C. *Brothers, Male Dominance and Technological Change*. London: Pluto Press, 1983.

Cockburn, C. *Machinery of Dominance: Women, Men and Technical Knowledge*. London: Pluto Press, 1986.

Cockburn, C. "The Gendering of Jobs: Workplace Relations and the Reproduction of Sex Segregation," in *Gender Segregation at Work*, S. Walby (ed.). Milton Keynes, UK: Open University Press, 1988.

Cockburn, C., and Omerod, S. *Gender and Technology in the Making*. London: Sage Publications, 1993.

Corby, S. "Opportunity 2000 in the National Health Service: A Missed Opportunity for Women," *Journal of Management in Medicine* (11:5), 1997.

Davies, C. *Gender and the Professional Predicament in Nursing*. Buckingham, UK: Open University Press, 1995.

Davies, C., and Rosser, J. "Gendered Jobs in the Health Service: A Problem for Labour Process Analysis," in *Gender and the Labour Process*, D. Knights and H. Willmott (eds.). Hampshire, UK: Gower Publishing Company Ltd., 1986.

Drummond, H. *Escalation in Decision-Making: The Tragedy of Taurus*. Oxford: Oxford University Press, 1996.

Dutton, W. H., Mackenzie, D., Shapiro, S., and Peltu, M. "Computer Power and Human Limits: Learning from IT and Telecommunications Disasters," Programme on Information and communications Technologies (PICT) Policy Research Paper No. 33, 1995.

Flowers, S. *Software Failure: Management Failure. Amazing Stories and Cautionary Tales*. Chichester, UK: John Wiley & Sons, 1996.

Flynn R (1992) Structures of Control in Health Management Routledge, London.

Fortune, J., and Peters, G. *Learning from Failure. The Systems Approach*. Chichester, UK: John Wiley & Sons, 1995.

Frenkel, K. A. "Women and Computing," *Communications of the ACM* (33:11), 1990, pp. 34-46.

Friedman, A. L., with Cornford, D. S. *Computer Systems Development: History, Organization and Implementation*. Chichester, UK: John Wiley & Sons, 1989.

Gable, G. G. "Interpreting Case Study and Survey Research Methods: An Example in Information Systems," *European Journal of Information Systems* (3:2), 1994, pp. 112-126.

Galliers, R. "Information and IT Strategy," in *Information Management in Health Services*, J. Keen (ed.). Buckingham, UK: Open University Press, 1994.

Grindley, C. B. B. (ed.). *Information Technology Review 1992/3*. London: Price Waterhouse, 1992.

Grundy, F. *Women and Computers*. Exeter, UK: Intellect, 1996.

Hirschheim, R. A., and Klein, H. K. "Four Paradigms of Information Systems Development," *Communications of the ACM* (32:10) 1989, pp. 1199-1216

Keen, J. "Information Policy in the NHS," in *Information Management in the Health Services*, J. Keen (ed.). Buckingham, UK: Open University Press, 1994.

Kelsey, T., and Brown, C. "Government Accused of Misleading Parliament over Wessex Health Authority's Computer Contracts: Former Tory Ministers Linked to Wasted £63m," *The Independent*, March 15, 1993.

Klawe, M., and Leveson, N. "Women in Computing: Where Are We Now?" *Communications of the ACM* (38:1), 1995, pp. 29-35.

Knights, D., and Murray, F. *Managers Divided*. Chichester, UK: John Wiley and Sons, 1994.

Knights, D., and Murray, F. "Markets, Managers, and Messages: Managing Information Systems in Financial Services," in *Information Technology in Organizations: Strategies, Networks, and Integration*, B. P. Bloomfield, R. Coombs, D. Knights, and D. Littler (eds.). Oxford: Oxford University Press, 1997..

Lander, R., and Adam, A. *Women in Computing*. Exeter, UK: Intellect, 1997.

Latour, B. "Ethnography of a 'High-tech' Case: About Aramis," in *Technological Choices: Transformations in Material Culture Since the Neolithic*, P Lemmonier (ed.). London: Routledge & Kegan Paul, 1993, pp. 372-398.

Laudon, K. C., and Laudon, J. P. *Management Information Systems: New Approaches to Organization and Technology*. Englewood Cliffs, NJ: Prentice Hall, 1998.

Law, J., and Bijker, W. E. "Postscript: Technology, Stability, and Social Theory," in *Shaping Technology/Building Society, Studies in Sociotechnical Change*, W. E. Bijker and J. Law (eds.). Cambridge, MA: MIT Press, 1992.

Law, J., and Callon, M. "The Life and Death of an Aircraft: A Network Analysis of Technological Change," in *Shaping Technology/Building Society, Studies in Sociotechnical Change*, W. E. Bijker and J. Law (eds.). Cambridge, MA: MIT Press, 1992, pp. 29-52.

LAS. *Report of the Inquiry into the London Ambulance Service London Ambulance Service*. London Ambulance Service, South West Thames Regional Health Authority, London, February 1993.

Lucas, H. "Organizational Power and the Information Services Department," *Communication of the ACM*, January 1984.

Lytinnen, K. "Stakeholders, Information Systems Failures and Soft Systems Methodology: An Assessment," *Journal of Applied Systems Analysis* (15), 1988, pp. 61-81.

Lyytinen, K., and Hirschheim, R. "Information Systems Failures: A Survey and Classification of the Empirical Literature," *Oxford Surveys in Information Technology* (4), 1987, pp. 257-309.

MacKenzie, D., and Wajcman, J. (eds.). *The Social Shaping of Technology*. Milton Keynes, UK: Open University Press, 1985.

Newman, M., and Wastell, D. G. "A Tale of Two Cities," in *Proceedings of the Fourth European Conference on Information Systems*, Lisbon, Portugal, 2-6 July 1996.

Oliver, I., and Langford, H. "Myths of Demons and Users: Evidence and Analysis of Negative Perceptions of Users," in *Information Analysis: Selected Readings*, R. Galliers (ed.). Wokingham, UK: Addison-Wesley, 1987.

Pinch, T. J., and Bijker, W. E. "The Social Construction of Facts and Artifacts or How the Sociology of Science and the Sociology of Technology Might Benefit One Another," in *The Social Construction of Technological Systems*, W. E. Bijker, T. P. Hughes, and T. J. Pinch (eds.). Cambridge, MA: MIT Press, 1987, pp. 17-50.

Quintas, P. "Software by Design," in *Communication by Design: The Politics of ICTs*, R. Mansell and R. Silverstone (eds.). Milton Keynes, UK: Open University Press, 1996.
Redmond, D. T. "An Overview of the Development of National Health Service Computing," in *The Impact of Computers on Nursing*, M. Scholes, Y. Bryant, and B. Barber (eds.). Amsterdam: North Holland IFIP-IMIA, 1983, pp. 59-69
Robinson, B. A. "And Treat Those Imposters Just the Same: Analyzing Systems Failure as a Social Process," Programme on Information and communications Technologies (PICT) Doctoral Conference, Edinburgh University, August, 1994.
Salvage, J. *The Politics of Nursing.* London: Heinman, 1985.
Sauer, C. *Why Information Systems Fail: A Case Study Approach.* Oxfordshire, UK: Alfred Waller, 1993.
Star, S. L. "The Politics of Formal Representations: Wizards, Gurus and Organizational Complexity,"in *Ecologies of Knowledge: Work and Politics in Science and Technology*, S. L. Star (ed.). Albany, NY: SUNY Press, , 1995, pp. 88-118.
Vaughan, D. *The Challenger Launch Decision: Risky Technology, Culture and Deviance at NASA.* Chicago: University of Chicago Press, 1996.
Wagner, I. "Women's Voice: The Case of Nursing Information Systems," *AI and Society* (7), 1993, pp. 295-310.
Wajcman, J. *Feminism Confronts Technology.* Cambridge, UK: Polity Press, 1991.
Watts, S. "The London Ambulance Service Crisis: Managers Created an Atmosphere of Mistrust," *The Independent*, November 8, 1992.
Webster, J. "Revolution in the Office? Implications for Women's Paid Work," in *Information and Communication Technologies: Visions and Realities*, W. H. Dutton (ed.). Oxford: Oxford University Press, 1996a.
Webster, J. *Shaping Women's Work: Gender Employment and Information Technology.* London: Longman, , 1996b.
Yin, R. K. *Case Study Research: Design and Methods.* Newbury Park, CA: Sage Publications, 1989.

About the Authors

Melanie Wilson is a lecturer in Management Information Systems in the School of Management, UMIST. For her Ph.d., she carried out an extensive study of nursing information systems, focusing on implementation issues and the role of gender as a factor leading to the rejection or acceptance of systems by users. She has also worked within the School as a research associate on a NHS-funded IT evaluation project. A trained linguist, she had previously worked in Materials Control within the export manufacturing sector, in the UK and abroad. Her interest in Information Systems derived from having experienced IS failure first hand in industry and she went on to obtain her Master's degree in Computation from UMIST in 1993. Melanie can be reached by e-mail at mcyssmw2@fs1.sm.umist.ac.uk.

Debra Howcroft is a lecturer at the Information Systems Institute and the Deputy Director of the Information Systems Research Centre at the University of Salford. She holds a Ph.D. from UMIST that focused on the nature and characteristics of Internet usage in practice. Her research interests include information society issues, evolving computer and telecommunications technologies, and the nature of methodological intervention, particularly participatory design practices. Debra can be reached by e-mail at D.A.Howcroft@salford.ac.uk.

27 LIMITATIONS AND OPPORTUNITIES OF SYSTEM DEVELOPMENT METHODS IN WEB INFORMATION SYSTEM DESIGN

Lars Bo Eriksen
Aalborg University
Denmark

Abstract

This article addresses the question of whether traditional object-oriented system development methods can be applied in the transformation of organizations toward web-based dissemination of information. System development methods have traditionally been argued as valuable tools in the process of capturing the complexity of information technology, organizations, and the interplay between the two. Relatively little is known about the application of traditional methods in web-based information system design. This is due partly to the limited history of web use, but also based in the normative argument that traditional methods do not play a significant role in WIS design. Through an action research project, the application of two system development methods (OMT and RMM) is evaluated. The research suggests that system development methods and a "traditional" IT-system perspective are applicable as a tool to capture central properties of a web information system. However, the use is limited to an interior organizational perspective: that of the information provider. From the audience context, the system development methods used did not provide proper guidance. The research suggests that future research should pursue the goal of providing better support for making and capturing design decisions about IT, individual, and organizational use from the perspective of the audience side.

Keywords: Web information system, system design, development methods, action research, emerging development contexts.

1. Introduction

System development methods are important tools for organizations in the process of introducing new information technology. The history of IS research has yielded a set of methods that guide developers in capturing and describing essential features of future systems. In a normative fashion, system development methods (SDM) provide developers with guidance on the design process through concepts, values, principles, and material resources (Lyytinen 1987). The application of system development methods in practice is a future-oriented means to change organizational and individual use of IT through a focus on the development of new IT solutions (Mathiassen 1997). For the novice developer, a SDM provides a path to follow and rules to play by. For the experienced designer, a SDM provides a framework for reflection on the development process (Mathiassen 1997). In any case, a method provides a mental tool for framing and discussing the design of the future. This paper is about the role of SDM in the context of web-based dissemination of information.

The paper departs from a dichotomy between arguments in favor and disfavor of SDM as tools in the design and discussion of web-based information systems (WIS). One stream of arguments is in favor of an evolution of SDM toward the web. In this stream of thought, a WIS is simply a specialized instance of the entire class of information systems (e.g. Isakowitz, Bieber, and Vitali 1998; Rice et al. 1996). In the light of this argument, SDM should be usable for web information system for the same reasons as they are usable in traditional contexts. The SDM of today may not be totally adequate for the special peculiarities of the web, but they are good starting points. The other stream of argument is that the web and its use is so different from what we have done traditionally that SDM in their current form not are adequate tools to apply (e.g., Braa and Sørensen 1998; Ehn 1998; Vessey and Glass 1998). The truth may very well lie somewhere within these extremes, but a common feature of both streams of argument is that they are primarily normative in nature and lack the support of empirical evidence.

This paper empirically investigates the proposition that system development methods are a valuable tool in the process of creating a web-based information system. This is done through the researcher's engagement in the development of a web information system. The remainder of the paper is structured as follows: Section 2 provides an overview on the research method. Section 3 presents the departing point, the development context. Section 4 in a rationalized manner describes the process of developing a WIS. As a reflection on the performed process, section 5 presents a model of an IS development context for WIS. Section 6 relates this research to other research contributions on the future of SDM. Finally, section 7 concludes the paper and suggests future research.

2. Research Method

A practice oriented intervention-driven research approach was selected as research method. This was selected in favor of an experimental or quasi-experimental approach. Given the hypothetico-deductive nature of the proposition from which the research departs, an experimental approach also seems an obvious choice (Patton 1990). An experiment, positivistic in nature, has traditionally proved a usable research approach in such cases. The experimental approach was disfavored because of the richness of system development in action. System design is as much reflection-in-action (Mathiassen 1997). With the desire to understand the complex social process of applying a system development method, the breakdowns, solutions, and their rationale, the AR approach was favored over the pure experimental research approach. An experimental approach was not believed to provide the desired qualitative insight into the process. Secondly, the establishment of an experiment would create an artificial situation, which could negatively influence the generality of the research. The definition and framing of action research is complicated by the absence clear theoretical frameworks and detailed guidelines for defining AR projects (Avison et al. 1999). In this section, the research will be defined in terms of previous work by Susman and Evered (1987), Baskerville and Stage (1996), and Lau (1997).

Action research (AR) is an intervention-driven approach to research (Susman and Evered 1978; Vidgen and Braa 1997). It is a future oriented and situational research approach that implies the collaboration between researcher and actors of the situation under study (Susman and Evered 1978). AR in nature is longitudinal as it applies a process focus. There are competing definitions and interpretations of the term action research (cf. Lau 1997). The approach selected is defined as an action research approach exploring the assumption that the application of a SDM was a strategy that would support the design of an information systems for course dissemination (ibid.). Baskerville and Stage define four defining characteristics interpreted from Susman and Evered. They are intervention, collaboration, creating understanding, and that the research should generate a solution to the immediate problem understanding.

These four criteria were met through the researchers participation in the development of a prototype of a web site for supporting the dissemination of course information for an open university setting. A problematic situation and the sketch of a desired solution had been stated by the responsible people of the organization *a priori* to the research project. The researcher volunteered to enter the development project and contribute with his knowledge on system development methods. The use of the method was conducted in collaboration with the participants of the development project. Over a period of four months, a team consisting of the lecturer responsible for the open university, a secretary, and the researcher applied various SDM activities in the OU context.

Lau suggests the four dimensions of type and focus, assumptions, process, and presentation as four dimensions of the design of AR projects. In the terminology of Lau, the conducted research is defined as classical action research project building on an interpretive assumption that seeks to evaluate the applicability of SDM in web design practice. The research process was designed as predetermined by the SD methods but still open to emerged factors requiring a change in process. The process included only a single group and the extent of change was limited to this context (and its audience). In the process, the researcher took the role as method expert, considering the participants as collaborators and to some extent co-researchers. The presentation style of the research is defined as an illustrative case presenting the rationalized process and the lessons learned.

The collection of research data was triangulated through the use of a researcher diary, discussions with stakeholders, and system development documentation. Along with the growing system documentation, notes taken by the researcher accumulated to an estimated 20 pages including figures. System documentation compromised 30 pages. A prototype consisting of more than 4,000 LOC in the Java programming language implemented part of the functionality required. Notes were taken as reflection-in-action and combined descriptive and interpretive elements. At the end of the project, the group held a final meeting where the researcher facilitated a discussion of findings and lessons on both the process and the product developed. This helped the OU staff define their future process and to clarify their goal further. It also validated the researcher's findings on the limitations of the application of system development methods. Two final outcomes of the research are identifiable: An evaluation of the usability of system development methods in WIS design, and a rationalized process proposal for others to explore or follow.

3. Point of Departure

The findings on method combination and application are based on a setting of an open university that wanted to create a web information system (WIS) to provide information for its students. Prior to the development project, the open university had 18 month of experience operating a "traditional" web system, based on use of the file system and standard web server. Over a this 18 month period, a "system" evolved into a coherent presentation in which the administration and lecturers of the open university provided material for the students. Over time, an increased awareness in the administration of the OU was established. The organization had problems in maintaining the information kept in the system. There were multiple causes and effects involved, of which the dominant were that information was not available as promised and that documents and links in the system were outdated and inconsistent with the reality on which the system tried to depict information. This awareness eventually lead to the formulation of the project that serves the role of empirical evidence of this research.

The existing Web system was operational, but as mentioned above, the solution was considered far less than optimal. The overall quality of the system was perceived as being too low. Rather than presenting a coherent updated structure, the system too often appeared to be inconsistent and giving promises of information not present. This lead to the definition of three overall design goals that the project should address.

The design goals expressed the desire to improve the operation of the web system in terms of *content, responsibility and automation*. The former needed an explicit address to reach a shared level of information between different courses. The responsibility for maintenance of data needed to be made explicit. Everyday life of operation was a negotiation about who should and who would not do what. Finally, the increased dynamic mass of data and the increase in pages called for automation of trivial routines, such as the removal of outdated information.

The researcher decided to rely on the general method OMT (Rumbaugh et al. 1991) and combine it with the dedicated hypermedia methods RMM (Isakowitz, Stohr, and Balasubramanian 1995). OMT was selected as primary method because it is assumed to represent "best-of-breed" in general object-oriented methods and because the method has seen some acceptance among professionals. Secondly, OMT is a generic method (Vessey and Glass 1998) that in theory should match many application domains and system development project. model. RMM was selected more on individual preference of the researcher than methodological difference from OOHDM (Schwabe and Barbosa 1994), which was considered as an alternative to RMM.

4. Design Process

In a rationalized view, the design process consisted of three activities: The decision of programming environment, the creation of a data model describing static and dynamic data properties, and the creation of the interface and interaction aimed at the audience of the web information system. In terms of reflection over the conducted process, the design process had two main activities:

- *Interior Perspective Design:* In this activity, the system is modeled and discussed from "within" the organization that eventually will operate the system. In this activity, attention is given to the description of objects, their static and dynamic relationships, and their behavior. Design ideas are motivated in a possible need in the exterior context of the system, e.g., is it of relevance for the audience, but decided upon in terms of feasibility within the operating organization, e.g., is it possible for us to provide and maintain this information.

- *Exterior Perspective Design:* In this activity, the system is modeled and discussed from the "outside," e.g., from the perspective of the intended audience of the system. What information is relevant for the audience, how should information be presented to audiences in terms of macro and micro information carrying units? Given the data model created in the interior perspective design activity, presentation units and their interrelationship is modeled.

Uncertainty in the early phase of the project toward the technical infrastructure of the Web led the project group to the decision that system architecture and system ambition had to be decided early. This was done to avoid the risk of designing a system and then at a late state in the project realize that it was not possible to implement the design due to misunderstandings or constraints of the Web infrastructure.

A four-leveled system structure extending the three layered model of Mathiassen et al. (1997) with a navigational layer to make the hypermedia interface explicit in the architecture resulted in the architecture depicted in Figure 1.

A first lesson of the project is that the choice of architecture and programming environment became a driving factor for the remaining design process. The decision on

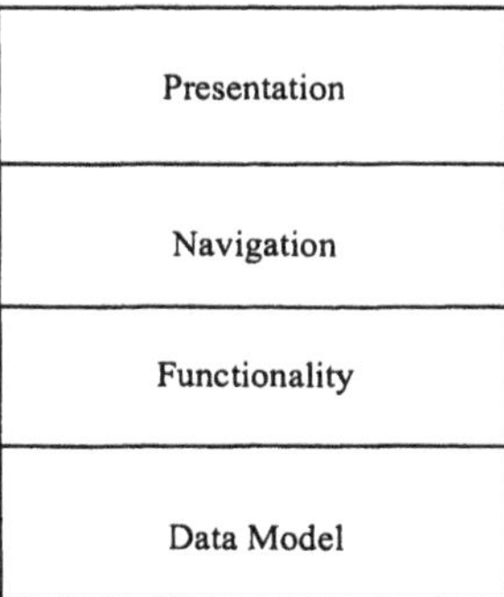

Figure 1. The Four-layered Architecture that Drove the Design of the WIS
(Adapted from Mathiassen et al. 1997)

architecture defined four separate design activities to perform: The elaboration of each of the layers in the architecture and the interplay between the layers. The perceived benefit of deciding early on system architecture was that, at any point in the process, there were no uncertainties on whether or not it was possible to implement a solution. Once the architecture had been decided upon, the remaining design task became to populate the architecture in more detail. In the spirit of OOA&D (Mathiassen et al. 1997), this was done in a bottom up fashion.

4.1 Interior Perspective

An object-oriented data model of static and dynamic relationships between objects and classes was created as prescribed by the OMT method. In addition, dynamics for the object classes were described in terms of events and actions. The model created was segmented into two components. One contains the general information about the education, another contains specific information about courses and classes. The final object model is depicted in Figure 2. Object dynamics were modeled using traditional state-transition diagrams (not depicted).

Creating the object model spawned a debate on just what objects were being modeled. The previous web solution was based on a document metaphor, as was the information circulated prior to "webification" of the information service. This lead to a debate in the development group as to whether the real world objects being modeled were documents or data entities. The researcher and senior lecturer perceived information as structured into classes and relations between these. The administrative staff perceived data as documents. Eventually the conflicting views were resolved as the architecture of the system was explained in details to the secretary and as the idea of separating data from navigation and presentation was understood. One immediate lesson to draw here is that to build an information system, one requires knowledge about what an information system is.

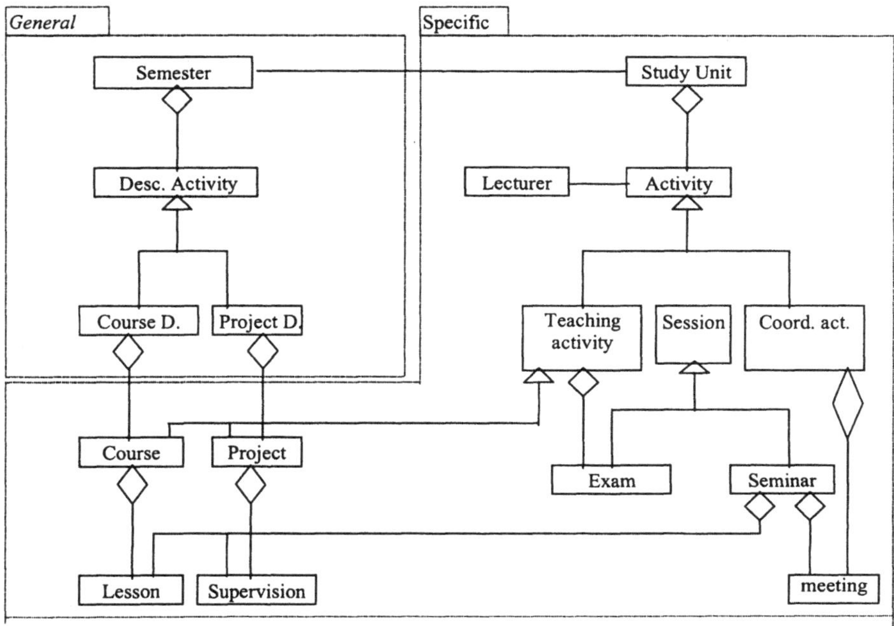

Figure 2. The Object Model Developed for the WIS

The increased focus on documents in web systems introduces confusion about what objects are being modeled. There is an extensive literature on document management systems and document repositories suggesting that documents are viable objects. In the object-oriented view of the world, a document presented to the user is a particular view on a set of object attributes. Dependent on what approach to documents to take, the project was facing decisions that would lead to very different solutions. Deciding on documents as objects would enable information providers to be in complete control of document design and a perception of the WIS as a simple document management system. Deciding on objects as classes in a traditional sense would enable the creation of functionality that extracted data from a model and generated documents in response to user queries. The project decided on the latter approach, documents as output rather than documents as data. This decision was made to enable reuse of data in different contexts.

A problem that arose in the process of dynamic modeling was that there is a large difference between "producer" users and "consumer" users of a web site. The former are the users responsible for maintaining the database, whereas the latter are the actual intended users of a WIS. The lesson to be learned here is that OMT in its foundation only supports the producer perspective. This relates to the observation that behavior of producers is expressible in terms of the objects. Once a data model is in place (e.g., Figure 1) it makes sense to discuss the behavior of producers in the light of the data model.

The project group benefitted from this activity because it led to a discussion about responsibility. With the consolidation of the data model, the project group had to decide on the interplay between objects and producer actors. This was a non-trivial balancing act between desired features in the data model and what was considered reasonable to require from producers of data (i.e., lecturers). As both the manager of the OU and the researcher were giving lectures on the open university, the requirements put forward were biased toward the limited resources available for producing and maintaining data.

4.2 Exterior Perspective

In this activity, a navigational model was constructed according to the prescriptions of RMM. A navigational model is an aggregation and abstraction over the data model. Whereas the data model reflects relationships between objects and classes, the navigational model reflects a design orientation toward the behavior of the audience. The navigational model defines an ordering of presentation nodes and the relationships between these. This allows designers to make navigational models different from the data model. The benefit of this is that the data model as perceived from the interior view can be abstracted into a different model reflecting the outside context. Over time, the data model can be kept stable and changes may independently be introduced to the navigational model (Mathiassen et al. 1997). Finally, the abstraction of the system into navigational and data model allows designers to create different navigational models for the same data model, in case there is a need to reflect different audience contexts for the same data model (Isakowitz, Stohr, and Balasubramanian 1995).

For each node in the navigational model where data could not be extracted directly from the model, e.g., in the case of a computation or the aggregation of different object attributes into new presentation units, a functional requirement was stated as an output and the required entities of the data model. Such "views" allowed designers to reduce the complexity of the data model and allowed designers to specify complex aggregations of data. A working draft model of the navigational model can be seen in Figure 3. Figure 4 illustrates how pseudo code was used to specify aggregates of information.

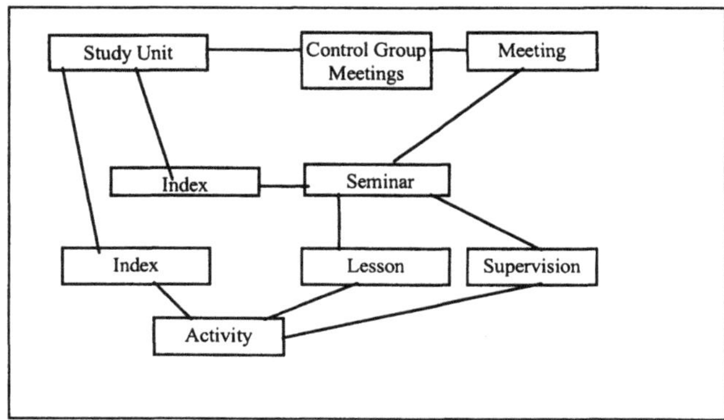

Figure 3. Navigational Model

```
View: Course overview
Classes: track, course
Input: track-id
-----------------------------------------
Function:
track tr = model.getCourse(track-id);
collection cou = agg.getCourses(tr);

output course.get("title");
for each c in cou{
 output link c.brief() to view.course(c.id);
}
```

Figure 4. Sample Pseudo Code for the Aggregation of Data from the Data Model

The project had a long and sound discussion on the navigational model. However, little guidance was found in RMM on how to actually make a model, except departing from the data model. Eventually the project by coincidence came across a book on the design of electronic documentation (Horton 1994). This provided examples of different structuring mechanisms of hypertext. Based on the inspiration from the book, the project team made its final decision on how to organize the navigational model.

The creation of the navigational model was the toughest challenge for the team and the project was not able to specify a solution using the guidelines of OMT and RMM. Eventually the project abandoned the model in Figure 4 and simply prototyped the desired interface directly on the web. By the end of the project, the group was still unsure whether this could be related to the inability of the developers to apply the navigational concepts or whether it was caused by the shift to object-orientation and in particular whether the use of aggregation and specialization caused problems for the project.

5. A Model for the Contexts of WIS

The project described in this paper illustrates why traditional system development methods are a usable tool in the design of web systems. The project departed from concerns on the operation and maintenance of a web site, concerns on the difficulty of dividing responsibility and maintaining the site with up to date information. As long as there is work going on "behind the service," work that is related to maintaining a database, there will be a need for system development methods like OMT.

However, the project also revealed problems in the design of the part of the system aimed at the actual users of the system, the target audience of the site. The absence of work tasks that can be described formally made OMT irrelevant in the design of the system properties aimed at the target audience. This was to some extent expected, and the dedicated hypermedia method RMM was introduced as a mean to deal with the specification of the interface aimed at the target audience.

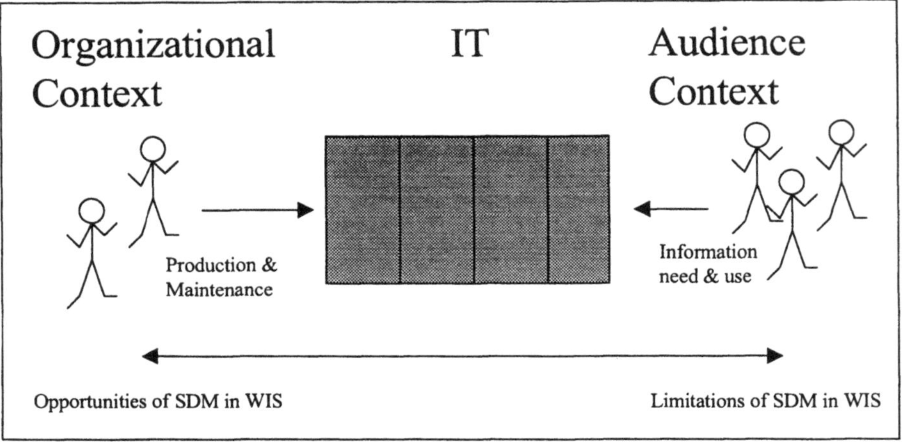

In the organizational context, the IT-based system is defined by the organization's ability to produce and maintain data. In the audience context, the system is defined through information need and use. Design in practice is the iterative balancing of requirements as perceived from the two contexts. The object-oriented SDM applied proved to be most supportive in the organizational context and less supportive in the audience context.

Figure 5. A Model of the Different Use Contexts of WIS

The research suggests that a traditional distinction between producer and consumer communities could be introduced to web information system design. Figure 5 depicts a model of a web-based information system in context. The *organizational context* represents the organization responsible for operating the WIS, for producing and maintaining data. The *audience context* represents the society of users for whom the WIS is intended.

Discussing the lessons learned on SDM in the light of Figure 5 suggests that the use of SDM is an opportunity to capture the often complex interplay between the content generating organization and IT system. Once a WIS is abstracted to an information system based on a database concept, a hard SDM approach to modeling static and dynamic data properties applies well. Once data properties have been defined, the IT system can be perceived as an administrative information system from the organizational context.

From the perspective of the audience, the concepts of OMT and RMM provided only limited support. In OMT and similar object-oriented methods, user behavior to a large extent is modeled through the dynamics of objects. As this is modeled for the organizational context, there is only a weak coupling between object dynamics and the audience community. The dedicated hypermedia design method RMM was introduced to handle the inability of OMT to model the system from the audience context. The concept of navigational model did provide designers with a notation for describing the navigational model. However even with the notation at hand, decisions were difficult to make.

6. Related Research

Method engineering has been advocated as a move to approach the diverse development contexts that occur today. In essence, method engineering is the idea that once a specific problem or situation is identified, a specific method can be "engineered" for this situation. This requires the existence of a library of method components and possibly a CASE environment that supports the method fragment selection and assembly (Harmsen, Brinkkemper, and Oei 1994). In some sense this is a formalization of what Mathiassen denotes as reflection-in-action: That practitioners do not follow a single method rigorously, but apply method fragments in response to the match between experience and a specific problem.

In the perspective of method engineering, this research suggest a model (Figure 5) of very different contexts that SDM should support. Although dating back to 1991, OMT is still a representative of the class of object-oriented methods, and there should be some generality in the observation that SDM are mainly applicable in the organizational context. This suggests that future research in method engineering should address issues of modeling a system from the perspective of the audience context.

There is a completely different chain of arguments based on communication research. In particular, the concept of genre is seeing increasing attention by IS researchers. There are competing definitions of genre, but a common definition is that a genre is a recurring type of communicative event. The introduction of genre theory to IS research is attributed to Yates and Orlikowski (1992). The concept has since then been elaborated, more recently in the context of web-based communication (e.g., Crowston and Williams 1997; Eriksen and Ihlström 2000; Watters and Shepard 1997). The power of genre is that it explains why design of (communicative) artifacts cannot be seen as independent of the established tradition of communication. Genre understanding reduces the infinite possibilities for structuring communication to a limited set of types of communication, thereby reducing the complexity of both creating and interpreting communication.

Genre theory explains why the project team experienced problems in modeling the system from the perspective of the audience context. As the web is a relatively new information infrastructure, a genre repertoire (Orlikowski and Yates 1994) has not been established in the organization. Over time this problem inevitably will be reduced as individuals and organizations develop a shared understanding of good design in terms of specific genres of WIS. It appears that the genre of course information dissemination has not reached a state of equilibrium yet. Until now, genre theory has primarily been an analytical tool. The research supports the argument of Brown and Duguid (1996) that genre may play a role as a tool in design within the audience context.

A third stream of argument that does not address the question of how to design web information systems is an artifact-oriented approach. Because of the almost infinite possibilities of restructuring practice, much research today is explorative in the sense that good ideas are presented, often along with their rationale. For instance, Sumner and Taylor (1998) present the design and design rationale of a novel approach to the dissemination of open university course information. From the perspective of SDM development and refinement, such contributions may be criticized for being to specific to provide support for the design process. However, if one takes the argument of genre design, such research contributions are exactly what is currently needed to disseminate and institutionalize new digital genres. In the perspective of Figure 5, the suggestion is that the audience context is as much understood in terms of the already installed base of similar systems. As little as students probably would accept a completely new style of course information for each and every course, students will expect similarities among different IT systems providing the same type of information.

7. Conclusion

The article departed from the question of whether the concepts and guidance in system development methods can play a role in the design of information systems in the web era. If so, there is an opportunity in SDM as a tool for describing, understanding, and designing the complex interrelationships between IT, information systems, organizations, and societies. The research suggests evidence of both opportunities and limitations of SDM in a web context. As the situation for which a system was designed is characterized by an organization disseminating dynamic information to an audience, the research and its lessons may possess generality beyond this case.

The application of system development methods in general was considered useful and worthwhile. The object model helped the organization to define the scope of the system in terms of objects and attributes. The modeling of the dynamics of the objects led to constructive discussion and solutions on how to define responsibility between the different actors in the organization. The shift from documents to objects as the basis data of the system and the implementation of a function layer made it possible for the organization to define presentation views as pseudo-queries on a database, thereby reducing the need for manual checking of links and outdated information. The methods clearly helped developers in structuring and organizing the development project into activities and in specifying outcomes. Yet there were some limitations in and lessons from method application that emerged during the project.

Emerged in response to complications in a design process of a WIS, the sharp distinction between organizational and audience contexts is proposed as a means to structure the design of web-based information systems. The different contexts are proposed as departing point for future WIS design. The organizational context is the context from which a WIS is operated in terms of production and maintenance of data. The audience context is the context in which the provided information is "consumed." System development methods proved a valuable tool to address problems of data standards, procedures for data maintenance, and static and dynamic properties of data, but only in the context of the producing organization.

System development methods failed to provide guidance for the design of the system in the audience context. Although not documented in the empirical data, the research suggests that standards, styles, and established practice in the form of specific genre characteristics play a stronger role in this than SDM. This suggest that WIS design for the audience context must be understood not in terms of the activities to perform, but in terms of similarities between WIS serving the same purpose. Future research could address the issue of how to make genre awareness and analysis more design oriented and how to integrate it with existing system development methods.

Acknowledgments

The researcher is grateful for the many constructive comments by the anonymous reviewers of the article. The researcher is also grateful to Kurt Nørmark and Lene Mogensen of the Open University of the Department of Computer Science at Aalborg University for their collaboration.

References

Avison, D., Lau, F., Myers, M., and Nielsen, P. A. "Action Research," *Communications of the ACM* (42:1), 1999, pp. 94-97.

Baskerville, R. L., and Stage, J. (1996). "Controlling Prototype Development Through Risk Analysis," *MIS Quarterly* (20), December 1996, pp. 481-504.

Braa, K., and Sørensen, C. "The Internet Factor," *Scandinavian Journal of Information Systems* (10:2), 1998, pp. 235-240.

Brown, J. S., and Duguid, J. "Keeping It Simple," in *Bringing Design to Software*, T. Winograd (ed.). New York: 1996, pp. 129-145.

Crowston, K., and Williams, M. "Reproduced and Emergent Genres of Communication on the World-Wide Web," in *Proceedings of Thirtieth Hawaii International Conference on System Sciences*. Los Alamitos, CA: IEEE Computer Society, 1997.

Ehn, P., and Malmborg, L. "The Design Challenge," *Scandinavian Journal of Information Systems* (10:2), 1998, pp. 211-218.

Eriksen, L. B., and Ihlström, C. "Evolution of the Web News Genre: The Slow Move Beyond the Print Metaphor," in *Proceedings of Thirty-third Hawaii International Conference on System Sciences*. Los Alamitos, CA: IEEE Computer Society, 2000.

Harmsen, F., Brinkkemper, S., and Oei, H. "Situational Method Engineering for Information System Project Approaches," in *Proceedings of IFIP WG8.1 Working Conference on Methods an Associated Tools for the Information System Life Cycle*, A. A. Verrijn-Stuart (ed.). Maastrict, The Netherlands: Elsevier, 1994.

Horton, W. *Designing and Writing Online Documentation*. New York: Wiley Technical Communication Library, 1994.

Isakowitz, T., Bieber, M., and Vitali, F. "Introduction to Special Issue on Web Information Systems," *Communications of the ACM* (41:7), 1998, pp. 79-80.

Isakowitz, T., Stohr, E. A., and Balasubramanian, P. "RMM: A Methodology for Structured Hypermedia Design," *Communications of the ACM* (38:8), August 1995, pp. 34-44.

Lau, F. "A Review on the Use of Action Research in Information Systems Studies," in *Information Systems and Qualitative Research*, A. S. Lee, J. Liebenau, and J. I. DeGross (eds.). London: Chapman & Hall, 1997, pp. 31-68.

Lyytinen, K. "A Taxonomy Perspective of Information Systems Development: Theoretical Constructs and Recommendations," in *Critical Issues in Information Systems Research*, R. J. Boland and R. Hirschheim (eds.). Chichester, UK: John Wiley, 1987.

Mathiassen, L. *Reflective System Development*. Doctoral thesis, Department of Computer Science, Aalborg University, Denmark, 1997.

Mathiassen, L., Munk-Madsen, A., Nielsen, P. A., and Stage, J. *Object Oriented Analysis and Design*. Aalborg: Marko (in Danish; to appear in English), 1997.

Orlikowski, W. J., and Yates, J. "Genre Repertoire: The Structuring of Communicative Practices in Organizations," *Administrative Science Quarterly* (39:4), 1994, pp. 541-574.

Patton, M. Q. *Qualitative Evaluation and Research Methods*, 2nd ed. Newbury Park, CA: Sage Publications , 1990.

Rice, J., Farquhar, A., Piernot, P., and Gruber, T. "Using the Web Instead of a Window System," in *Proceedings of CHI'96*, Vancouver, British Columbia, Canada. New York: ACM Press, 1996.

Rumbaugh, J., Blaha, M., Premerlani, W., Eddy, F., and Lorensen, W. *Object-Oriented Modeling and Design*. Englewood Cliffs, NJ: Prentice Hall, 1991.

Schwabe, D., and Barbosa, S. D. J. "Navigation Modeling in Hypermedia Applications," Internal Paper, PUC-rio, Departamento de Informatica, 1994.

Sumner, T., and Taylor, J. "New Media, New Practices: Experiences in Open Learning Course Design," in *Proceedings of CHI'98*, Los Angeles. New York: ACM Press, 1998, pp. 432-439.

Susman, G. I., and Evered, R. D. "An Assessment of the Scientific Merits of Action Research," *Administrative Science Quarterly* (23:4), 1978, pp. 582-603.

Vessey, I., and Glass, R. "Strong vs. Weak Approaches to System Development," *Communications of the ACM* (41:4), April 1998, pp. 99-102.

Vidgen, R., and Braa, K. "Balancing Interpretation and Intervention in Information System Research: The Action Case Approach," in *Information Systems and Qualitative Research*, A. S. Lee, J. Liebenau, and J. I. DeGross (eds.). London: Chapman & Hall, 1997.

Watters, C., and Shepard, M. "The Role of Genre in the Evolution of Interfaces for the Internet," in *Proceedings of NET'97*, 1997 (http://net97.dal.ca).

Yates, J., and Orlikowski, W. J. "Genres of Organizational Communication: A Structurational Approach to Studying Communication and Media," *Academy of Management Review* (17:2), 1992, pp. 299-326.

About the Author

Lars Bo Eriksen is an assistant professor at the InterMedia-Aalborg , Computer Science Department of Aalborg University. Since obtaining his Computer Science Master's degree in 1995, he has done empirical research on CASE tools, WIS development, and newspapers' use of the web. He is currently responsible for the usability laboratory at Aalborg University. Lars can be reached by e-mail at lbe@intermedia.auc.dk.

28 LESSONS FROM A DINOSAUR: MEDIATING IS RESEARCH THROUGH AN ANALYSIS OF THE MEDICAL RECORD

Marc Berg
Erasmus University of Rotterdam
The Netherlands

Abstract

Many approaches critical of traditional ISD make an important claim: they argue that an interpretative approach to human work and organizations is a sine qua non *for proper ISD and that ISD is, therefore, as much—or even more—a social science as a technical one. The argument presented in this paper is that actor-network theory might help this science to better understand and design the important phenomena taking place in and through the interrelation of human and non-human elements in a work practice.*

Through a brief analysis of the paper-based and electronic patient record, this paper demonstrates how such tools can completely transform work practices through mediating the activities of doctors and nurses: by accumulating inscriptions and coordinating events. These interrelations cannot be captured in terms of the tool serving health care workers or automating part of their tasks. Neither concept sufficiently highlights the transformations of work practice and *tasks that ensue from a synergistic interrelation. Synergy, rather, lies in mutually affording bringing out new capacities in each other. The record can only transform the activities of doctors and nurses, however, if they concurrently afford the tool to do so by partially submitting to the prerequisites of its operation. This submission is neither being deskilled nor being served—it is a form of coexisting with artifacts for which a theoretical vocabulary is as yet in its infancy.*

1. Introduction

In recent years, there has been a steady increase in critical alternatives for traditional information system development (ISD) and research. Stemming from the large number of failed ISD projects, and from concerns about the unexamined social and organizational aspects of ISD, the theoretical framework that drives most ISD is more and more under attack. This framework has been shown to be highly technology-centered and functionalist, approaching the organization and the human interactions in which the information technology (IT) will operate in the same formal and schematic terms as the technology itself (see Bowker et al. 1997; Sauer 1993; Winograd and Flores 1986).

According to the critics, there is an important reason why traditional ISD projects so often fail. In essence, the dominant, functionalist approach to ISD commits a category mistake by conceiving that *both* work practice and technology operate according to the same instrumental, techno-centric logic. Traditional ISD, in other words, mistakenly sees human work as describable by the logic that belongs to the realm of technology: as consisting of clear-cut, well-circumscribed tasks, executable in a predictable and pre-designed sequence. In this depiction of human work practices, humans are just cogs in the wheel of the larger technological system, whose work tasks are precisely describable and (no coincidence intended) fit perfectly in an authoritarian chain of command. What traditional systems design does not see, according to the critics, is that work as it actually takes place follows a fundamentally different logic: a logic of fluid interactions, of situated action, of contingencies, and local circumstances (Forsythe 1993; Star 1995a; Suchman 1994). According to these authors, information systems are primarily *social* entities: information exists only by means of constant processes of meaning-attribution and negotiation, and an information system is ultimately comprised of people communicating and interpreting data (Checkland and Holwell 1998).

The emergence of this "socially informed" or "interpretative" view of IS (Hirschheim, Klein, and Lyytinen 1995, p. 1; Walsham 1993) is an important development. It promises a reorientation of ISD methods and it poses a novel challenge to interpretative and constructivist approaches within sociology, anthropology, and social philosophy to contribute to information system development. Yet notwithstanding its credits, the theoretical framework espoused by many socially informed approaches contains a blind spot that may ultimately limit its capacity to contribute to ISD and to understand the profound nature of the changes in the workpractices. Put concisely, as a counter-reaction to the dominant, technology-driven view which "forgets" the social, these approaches generally draw upon sociological traditions that maintain the categorical distinction between the realms of technology and of human work and forget their interrelation. Aiming to emphasize the importance of human communication and interactions that constitute work practices, traditions such as symbolic interactionism and social theorists such as Habermas, Berger, Luckmann, Giddens, and Searle are brought in. Important as these innovations are, however, these theoretical traditions are stronger in emphasizing the distinctiveness and primacy of human interaction than in illuminating the dynamics of the interrelation of human action and technologies. The social-relativist theorists treat the technical as a field of meanings constructed by human interaction, while the more realist theorists tend to argue that we need two wholly different vocabularies for the two realms: the vocabulary of the natural sciences for the technical realm and an interpretative vocabulary for the social realm (see the discussion in Hirschheim, Klein, and Lyytinen 1995, pp. 46-57). In both cases, the tools to look at the shape and dynamics of the interrelation of human workers and IT are underdeveloped.[1]

Recent developments in science and technology studies (in particular those approaches loosely labeled actor-network theory) can be of help here. As several authors in the domain of IS research have already pointed out, actor network theory is particularly (in)famous for its symmetrical treatment of human and non-human actors (e.g., Bloomfield 1991; Bowers 1992; Hanseth and Monteiro 1997; Orlikowski et al. 1996; Walsham 1997). Starting from the position that *any* category we maintain in our theories should be seen as historically emerged, actor network theory is strongly opposed to theories that start from preset distinctions between the social and the technical. Rather, it studies how the distinction between the two came into being, or how what we call social is in fact shot through with, and dependent upon a wide variety of, artifacts (Latour 1996; Law and Hassard 1999). In actor network theory, a given work practice would be considered as a network of interrelated people, machines, paperwork, and architectures, which *together* produce the work practice's output. The assumption is not that a machine is a true actor in a humanist sense; rather, the assumption is that only by taking the *active* roles of all these entities into account can we hope to understand the functioning of the work practice and the interrelations between its constituents.

This paper will focus on one specific point that can be drawn out of this approach: the transformation of work practices that can emerge when the entities composing an IS—including artifacts and humans—interrelate synergistically.[2] To do so, I will draw upon the example of a very successful information technology: the paper-based patient-centered record, which emerged at the beginning of the 20th century and has remained virtually unchanged over the years. I will briefly argue how the interrelation of the activities of this paper-based technology with the activities of doctors, nurses, and clerical personnel has afforded (and survived!) the unprecedented organizational changes that typified medical organizations in the 20th century, and the changing nature and content of medical work. Doing so, I will tease out the activities that seem to characterize this technology and that are responsible for its transformative power: *accumulating inscriptions* and *coordinating events*.

In recent years, however, the paper-based patient-centered record has been called a dinosaur and its extinction—although often prematurely announced—now seems near. Subsequently, I will discuss the emergence of the electronic patient record (EPR): the electronic version of the medical record, in which many new ICT features will be integrated. Strictly speaking, this application will be much more than a record (including, for example, telemedicine applications), but since its function as the patient record will remain a core feature, we will discuss it under this name. By discussing how the

---

[1]Several authors have attempted to undo these asymmetries from within ethnomethodology, activity theory, and symbolic interactionism (see Engeström 1995; Star 1995b). For a detailed discussion of these issues, see Berg (1998).

[2]For introductory texts about actor network theory, see Latour (1987) and Callon (1987). For an introduction to actor network theory in the specific field of IS research, see Walsham (1997). The analysis presented here also draws upon Hutchins' work (1995). I remain close to the more socially informed definitions of IS as including also humans, storage cabinets, and organizational routines (see Checkland and Holwell 1998; Kling 1989). IT refers to the technological artifacts (the hardware and software).

coordinating and accumulating functions of the patient record might be transformed by moving toward such an IS, we can discuss some of the new work practices that might emerge. In this way, we can see if the analytic framework developed here can also be of use to *inform* design.

2. A Venerable Information Technology: The Paper-based Patient-centered Record

At the end of the 19th century, Western physicians kept their records in the form of casebooks, a kind of log or diary, in which they would write down (often at the end of the day) what patients they had seen, the diagnostic findings, and the actions they had recommended. These casebooks were kept in the doctor's private office and were used mainly for administrative, research, or teaching purposes. Only rarely, when later developments made that necessary, would they be used to look up an earlier entry about a patient. This was not easy, however: only very few doctors kept indexes to their casebooks and one needed to know the date of a visit to find that entry. This form of record keeping was well adapted to the exigencies of medical work in that era: doctors worked largely on their own, so there was no need to keep records for other practitioners to consult. In addition, the variety of diagnostic and therapeutic activities that a doctor could undertake was rather limited and doctors could often rely on their memories to keep track of a particular case.

The patient-centered record emerged in the United States in the first decades of the 20th century, and quickly became the standard of record keeping in Western medicine. From a bound casebook in the physician's private office, with handwritten notes gradually and consecutively filling the empty pages, the record became a patient-centered case*file*. Each patient now had his/her own record, which usually consisted of a binder or folder. In this folder, we would now find the doctor's and nurses progress notes (handwritten) and, in consecutive sections, correspondence with the patient or about the patient and standardized forms and graphs from the different laboratories and other auxiliary services. The record would be empty at the beginning and slowly fill up with loose sheets. Each new sheet was added to its own section in chronological order. As the 20th century progressed, this format remained basically the same, although the average record became much thicker. With the increased diversification of medical specialisms, laboratories, visualizing technologies, and so forth, many more sections have been added, many more standardized forms are used, and many more entries fill each individual section.

The introduction of the patient-centered record had all the characteristics of the construction of a novel IS. It did not merely consist of the replacement of casebooks by casefiles: it went hand in hand with architectural changes to hospital buildings (due to the need to have a centrally located, easily accessible record room), with the emergence of a wholly new profession (the medical record professional, who became responsible for the storage, retrieval, and quality of records), and with a novel delineation of tasks between nurses, physicians, and the new professionals. Doctors' notes now became the responsibility of the hospital. This implied that they had to thoroughly adjust their work routines: standardized forms and medical terminologies were introduced to facilitate a common use of the same file and doctors were sometimes chastised by the record professionals for not keeping their records complete. These developments did not occur without major conflicts (see Reiser 1984; Stevens 1989), but within two or three decades, the patient-centered record had become the unquestioned standard for medicine (and would largely remain so until the 1960s, see further).

In the next three subsections, I will briefly delineate some of the ways in which the patient-centered record has transformed (and become an integral part of) the activities of health care workers. In each subsection, the emphasis is on the *emergence* of new levels of complexity, or wholly new modes of working and/or reasoning. These new worlds could and can only exist in the interrelation of the activities of humans and non-humans; their characteristics cannot be reduced to the contributions of one or the other.

2.1 Transforming the Work of Individual Doctors and Nurses

In the daily work of a doctor or nurse, a huge number of documents are looked up, scanned, checked, checked off, filled in, handed over, and mailed. Many of these documents are *external memories* to the nurse and doctor. The documents provide a constantly updated record of what has been done, said, decided, or occurred. Compared to the casebooks of the previous century, the design of the patient-centered record makes central retrieval much easier. As the historian Barbara Craig (1990, p. 25) nicely puts it, "the file itself became the index": that what binds the folder—the patient—is the entity to which the search is oriented. Moreover, once the right record lies before them, experienced health care workers draw upon cues like thickness of the section, colors of different forms, handwriting of colleagues, and so forth, and find their previous entries with amazing speeds (Nygren and Henriksson 1992).

The patient-centered record carries the burden of recollecting and retelling what has happened with a patient for a wide variety of health care workers. For all of these individual practitioners, the record makes it possible to deal with potentially large numbers of individual patients and to keep an overview of their own activities and of the highly complex trajectories that their patients might have traveled.

In addition, in working with a check list, or in filling in a pre-structured form during history taking, the doctor's or nurse's work is *structured* and *sequenced*. Checking off the questions or actions, the activities of the health care worker are given shape and ordered in a pre-specified way. In this *mediation* of the health care worker's tasks by the form, some of the structuring and sequencing of his/her work tasks is delegated to this form (Latour 1996). The cognitive demands of the individual, as Hutchins (1995, pp. 151-155) would phrase it, are simplified and transformed by having the checklist or the pre-structured form do part of the thinking. In these forms, a preferred organization of questions to ask and activities to undertake is already embedded and, by drawing upon these forms, the health care worker automatically integrates this organization into his/her activities. Because of the reduction of the individual's cognitive load that occurs in this delegation, pre-structured forms and checklists afford an increase in the overall *complexity* of the work. When drawing upon such forms, for example, doctors and nurses can handle very complexly structured therapeutic treatment modalities or intricate diagnostic schemes.

With only marginal changes, then, the patient-centered record has, throughout the 20th century, cooperated with physicians and nurses in handling increasingly complex work tasks. While the overall cognitive load of these individuals probably did not change significantly, the patient-centered record took over gradually more and more memorizing, structuring, and sequencing tasks. This transformed the content of the work tasks of the health care worker, who—as one element of the information system—is now able to oversee more patients, cover longer time spans, and handle unprecedented finesses in therapeutic and diagnostic know-how.

2.2 Transforming the Collective Nature of Health Care Work

The documents that health care workers handle every day are maybe even more important for the organization and distribution of work tasks *between* health care workers and between organizational units. As we said earlier, the patient-centered record developed as a response to the problem of coordinating a growing number of people and events. More and more health care workers have become involved in the care of a single patient. Cooperation between doctors has slowly become the norm rather than the exception and the number of professions and auxiliary services has increased manifold. In addition, the modernization of hospital administration has required a streamlining of billing and throughput control, while simultaneously an increasing number of third parties are more and more interested in what takes place in what we now call the primary care process.

More and more, then, the external memory function of the patient-centered record has become geared at informing *others* of the medical history and the diagnostic and therapeutic activities undertaken. The capacities of the record that made it easy to find one's own entries also made it easy to find the entries of each other. In addition, health care workers know to find the referral and discharge letters and their colleagues' progress notes, which they know contain summaries of the case. The record affords close cooperation between two professionals without a need for personal communication between the two. The organizational memory thus created served administrators and researchers as well: compiling the data in records in specific ways (patients with a certain diagnosis or patients admitted in a particular year) has become the source upon which clinical science and hospital administration has started to feed.

The structuring and sequencing of individual health care worker's tasks that checklists and pre-structured forms achieve is equally crucial for the coordination of their tasks with each other, and with the demands of administrators, researchers, and others. By, for example, demanding a physician's signature on a medication order slip, this simple form ensures that a physician underwrites a certain change in medication, whether a nurse initiated this change or not. When different individuals work with the same pre-structured forms or checklists, they can anticipate each other's past, current, and future activities—and track each other's activities through the changes made on the form. Here, the work that would otherwise be necessary to coordinate an individual's working routines is delegated to these forms: the simplification and transformation that Hutchins speaks of is here found in the reduced work of *articulating* distributed work tasks (Mambrey and Robinson 1997; Schmidt and Simone 1996). The use of the form presets the content and shape of a particular interaction. The correct nature and sequence of steps is built into the forms, "and incorrect relations are 'built out'" (Hutchins 1995, p. 151): it is not possible to send off a medication order without a doctor's signature.

The way the record as a whole is set up, then, affords an unprecedented level of organizational complexity. At the same time, however, it allows the work around individual patient trajectories to be highly varied, ad hoc, and adapted to the particular needs that a patient might have. As a binder organized around the patient, each organizational subsection of the hospital has its own dedicated place within the record (either a whole section, or a separate form, or merely a few parts of a single form). Because the binder travels with the patient through these organizational subsections, all of the information from all of these different activities are automatically "filed as they [are] created (Craig 1990, p. 28): in its proper place and, by merely putting it in front of a section, in chronological order. In all its simplicity and matter-of-factness, organizing the plethora of data generated and retrieved by an ever-increasing array of services and professions around the least permanent participant in the network (i.e., the individual patient) is a brilliant conception. For the organization of medical work within a professional bureaucracy such as a hospital, having a patient-centered record as the core of one's information system is highly efficient and economical. In medicine, work flows can never be fully predetermined: what patients want and require varies with each patient, therapeutic interventions can yield unexpected results, and the organizational complexity of the institution itself guarantees a never ending stream of contingencies that have to be acted upon immediately (Strauss et al. 1985). In such a situation, tying the institution's core coordinating mechanism to the actual object of action is much more efficient than centralizing it by organizational subunits, for example. It is the perfect tool for this professional bureaucracy: it is a coordination device that anticipates the limits to predetermined coordination.[3]

2.3 Redrawing the Temporal Organization of Medicine

The emergence of the patient-centered record also played an important role in the way medicine became restructured *temporally*. At the end of the 19th century, the emergence of physiology gradually started to replace medicine's focus on anatomical *structure* by a temporal and quantitative focus on bodily *processes*. Several authors have argued how the emergence of *graphs* was a vital part of this transition: graphs were life—redefined as process—"'inscribing itself on paper (Braun 1992, p. 83; cf. Cartwright 1995). A graph transformed the subjective, fleeting experience of a physical event into "an objective, visual, graphic representation that was a permanent record of the transient event, a record that could be studied and criticized by a single physician or by a group" (Braun 1992, p. 18).

Such developments in medical theory, however, did not always immediately affect the way medicine was practiced (Warner 1986). In the everyday practice of medicine, the focus on anatomy as the site of disease and the anatomical lesion as the focus of interest lasted well into the 20th century (Armstrong 1985). A thoroughly temporal focus,

[3]There are limits to this effectiveness, however, that are partly inherent to its current incarnation; see further.

historizing the patient's body in a new, unilinear, and universalizing way only emerged *after* the appearance of the patient-centered record. In the everyday practice of medicine, this latter technology materialized, in a more general way, the logic of the graph. The patient-centered record would generally contain several graphs (pulse, temperature) and it would contain temporally organized tables for quantifiable urinalysis results, blood test results, and so forth. In addition, its very functioning created a history for the patient. Its sections were all organized chronologically, so that leafing through the record would become a journey through *time*. More precisely, it would become a journey through physiological time. The repetition of pages, the sequences of tests, and the grids of the graphs transport the reader into the time zones of the rise and fall of temperature peaks, the increase and decline of blood cell levels, or the steady or irregular growth of a tumor (Berg and Bowker 1997).

Compared to the series of brief narratives in a casebook, the patient-centered record affords a physiological historization of the body in several ways. Its thickness is already a measure of time—thick files emphasizing chronicity—and its standardized, pre-formatted, and serially stored forms structure and unify time much like the graph's grid does. In the further history of medical practice, this historicity has generalized beyond the emphasis on physiology—and this generalization is only thinkable from a practice that has already incorporated the patient-centered record in its core. Early in the 20th century, reformers had already emphasized the importance of *follow-up*, which was only possible when a proper record system was in place: checking and registering the outcome of one's therapeutic activities, they argued, was a *sine qua non* for the development of clinical science (Reiser 1984; Reverby 1981). This emphasis on follow-up was followed later by an extension into the other temporal direction of the linear time grid: early diagnosis and intervention. When both life and illness are seen as a temporal continuity, intervening early to prevent later disease has become a logical possibility. "[Pushing] the identification of illness or its precursors back in time" was part and parcel of a logic of medical intervention that had shifted from the lesion to a "temporal space of possibility" (Armstrong 1985). This temporal inversion between the moment of intervention and the onset of symptomatic disease—the notions of early diagnosis and prevention—is, of course, unthinkable without a record in which these long time stretches are brought together and made over seeable. This temporal space that now typifies medical work, then, did not emerge as some ephemeral discursive notion or idea: it is partly performed in the very structure and functioning of the patient-centered record.

2.4 Mediating Work: Accumulating Inscriptions and Coordinating Events

In the previous three subsections, I have attempted to illustrate some of the ways the mediation of nurses' and doctors' activities with a specific information technology may fundamentally transform the work practices involved. The claim is not that the patient-centered record caused all these changes; the claim is that this device was a *sine qua non* for these changes to occur and that it is, therefore, a constitutive element of modern Western medicine. It does so in two, closely intertwined ways (Berg 1999). First, through structuring and sequencing the work of health care workers, the record *coordinates* activities and events at various locations and times. This coordination function affords highly complex decision making by individual health care workers, the cooperation of several specialists around a single patient, and the practice of follow-up. Second, the patient record *accumulates* all inscriptions[4] that are gathered during the course of a patient trajectory, resulting in the external memory referred to above. The accumulation of data from a plethora of sources in the different parts of the patient-centered record afford the individual health care worker to keep an overview of increasingly complex trajectories. It likewise generates the possibility of the secondary use of health care data for research and administration. It is the specific form that this accumulation takes in the patient-centered record, also, that helps produces the particular temporal space of possibility typifying late 20th century medicine.

For the record to afford these transformations, the health care workers interacting with the record will have to align themselves to the demands of the tool. They have to let the tool structure and sequence their activities for the linkages to persist and they have to follow the preset classifications that the sections of the record offer lest their entries will not be retrievable at some later date. They have to do the work of articulating the preset work flows in the checklists and forms to actual work tasks, whose details always vary, and that are always structured by many more pressing issues and needs than could have been foreseen (Collins 1990; Suchman 1987). Likewise, they have to do the work of translating locally generated information into data-items that will keep their meaning once they are transported out of the local setting. The health care worker's inscriptions need to be standardized so that different professions can understand each other and so that administrators and researchers can find what they require. Health care workers, in other words, have to "disentangle" the information from the local networks in which their meaning was self-evident, and "frame" it in such a way that the record indeed accumulates comparable entities (Callon 1999).

In modern medical work, then, health care workers truly act *with* the record. Traditional ISD cannot conceptualize this interrelation properly, since for this approach the human entity is just an extension of the larger technical system. Most theoretical approaches drawn upon by the socially informed alternatives, however, equally do not have the tools to look at this interrelation, or to study the historical transformation of the emerging hybrids of people and technologies (Latour 1993; Walsham 1997). The record does not merely support the work of doctors and nurses: such a phrasing downplays the constitutive role of the former and presupposes a subordination that does not do justice to the way the record's and worker's activities are interrelated and interdependent.

3. What Electronic Patient Records Could Bring

Notwithstanding all its achievements, the patient-centered medical record has increasingly come under fire during the last few decades. The incessant growth of the record is now more and more seen to hamper its functioning. With the increase in sources generating information for the record, and with the increase in the amount of information each source

[4]The term inscriptions refers to the marks left in records by both people (doctors, nurses) and machines (laboratory equipment, monitoring devices). It emphasizes the importance of the activities of reading and writing in the scientific and science-based practices (Latour 1987).

generates, weighty records of over 500 pages are not unusual in current hospitals. This threatens the effect of integration that the binder, when it was introduced, so nicely achieved: the "pastiche" of reports, statements, and narratives has become so large that it is no longer over seeable (Reiser 1978, p. 209). In addition, with the growth of the record, and with the increasing use by an increasingly wide circle of professionals and services, the costs of storing, maintaining, retrieving, and transporting the records become momentous (Dick, Steen, and Detmer 1997).

One interesting line of criticism on the record as we described it is that, in its current functioning, it is not nearly as patient-centered as its name would have it. It might have been a brilliant invention when compared to, for example, the earlier logbook, yet its current implementation leaves much to be improved. The binder may have the name of a single patient on it, for example, but its separation in sections reflects the organizational design and needs of the medical institution more than the perspective of the patient's trajectory. Moreover, the patient-centered record would only truly live up to its name if only *one* record would exist per patient. Yet even within one hospital, several patient-centered records of one patient often exist side by side. As is still the case in most current hospitals, separate records exist for each outpatient specialty and for inpatients.

An electronic patient record (EPR), many authors argue, holds the potential to overcome many of these problems to create a truly unified patient record that would integrally cover the patient's (para-)medical history from the cradle to the grave (Dick, Steen, and Detmer 1997). So far, however, truly integrated record systems that have eradicated paper are extremely rare (Levitt 1994; Mohr *et al.* 1995). The design and implementation of such an IS runs into all of the problems and issues that have been already painfully experienced in other information-dependent organizations (Kumar, van Dissel, and Bielli 1997; Sauer 1993). What, then, could EPRs bring to medicine? Drawing upon the analysis made above, the following two subsections will discuss some of the new ways an EPR could perform its coordinating and accumulating functions. In each subsection, I will discuss some of the ways these novel functions could be employed so as to mediate health care workers' activities in ways that benefit them as well as the central actor of health care work: the patient.

3.1 Electronic Accumulation

In a paper-based record, accumulation of inscriptions occurs through the addition of entries by health care workers, combined with the way the forms and the binder order these inscriptions. The powerful way that graphs or tables can create overview and achieve a meaningful link between the individual inscriptions is easily overlooked: after all, these are mere sheets of paper, and conceptualizing their role as an active entity might be counter-intuitive at first glance (Hutchins 1995). Yet it is evident that the computational powers of ICT far outstrip those of a paper sheet. An EPR could change the accumulation function in four ways:

- It could *draw upon larger databases*: The ability of ICT to access records in a (distributed) database is incomparable with the effort it takes to physically access a paper-based medical record.

- It could perform *more powerful operations* on these data: An integrated EPR could synthesize overviews of workload, monthly throughput overviews, but also run checks on data entered to flag incompatible drug combinations, for example.
- It could *more easily allow for changing the logic of accumulation*: Whereas in the paper-based record, the information is always presented per patient and by source, an EPR could facilitate searching the database by disease-category, by treating specialist, and so forth.)
- It could *make real-time accumulation possible*: The EPR could perform all of the above mentioned operations in real-time (give immediate feedback upon the entry of a wrong medication, for example).

For all of this potential to become true, however, the nature of the information fed into the record is crucial (Burnum 1989; van der Lei 1991). The entries that health care workers make in today's paper records are mostly made for the purpose of the proper unfolding of the primary care process. The notes made and orders entered are meant to be understood by colleagues and other "insiders." For these purposes, omitting issues that are self-evident and heavily dwelling upon local dialects are the rule rather than the exception. Such features are actually highly functional for an optimal and flexible performance of common tasks (Garrod 1998; Heath 1982), but the information thus produced is strongly context-dependent. In more general terms, information is not a kind of fixed substance that you can store and retrieve: it should rather be conceptualized as a *flow* that is always directed to somewhere and coming from somewhere else and can only be understood in the light of this purposeful directionality (cf. Agre 1995; Nunberg 1996).

For any automatic operation on these data entries, then, the need to disentangle them and frame them for the accumulation purpose at hand becomes acute. This need to be meticulous; to code data entered or to use even more pre-structured forms is a demand that is inevitable when EPRs are developed—but that should not be taken lightly. This further alignment to the demands of the novel IS can easily become just an additional burden for the health care workers who have to work with this system. For this to be avoided, the articulation between the activities of health care workers and the electronic accumulation features of the EPR is precarious and should be sought in a combination of strengths.

So how do we achieve this synergy? It should not be sought in an *imitation* of human activities by the computer. In the 1970s and 1980s, attempts to build expert systems mimicking the way humans supposedly reason (in order to outperform them, or to have electronic "experts" available where human experts might be absent; Clancey and Shortliffe 1984) has never had much impact on the practice of medicine.[5] Albeit of academic interest, from an ISD point of view it is a certain dead-end to start with putting most of the computational potential of a tool in the attempt to make it do everything that the user already does without it.

A system that generates *reminders* or *critiques* on the basis of data that are entered in a EPR, on the other hand, needs only a simple rule-base to alert health care workers about, for example, potential contraindications for suggested medications (e.g. McDonald

[5]For this story, see Berg (1997).

et al. 1984; van der Lei et al. 1993). Such a tool employs the EPR's accumulation function to link some crucial bits of information together that may be located in different parts of a record, and thus frequently overlooked by the health care worker. At the same time, such a tool requires the precise entry of only those data-items needed for the generation of the reminders or critiques. In a well-crafted design, this precise entry fulfills more functions at once: in a system checking medication patterns, for example, the medication data it can draw upon would have been already entered through a coded list to facilitate the printing of prescriptions.

Synergy will likewise not be found in the attempt to create "information superhighways" (Dick, Steen, and Detmer 1997). The disentangling and framing of data will allow data to be transferred from one specific context to another—but the idea to create context-*free* data, usable by everyone for whatever purposes, is based upon a fundamentally mistaken conception of the nature of information. Moreover, when the benefits of the disentangling and framing work do not return to those who do this work, synergy cannot be achieved either (Robinson 1993). The head of a nursing unit might find a precise workload measuring system very useful, for example, but it would be problematic to ask health care workers to meticulously register their activities when those precise data do not benefit the primary care process itself. In this sense, systems that help general practitioners organize preventive activities *do* function synergistically. Such systems may generate lists of patients requiring a vaccination, for example, and automatically print out letters for these patients. Here, doctors will need to add additional codes to their files that subsequently help them attain a level of preventive care impossible beforehand.

3.2 Electronic Coordination

Paper-based patient records contain, and are themselves, powerful coordinating artifacts: checklists, forms, medication slips, the binder separated in sections, and so forth. Here again, however, the computational powers of ICT can change the coordination function of the record in four ways:

- It could *track* events and *send messages* to trigger these, and so coordinate them more powerfully.
- It could *sequence and structure activities more powerfully* (for example, by not letting a health care worker proceed to a next step before a previous step is completed).
- It could make *synchronous coordination* possible (the speed of electronic communication makes possible the simultaneous coordination of activities in different geographical sites).
- It could facilitate *coordination between more locations and/or more entities* (once an infrastructure is installed, all the above mentioned functions can be distributed over larger numbers of recipients).

As with electronic accumulation, the more powerful coordinating functions of an EPR can only properly articulate with health care workers' activities if the latter become more strongly aligned to the record's demands. As the critics of traditional ISD have well established, complex work activities such as health care work do not come structured in a logic of flowcharts and unequivocal decision rules (Star 1995a; Suchman 1987). Health care work is characterized by a constant "making relevant" of organizational routines and technical possibilities for individual patient trajectories, whose courses are never fully predictable (Berg 1997; Strauss et al. 1985). As these critics have argued, the mechanical representations of work embedded in traditionally developed information technologies clash continuously with the contingent and complex logic of actual work practices. It is then up to the health care workers to "repair" the cracks that this causes in the flow of their activities (Button 1993). This generates the failures of so many IT applications: the technology creates difficulties in "good working practices...because it is insensitive to the contextual reasons for the existence of those practices" (Button and Harper 1993).

So how do we avoid these failures and achieve synergy? Within medicine, many examples can indeed be seen when ICT is employed in an attempt to structure and sequence health care work as if it indeed *is* or *should* proceed with the mechanical logic of a flowchart. Several attempts to create clinical pathways,[6] for example, structure and sequence the work of nurses and doctors in such detail that the central importance of articulating such general pathways to individual trajectories is denied. The latter becomes noise upsetting the smooth flow of the preset categories of trajectories rather than the core activity typifying health care work. Here, designers have indeed collapsed the nature of their tool with the nature of the work practices in which that tool is to function.

Yet examples of synergistic interrelations of ICT in health care work practices do exist—and in these circumstances, it is in the alignment of their activities to the ICT that the latter affords new capacities to arise at the level of the work practice. As a simple example, consider a program that is being tested in Dutch general practitioner practices to "change the behavior of general practitioners ordering laboratory tests."[7] Ordinarily, practitioners use paper order forms that list all of the tests that they can order from the local laboratories. To have tests performed for a specific patient, they only have to check off the tests and mail the form. The new software module replaces the paper forms. Now, a practitioner first has to enter the patient problem for which the tests are ordered, upon which s/he is presented a shortlist of tests that are appropriate for such a problem (these recommendations are based on Dutch General Practitioners' standards). This is a subtle enforcement of a structure to the practitioner's test selection process: the practitioner can add or remove tests at will, but having to go through an explicit problem-selection step and then having to decide whether to modify a preset list (which has the weight of being derived from a professional standard) turns test selection into a rather different process. The delegation of tasks to the software program, here, simplifies *and* transforms the doctor's cognitive task: the doctor can start from an already assembled list, yet by doing so, and by enforcing an additional step of problem-selection, the program integrates its preferred list of tests into the doctor's thinking process and enforces a critical reflection on the test-ordering activity. The overall effect has been a significant reduction in the number of tests.

The coordinating functions of EPRs are already referred to as basic prerequisites for the transformation from care institutions that are organized around "functions" (the

---

[6]Clinical pathways are protocols that outline the diagnostic and therapeutic steps to be taken for a certain category of patients.

[7]A discussion of the project can be found at http://www.eur.nl/fgg/mi/annrep/p15.html. The mode of working is one of the two investigated.

specialisms, the labs) to institutions that are organized around patient trajectories (Dick, Steen, and Detmer 1997). In the former, activities around the patient are coordinated first and foremost according to the internal working schemes of the different functions: if a patient arrives at 3:00 p.m. at the blood puncture service, s/he is put on their worksheet at that time and processed whenever the next empty slot occurs in their schedule. In his/her trajectory past several such laboratories and services, the patient might have to spend hours to undergo interventions that, taken together, maybe take a few minutes. Similarly, when a patient arrives at the desk of a specialist, s/he is treated according to their logic and only secondarily will attention be paid to the parts of the patient's trajectory that fall under the domain of other specialisms.

Organizing events around patient trajectories, however, would necessitate a level of coordination *between* functions that will require a thorough reorganization and integration of these functions in conjunction with an IS whose coordinating activities will take the individual patient's trajectory as its point of departure. What this will look like is hard to predict: the organizational changes that will need to occur hand in hand with the development of such an ICT application can only take place iteratively, in a process geared toward learning from all the unforeseen consequences and unanticipated uses that occur at each step (Atkinson and Peel 1998). A system that will help specially designated doctors or nurses to function as "case managers" might help overcome the lack of continuity of care between different specialisms and services. By pooling information and coordinating events, the application might afford the case manager to help integrate the overall distribution of tasks over different functions in unprecedented ways. The redistribution of responsibilities that would be implied by the use of such a system would ensure a conflict-rife implementation process; yet it could be an important part of a more patient-oriented care process.

4. Conclusion

As a critique on traditional ISD, the alternative approaches discussed in this paper make an important claim: they argue that an interpretative approach to human work and organizations is a *sine qua non* for proper ISD and that ISD is, therefore, as much—or even more—a *social* science as a technical one. In this paper, I have argued that actor-network theory might help this science focus at the important phenomena taking place in and through the interrelation of human and non-human elements in a work practice.

The brief analysis of the paper-based and electronic patient record demonstrated how such tools can completely transform work practices through mediating the activities of doctors and nurses by accumulating inscriptions and coordinating events. These interrelations cannot be captured in terms of the tool serving health care workers or automating part of their tasks. Both concepts do not sufficiently highlight the transformations of work practice *and* tasks that ensue from a synergistic interrelation. Similarly, a synergistic interrelation only exists where the tool's operation goes beyond serving or automating. If the EPR simply imitates what humans or the paper record did before, I argued, its computational powers are exhausted in imitating the strengths of other entities in the network and in developing its own potential contribution. The synergy lies in mutually affording bringing out new capacities in each other: doctors handling more complex diagnostic categories, nurses overseeing longer parts of their patients' trajectories, and records allowing secondary usage for research and administration. The record can only transform the activities of doctors and nurses, then, if they concurrently afford the tool to do so by partially submitting to the prerequisites of its operation. This submission is neither being de-skilled nor being served—it is a form of coexisting with artifacts for which a theoretical vocabulary is yet in its infancy (Knorr-Cetina 1999; Latour 1996; Star 1995b).

The discussion of the EPR examined whether the elementary framework discussed here could also be used to explore the normative question how such an EPR should be developed. Through mapping out the different ways in which the EPR's functioning differed from its paper predecessors, it was possible to discuss at least the contours of fruitful design directions. The important trick will be to learn from the paper-based record, a simple and robust device, very flexible in its use, yet affording health care workers the ability to handle elaborately structured diagnostic and therapeutic know how and sustaining an awesome level of organizational complexity. The danger lies in the traditional ISD notion that everything should be modeled, planned, and controlled top-down. Apart from the inflexibility that ensues from such an approach, it assumes that the intelligence present in the IS has to be preset in the design and come from the designer. This underestimates the collective intelligence that exists in the articulations of already existing tools, skills, and organizational routines, upon which a novel application has to build (Henderson 1998). These have evolved together over time, their articulations emerging in a learning process to which designer, health care worker, and new tool contribute interatively, sparking off insights from expected and unexpected interactions with each other (Suchman 1994). On the other hand, finding synergy will require that we allow our sociological categories to be trespassed by technology: to interfere in the user's communications and to let them be affected by their tools.

Acknowledgments

I would like to thank Jos Aarts, Els Goorman, Pieter Toussaint, Berti Zwetsloot, my colleagues within the members of the Social Medical Sciences Department, and the two anonymous referees for their critical comments and stimulating discussions.

References

Agre, P. E. "Institutional Circuitry: Thinking about the Forms and Uses of Information," *Information Technology and Libraries* (14), 1995, pp. 225-230.

Armstrong, D. "Space and Time in British General Practice,"*Social Science and Medicine* (20), 1985, pp. 659-666.

Atkinson, C., and Peel, V. J. "Growing, Not Building, the Electronic Patient Record System," *Methods of Information in Medicine* (37), 1998, pp. 206-310.

Berg, M. *Rationalizing Medical Work. Decision Support Techniques and Medical Practices.* Cambridge, MA: MIT Press, 1997.

Berg, M. "Accumulating and Coordinating: Occasions for Information Technologies in Medical Work," *Computer Supported Cooperative Work* (8), 1999, pp. 373-401.

Berg, M. "The Politics of Technology: On Bringing Social Theory in Technological Design," *Science, Technology and Human Values* (23), 1998, pp. 456-490.

Berg, M., and Bowker, G. "The Multiple Bodies of the Medical Record: Towards a Sociology of an Artefact," *The Sociological Quarterly* (38), 1997, pp. 511-535.

Bloomfield, B. P. "The Role of Information Systems in the UK National Health Service: Action at a Distance and the Fetish of Calculation," *Social Studies of Science* (21), 1991, pp. 701-34.

Bowers, J. "The Politics of Formalism," in *Contexts of Computer Mediated Communication,* M Lea (ed.). Hassocks, UK: Harvester, 1992.

Bowker, G. C., Star, S. L., Turner, W., and Gasser, L. (eds.). *Social Science, Technical Systems, and Cooperative Work: Beyond the Great Divide.* Mahwah, NJ: Lawrence Erlbaum, 1997.

Braun, M. *Picturing Time: The Work of Etienne-Jules Marey (1830-1904).* Chicago: University of Chicago Press, 1992.

Burnum M. "The Misinformation Era: The Fall of the Medical Record," *Annals of Internal Medicine* (110), 1989, pp. 482-484.

Button, G. (ed.). *Technology in Working Order: Studies of Work, Interaction, and Technology.* London: Routledge, 1993.

Button, G., and Harper, R. H. R. "Taking the Organization into Accounts," in *Technology in Working Order: Studies of Work, Interaction, and Technology,* G. Button (ed.). London: Routledge, 1993.

Callon, M. "Society in the Making: The Study of Technology as a Tool for Sociological Analysis," in *The Social Construction of Technological Systems,* W. E. Bijker, T. P. Hughes, and T. Pinch (eds.). Cambridge, MA: MIT Press, 1987.

Callon, M. "Actor-Network Theory: The Market Test," in *Actor Network Theory and After,* J. Law and J. Hassard (eds.). Oxford: Blackwell, 1999.

Cartwright, L. *Screening the Body: Tracing Medicine's Visual Culture.* Minneapolis: University of Minnesota Press, 1995.

Checkland, P., and Holwell, S. *Information, Systems and Information Systems: Making Sense of the Field.* Chichester: Wiley, 1998.

Clancey, W. J., and Shortliffe, E. H. (eds.). *Readings in Medical Artificial Intelligence: The First Decade.* Reading, MA: Addison-Wesley, 1984.

Collins, H. M. *Artificial Experts: Social Knowledge and Intelligent Machines.* Cambridge, MA: MIT Press, 1990.

Craig, B. L. "Hospital Records and Record-Keeping, c. 1850 - c. 1950. Part II: The Development of Record-Keeping in Hospitals," *Archivaria* (30), 1990, pp. 21-38.

Dick, R. S., Steen, E. B., and Detmer, D. E. (eds.). *The Computer-Based Patient Record: An Essential Technology for Health Care.* Washington, DC: National Academy Press, 1997.

Engeström, Y. "Objects, Contradictions and Collaboration in Medical Cognition: An Activity-theoretical Perspective," *Artificial Intelligence in Medicine* (7), 1995, pp. 395-412.

Forsythe, D. E. "The Construction of Work in Artificial Intelligence," *Science, Technology and Human Values* (18), 1993, pp. 460-479.

Garrod, S. "How Groups Coordinate Their Concepts and Terminology: Implications for Medical Informatics," *Methods of Information in Medicine* (37), 1998, pp. 471-6.

Hanseth, O., and Monteiro, E. "Inscribing Behavior in Information Infrastructure Standards," *Accounting, Management and Information Technologies* (7), 1997, pp. 183-211.

Heath, C. "Preserving the Consultation: Medical Record Cards and Professional Conduct," *Sociology of Health and Illness* (4), 1982, , pp. 56-74.

Henderson, K. "The Role of Material Objects in the Design Process: A Comparison of Two Design Cultures and How They Contend with Automation," *Science, Technology and Human Values* (23), 1998, pp. 139-174.

Hirschheim, R., Klein, H. K., and Lyytinen, K. *Information Systems Development and Data Modeling. Conceptual and Philosophical Foundations.* Cambridge, UK: Cambridge University Press, 1995.

Hutchins, E. *Cognition in the Wild.* Cambridge, MA: MIT Press, 1995.

Kling, R. "Theoretical Perspectives in Social Analyses of Computerization," in *Perspectives of the Computer Revolution,* 2nd ed., Z. W. Pylyshyn and L. J. Bannon (eds.). Exeter, UK: Intellect, 1989.

Knorr-Cetina, K. *Epistemic Cultures: How the Sciences Make Knowledge.* Boston: Harvard University Press, 1999.

Kumar, K., van Dissel, H. G., and Bielli, P. "The Merchant of Prato—Revisited: Toward a Third Rationality of Information Systems," *MIS Quarterly* (22), 1998, pp. 199-226.

Latour, B. *Science in Action.* Cambridge,UK: Cambridge University Press: Cambridge, 1987.

Latour, B. *We Have Never Been Modern.* Boston: Harvard University Press, 1993.

Latour, B. "On Interobjectivity," *Mind, Culture and Activity* (3), 1996, pp. 228-245.

Law, J., and Hassard, J. (eds.). *Actor Network Theory and After.* Oxford: Blackwell, 1999.

Levitt, J. I. "Why Physicians Continue to Reject the Computerized Medical Record," *Minnesota Medicine* (77:8), 1994, pp. 17-21.

Mambrey, P., and Robinson, M. "Understanding the Role of Documents in a Hierarchical Flow of Work," *ACM Conference on Supporting Group Work (GROUP'97).* Phoenix, Arizona, November 16-19, 1997.

McDonald, C. J., Hui, S. L., Smith, D. L., Tierney, W. M., Cohen, S. J., Weinberger M, and McCabe, G. P. "Reminders to Physicians from an Introspective Computer Medical Record: A Two-year Randomized Trial," *Annals of Internal Medicine* (100), 1984, pp. 130-8.

Mohr, D. N., Carpenter, P. C., Claus, P. L., Hagen, P. T., Karsell, P. R., and Van Scoy, R. E. "Implementing an EMR: Paper's Last Hurrah," in *Proceedings of the Nineteenth Annual Symposium on Computer Applications in Medical Care,* R. M. Gardner (ed.). New Orleans: American Medical Informatics Association, 1995.

Nunberg, G. "Farewell to the Information Age," in *The Future of the Book,* G. Nunberg (ed.). Berkeley: University of California Press, 1996.

Nygren, E., and Henriksson, P. "Reading the Medical Record. I. Analysis of Physicians' Ways of Reading the Medical record," *Computing Methods and Programs in Biomedicine* (39), 1992, pp. 1-12.

Orlikowski, W. J., Walsham, G., Jones, M. R., and DeGross, J. I. (eds.). *Information Technology and Changes in Organizational Work.* London: Chapman & Hall, 1996.

Reiser, S. J. *Medicine and the Reign of Technology.* Cambridge, England: Cambridge University Press, 1978.

Reiser, Stanley J (1984): Creating Form Out of Mass: The Development of the Medical Record. In *Transformation and Tradition in the Sciences: Essays in Honor of I.* Bernard Cohen, ed. E. Mendelsohn. Cambridge University Press: New York.

Reverby, S. "Stealing the Golden Eggs: Ernest Amory Codman and the Science and Management of Medicine," *Bulletin of the History of Medicine* (55), 1981, pp. 156-71.

Robinson, M. "Computer Supported Cooperative Work: Cases and Concepts," in *Readings in Groupware and Computer-Supported Cooperative Work: Assisting Human-Human Collaboration,* R. M. Baecker (ed.). San Mateo, CA: Morgan Kaufman, 1993.

Sauer, C. *Why Information Systems Fail: A Case Study Approach.* Henley-on-Thames, England: Alfred Waller, 1993.

Schmidt, K., and Simone, C. "Coordination Mechanisms: Towards a Conceptual Foundation of CSCW Systems Design," *Computer Supported Cooperative Work* (5), 1966, pp. 155-200.

Star, S. L. (ed.). *The Cultures of Computing.* Oxford: Blackwell, 1995a.

Star, S. L. (ed.). *Ecologies of Knowledge: Work and Politics in Science and Technology.* New York: State University of New York Press, 1995b.

Stevens, R. *In Sickness and in Wealth: American Hospitals in the Twentieth Century.* New York: Basic Books, 1989.

Strauss, A., Fagerhaugh, S., Suczek, B., and Wieder, C. *Social Organization of Medical Work.* Chicago: University of Chicago Press, 1985.

Suchman, L. *Plans and Situated Actions: The Problem of Human-machine Communication.* Cambridge, England: Cambridge University Press, 1987.

Suchman, L. "Working Relations of Technology Production and Use," *Computer Supported Cooperative Work* (2), 1994, , pp. 21-39.

van der Lei, J. "Use and Abuse of Computer-stored Medical Records" (Editorial), *Methods of Information in Medicine* (32:2), 1991, 2, pp. 79-80.

van der Lei, J., van der Does, E., Man in 't Veld, A. J., Musen, M. A., and van Bemmel, J. H. "Response of General Practitioners to Computer-generated Critiques of Hypertension Therapy," *Methods of Information in Medicine* (32:2), 1993, pp. 146-53.

Walsham, G *Interpreting Information Systems in Organizations*. Chichester: J. Wiley & Sons, 1993.

Walsham, G. "Actor-Network Theory and IS Research: Current Status and Future Prospects," in *Information Systems and Qualitative Research,* A. S. Lee, J. Leibenau, and J. I. DeGross (eds.). London: Chapman & Hall, 1997.

Warner, J. H. *The Therapeutic Perspective: Medical Practice, Knowledge and Identity in America, 1820-1885*. Boston: Harvard University Press, 1986.

Winograd, T , and Flores, F. *Understanding Computers and Cognition: A New Foundation for Design*. Norwood, NJ: Ablex, 1986.

About the Author

Marc Berg is an associate professor at the Institute of Health Policy and Management, Erasmus University, Rotterdam, The Netherlands, where he leads a research group focusing on ICT in health care. He is the author of, among others, *Rationalizing Medical Work: Decision Support Techniques and Medical Practices* (MIT Press, 1997). He is currently finishing a book on standardization in medicine with Stefan Timmermans. Marc can be reached by e-mail at M.Berg@bmg.eur.nl.

Part 10:

Panels on Research Methods and Distributed Organizations

29 ADDRESSING THE SHORTCOMINGS OF INTERPRETIVE FIELD RESEARCH: REFLECTING SOCIAL CONSTRUCTION IN THE WRITE-UP

Ulrike Schultze
Southern Methodist University
U.S.A.

Michael D. Myers
University of Auckland
New Zealand

Eileen M. Trauth
Northeastern University
U.S.A.

Recent critical discussion of interpretive field research suggests that there exist two major concerns with the state of the art use of interpretive field methods in Information Systems. First, many interpretive researchers pay insufficient attention to critical reflection on how the research materials were socially constructed through interaction between the researchers and the participants in their field studies. Second, it appears that the preferred pattern for interpreting field data is to begin with an *a priori* theory. This raises concerns not only about unwarranted theoretical bias in researchers' interpretations, but also about the researcher's propensity to question his/her assumptions in the process of making sense of the data. The implication of these two shortcomings is that "we are given little understanding of how the researchers' analysis developed over the course of the project. As it stands, we are presented with a finished piece of interpretive research with few indications of its emergent nature" (Klein and Myers 1999, p. 25). The IS community is thus not given enough insight into how interpretive researchers resolve the contradictions between the theoretical preconceptions guiding their research and the results that are generated.

In this panel, we wish to explore how widespread these two shortcomings are in interpretive IS research, what accounts for their existence, whether they serve a purpose and how they can be remedied. Do the politics of publishing play a role in generating certain shortcomings in current interpretive field research? What does it take to give readers a better understanding of how the researcher and his/her analysis developed over the course of the research? Is it a problem that can be addressed by special methodological strategies and techniques and if so, by which? Or could it be that authors simply need to adopt a different style of writing up the research? What would such writing styles look like? Is the confessional style of writing (Van Maanen 1988; Trauth 1997; Schultze 1999) the only alternative? If so, what are the consequences of an author making him/herself vulnerable by including autobiographical details in his/her "scientific" writings? Can this style of writing be taught? How will it be viewed and received in different cultures?

These are the kinds of questions we wish to explore with the audience in this panel. We plan to proceed in the following way:

Ulrike Schultze will introduce the panel by outlining the motivation for the discussion about the challenge of living up to the expectation of high-quality interpretive research. She will then introduce the panel members' research and relate their expertise to the purpose of the panel. This should orient the audience to how each panel member will approach the questions and issues that the panel seeks to address.

Michael Myers will highlight some of the major shortcomings identified in interpretive field research in IS and the importance of correcting them. He will focus particularly on two shortcomings, namely the lack of critical reflection on how the research materials were socially constructed through interaction between the researcher and the participants, and the preference for theory-driven interpretation. He will then discuss how and why some of the principles that he advocated in Klein and Myers were not applied in the writing up of some of his earlier research work (e.g., Myers 1994). He will argue that it is possible to satisfy both shortcomings in interpretive IS research without necessarily adopting a confessional style of writing.

Eileen Trauth will discuss the same two shortcomings in interpretive IS research with respect to her own research and publishing experiences. Her own interpretive work places a priority on establishing a connection between the researcher and the participants in the field study. She will show how a data-driven (vs. a theory-driven) approach in conducting the research led her to a confessional style of writing the results. She will also discuss the political difficulties she had in trying to get interpretive work that is "in her own voice" published. She will raise questions about our own academic culture and its effect on the audience for self-reflexive writing. Finally, she will discuss the joys of conducting self-reflexive research and the liberation of the author through the more personalized styles of writing

Ulrike Schultze will explain how she addressed the same two shortcomings of interpretive IS research in her work. She will outline her use of a confessional genre of representation and the implications such a genre has for both the fieldwork and the subsequent deskwork. In contrast to Eileen Trauth, who will address the joys and liberation of self-revealing writing, Ulrike will focus on the concerns about professional identity that are related to writing in a confessional style.

Each panelist's presentation will be timed in such a way that at least 30 minutes remain at the end for the question and answer session and audience participation.

References

Klein, H. K., and Myers, M. D. "A Set of Principles for Conducting and Evaluating Interpretive Field Studies in Information Systems," *MIS Quarterly* (23:1), 1999, pp. 67-93.

Myers, M. D. "A Disaster for Everyone to See: An Interpretive Analysis of a Failed IS Project," *Accounting, Management and Information Technologies* (4:4), 1994, pp. 185-201.

Schultze, U. "A Confessional Account of an Ethnography about Knowledge Work," *MIS Quarterly* (forthcoming, 1999, electronic pre-print available at www.cox.smu.edu/faculty/uschultz/research/).

Trauth, E. M. "Achieving the Research Goal with Qualitative Methods: Lessons Learned Along the Way," in *Information Systems and Qualitative Research*, A. S. Lee, J. Liebenau, and J. I. DeGross (eds.). London: Chapman and Hall, 1997, pp. 225-245.

Van Maanen, J. *Tales of the Field: On Writing Ethnography*. Chicago: University of Chicago Press, 1988.

About the Panelists

Ulrike Schultze is an assistant professor in the Information Systems and Operations Management Department at Southern Methodist University. Her research focuses on knowledge work, particularly informing practices, i.e., the social processes of creating and using information in organizations. Ulrike has written on hard and soft information genres, information overload, knowledge management and knowledge workers' informing practices. Her dissertation research, which was completed in Fall 1997, is an ethnographic field study of knowledge work. Her first publication from this study is written in a confessional genre of representation. Ulrike can be reached by e-mail atuschultz@mail.cox.smu.edu.

Michael D. Myers is an associate professor in the Department of Management Science and Information Systems at the University of Auckland, New Zealand. Michael has published widely in the areas of qualitative research methods, interpretive research, ethnography, and information systems implementation. He is editor of the *ISWorld Section on Qualitative Research*, an associate editor of *MIS Quarterly,* and an associate editor of *Information Systems Journal.* Michael can be reached by e-mail at m.myers@auckland.ac.nz.

Eileen M. Trauth is an associate professor of Management Information Systems in the College of Business Administration at Northeastern University. Her research interests center around the cultural, societal, organizational, and educational impacts of information technology. She has published articles recently about the global information economy, information policy, IT education, and information management. Her research, which employs qualitative methods, is primarily directed at uncovering contextual influences and impacts on information technology and information professionals. While she has done research in several countries, her most extensive piece of qualitative work has been a multiyear, ethnographic investigation of socio-cultural influences on an emerging information economy. She has published several papers from this research and a forthcoming book about it is written in the confessional genre. Eileen can be reached by e-mail at trauth@neu.edu.

30 LEARNING AND TEACHING QUALITATIVE RESEARCH: A VIEW FROM THE REFERENCE DISCIPLINES OF ANTHROPOLOGY AND HISTORY

Bonnie Kaplan
Yale University School of Medicine
U.S.A.

Jonathan Liebenau
London School of Economics and Political Science
United Kingdom

Michael D. Myers
University of Auckland
New Zealand

Eleanor Wynn
University of Oregon
U.S.A.

1. Introduction

As interest grows in qualitative methods, more attention is being focused on how to teach others to carry out qualitative research projects. Qualitative research has roots in social science and humanities disciplines. Over the years, individuals in these disciplines have honed data collection and analysis techniques and developed approaches to understanding through a variety of descriptive and interpretive methods. Their methods include observation, participation, close reading of texts, and, more recently, data analysis techniques such as discourse analysis, hermeneutic interpretation, and grounded theory. In many cases, these researchers articulated and practiced ways of understanding the points of view of the individuals they study (be this focused on a text, an historical development, or a ritual practice). They then went deeper by using additional theoretical knowledge together with additional evidence to interpret and explain the situation under study. They also may have incorporated methods and epistemological stances from other fields, developing such areas as cliometrics in history, ethnoscience in anthropology, and the widespread use of sophisticated statistical analysis in sociology and psychology.

Just as we in IS have learned from methodological work in these other fields, we also can learn about teaching these methods by examining the educational approaches they employ. By adapting not only the methods, but also how they are taught to graduate students, we may both deepen our understanding of these forms of inquiry and how to teach them to others.

This panel is comprised of IS researchers who hold doctorates in reference disciplines for qualitative research. With credentials and publications in both IS and other fields, the panelists may speak to issues of teaching and evaluating qualitative research in IS in ways that bring perspectives from reference disciplines and that are well grounded in personal experiences in both the reference discipline and in IS. The focus in this panel is on teaching, and particularly on educating graduate students in qualitative research. Panelists will discuss how this material has been taught in their base disciplines and raise issues for how it should be taught in IS in the future. They will describe how they and other students learned the methods, the research questions, and the standards for good work in their doctoral fields. They will discuss how they themselves adapted their training for doing IS research, how they teach research methods in IS, and what they recommend for teaching qualitative research in IS in the future. Following the panelists' remarks, the chair will invite the audience to discuss with panelists what we may learn by looking at ways of adapting reference discipline research and teaching methods when teaching IS research.

The panel is comprised of historians and anthropologists. Between them, these two disciplines use a wide range of qualitative and interpretive research approaches. Historians have, for centuries, been experts in analyzing texts and historical artifacts to develop explanatory narratives of events as they unfold over time. Anthropologists have developed well established expertise in studying and explaining culture by immersing themselves in it. History and anthropology represent disciplines that do research involving immersion in other settings and vast amounts of textual, verbal, and observational data to make sense of dynamic situations. In both fields, too, newer quantitative approaches have been combined with more traditional qualitative research approaches. Having two panelists from each of these disciplines provides a more thorough overview of each of the reference disciplines: how they evolved over time, how they are taught in different universities, how each panelist has used this knowledge to contribute to IS, and changes each of them have seen in these approaches as used in IS. Together, these two disciplines provide a view of how research in IS, and particularly qualitative research, may be taught, practiced, and evaluated.

2. Panelist Statements

Bonnie Kaplan: I took a faculty position in information systems immediately upon completing a doctorate in history at a school that values interdisciplinarity, and where history is considered as a field in both the social science and humanities. My work draws on the kinds of questions social scientists and humanists ask, and on methods used by them for answering such questions. My graduate program emphasized traditional historical narrative based on original documents as primary source material. Methodological and theoretical issues were implicit and interpretive approaches were such an established tradition that they were more assumed than discussed in graduate courses and departmental seminars. Graduate students read numerous (several each week) historical books and papers together with original documents in several fields within the discipline, covering different time periods, places, issues, and interpretations. They then spend many years (at my school, it was not unusual for a Social Sciences Division student to take over 10 years) doing the research for a dissertation. I worked on the history of computer use in clinical medicine by analyzing original source material of medical computing publications, historical and sociological literature pertaining to medicine and to technology, and by engaging in participant observation as a programmer/analyst at two academic hospitals. I since have conducted studies of information systems in clinical settings by using a combination of methods: ethnographic interviewing, participant observation, observation, surveys, and document analysis.

Today, methodological approaches are addressed more explictly both in graduate history courses and in historical writing than when I was a graduate student. I will discuss how I learned to do history and how students learn it today. I will talk about how, later, as a researcher and teacher in both information systems and medical informatics, I learned how to make methods explicit and how to combine them with quantitative methods to achieve robust, rigorous, and relevant research results. Based on these experiences, I believe that what makes for good research, whether qualitative or quantitative, must be partially self taught and internalized. While helpful, research methods courses, checklists, and guidelines are secondary to steeping oneself in models of excellent research through apprenticeship and emulation and then personally undertaking a significant project to develop both mastery and originality.

Jonathan Liebenau: As a student of history in the 1970s, I was much influenced by the renewed debate on the use of quantitative methods (dubbed "cliometrics" at the time, but in principle not different from much of the economic history of the 1930s and some of the social history of the 1960s). All historians who were interested in methodology had to engage in that debate, and many of us developed work that brought out the relationship between the qualitative methods, which had been very well crafted by historical scholarship, and "new" quantitative methods. I find it easy to defend either approach, and I believe very strongly that both are legitimate (and that those who attempt to de-legitimate qualitative methods are very seriously misguided). Furthermore, I find it easy to demonstrate the great value of carefully executed qualitative methods used in many social sciences, especially history, anthropology, and sociology.

Michael D. Myers: I will discuss how I learned to do textual analysis and observation. After majoring in anthropology and psychology at the undergraduate level, I went on to specialize in Social Anthropology for my Master's and Ph.D. I will discuss how a doctoral student in anthropology was expected to have read and understood the ethnographic works of anthropologists from many different schools (e.g., Malinowski, Geertz, Radcliffe-Brown, Boaz, Benedict, Godelier, Levi-Strauss, Firth, Turner, etc). The differences between these schools were regularly discussed and debated (e.g., the functionalism of Radcliffe-Brown versus the structuralism of Levi-Strauss). Ph.D. students, however, would typically become disciples of one particular approach and use that approach in their fieldwork. In my case, I used critical hermeneutics to analyze and interpret my ethnographic materials. I spent 14 months doing fieldwork studying the independence movement in Vanuatu (a group of islands in the South-West Pacific).

I will argue that one of the key challenges for qualitative researchers in information systems is that of superficiality. How can we expect doctoral students in IS to produce good qualitative research if they have completed just one or two courses in qualitative methods? I will argue that this superficiality is one of the main reasons why so many qualitative research articles are rejected in our top journals (although this is not to say that the rejection rate for quantitative articles is any lower). I will suggest a few ways in which this challenge might be overcome.

Eleanor Wynn: When I went to Xerox PARC as an intern in 1976, I was both fascinated with the altogether new-to-me world of computing as well as struck with how totally my world-view differed from the one around me. It took many years of acculturation before I was able to tell people they were asking the wrong question when they asked me "What is your instrument?" There actually is an answer: my instrument is my brain and the repeated exposure to a form of viewing and analyzing that comes with an apprenticeship in anthropology and conversation analysis. But in the beginning I had felt compelled to come up with some "technic." I was struck by how most people consider themselves social analysts (they are, but not at an academic level) and with the fact that once you point out something previously unarticulated, it is, of course, obvious. It therefore quickly becomes "not special." Everyone possesses a sort of lumpen social science. This is the premise of ethnomethodology, the theoretical view I specialized in, based in phenomenology.

What is different about specializing in social analysis is the review of theories and results of others' studies and formation of patterns and arguments based in this. Part of my work as a person bridging the worlds of social science, technology, and business, was to grasp how powerful these tools are for practical application, something that wasn't a goal in graduate school. By taking the everyday and relatively unactionable insights and implicit knowledge of organizational participants and articulating them, these common sense understandings can be used to guide designs and decisions about appropriate technology. I will discuss first the basic premise of anthropology (members' competence in organizing a workable world) and then talk about how language studies reveal this.

About the Panelists

Bonnie Kaplan is a lecturer at the Yale School of Medicine's Center for Medical Informatics, a senior scientist at Boston University's Medical Information Systems Unit, and president of Kaplan Associates. She has published research in change management, organizational issues surrounding the introduction and use of information technologies, benefits realization, and evaluating organizational and user acceptance issues pertaining to clinical applications of computer information systems in medicine and health care. She is on the Informatics Advisory Board for a major health care company, serves as Chair of both the American Medical Informatics Association People and Organizational Issues Working Group and the International Medical Informatics Association WG13: Organizational Impacts. She has taught a variety of information systems courses in business administration and hospital administration programs. She holds a Ph.D. in history with a specialization in history of science and medicine from the University of Chicago. Bonnie can be reached by e-mail at bonnie.kaplan@yale.edu.

Jonathan Liebenau is a senior lecturer in Information Systems at the London School of Economics with a background in social studies of science, technology, and medicine. He has published on the application of social science theory to the study of information systems and he teaches a large, year-long doctoral seminar on "Theory and Research Methods in Information Systems." He was a program chair of a recent IFIP 8.2 meeting on "Information Systems and Qualitative Research" and is an editor of the associated volume. His doctorate in history is from the University of Pennsylvania. Jonathan can be reached by e-mail at liebenau@LSE.AC.UK.

Michael D Myers is an associate professor in the Department of Management Science and Information Systems at the University of Auckland. He has a Ph.D. in Social Anthropology from the University of Auckland. Michael has published widely in the areas of qualitative research methods, interpretive research, ethnography, and information systems implementation. He is Editor of the *ISWorld Section on Qualitative Research*, an Associate Editor of *MIS Quarterly,* and an Associate Editor of *Information Systems Journal.* Michael can be reached by e-mail at m.myers@auckland.ac.nz.

Eleanor Wynn is in Applied Information Management at the University of Oregon. She has worked for the past 25 years within the IS, telecommunications, and computer science communities, applying contemporary linguistic and social theories to communications issues surrounding technology development and use. Her interests evolved from a concern with tacit work practices and knowledge sharing to organizational decision-making about technology and then to assumptions embedded in software development methods and practice. Returning to an earlier interest in Latin America, where she was raised, she is now working on a framework for research on globalization and information technology for less-developed countries that follows from all these prior concerns. She received her Ph.D. in 1979 from the University of California Department of Anthropology with a concentration in social interaction. Her doctoral thesis, *Office Conversation as an Information Medium*, was widely read during the early 1980s when system developers, especially in Europe, were searching for way of studying the workplace that would support a labor-oriented approach. After an internship at Xerox PARC, she worked as a Senior Scientist at Bell-Northern Research and eventually went into private consulting until returning to academic life. She is editor-in-chief of *Information Technology & People*. Eleanor can be reached by e-mail at wynn@zephyr.net.

31 SUCCESSFUL DEVELOPMENT, IMPLEMENTATION, AND EVALUATION OF INFORMATION SYSTEMS: DOES HEALTHCARE SERVE AS A MODEL FOR NETWORKED ORGANIZATIONS?

Jos Aarts
Erasmus University Rotterdam
The Netherlands

Els Goorman
Erasmus University Rotterdam
The Netherlands

Heather Heathfield
Manchester Metropolitan University
United Kingdom

Bonnie Kaplan
Center for Medical Informatics
Yale University
U.S.A.

The use of information systems in healthcare lags behind other business sectors. One of the possible reasons is that healthcare traditionally has been a "networked" organization with no "central command." The introduction and use of IS has been shaped by the different powerful actors determining the delivery of care: professional groups (such as physicians, pharmacists, and, to a lesser extent, nurses), healthcare organizations, insurance companies (private and/or public) and regulating bodies (including governments). Each of the actors has different knowledge reference frameworks and knowledge that seems difficult to integrate. Private businesses are more rooted in the concept of "single line of command" and are moving swiftly into networked knowledge based organizations to meet the demands of a dynamic marketplace. What has happened in healthcare suggests that the development and implementation of IS will become increasingly difficult as businesses move toward networked organizations. The four panelists are researching IS in healthcare from different theoretical angles. They propose that healthcare as an example of a networked organization bears important lessons for the IS field.

Jos Aarts is researching the design and implementation process of a large hospital information system. A major feature of the system is the order entry functionality, which allows physicians to generate orders (such as lab, radiology, and medication), schedule patient appointments for these orders, and get results. The university hospital in the study bears the characteristics of a networked organization. Medical activities are grouped in clusters (internal medicine, surgery, pediatrics, neurology, clinical psychiatry, etc.) and support groups, such as radiology and pathology, which are actually independent entities, that are continuously negotiating about patients, services, and utilization of resources. Actor-network theory provides a theoretical framework to study information systems in networked organizations (Walsham 1997). Jos will argue that the order entry functionality of the system is inscribed by a development team representing the interests of the top level management in the hospital and this inscription causes problems for physicians to accept the system (Akrich 1992). The necessary translation of the system into medical work, therefore, is problematic and a stable network of aligned interests is not yet developing.

Els Goorman studies the connection between health care providers and information and communication technology (ICT) devices in care processes. She argues that the way synergy occurs and relevant factors influencing this synergy can give understanding in the way ICT supports health care providers in their activities (Luff and Heath 1993). Studying this possible synergy between care providers and ICT can give recommendations for future development of ICT applications and how to promote this synergy. Furthermore, she is studying the (un)expected consequences of the use of these ICT devices for the work of care providers and the transparency of the care process. Data gathered by means of participatory observations and interviews will be presented.

Heather Heathfield has evaluated the use of electronic patient records in UK hospitals. The rate of success of using the electronic patient record depends very much on the arrangements made between different "networks" in a hospital, including physicians, ICT-staff, and hospital boards. In order to make sense of the evaluation results, the complex social and organizational issues that surround the use of IS need to be understood. She argues that a useful way of analyzing evaluation results is to use models from the evaluation of social programs that differentiate between the "mechanism" of an IS and the context in which it is used (Pawson and Tilley 1997). This helps to identify what "works" for "whom" in different circumstances.

Bonnie Kaplan has been actively involved in both medical informatics and in Information Systems over a long period. She will argue that understanding healthcare organizations as networked does not provide predictors for newly networking organizations because healthcare organizations have a different organizational structure, history, and professional legacy from business organizations. Based on research evaluating both clinical and consumer reactions to information technologies, she will describe other ways in which what has been learned in healthcare settings is applicable to other organizations. In particular, she will draw on a social interactionist theoretical framework to present both methodological approaches and empirical findings that illuminate more general concerns in IS (Anderson and Aydin 1994; Kaplan 1997).

References

Akrich, M. "The De-scription of Technical Objects," in *Shaping Technology/Building Society: Studies in Sociotechnical Change*, W. E. Bijker and J. Law (eds.). Cambridge, MA: The MIT Press, 1992, pp. 205-224.

Anderson, J. G., and Aydin, C. E. "Overview: Theoretical Perspectives and Methodologies for the Evaluation of Health Care Information Systems," in *Evaluating Health Care Information Systems: Approaches and Applications*, J. G. Anderson, C. E. Aydin and S. J. Jay (eds.). Thousand Oaks, CA: Sage, 1994, pp. 5-29.

Kaplan, B. "Addressing Organizational Issues into the Evaluation of Medical Systems," *Journal of the American Medical Informatics Association* (4:2), 1997, pp. 94-101.

Luff, P., and Heath, C. "System Use and Social Organization: Observations on Human-computer Interaction in an Architectural Practice," in *Technology in Working Order: Studies of Work, Interaction and Technology*, G. Button and H. Harper (eds.). London: Routledge, 1993.

Pawson, R., and Tilley, N. *Realistic Evaluation.* Thousand Oaks, CA: Sage, 1997.

Walsham, G. "Actor-network Theory and IS Research: Current Status and Future Prospects," in *Information Systems and Qualitative Research,* A. S. Lee, J. Liebenau and J. I. DeGross (eds.). London: Chapman & Hall, 1997, pp. 466-480.

About the Panelists

Jos Aarts is an assistant professor of information management in healthcare at the Department of Health Policy and Management of Erasmus University Rotterdam in the Netherlands. He has a background in physics and is currently pursuing a Ph.D. on successful design and implementation of information systems in healthcare. He has published several articles about a model to teach and study the organizational impact of health informatics. He is course director of the information management program at the department and coordinates the Erasmus part in the joint University of Surrey, University of Manchester and Erasmus University Rotterdam MSc Health Informatics course. He is chairman of European Federation for Medical Informatics WG9 "Human and Organizational Issues of Medical Informatics." Jos can be reached by e-mail at j.aarts@bmg.eur.nl.

Els Goorman is a Ph.D. student at the Department of Health Policy and Management of Erasmus University Rotterdam. The title of her research is "Synergy and Transition: Information and Communication Technology, Healthcare Providers and the Primary Care Process." She teaches qualitative research methods. She is a registered nurse and has a MSc in Health Sciences from Maastricht University. Her dissertation dealt with a sociological perspective on the use of clinical information systems. Els can be reached by e-mail at goorman@bmg.eur.nl.

Heather Heathfield is a senior lecturer in the Department of Computer Science and Mathematics of Manchester Metropolitan University in the UK. She holds a Ph.D. from the University of Brighton in computer science on decision support systems in healthcare. Currently she is involved in the evaluation of decision support systems for general practitioners and alerting systems for chronic diseases. She was part of a multi-disciplinary research team conducting an NHS commissioned evaluation of electronic patient records in use. Heather can be reached by e-mail at h.heathfield@doc.mmu.ac.uk.

Bonnie Kaplan is a lecturer at the Yale School of Medicine's Center for Medical Informatics, a Senior Scientist at Boston University's Medical Information Systems Unit, and President of Kaplan Associates. A recognized expert in evaluating organizational and user acceptance issues pertaining to clinical applications of computer information systems in medicine and health care, she is the author of over 30 refereed papers as well as numerous other articles and publications. She is on the Informatics Advisory Board for a major health care company, serves as Chair of the American Medical Informatics Association People and Organizational Issues Working Group, and is Chair of the International Medical Informatics Association WG13 "Organizational Impact of Medical Informatics" as well as editor of their newsletter. She has taught a variety of information systems courses in business administration and hospital administration programs at several universities. Her Ph.D. is from the University of Chicago. Bonnie can be reached by e-mail at bonnie.kaplan@yale.edu.

32 STANDARDIZATION, NETWORK ECONOMICS, AND IT

Esben S. Andersen
Aalborg University
Denmark

Jan Damsgaard
Aarhus University
Denmark

Ole Hanseth
University of Oslo
Norway

John L. King
University of California, Irvine
U.S.A.

M. Lynne Markus
Claremont Graduate University
U.S.A.

Eric Monteiro
Norwegian University of Science and Technology
Norway

1. Introduction

Standard software systems pervade our lives, both as professionals working in an organization and as private persons. Organizations and individuals alike have to make multiple decisions. What browser to use (Netscape or Explorer)? What office suite to purchase? What instant messenger to use (preferably, the one most of your friends or colleagues use; AOL, Microsoft, and Yahoo represent three alternatives)? What operating system to purchase in order to protect the current IT infrastructure? What Internet search engine to purchase for the company's intranet? Whether to buy a single, double, or triple band cellular phone depending on where you live in the U.S. and how much you travel?

The answers to these questions are obviously related to the technological and social networks in which we are embedded, but it is not always easy to specify how.

In the IT section of ordinary newspapers, we can daily follow the battle between such standard systems: how lawsuits are being filled against companies using an established standard-based monopoly to bridgehead the establishment of a new; how billions of dollars are being made from introducing IT companies, that have not shown a profit yet, to the stock exchange; and how sworn enemies suddenly work together to define new standards in common. This has stirred quite an interest in the business world, specifically in high tech marketing (Moore 1995, 1999).

Underlying these strategic games are, to a large extent, reflections about network economies and their exploitation.

These systems all rely on or are built around some software or hardware standard, either piggybacking some existing open standard, a proprietary standard, or desperately seeking to set the standard in some new area. The advanced IT software systems operating behind the scene are costly to develop, but have a very limited life-cycle, usually measured in months, and their success or endurance are by no means guaranteed. To make the picture even more confusing, some can be downloaded for free, while others are very costly. IS scholars have so far had little advice to companies on how to maneuver in the seemingly erratic world of standard-based IS systems and networks.

2. IS Research

When examining the IS literature, it becomes clear that there has been relatively little attention given to these phenomena, which have such a profound impact on organizational processes and IT investments. A few that are close to the IFIP Working Group 8.2 are presented below.

Markus (1987) describes the diffusion and implementation of interactive media within the organizational boundary. Damsgaard and Lyytinen (1997, 1998) consider the impact of network externalities in their study of EDI diffusion and adoption in Finland and Hong Kong. Ciborra and Hanseth (1998) examine agendas for managing the information infrastructure, while Knuuttila, Lyytinen, and King (1996) examine the standardization process leading to the successful NMT and GSM standards for mobile telephony. Monteiro (1998) describes the lock-in effect of earlier standard adoption decisions (Internet Protocol version 5) and their implications for defining and launching IPv6. Hanseth (1996) examines IT as infrastructure and Damsgaard and Truex (2000) examine the limits of EDI standards to facilitate meaningful binary interorganizational communication.

3. Network Economics

There exists a whole body of research commonly referred to as network economics that offers a theoretical vehicle to explain and understand the phenomenon. The research tradition emanates from a desire to understand standards (Farrell and Saloner 1985, 1988), technological innovation (Dosi et al. 1988; Edquist 1997; Nelson 1994) and associated business strategy (Besen and Farrell 1994; Farrell and Saloner 1986; Oliva 1994; Shapiro and Varian 1999). Its point of departure is often historical accounts of technology innovation processes (Arthur 1989; David and Bunn 1988). For example, network economics offers explanations of the static QWERTY layout of the keyboard (David 1985) or the battle between VHS and Betamax (Katz and Shapiro 1986), but also addresses broader issues such as policy making, regulation, and maintaining and sustaining systems of innovation.

These studies are quite important for recent IT discussions and regulations, so it is understandable why some researchers have tried to articulate a more harmonious picture of the market-based evolution of IT-related networks (Liebowitz and Margolis 1999). Still, the ideas of strategic games and suboptimal results are gaining strength.

4. Objectives

The panel participants have all addressed key aspects of network economics and have all referenced network economics literature. The panel discusses key concepts of network economics suitable for IS research, assesses the experiences with applying network economics to the IS field, and suggests future directions for IS research based on a networked perspective.

Specifically Jan Damsgaard will chair and moderate the panel. He will make some introductory remarks and tie together the panelist' statements. Esben S. Andersen will provide an overview of network economics by summarizing its history and main contributions. Ole Hanseth will describe his experiences with applying network economics to study the evolution IT infrastructure for global companies. John L. King will describe how network economics can inform us when seeking to understand the development of national information infrastructure. M. Lynne Markus will demonstrate how she has applied network economics to analyze the evolution of IT systems within organizations. Finally, Eric Monteiro will explain how network economics can be applied to understand the evolution of software standards.

References

Arthur, W. B. "Competing Technologies, Increasing Returns and Lock-in by Historical Events," *Economic Journal* (99), 1989, pp. 116-131.

Besen, S. M., and Farrell, J. "Choosing How To Compete: Strategies and Tactics in Standardization," *Journal of Economic Perspectives* (8:2), 1994, pp. 117-131.

Ciborra, C. U., and Hanseth, O. "Toward a Contingency View of Infrastructure and Knowledge: An Exploratory Study," in *Proceedings of the Nineteenth International Conference on Information Systems*, R. Hirschheim, M. Newman, and J. I. DeGross (eds.), Helsinki, Finland, 1998, pp. 263-272.

Damsgaard, J., and Lyytinen, K. "Hong Kong's EDI Bandwagon: Derailed or on the Right Track?" in *Facilitating Technology Transfer Through Partnership: Learning from Practice and Research,* T. McMaster, E. Mumford, E. B. Swanson, B. Warboys, and D. Wastell (eds.). London: Chapman & Hall, 1997, pp. 39-63.

Damsgaard, J., and Lyytinen, K. "Contours of Electronic Data Interchange in Finland: Overcoming Technological Barriers and Collaborating to Make It Happen," *The Journal of Strategic Information Systems* (7:4), 1998, pp. 275-297.

Damsgaard, J., and Truex, D. "The Procrustean Bed of Standards: Binary Relations and the Limits of EDI Standards," 2000 (submitted for publication).

David, P. A. "Clio and the Economics of QWERTY," *The American Economic Review* (75:2), 1985, pp. 332-337.

David, P. A., and Bunn, J. A. "The Economics of Gateway Technologies and Network Evolution: Lessons from Electricity Supply History," *Information Economics and Policy* (3), 1988, pp. 165-202.

Dosi, G., Freeman, C., Nelson, R., and Soete, L. (eds.). *Technical Change and Economic Theory.* London: Pinter Publishers, 1988.

Edquist, C. (ed.). *Systems of Innovation: Technologies, Institutions, and Organizations.* London: Pinter Publishers, 1997.

Farrell, J., and Saloner, G. "Standardization, Compatibility, and Innovation," *Rand Journal of Economics* (16:1), 1985, pp. 70-83.

Farrell, J., and Saloner, G. "Installed Base and Compatibility: Innovation, Product Preannouncements, and Predation," *The American Economic Review* (76:5), 1986, pp. 940-955.

Farrell, J., and Saloner, G. "Coordination Through Committees and Markets," *Rand Journal of Economics* (19:2), 1988, pp. 235-252.

Hanseth, O. *Information Technology as Infrastructure.* Gothenburg: Ph.D. Thesis, Report 10, Department of Informatics, Gothenburg University, 1996.

Katz, M. L., and Shapiro, C. "Technology Adoption in the Presence of Network Externalities," *Journal of Political Economy* (94:4), 1986, pp. 822-841.

Knuuttila, J., Lyytinen, K., and King, J. L. *The Parturition of Mobile Telephony Revisited: The Case of Standardization and Institutional Intervention in the Nordic Countries.* Jyväskylä, Finland: University of Jyväskylä, 1996.

Liebowitz, S. J., and Margolis, S. E. *Winners, Losers and Microsoft: Competition and Antitrust in High Technology.* Oakland, CA: Independent Institute, 1999.

Markus, M. L. "Toward a 'Critical Mass' Theory of Interactive Media: Universal Access, Interdependence and Diffusion," *Communication Research* (14:5), 1987, pp. 491-511.

Monteiro, E. "Scaling Information Infrastructure: The Case of Next-generation IP in the Internet," *The Information Society* (14), 1998, pp. 229-245.

Moore, G. A. *Inside the Tornado: Marketing Strategies from Silicon Valley's Cutting Edge.* New York: Harper Business, 1995.

Moore, G. A. *Crossing the Chasm: Marketing and Selling High-Tech Products to Mainstream Customers.* New York: Harper Business, 1999.

Nelson, R. R. "The Co-evolution of Technology, Industrial Structure and Supporting Institutions," *Industrial and Corporate Change* (3:1), 1994, pp. 47-63.

Oliva, T. A. "Technological Choice Under Conditions of Changing Network Externality," *The Journal of High Technology Management Research* (5:2), 1994, pp. 279-298.

Shapiro, C., and Varian, H. R. "The Art of Standards Wars," *California Management Review* (41:2), 1999, pp. 8-32.

About the Panelists

Esben Sloth Andersen is an associate professor at the Department of Business Studies, Aalborg University. He teaches general economics, economic simulation, and IT and network economics. His primary research is in the economics on knowledge creation, specialization and networks, evolutionary modeling and simulation, and the history of economic analysis. Esben is a member of the IKE Group on Innovation Studies, Aalborg University and the Danish Research Unit of Industrial Dynamics (DRUID). He can be reached by e-mail at esa@business.auc.dk.

Jan Damsgaard is an associate professor at the Department of Computer Science, Aalborg University, Denmark. His research focuses on the diffusion and implementation of networked technologies such as Intranet, e-commerce, EDI, Extranet, and Internet technologies. He has presented his work at international conferences (ICIS, ECIS, HICSS, IFIP 8.2. and 8.6) and in international journals (*Journal of Strategic Information Systems*, *International Journal of IT*, *Information Technology and People*, and *Information Infrastructure and Policy*). He holds visiting positions at University of Pretoria, University of Jyväskylä, and Copenhagen Business School. He has a Master's degree in Computer Science and Psychology and a Ph.D. in Computer Science. Jan can be reached by e-mail at damse@cs.auc.dk.

Ole Hanseth has worked most of his career in industry and applied research before moving to his current position as an associate professor at the Department of Informatics at the University of Oslo. He also held a part time position at Göteborg University. The main focus of his research has been comprehensive, integrated, and geographically dispersed information systems—information infrastructures—in both the private and public sectors. Recent journal publications on these issues have appeared in *Computer Supported Cooperative Work; Information Technology and People; Accounting, Management, and Information Technology; Science, Technology, and Human Values; Systèmes d'Information et Management; and Scandinavian Journal of Information Systems.* Ole can be reached at ole.hanseth@ifi.uio.no.

John Leslie King is Dean and Professor in the School of Information at the University of Michigan. His research focuses on high-level requirements for information systems design and implementation in strongly institutionalized production sectors. He served as Editor-in-Chief of the journal *Information Systems Research* from 1993-1998, and served as Marvin Bower Fellow and Visiting Professor at the Harvard Business School in 1990. From 1980-1999, he was Professor of Information and Computer Science and Management at the University of California, Irvine. He holds a B.A. degree in Philosophy and a Ph.D. in Administration from the University of California, Irvine. John can be reached by e-mail at jlking@umich.edu.

M. Lynne Markus is Professor of Management at the Peter F. Drucker Graduate School of Management and Professor of Information Science at the School of Information Science, Claremont Graduate University. She has also taught at the Sloan School of Management (MIT), the Anderson Graduate School of Management (UCLA), the Nanyang Business School, Singapore (as Shaw Foundation Professor), and Universidade Tecnica de Lisboa, Portugal (as Fulbright/FLAD Chair in Information Systems). She began thinking about IT-related risks while conducting research for the Financial Executives Research Foundation. Lynne can be reached by e-mail at m.lynne.markus@cgu.edu.

Eric Monteiro is professor at the Department of Computer and Information Science at the Norwegian University of Science and Technology and the University of Oslo. He has for some years researched in and published on issues relating to the development and diffusion of infrastructure technologies in various areas including the health care sector, the Internet, and within business organizations. Eric can be reached by e-mail at ericm@idi.ntnu.no.

Index of Contributors

GPSR Compliance
The European Union's (EU) General Product Safety Regulation (GPSR) is a set of rules that requires consumer products to be safe and our obligations to ensure this.

If you have any concerns about our products, you can contact us on

ProductSafety@springernature.com

In case Publisher is established outside the EU, the EU authorized representative is:

Springer Nature Customer Service Center GmbH
Europaplatz 3
69115 Heidelberg, Germany

www.ingramcontent.com/pod-product-compliance
Ingram Content Group UK Ltd.
Pitfield, Milton Keynes, MK11 3LW, UK
UKHW021333070726
13610UKWH00011B/49

* 9 7 8 1 4 7 5 7 6 1 0 6 1 *